Graphs and Patterns
in Mathematics
and Theoretical Physics

Proceedings of Symposia in PURE MATHEMATICS

Volume 73

Graphs and Patterns in Mathematics and Theoretical Physics

Proceedings of the Conference on Graphs and Patterns in Mathematics and Theoretical Physics
Dedicated to Dennis Sullivan's 60th birthday
June 14–21, 2001
Stony Brook University, Stony Brook, New York

Mikhail Lyubich
Leon Takhtajan
Editors

American Mathematical Society
Providence, Rhode Island

Proceedings of the conference on Graphs and Patterns in Mathematics and Theoretical Physics held at Stony Brook University, Stony Brook, New York, June 14–21, 2001.

2000 *Mathematics Subject Classification*. Primary 81Txx, 57-XX 18-XX 53Dxx 55-XX 37-XX 17Bxx.

Library of Congress Cataloging-in-Publication Data
Stony Brook Conference on Graphs and Patterns in Mathematics and Theoretical Physics (2001 : Stony Brook University)
 Graphs and Patterns in mathematics and theoretical physics : proceedings of the Stony Brook Conference on Graphs and Patterns in Mathematics and Theoretical Physics, June 14–21, 2001, Stony Brook University, Stony Brook, NY / Mikhail Lyubich, Leon Takhtajan, editors.
   p. cm. — (Proceedings of symposia in pure mathematics ; v. 73)
 Includes bibliographical references.
 ISBN 0-8218-3666-8 (alk. paper)
  1. Graph Theory. 2. Mathematics–Graphic methods. 3. Physics–Graphic methods. 4. Manifolds (Mathematics).   I. Lyubich, Mikhail, 1959–   II. Takhtadzhyan, L. A. (Leon Armenovich)   III. Title. IV. Series.

QA166.S79 2001
511′.5–dc22   2004062363

Copying and reprinting. Material in this book may be reproduced by any means for educational and scientific purposes without fee or permission with the exception of reproduction by services that collect fees for delivery of documents and provided that the customary acknowledgment of the source is given. This consent does not extend to other kinds of copying for general distribution, for advertising or promotional purposes, or for resale. Requests for permission for commercial use of material should be addressed to the Acquisitions Department, American Mathematical Society, 201 Charles Street, Providence, Rhode Island 02904-2294, USA. Requests can also be made by e-mail to reprint-permission@ams.org.
 Excluded from these provisions is material in articles for which the author holds copyright. In such cases, requests for permission to use or reprint should be addressed directly to the author(s). (Copyright ownership is indicated in the notice in the lower right-hand corner of the first page of each article.)

© 2005 by the American Mathematical Society. All rights reserved.
The American Mathematical Society retains all rights
except those granted to the United States Government.
Copyright of individual articles may revert to the public domain 28 years
after publication. Contact the AMS for copyright status of individual articles.
Printed in the United States of America.

∞ The paper used in this book is acid-free and falls within the guidelines
established to ensure permanence and durability.
Visit the AMS home page at http://www.ams.org/

10 9 8 7 6 5 4 3 2 1   10 09 08 07 06 05

This Volume is Dedicated to Dennis Sullivan's 60th Birthday

Contents

Preface	ix
Dennis Sullivan–A short history	xiii
Dennis Sullivan's List of publications	xv
Sigma models and string topology DENNIS SULLIVAN	1

Feynman diagrams

Feynman diagrams for pedestrians and mathematicians MICHAEL POLYAK	15
Structures in Feynman graphs: Hopf algebras and symmetries DIRK KREIMER	43

Algebraic structures

Notes on universal algebra ALEXANDER A. VORONOV	81
The ring of differential operators on forms in noncommutative calculus DMITRI TAMARKIN and BORIS TSYGAN	105
Twisted chiral de Rham algebras on $\mathbb{P}^1$ VASSILY GORBOUNOV, FYODOR MALIKOV, and VADIM SCHECHTMAN	133

Manifolds: invariants and mirror symmetry

Invariants of tangles with flat connections in their complements RINAT KASHAEV and NIKOLAI RESHETIKHIN	151
Tree-level invariants of three-manifolds, Massey products and the Johnson homomorphism STAVROS GAROUFALIDIS and JEROME LEVINE	173
Multivalued Morse theory, asymptotic analysis and mirror symmetry KENJI FUKAYA	205

Combinatorial aspects of dynamics

Some applications of combinatorial differential topology   281
    Robin Forman

Extensions, quotients and generalized pseudo-Anosov maps   315
    André de Carvalho

Unimodal maps and hierarchical models   339
    Michael Yampolsky

Physics

Quantum geometry in action: big bang and black holes   361
    Abhay Ashtekar

Supersymmetry, supergravity, superspace and BRST symmetry in a simple model   381
    Peter van Nieuwenhuizen

Graphs and Patterns
in Mathematics and Theoretical Physics
Stony Brook   June 14-21, 2001

Graphs occur as organizing objects in several fields of mathematics and theoretical physics. In difficult areas like dynamics or quantum theory, they provide a natural point of access. But each field interprets them in its language. As a consequence, parallel developments are often played out in isolation from each other. The conference will be centered around five minicourses and about ten related invited lectures, presented by experts in each of the areas and aimed at an audience including specialists in other areas as well as graduate students. There will be evening sessions for additional lectures and informal discussions.

Themes to be included:
- Graphs and universal algebra
- Graphs and discrete Riemannian geometry or gauge theory
- Graphs and bifurcation patterns in dynamical systems
- Graphs and quantum field theory - topology

Speakers Include: **Abhay Ashtekar, John Baez, André de Carvalho, Robin Forman, Kenji Fukaya, John Hubbard, Dirk Kreimer, Curtis McMullen, Yair Minsky, Nikolai Reshetikhin, Misha Polyak, Dennis Sullivan, Dylan Thurston, Boris Tsygan, Peter van Nieuwenhuizen, Alexander Voronov.**

A Conference to Celebrate Dennis Sullivan's 60th Birthday

Organizing Committee: André de Carvalho, Detlef Gromoll, Alexander Kirillov, Jr., Blaine Lawson, Mikhail Lyubich, John Milnor, Anthony Phillips, Santiago Simanca, Scott Sutherland, Leon Takhtajan

The background image symbolizes, among other things, the fundamental interconnectedness of all of mathematics.

Preface

This volume offers a variety of papers based on the talks delivered at the Stony Brook Conference (June 2001) "Graphs and Patterns in Mathematics and Theoretical Physics" dedicated to Dennis Sullivan's 60th birthday.

At the conference, whose scientific content was suggested by Sullivan, an attempt was made to overcome conceptual barriers between experts in various branches of mathematics and theoretical physics who encounter graphs in their research. To achieve this goal, the conference was largely based on mini-courses and survey lectures directed to experts in other areas as well as graduate students.

The above idea is reflected in this volume as well: along with research papers, it contains a number of surveys aimed to introduce a reader to the corresponding fields.

The volume opens with an article by Dennis Sullivan "Sigma models and string topology". It describes a background algebraic structure for the sigma model based on algebraic topology and transversality. The background allows one to write the quantum master equation of Batalin and Vilkovisky. It is shown that the Gromov moduli space of all J-holomorphic curves gives a solution to this equation, thus providing a mathematical foundation of the quantum field theory aspects of the Gromov-Witten theory.

The other contributions are organized into five sections: Feynman Diagrams, Algebraic Structures, Manifolds: Invariants and Mirror Symmetry, Combinatorial Aspects of Dynamics, and Physics. This classification is rather conventional as there is a strong interplay between ideas and methodology presented in different contributions which have a common origin in topology and quantum physics.

The articles in the volume are ordered in such a way that survey-style papers are followed (within a given section) by more special research contributions.

Feynman diagrams

This section opens with a survey "Feynman diagrams for pedestrians and mathematicians" by M. Polyak. It introduces the reader to the machinery of Feynman diagrams (starting with a finite dimensional model), and then explains how it can be combined with the Chern-Simons theory to produce "quantum" invariants of knots and three-manifolds.

The next paper, "Structures of Feynman graphs: Hopf algebras and symmetries" by D. Kreimer describes algebraic structures on the space of Feynman diagrams that puts the classical renormalization procedure on a regular basis.

Algebraic structures

The survey "Notes on universal algebra" by A. Voronov introduces the reader to Hochschild cohomology and deformation quantization, operad theory, and graph homology. In particular, it contains a sketch of Kontsevich's proof of the celebrated Formality Theorem, as well as ideas of its proof by Cattaneo and Felder motivated by quantum field theory (based, again, on the machinery of Feynman diagrams).

The paper "The ring of differential operators on forms in non-commutative calculus" by D. Tamarkin and B. Tsygan gives a non-commutative generalization of the Hochschild-Kostant-Rosenberg Theorem that identifies the Hochschild cohomology of the commutative algebra of smooth functions on a manifold M with the algebra of polyvector fields on M. In particular, it gives an extension of the Formality Theorem to the case of an arbitrary associative algebra.

In the paper "Twisted chiral de Rham algebras on $\mathbb{P}^1$" by V. Gorbounov, F. Malikov and V. Schechtman, the authors continue their exploration of chiral de Rham complexes, namely, they construct a twisted chiral de Rham complex of vertex algebras over $\mathbb{P}^1$.

Manifolds: invariants and mirror symmetry

In the paper "Invariants of tangles with flat connections" by R. Kashaev and N. Reshetikhin, the authors define invariants of tangles with flat connections in their complements extending the classical work of Reshetikhin and Turaev.

In the next paper, "Tree-level invariants of three-manifolds, Massey products and the Johnson homomorphism" by S. Garoufalidis and J. Levine, the tree-level part of the finite type invariants for 3-manifolds is expressed in terms of classical algebraic topology.

In the paper "Multivalued Morse theory, asymptotic analysis, and mirror symmetry", K. Fukaya proposes a new geometric approach to the homological mirror symmetry conjecture for Calabi-Yau manifolds represented as dual torus fibrations with the same base.

Combinatoral aspects of dynamics

In the first paper of this section, "Some applications of combinatorial differential topology", R. Forman develops Morse theory for cell complexes based on the dynamics of discrete vector fields, and discusses its applications in topology, geometry, and other fields.

The paper "Extensions, quotients and generalized pseudo-Anosov maps" by A. de Carvalho describes the interplay between one- and two-dimensional dynamics (on graphs and on surfaces). This leads the author to a notion of a generalized pseudo-Anosov map whose stable/unstable foliations may have infinitely many singularities.

The survey "Unimodal maps and hierarchical models" by M. Yampolsky overviews ideas unifying the notions of renormalization in statistical mechanics and one-dimensional dynamics. "Microscopic renormalization" acting on Hamiltonians on binary trees plays a prominent role in this discussion.

Physics

The last section contains two contributions written from the physicist's point of view. A non-technical survey "Quantum geometry in action: big bang and black holes" by A. Ashtekar describes the "spin-network model" approach to quantum gravity and provides an extensive bibliography on the subject. The paper "Supersymmetry, supergravity, superspace and BRST symmetry in a simple model" by P. van Nieuwenhuizen works out ideas of the supersymmetry in the simplest one-dimensional example.

Acknowledgment. We would like to thank S. Gelfand, A. Kirillov, Jr., and all the referees for their help in editing this volume, and C. Cabrera-Ocanas, M. Milea, S. Sutherland, and S. Wilson for technical help. The poster of the Conference "Graphs and Patterns in Mathematics and Theoretical Physics" was designed by A. Phillips (with technical help by S. Simanca and S. Sutherland). The Conference was supported by the Simons Math-Physics Endowment, the NSF, and the Stony Brook Research Foundation.

Mikhail Lyubich and Leon Takhtajan

Dennis Sullivan - A Short History

Dennis Sullivan was born on February 12, 1941 in Port Huron, Michigan; but he spent his formative years in Houston and has since presented himself as a Texan. He graduated from Rice in 1963 and came to Princeton, where he completed his Ph.D. in two years under the direction of William Browder.

Browder's students at that time were mining surgery theory for topological gold. Dennis' thesis was in this vein and led to his work on the Hauptvermutung (1967). After graduation he held fellowships at Warwick and at Berkeley; he then spent two years on the Princeton faculty and four at MIT. During this time his work focused on what he named geometric topology (in particular the study of Galois symmetry) and on the construction of minimal models for the rational homotopy type of manifolds, using differential forms.

In 1973-74, Dennis visited the University of Paris-Orsay. He remained in France as *Professeur permanent* at the IHES, full-time until 1981, when he was named Einstein Professor at CUNY, and half-time after that until 1996, when he joined the Mathematics Department and the IMS at Stony Brook.

During his years in France, his interests expanded first towards dynamical systems, including ergodic theory, foliations, Kleinian groups, and renormalization, and then, motivated originally by problems in conformal dynamics, towards Teichmüller theory. Dennis's proof of the No Wandering Domains Theorem (1982) led to revival of holomorphic dynamics after 60 years of stagnation, and his ideas in the renormalization theory linking real and complex dynamics (1980's) left a long-lasting trace in the field.

His recent work addresses problems of algebraic topology related to the quantum field theory and the string theory.

The reader can go through the list of Dennis's publications to see a broader picture of his scientific life.

Anthony Phillips

Dennis Sullivan's List of Publications

[1] *Triangulating homotopy equivalences.* Ph.D. Thesis, Princeton, (1966) 187 pages. (See [105]).

[2] *Triangulating homotopy equivalences.* Preprint, Univ. of Warwick (1967), 10 pages.

[3] *Smoothing homotopy equivalences.* Preprint, Univ. of Warwick (1967), 12 pages.

[4] *On the Hauptvermutung for manifolds.* Bull. Amer. Math. Soc. **73** (1967), 598-600.

[5] *On the regular neighborhood of a two-sided submanifold* (with M. Cohen). Topology **9** (1970), 141-147.

[6] *Geometric topology I. Localization, periodicity and Galois symmetry.* MIT, 1970, Mimeogr. notes, 432 pages. (See [20] and [117]).

[7] *Combinatorial invariants of analytic spaces.* Proc. Liverpool Singularities Symposium, I (1969/70), 165-168, Lecture Notes in Math. **192** Springer, Berlin, 1971.

[8] *Geometric periodicity and the invariants of manifolds.* Manifolds – Amsterdam 1970, 44-75, Lecture Notes in Math. **197** Springer, Berlin, 1971.

[9] *Galois symmetry in manifold theory at the primes.* Actes du Congrès International des Mathèmaticiens (Nice, 1970), Tome 2, pp. 169–175. Gauthier-Villars, Paris, 1971. (See [117]).

[10] *Singularities in spaces.* Proc. of Liverpool Singularities Symposium, II, 196-206, Lecture Notes in Math. **209** Springer, Berlin, 1971.

[11] *On the Kervaire obstruction* (with C. Rourke). Ann. of Math. (2) **94** (1971), 397-413.

[12] *The signature, geometric cycles, and K-theory.* Manuscript 1973/74. (See P. Siegel, *Witt spaces: a geometric cycle theory for K_0 homology at odd primes.* Amer. J. Math. **105** (1983), 1067-1105).

[13] *The transversality characteristic class and linking cycles in surgery theory* (with J. Morgan). Ann. of Math. (2) **99** (1974), 463-544.

[14] *Genetics of homotopy theory and the Adams conjecture.* Ann. of Math. (2) **100** (1974), 1-79.

[15] *A remark on the Lefschetz fixed point formula for differentiable maps* (with M. Shub). Topology **13** (1974), 189-191.

[16] *Inside and outside manifolds.* Proceedings of the International Congress of Mathematicians (Vancouver, B. C., 1974), Vol. 1, pp. 201–207. Canad. Math. Congress, Montreal, 1975.

[17] *Differential forms and the topology of manifolds.* Manifolds – Tokyo 1973 (Proc. Internat. Conf., Tokyo, 1973), pp. 37–49. Univ. Tokyo Press, Tokyo, 1975.

[18] *Homology theory and dynamical systems* (with M. Shub). Topology **14** (1975), 109-132.

[19] *On the intersection ring of compact three manifolds.* Topology **14** (1975), 275-277.

[20] "Геометрическая топология. Локализация, периодичность и симметрия Галуа". Библиотека Сборника "Математика". Издат. "Мир", Москва, 1975, 284 стр. (Russian) "Geometric topology. Localization, periodicity and Galois symmetry". Library of the Journal "Matematika", translated from the English and edited by D.B. Fuks.

[21] *Genericity theorems in topological dynamics* (with J. Palis, C. Pugh, M. Shub). Dynamical systems – Warwick 1974 (Proc. Sympos. Appl. Topology and Dynamical Systems, Univ. Warwick, Coventry, 1973/1974), 241–250, Lecture Notes in Math. **468**, Springer, Berlin, 1975.

[22] *A topological invariant of flows on 1-dimensional spaces* (with B. Parry). Topology **14** (1975), 297-299.

[23] *Currents, flows and diffeomorphisms* (with D. Ruelle). Topology **14** (1975), 319-327.

[24] *Real homotopy theory of Kähler manifolds* (with P. Deligne, P. Griffiths and J. Morgan). Invent. Math. **29** No. 3 (1975). 245-274. See also Seminaire Bourbaki, **1975/76**, Exp. No. 475, 69-80. Lecture Notes in Math. **567**, Springer, Berlin, 1977.

[25] *The Euler characteristic of an affine space form is zero* (with B. Kostant). Bull. Amer. Math. Soc. **81** (1975), 937-938.

[26] *La classe d'Euler reelle d'un fibre vectoriel a groupe structural* $SL_n(Z)$ *est nulle.* C.R. Acad. Sci. Paris Sér. A-B. **281**, (1975), A17-A18.

[27] *Geometrie differentielle. Sur l'existence d'une infinite de geodesiques periodiques sur une variete Riemannienne compacte.* C.R. Acad. Sci. Paris Sér. A-B, **281** (1975), A289-A291.

[28] *Complex vector bundles with discrete structure group* (with P. Deligne). C.R. Acad. Sci. Paris Sér. A-B. **281**, (1975), A1081-A1083.

[29] *A semi-local combinatorial formula for the signature of a 4k-manifold* (with A. Ranicki). J. Differential Geom. **11** (1976), 23-29.

[30] *On the homology of attractors* (with R.F. Williams). Topology **15** (1976), 259-262.

[31] *A generalization of Milnor's inequality concerning affine foliations and affine manifolds.* Comment. Math. Helv. **51** (1976), 183-189.

[32] *A counterexample to the periodic orbit conjecture.* Inst. Hautes Études Sci. Publ. Math. **46** (1976), 5-14.

[33] *Cycles for the dynamical study of foliated manifolds and complex manifolds* (dedicated to Jean-Pierre Serre). Invent. Math. **36** (1976), 225-255.

[34] *The homology theory of the closed geodesic problem* (with M. Vigué Poirrier). J. Differential Geom. **11** (1976), 633-644.

[35] *Cartan-de Rham homotopy theory.* Colloque "Analyse et topologie" en l'honneur de Henri Cartan (Orsay 1974), Asterisque **32-33** (1976), 227-254.

[36] *A new flow.* Bull. Amer. Math. Soc. **82** (1976), 331-332.

[37] *Real homotopy theory of Kähler manifolds* (with F. Griffiths, P. Deligne, P. Morgan) (Russian). Uspehi Mat. Nauk **32** (1977), 119–152.

[38] *Foliations with all leaves compact* (with R. Edwards, K. Millett). Topology **16** (1977), 13-32.

[39] *Infinitesimal computations in topology.* Inst. Hautes Études Sci. Publ. Math. **47** (1977), 269-331.

[40] *A foliation of geodesics is characterized by having no "tangent homologies"* (George Cooke memorial volume). J. Pure Appl. Algebra **13** (1978), 101-104.

[41] *On complexes that are Lipschitz manifolds* (with L. Siebenmann). Geometric topology (Proc. Georgia Topology Conf., Athens, Ga., 1977), pp. 503–525, Academic Press, New York, London, 1979.

[42] *Hyperbolic geometry and homeomorphisms.* Geometric topology (Proc. Georgia Topology Conf., Athens, Ga., 1977), pp. 543–555, Academic Press, New York, London, 1979.

[43] *A homological characterization of foliations consisting of minimal surfaces.* Comment. Math. Helv. **54** (1979), 218-223.

[44] *On the density at infinity of a discrete group of hyperbolic motions.* Inst. Hautes Études Sci. Publ. Math. **50** (1979), 171-202.

[45] *Rufus Bowen (1947-1978).* Inst. Hautes Études Sci. Publ. Math. **50** (1979), 7-9.

[46] *Travaux de Thurston sur les groupes quasifuchsiens et les varietes hyperboliques de dimension 3 fibrees sur S^1.* Seminaire Bourbaki, 1979/80, 196-214, Lecture Notes in Math. **842**, Springer, Berlin, 1981.

[47] *For $n > 3$, there is only one finitely additive measure on S^n rotationally invariant and defined on all Lebesgue measurable sets.* Bull. Amer. Math. Soc. (N.S.) **4** (1981), 121-123.

[48] *A finiteness theorem for cusps.* Acta Math. **147** (1981), 289-299.

[49] *Growth of positive harmonic functions and Kleinian group limit sets of zero planar measure and Hausdorff dimension two.* Geometry Symposium (Nicolaas Kuiper volume), Utrecht 1980 (Utrecht, 1980), pp. 127–144, Lecture Notes in Math. **894**, Springer, Berlin-New York, 1981.

[50] *Zariski dynamics of a homotopy type* (with R. Body). Manuscript 1978. (See [112]).

[51] *Geometry of leaves* (with A. Phillips). Topology **20** (1981), 209-218.

[52] *On lifting automorphisms of monotone σ-complete C^*-algebras* (with J.D. Maitland Wright). The Quart. J. Math. Oxford Ser. (2) **32** (1981), 371-381.

[53] *On the ergodic theory at infinity of an arbitrary discrete group of hyperbolic motions.* Riemann surfaces and related topics: Proceedings of the 1978 Stony Brook Conference (State Univ. New York, Stony Brook, N.Y., 1978), pp. 465–496, Ann. of Math. Stud., **97**, Princeton Univ. Press, Princeton, 1981.

[54] *Disjoint spheres, Diophantine approximation, and the logarithm law for geodesics.* Acta Math. **149** (1982), 215-237.

[55] *Discrete conformal groups and measurable dynamics.* Bull. Amer. Math. Soc. (N.S.) **6** (1982), 57-73.

[56] *Seminar on conformal and hyperbolic geometry*, notes by M. Baker and J. Seade. Preprint IHES, 1982, 92 pp.

[57] *Iteration des fonctions analytiques complexes*, C.R. Acad. Sci. Paris Sér. I Math. **294** (1982), 301-303.

[58] *On the dynamics of rational maps* (with R. Mané, P. Sad). Ann. Sci. École Norm. Sup. (4) **16** (1983), 193-217.

[59] *Conformal dynamical systems.* Geometric dynamics (Rio de Janeiro, 1981), 725–752, Lecture Notes in Math. **1007**, Springer, Berlin, 1983.

[60] *Manifolds with canonical coordinates charts: some examples* (with W. Thurston). Enseign. Math. (2) **229** (1983), 15-25.

[61] *Division algebras and the Hausdorff-Banach-Tarski paradox* (with P. Deligne). Enseign. Math. (2) **29** (1983), 145-150.

[62] *The Dirichlet problem at infinity for a negatively curved manifold.* J. of Differential Geom. **18** (1983), 723-732.

[63] *An analytic proof of Novikov's theorem on rational Pontrjagin classes* (with N. Teleman). Inst. Hautes Études Sci. Publ. Math. **58** (1983), 79-81.

[64] *On the iteration of a rational function: computer experiments with Newton's method* (with J.H. Curry and L. Garnett). Comm. Math. Phys. **91** (1983), 267-277.

[65] *Entropy, Hausdorff measures old and new, and limit sets of geometrically finite Kleinian groups*. Acta Math. **153** (1984), 259-277.

[66] *Function theory, random paths, and covering spaces* (with T. Lyons). J. Differential Geom. **19** (1984), 299-323.

[67] *Rational ergodicity of geodesic flows* (with J. Aaronson). Ergodic Theory and Dynam. Systems **4** (1984), 165-178.

[68] *Brownian motion and harmonic functions on the class surface of the thrice punctured sphere* (with H. McKean). Adv. in Math. **51** (1984), 203-211.

[69] *Nilpotent bases for distributions and control systems* (with H. Hermes, A. Lundell). J. of Differential Equations **55** (1984), 385-400.

[70] *Quasiconformal homeomorphisms and dynamics. I. Solution of the Fatou-Julia problem on wandering domains*. Ann. of Math. (2) **122** (1985), 401-418.

[71] *Quasiconformal homeomorphisms and dynamics. II. Structural stability implies hyperbolicity for Kleinian groups*. Acta Math. **155** (1985), 243-260.

[72] *Expanding endomorphisms of the circle revisited* (with M. Shub). Ergodic Theory and Dynam. Systems **5** (1985), 285-289.

[73] *On the measurable dynamics of the exponential map* (with E. Ghys and L. Goldberg). Ergodic Theory and Dynam. Systems **5** (1985), 329-335.

[74] *On the dynamical structure near an isolated completely unstable ellipitic fixed point*. Atas do 14° Colóquio Brasileiro de Matemática, vol. II, Publ. do IMPA, 1985, 529–542.

[75] *A relation between probability, geometry and dynamics – the random path of the heat equation on a Riemannian manifold*. Atas do 14° Colóquio Brasileiro de Matemática, vol. II, Publ. do IMPA, 1985, 553–559.

[76] *Related aspects of positivity: λ-potential theory on manifolds, lowest eigenstates, Hausdorff geometry, renormalized Markov processes ....* Aspects of Mathematics and its Applications. North-Holland, Math. Library **34** 747-779, North-Holland, Amsterdam, 1986. (Same as [82] with longer introduction).

[77] *Generic dynamics and monotone complete C^*-algebras* (with B. Weiss, J. Wright). Trans. Amer. Math. Soc. **295** (1986), 795-809.

[78] *Ground state and lowest eigenvalue of the Laplacian for noncompact hyperbolic surfaces* (with T. Pignataro). Comm. Math. Phys. **104** (1986), 529-535.

[79] *Quasiconformal homeomorphisms in dynamics, topology, and geometry*. Proceedings of the International Congress of Mathematicians, Vol. 1,2, (Berkeley, Calif., 1986), 1216–1228, Amer. Math. Soc., Providence, RI, 1987. (See [92]).

[80] *Extending holomorphic motions* (with W. Thurston). Acta Math. **157** (1986), 243-257.

[81] *The convergence of circle packings to the Riemann mapping* (with B. Rodin). J. Differential Geom. **26** (1987), 349-360.

[82] *Related aspects of positivity in Riemannian geometry*. J. Differential Geom. **25** (1987), 327-351.

[83] *Estimating small eigenvalue of Riemann surfaces* (With J. Dodziuk, T. Pignataro and B. Randol). The legacy of Sonya Kovalevskaya (Cambridge, Mass., and Amherst, Mass., 1985), 93–121, Contemp. Math. **64**, Amer. Math. Soc., Providence, RI, 1987.

[84] *The spinor representation of minimal surfaces in space*. Preprint Texas 1986 (a letter to H. Rosenberg).

[85] *On negative curvature, variable, pinched and constant*. Preprint Texas 1986. (See [79]).

[86] *Differentiable structures on fractal-like sets, determined by intrinsic scaling functions on dual Cantor sets*. Proceedings of the Herman Weyl Symposium, Duke University, Proc. Sympos. Pure Math. **48** (1988), 15-23. See also: "Nonlinear

Evolution and Chaotic Phenomena" (Noto, 1987), 101-110, NATO Adv. Sci. Inst. Ser. B Phys., 176, Plenum, New York, 1988.

[87] *Variation of the Green function on Riemann surfaces and Whitney's holomorphic stratification conjecture* (with R. Hardt, René Thom volume). Inst. Hautes Études Sci. Publ. Math. **68** (1988), 115-137.

[88] *Quasiconformal 4-manifolds* (with S. K. Donaldson). Acta Math. **163** (1989), 181-252.

[89] *Bounded structure of infinitely renormalizable mappings.* (The quasiconformal explanation of the universal constants in one dimensional dynamical bifurcations), Cvitanović, Universality of Chaos, 2nd Edition, Adam Hilger, Bristol, 1989. (See [92]).

[90] *On conformal welding homeomorphisms associated to Jordan curves* (with Y. Katznelson and S. Nag). Ann. Acad. Sci. Fenn. Ser. A I Math. **15** (1990), 293-306.

[91] *Renormalization, Zygmund smoothness and the Epstein class.* Chaos, order, and patterns. (Lake Como, 1990), 25-34, Nato Adv. Sci. Inst. Ser. B Phys. **280**, Plenum, New York, 1991.

[92] *Bounds, quadratic differentials, and renormalization conjectures.* American Mathematical Society Centennial Publications, **II** (Providence, RI, 1988), 417-466, Amer. Math. Soc., Providence, RI, 1992.

[93] *Quantized calculus on the circle and quasi fuchsian groups* (with A. Connes). Preprint IHES. (See A. Connes "Non-commutative geometry", Academic Press, Inc., San Diego, CA, 1994).

[94] *Expanding directions of period doubling operator* (with Y. Jiang and T. Morita). Comm. Math. Phys. **144** (1992), 509-520.

[95] *Symmetric structures on a closed curve* (with F. Gardiner). Amer. J. Math. **114** (1992), 683-736.

[96] *Formules locales pour les classes de Pontrjagin topologiques* (with A. Connes and N. Teleman). C.R. Acad. Sci. Paris Sér. I Math. **317** (1993), 521-526.

[97] *Linking the universalities of Milnor-Thurston, Feigenbaum and Ahlfors-Bers.* Topological Methods in Modern Mathematics, 543-564 (The Proceedings of Symposium held in honor of John Milnor's 60th Birthday, SUNY at Stony Brook, 1991), Publish or Perish, Houston, TX, 1993.

[98] *Infinite cascades of braids and smooth dynamical systems* (with J.-M. Gambaudo and C. Tresser). Topology **33** (1994), 85-94.

[99] *Quasiconformal mappings, operators on Hilbert space, and local formulae for characteristic classes* (with A. Connes and N. Teleman). Topology **33** (1994), 663-681.

[100] *Lacunary series as quadratic differentials in conformal dynamics* (with F. Gardiner). The mathematical legacy of Wilhelm Magnus: groups, geometry and special functions (Brooklyn, NY, 1992), 307-330, Contemp. Math. **169**, Amer. Math. Soc., Providence, RI, 1994.

[101] *Exterior d, the local degree, and smoothability.* Prospects in topology (Princeton, NJ, 1994), 328-338, Ann. of Math. Stud. **138**, Princeton Univ. Press, Princeton, NJ, 1995.

[102] *Teichmüller theory and the universal period mapping via quantum calculus and the $H^{1/2}$ space on the circle* (with S. Nag). Osaka J. Math. **32** (1995), 1-34.

[103] *Wandering domains and invariant conformal structures for mappings of the 2-torus* (with A. Norton). Ann. Acad. Sci. Fenn. Math. **21** (1996), 51-68.

[104] *Determinant bundles, Quillen metrics and Mumford isomorphisms over the universal commensurability Teichmuller space* (with I. Biswas and S. Nag). Acta Math. **176** (1996), 145-169.

[105] *Triangulating and smoothing homotopy equivalences and homeomorphisms.* The Hauptvermutung book, 69-103. A collection of papers of the topology of manifolds. K-Monographs in Mathematics **1**. Kluwer Academic Publishers, Dordrecht, 1996.

[106] *Dynamics of geometrically finite rational maps* (with G. Cui and Y. Jiang), Manuscript 1996.

[107] *Topological conjugacy of circle diffeomorphisms* (with Jun Hu). Ergodic Theory and Dynam. Systems **17** (1997), 173-186.

[108] *Quasiconformal homeomorphisms and dynamics. III. The Teichmüller space of a holomorphic dynamical system* (with C. McMullen). Adv. Math. **135** (1998), 351-395.

[109] *P-universal spaces and rational homotopy types* (with R. Body, M. Mumura, and H. Shiga). Comment. Math. Helv. **73** (1998), 427–442.

[110] *On the foundation of geometry, analysis, and the differential structure for manifolds.* Topics in low-dimensional topology (In Honor of Steve Armentrout. Proceedings of the Conference on low-Dimensional Topology, University Park, PA, 1996), 89–92, World Sci. Publishing, River Edge, NJ, 1999.

[111] *String topology* (with M. Chas). math.GT/9911159.

[112] *Reminiscences of Michel Herman's first great theorem.* In "Michael R. Herman" Gaz. Math. No. 88 (2001). Société Mathématique de France, Paris, 2001, 91–94.

[113] *On the locally branched Euclidean metric gauge* (with J. Heinonen). Duke Math. J. 114 (2002), 15–41.

[114] *Closed string operators in topology leading to Lie bialgebras and higher string algebra* (with M. Chas). Proceedings of the Abel Bicentenniel Symposia, Oslo Norway 2002.

[115] *Open and closed string field theory interpreted in classical algebraic topology.* Proceedings of Oxford Symposium for Graeme Segal "Geometry, Topology, and Quantum Field Theory", 2003, ed. U. Tillman, Cambridge University Press, 2004, 344-357.

[116] *René Thom's work on geometric homology and bordism.* Bull. Amer. Math. Soc. (N.S.) **41** (2004), 341–350.

[117] *Geometric topology, Part I: Localization, periodicity, and Galois symmetry.* K-Monographs in Math, Kluwer Academic Publishers, Dordrecht, 2004.

[118] *Higher genus string topology* (with M. Chas). Manuscript 2004.

[119] *Dynamical systems applied to asymptotic geometry* (with A. Pinto). IMS at Stony Brook University, Preprint No. 6, 2004.

[120] *Sigma models and string topology.* These Proceedings.

Sigma models and string topology

Dennis Sullivan

ABSTRACT. Gromov's homology class of J-holomorphic curves is shown to satisfy at the chain level the quantum master equation $\partial S = \Delta S + \frac{1}{2}\{S,S\}$ in a background described by algebraic topology and transversality. Conjecturally, this provides a mathematical interpretation of Witten's discovery of the corresponding Gromov homology periods as the correlations of a quantum field theory. The basic idea is due to Zwiebach. A new concept of unbounded algebraic structures arising from transversality is used.

Contents

1. Short introduction
2. Longer introduction
3. The modular operad and BV algebra structure
4. The Gromov chain of perturbed J-holomorphic curves
References

1. Short introduction

Associated to any oriented smooth manifold M of even dimension d, the target manifold of the sigma model[1], there is a modular operad structure (C_{ij}, S_{ij}) on one version, and a related BV algebra structure[2] $(\delta, \{\,,\,\})$ on another version of an equivariant singular chain complex associated to the mapping spaces $\mathrm{Map}(\Sigma, M)$ of the corresponding sigma model. These structures are in the unbounded sense

---

2000 *Mathematics Subject Classification.* 55N, 57R, 17B.

Key words and phrases. BV algebra, quantum master equation, Gromov chain.

[1]Sigma models are putative quantum field theories where the classical fields are based on all maps of all Riemann surfaces probing a target Riemannian manifold M, and a classical action on these is given by the energy of the map.

[2]Used by Batalin and Vilkovisky [**2**] in an algebraic quantization scheme for classical field theories with constraints. In the sigma model the defining equations of the target M could be considered to be constraints. A BV algebra is a graded commutative algebra with an operator Δ of degree -1 so that $\Delta \cdot \Delta = 0$ and the deviation of Δ from being a derivation, $\{\,,\,\}$, is itself a derivation in each variable. It follows $\{\,,\,\}$ is a Lie bracket of degree -1 and Δ is a derivation of $\{\,,\,\}$.

© 2005 American Mathematical Society

(see Section 3). Here Σ runs over diffeomorphism types of closed oriented connected surfaces of genus g with n marked points (i.e., there is one Σ for every (g,n)). After an even shift of gradings by $(d-6)(1-g) + 2n + c(\beta)$, $\beta \in H_2(M)$, c linear and even, the degrees of the chain complex ∂ operator and transversal gluing operations C_{ij}, S_{ij}, δ, Δ, and $\{\ ,\}$ are all equal to -1. A solution S in degree zero of the quantum master equation, or $\Delta S + \frac{1}{2}\{S,S\} = \partial S$ in the BV algebra, will be treated homologically in a future publication.

When M is almost complex, the dimension shift $(d-6)(1-g) + 2n + 2c_1(\beta)$ gives the formal real dimension of the piece of the moduli space corresponding to J-holomorphic curves of genus g and n marked points in the homology class of β in $H_2(M)$. A J-holomorphic curve in M is an equivalence class of a pair (complex structure on Σ, map of Σ into M) such that $\bar{\partial}(\text{map}) = 0$.

When M is closed symplectic, one can introduce constraints in the complex structure of Σ and the concentration of energy in the map defining a J-holomorphic curve or a perturbed version. Then for each g, n, and β, the set of perturbed J-holomorphic curves defines a compact set and, assuming enough transversality, an oriented chain. These chains with boundary yield a degree zero solution of the quantum master equation in the BV algebra using the dual pictures of Gromov [8] and Sen-Zwiebach [12]. The homological treatment will determine Gromov-Witten invariants.

The transversality and parametrized surgery defining the unbounded structure maps are just like those in String Topology [3, 14]. Underlying both cases, here and [3], is the structure of an unbounded modular operad with Δ and $\{\ ,\}$ added in. See the longer introduction below for the definition of modular operad and more detailed descriptions.

The discussion here for general M may be viewed as a chain level or off-shell background for the closed string A-model or Gromov-Witten theory, defined when the target M is closed and symplectic, but which may be formulated more generally in terms of the quantum master equation of the BV algebra. One can ask if there is also a relation with the B-model defined when M is complex with holomorphic volume form.

Whereas the above discussion uses only closed surfaces Σ with finite subsets I of interior marked points, there is also a construction for surfaces with boundary mapping to M where pieces of the boundary between marked boundary points could be required to land in various submanifolds of M. Using parametrized interior and ∂ connected sum one obtains an (unbounded) generalized structure to the one presented here. This generality is needed to discuss open and closed strings and D-branes, see [14] for a beginning.

2. Longer introduction

A singular chain complex of the smooth mapping spaces $\text{Map}(\Sigma, M)$ of closed oriented surfaces Σ into a target manifold M of even dimension d, taken altogether as Σ varies and equivariantly with respect to diffeomorphism groups of the Σ's (see Subsection 3.1) has an algebraic structure called *unbounded modular operad*. This means [7] there are non-negative graded chain complexes $(C(I), \partial)$ with an additional direct sum decomposition over the genus $C(I) = \bigoplus_{g=0}^{\infty} C(g, I)$ functorially attached to finite sets $I, J, \ldots$, with unbounded chain maps

$$C(I) \otimes C(J) \xrightarrow{C_{ij}} C(I \cup J - \{i,j\})$$

with $i \in I, j \in J$, and unbounded chain maps $C(I) \xrightarrow{S_{ij}} C(I - \{i,j\})$ with $\{i,j\} \subset I$. C_{ij} is defined for each pair of families of maps of Σ into M as a parametrized connected sum at i and j of different surfaces along the locus where the maps of the family transversally (over the base) identify i and j. S_{ij} is defined for each family of maps as a parametrized self-connected sum at i and j along the locus where the map of the fibre transversally identifies i and j. These structure maps act on bidegrees by

$$(g_1, k) \otimes (g_2, l) \mapsto (g_1 + g_2, k + l - d + 1)$$

and $(g, k) \mapsto (g + 1, k - d + 1)$ where $d = \dim M$, $g = \text{genus } \Sigma$, and k, l are the geometric dimensions of the equivariant chains. The structure maps satisfy the relations that any compositions are associative and anticommute (see Subsection 3.3). The structure maps C_{ij} and S_{ij} are unbounded in the sense they are only defined on dense core domains which are subcomplexes of $\otimes_\alpha C(I_\alpha)$. These subcomplexes satisfy inclusion relations and have isomorphic homology to the entire complexes (see Subsection 3.2). These unbounded (or partial) operations are sufficient for our homological or derived category purposes because there are functorial quasi-isomorphic completion constructions (in two senses, see [**15**] motivated by [**9**]). The core domains are defined by transversality conditions, and the structure maps are defined by parametrized surgery both analogous to String Topology [**3**]. The geometric operations C_{ij} and S_{ij} for d odd will not be discussed further here.

There is a further direct sum decomposition of $C(g, I) = \oplus_\beta C(\beta, g, I)$ where β ranges over all elements in $H_2(M)$ which may be represented by maps of a genus g surface into M.

THEOREM 1. *When d is even, the chain complexes $C(I) = \oplus_g \oplus_\beta C(g, \beta, I)$ form a modular operad (in the unbounded sense explained above). After an even grading shift down by $(d-6)(1-g) + 2n + c(\beta)$, where c is linear in β and even, all the operations ∂, C_{ij}, and S_{ij} have degree -1.*

Now we add more structure to the modular operad. We let $C = C(\emptyset)$, and we define $\{\,,\,\}$ and δ (Subsection 3.4) by the compositions

$$C \otimes C \xrightarrow{M_1 \otimes M_1} C(\{i\}) \otimes C(\{j\}) \xrightarrow{C_{ij}} C$$

and

$$C \xrightarrow{M_2} C(\{i,j\}) \xrightarrow{S_{ij}} C$$

by adding marked points via operators M_1 and M_2. An anomaly appears in the structure of δ, which is discussed in Remark 3.10.

THEOREM 2. *When d is even and c is even, $(C, \{\,,\,\}, \delta)$ is defined and is an unbounded differential Lie algebra of degree -1. (See Subsection 3.4)*

REMARK 2.1. $(C, \delta, \{\,,\,\})$ is a Lie algebra of degree -1 means $[a, b] = (-1)^{|a|}\{a, b\}$ is the bracket of a differential Lie algebra (of degree zero) on the graded space C shifted by one.

We can consider the Maurer-Cartan equation $\delta S + \frac{1}{2}\{S,S\} = \partial S$ where S has degree zero. A solution here implies a solution to the equation $\Delta\, e^S = \partial e^S$ in the graded symmetric algebra ΛC generated by C. Here Δ is the (second order) derivation on ΛC obtained by adding together both δ and $\{\ ,\}$ extended to be coderivations of ΛC. The equation $\Delta \cdot \Delta = 0$ summarizes the differential Lie algebra properties of $(C, \{\ ,\}, \delta)$. The second order nature and nilpotence of Δ comes by viewing Δ as the extension of δ plus $\{\ ,\}$ to all disconnected surfaces by summing over the possible gluings (i,j). Then since $a \wedge b$ can be viewed in terms of disjoint union of surfaces and maps over the product of the bases for a and for b,

$$\Delta(a \wedge b) = (\Delta a) \wedge b + (-1)^{|a|} a \wedge \Delta b + \{a,b\},$$

where $\{a,b\}$ denotes the extension of $\{\ ,\}$ on C to all of ΛC to a binary operation which is a graded derivation in each variable. The equation for S in terms of Δ is called the quantum master equation and can be written either as $\Delta S + \frac{1}{2}\{S,S\} = \partial S$, or $\Delta\, e^S = \partial e^S$.

THEOREM 3. *When d is even and $c(\beta)$ is even, $(\Lambda C, \Delta)$ is defined and is an unbounded BV algebra attached to the spaces $\mathrm{Map}(\Sigma, M)$ modulo diffeomorphisms of the source. Δ is a second order derivation and a coderivation. $(\Lambda C, \Delta, \partial)$ is a BV background for the sigma model in physics[3] constructed by algebraic topology and transversality.*

(See Subsection 3.4 for the proof).

2.1. Application. As mentioned above, the even integer $(d-6)(1-g) + 2n + c(\beta)$ used to shift the chain complexes $C(g, I)$ above, when $c(\beta) = 2c_1(\beta)$ yields the formal dimension of the moduli space of J-holomorphic maps of Σ into an almost complex manifold (M, J) representing the homology class β in $H_2(M)$, $n = $ cardinality of I.

If the target manifold M is a closed symplectic manifold with 2-form ω, and J is a compatible almost complex structure ($\omega(x, Jy)$ is symmetric and positive definite), Gromov studied the J-holomorphic mappings $\Sigma \xrightarrow{f} M$ of Riemann surfaces into M and analyzed the non-compactness of the space of these in the given homology class β in $H_2(M)$. He found exactly the familiar non-compactness in the complex structure of Σ corresponding to pinching curves, together with Freed-Uhlenbeck bubbling off of 2-spheres in the map into M. Bubbling-off or pinching-off an essential separating curve in the complex structure of Σ will be seen to be inverse to an operation like C_{ij}. Similarly, pinching-off a nonseparating curve is inverse to an operation like S_{ij}. Thus, if one introduces cut-off inequalities to prevent these pinching and bubbling phenomenon from happening (see Section 4) one obtains for each $C(\beta, g)$ a compact family of J-holomorphic curves. Using $\mathbb{Q}$ coefficients and

---

[3]In the physics discussions the ordinary maps of surfaces Σ into the target M are augmented by (contractible) odd degree components. A (formal) symplectic manifold of odd degree appears, whose functions define (formally) an odd version of Poisson algebra with bracket derived from a BV operator [**2**]. The functions are closely related to differential forms on the original mapping spaces. New variables are added to form the Borel-Cartan version of equivariant forms relative to diffeomorphisms of Σ, the BV algebra structure extends formally and a nondegenerate solution of the quantum master equation is sought where the leading term is a classical action in the original superfields. The structure is formal. However, see [**1**] for a rigorous finite dimensional analogue.

perturbing the equation the (relative) cycle is defined [**10**], [**6**] and has a boundary essentially described by a sum over the operations $\{\,,\}$ and δ. In this way (after shifting each $C(\beta, g)$ down by $(d-6)(1-g) + 2c_1(\beta)$ we obtain a solution S to the quantum master equation of degree zero, where S assigns to each β and each genus of Σ the oriented Gromov moduli space compactified by cut-off, plus small collars.

THEOREM 4 (Gromov, Eliashberg, Fukaya, Hofer, Kontsevich, Li, McDuff, Ono, Ruan, Salomon, Sullivan, Tian, ..., Zwiebach). *For a closed symplectic manifold M the set of all cut-off oriented moduli spaces of perturbed connected J-holomorphic curves defines a Q-chain which provides a degree zero solution to the quantum master equation $\partial S = \Delta S + \frac{1}{2}\{S, S\}$ in an unbounded BV algebra attached to the sigma model of mapping spaces with target M.*

(See Section 4 for the proof.)

REMARK 2.2. The dots in the theorem refer to anyone omitted who has contributed to the formidable task of making the picture of the Gromov chain or homology class into rigorous mathematics.

Let us look at the quantum master equation in Zwiebach's form [**12**], which directly inspired this paper,

$$\partial S = \delta S + \frac{1}{2}\{S, S\}.$$

Note that we can write δS instead of ΔS because S has monomial grading one. We see the geometry of Gromov's virtual cycle where the first term on the right-hand side corresponds to approaching the boundary of S by pinching a nonseparating curve, while the second term on the right-hand side corresponds to approaching the boundary of S by either pinching-off an essential separating curve in the complex structure, or by bubbling off a 2-sphere in the map.

3. The modular operad and BV algebra structure

3.1. The equivariant chain complex. We first define the equivariant chain complexes $C(I)$ for the mapping spaces $\mathrm{Map}(\Sigma, M)$. Let I be a finite set and for each oriented connected pseudomanifold σ (see Remark 3.1), consider bundles η over σ with fibre the pair (Σ, I), where Σ is a closed connected piecewise smooth oriented surface of genus g and $I \subset \Sigma$ is an embedding of I into Σ. We also need piecewise smooth maps $f : \eta \to M$. Two such pairs (η, f) and (η', f') are equivalent iff there is an oriented bundle isomorphism b which is an oriented piecewise diffeomorphism between η and η', sends I to I on each fibre by the identity, and relates f to f', namely $f' \circ b = f$. We take the vector space $C(g, I)$ over $\mathbb{Q}$ on these equivalence classes for σ connected and oriented. We add the relation:

$$-(\eta, f, orientation\ \sigma) = (\eta, f, opposite\ orientation\ \sigma).$$

REMARK 3.1. We work in the piecewise differentiable category of spaces built by gluing together compact manifolds with corners (e.g., curvilinear polyhedra) and piecewise smooth maps. A pseudomanifold of dimension k by definition is an object in this category which admits such a decomposition with each $(k-1)$-dimensional part lying in the face of one or two k-dimensional pieces. The ∂ is the $(k-1)$-dimensional part made of pieces with only one k-dimensional face.

REMARK 3.2. A self-equivalence of (σ, η, f) which is orientation reversing on σ forces that generator to be zero.

The sum of the oriented geometric boundary components of σ defines the boundary operator ∂. The direct sum of all these chain complexes for the equivalence relation above requiring the identity map on I over all genera $g = 0, 1, 2, \ldots$, is the chain complex $C(I)$ over $\mathbb{Q}$ functorially attached to the finite set I.

3.2. Core domains of unbounded structures.
Now we define core domains $D(I) \subset C(I)$ and $D(I_1, I_2, ..., I_n) \subset \bigotimes_{\alpha=1}^n C(I_\alpha)$ by transversality. In each case the constraint is imposed on the connected basis elements and then the core domain is closed under $\mathbb{Q}$-linear combinations. For one generator (η, f) the constraint is that the map $\sigma \to M^I$ which evaluates f at the subset I of the fibre is in general position relative to all the subdiagonals of M^I. For a collection $(\eta_\alpha, f_\alpha, \sigma_\alpha)$ the constraint is that the product evaluation map is in general position relative to all the subdiagonals of the big product, and this is also true for the restriction to all the natural product strata in the Cartesian product of the $(\sigma_\alpha, \partial \sigma_\alpha)$.

PROPOSITION 3.3. *The core domains $D(I_1, I_2, ..., I_n)$ are subcomplexes and satisfy the following two properties (compare [9] and [11]).*
i) For each partition $\alpha_1, \alpha_2, ..., \alpha_k$ of $\{1, 2, ..., n\}$ with $\alpha_i = \{j_i, j_i + 1, \ldots, j_i + m_i\}$, we have

$$D(I_1, I_2, ..., I_n) \subseteq D(\{I_\alpha\}_{\alpha \in \alpha_1}) \otimes \cdots \otimes D(\{I_\alpha\}_{\alpha \in \alpha_k}) \subset \bigotimes_{\gamma=1}^n C(I_\gamma);$$

ii) Each inclusion in i) induces an isomorphism on homology.

PROOF. They are subcomplexes because of the definition of general position. The inclusions of $i)$ exist because more constraints are added by the definitions passing from right to left. To see that any such inclusion is onto in homology observe that any cycle in the range may be perturbed slightly to satisfy the constraints or the further constraints in the inclusions of $ii)$. Injectivity follows because a homology in the range between cycles in general position is also in general position near its boundary and by a small deformation relative to its boundary may be put in general position everywhere. $\square$

REMARK 3.4. The set of generators which are in general position is dense for a natural topology. The "denseness" of the core domain motivates the adjective "unbounded" in describing these structures.

3.3. The operations of the modular operad.
Now we define the unbounded operations of the modular operad. First

$$C(I) \otimes C(J) \xrightarrow{C_{ij}} C(I \cup J - \{i, j\}).$$

On the core domain subcomplex (see Subsection 3.2) $D(I, J) \subset C(I) \otimes C(J)$ for each generator $\sigma_I \times \sigma_J$ in $D(I, J)$ the preimage of the diagonal by $(f(i), f(j))$ (called the locus (i, j)) has a normal bundle by transversality. Thus, the normal bundle and then the locus may be oriented when M is oriented to define a relative cycle whose ∂ lies in $\partial(\sigma_I \times \sigma_J)$ in a manner compatible with the natural pieces $\partial \sigma_I \times \sigma_J$ and $\sigma_I \times \partial \sigma_J$. Note this orientation of the normal bundle is independent of the order of i and j because d is even.

Along the locus (i,j) we have a well defined mapping of the one point union of the fibre over σ_I and the fibre over σ_J into M. Now for each surface we replace the marked point by the set of tangent directions at that point and glue these together along the boundary by a rigid map. The ambiguity in such a map is a circle parametrizing the set of orientation reversing isometries between these two boundaries (after choosing metrics on the boundaries). We can map any of the connected sums to the one point union by collapsing the glued up circles to the one common point.

Combining all this we get a fibration of connected sum surfaces over a circle bundle (of all rigid gluings) over the (i,j) locus and a canonical map into M of the total space of the surface bundle. Up to our equivalence the result is a well defined map into M. We combine the natural orientation of the circle (see Remark 3.5) with the orientation of the locus (i,j) to orient the circle bundle. This is the output of the operation between two elements C_{ij} which commutes with the ∂ operator.

The self-connected sum construction of $C(I) \stackrel{S_{ij}}{\to} C(I - \{i,j\})$ on the domain $D(I) \subset C(I)$ is done essentially the same way as in the case of C_{ij}, and S_{ij} also commutes with ∂.

REMARK 3.5 (Orientation of the circle). The surfaces Σ are oriented. Thus, each small circle around a marked point is oriented. In the gluing an orientation reversing isometry between boundary circles is used so that the glued up surfaces has a natural orientation. There is a circle of such orientation reversing isometries and this circle needs to be given an orientation. The family of all the glued up surfaces is a standard Dehn twist family over the circle of gluings. We orient that base circle by declaring the monodromy around that direction to be a right Dehn twist.

PROPOSITION 3.6. *When d is even, S_{ij} only depends on the unordered pair $\{i,j\}$, and C_{ij} is graded symmetric. The gluing operations have odd degree, and compositions are associative and anticommute.*

PROOF. This follows from the construction, Remark 3.5, and Remark 3.2. □

Now let us shift down the grading in the (g,β) component of $C(I)$, $n =$ cardinality of I, by $(d-6)(1-g) + 2n + c(\beta)$, where c is linear in β. In the new grading we have

PROPOSITION 3.7. *When d is even and $c(\beta)$ is even, the dimension shift is even and ∂, C_{ij} and S_{ij} each has degree -1.*

PROOF. We check the degrees. Before C_{ij} sent $(g_1, n_1, k_1) \otimes (g_2, n_2, k_2) \mapsto (g_1 + g_2, n_1 + n_2 - 2, k_1 + k_2 - d + 1)$ and S_{ij} sent $(g,n,k) \mapsto (g+1, n-2, k-d+1)$. Using $c(\beta_1 + \beta_2) = c(\beta_1) + c(\beta_2)$ the degree of C_{ij} in the new grading is thus

$$((d-6)(1-g_1) + 2n_1 + (d-6)(1-g_2) + 2n_2) + (-d+1)$$
$$- ((d-6)(1-g_1-g_2) + 2(n_1+n_2-2)) = -1$$

by direct calculation.

Similarly, the degree of S_{ij} in the new grading is $((d-6)(1-g) + 2n) + (-d+1) - ((d-6)(1-g-1) + 2(n-2)) = -1$. □

REMARK 3.8. The number $(d-6)(1-g) + 2n + c(\beta)$ used to shift the grading in the above proposition when $c(\beta) = 2c_1(\beta)$ gives the formal (real) dimension of the moduli space of J-holomorphic curves of genus g with n marked points in the homology class β of any almost complex manifold (M, J) of real dimension d. In other words, the formal complex dimension is $(d_{\mathbb{C}} - 3)(1 - g) + n + c_1(\beta)$ where $d_{\mathbb{C}}$ is the complex dimension of (M, J). Thus, if $c_1 = 0$ (e.g., in the Calabi-Yau case), $d_{\mathbb{C}} = 3$ (or $d = 6$) is the critical dimension when there is a discrete number of robust J-holomorphic curves of every genus.

3.4. The operations of the BV algebra.
We define δ and $\{\,,\}$ on the chain complex C of Subsection 3.1 and on an enlargement NC which includes maps of surfaces with nodes. There is a map $C \xrightarrow{M_1} C(\{*\})$ which adds a marked point in all possible positions to the surfaces in the family defining an element in C (that is, the bundle $\eta \to \sigma$ is pulled back to η and η becomes the new parameter space). M_1 has degree $+2$ because the base of $M_1(x)$ in $C(\{*\})$ is the total space of x in C. The bracket $\{\,,\}$ in C is the composition

$$C \otimes C \xrightarrow{M_1 \otimes M_1} C(\{i\}) \otimes C(\{j\}) \xrightarrow{C_{ij}} C$$

where i is the marked point in the first factor and j in the second. This bracket has degree $4 - d + 1 = -d + 5$ and is analogous to the generalized Goldman bracket, or string bracket in [**3**] of degree $-d + 2$.

We will define δ analogously as the composition $C \xrightarrow{M_2} C(\{i,j\}) \xrightarrow{S_{ij}} C$, where M_2 replaces a family of surfaces with no marked points with the enlarged family with a pair of distinguished distinct points in all possible positions. The base of $M_2(x)$ is the total space of the fibrewise 2-point configuration space over the base of x. There is a compactness issue which is discussed in Remark 3.10. Since M_2 has degree 4, the composition δ also has degree $4 - d + 1 = -d + 5$.

Now form ΛC, the free graded commutative algebra generated by C, which is also a graded cocommutative coalgebra and a bialgebra with the diagonal uniquely defined by $x \mapsto x \otimes 1 + 1 \otimes x$, $x \in C$. Extend $-\partial$, δ, and $\{\,,\}$ to coderivations of ΛC. The sum of the extensions of δ and $\{\,,\}$ is called Δ. Extend $\{\,,\}$ to a bracket on ΛC so that the Leibniz rule holds, and denote it $\{\,,\}$.

THEOREM 5. *ΛC with its multiplication and the operator Δ is a BV algebra with derived Lie bracket $\{\,,\}$ of odd degree. Namely, $\Delta \cdot \Delta = 0$ and the deviation of Δ from being a derivation is $\{\,,\}$. Furthermore, Δ is a coderivation of the natural coalgebra structure on ΛC. (All in the unbounded sense).*

PROOF. A general point on the locus where each of two pairs is identified in M contributes two points (by definition) to the chain representing the output of $\Delta \cdot \Delta$. Since the circles of gluings for each pair is taken in opposite orders for the two points, the orientations at the two points are opposite. Thus, the involution interchanging the two points is orientation reversing and this output chain is equivalent to zero by Remark 3.2. The rest follows directly from the definitions and the gluing picture. □

REMARK 3.9. We may extend all the above to the chain complex NC corresponding to maps of connected nodal surfaces into M — namely, disconnected surfaces glued together at distinct pairs of points to make the entire collection connected. These are called *garlands* in Chernov-Rudyak [**4**] that studied a factor of $\{\,,\}$. Now there is also a compactness issue for $\{\,,\}$, when a point of a gluing pair we add comes close to a nodal point. This also concerns δ as well as the issue for δ of the two points of the gluing pair approaching each other in the surface.

REMARK 3.10. The locus for the compactified definition of δ will be a subspace of the union $\mathcal{C}$ over the base of the direction blow-up of the diagonal in the 2-point configuration spaces of the fibres. The locus intersected with the interior of $\mathcal{C}$ consists of distinct pairs on the fibre transversally (over the base) identified by the map.

For two points x and y in a surface Σ the operation $g(x,y)$ of forming a self-connected sum between x and y extends continuously to the diagonal blow-up of the 2-point configuration space. Now blow-up each point in an antipodal pair of boundary points which separate the boundary into antipodal arcs. This adds two new arcs to the boundary. Identify the mentioned antipodal arcs (by the old antipodal map) to obtain a surface with two boundary components. These circles may be identified by orientation reversing maps to extend $g(x,y)$.

We define the domain of δ by: i) at an interior point (x,y) of $\mathcal{C}$ we have the transversal coincidence of the map at x and y, a codimension d locus; ii) on the boundary of $\mathcal{C}$ (assume the base has no boundary for simplicity now) the points of i) tend generically to pairs $(x = y, l)$ where the differential of the map at x transversally drops rank by one with kernel l; iii) this picture is completed by the transversal locus when the rank of differential of the map drops by two of codimension $2d$, pulled back to the boundary of $\mathcal{C}$ by the projection.

Then δ is defined on this closed locus using the gluing extension of $g(x,y)$ above. Note that even though (in this special case) the base has no boundary, the output of δ does have boundary. In the general case, where the base has boundary, the output of δ will have extra boundary, called the *anomalous* boundary.

In the next section when we apply δ to a family of perturbed J-holomorphic curves, we will trim away a collar neighborhood of the anomalous boundary corresponding to these pairs (x,y) which are so close that the gluing operation creates a Riemann surface violating the cut-off constraints imposed in the definition of the cut-off Gromov chain.

4. The Gromov chain of perturbed J-holomorphic curves

4.1. Perturbed J-holomorphic curves. We will refer to the formalism of [**6**], especially Chapter 1, Section 6, for discussing the Gromov chain associated to all J-holomorphic curves in a symplectic almost complex manifold (M, J, ω). Fixing a stable combinatorics and homology data (Σ, β), there is a set of J-holomorphic curves with this type which is compact if we add the J-holomorphic curves of all finitely many stable combinatorial types (Σ', β') obtained by degenerating. (see Subsection 4.2 and Subsection 4.3 below). This compactness is the achievement of Gromov to which many have added. See [**10**], [**6**], and the introduction and references. This compact set of J-holomorphic curves admits a finite stratification into types but the dimension and the regularity of the various strata is not necessarily what we want because the solution set $\{(\text{complex structure, map}) : \overline{\partial}(\text{map}) = 0\}$ may not be defined transversally.

The idea is to locally perturb the equation $\overline{\partial}(\text{map}) = 0$ near the compact set and consider the transversal zeros which will be in some neighborhood of the compact set. Because of the presence of finite automorphism groups, orbifolds, orbibundles, and multisections must be used (see [**6**] Chapter 1, Section 6) to construct the

Gromov chain with ℚ-coefficients. It is a tour de force challenge and achievement. All we need to know here is that the objects are produced by local perturbations and transversality, that *all* perturbed J-holomorphic curves are included, and the non-compactness is described by degeneration as in Subsection 4.3 below.

4.2. Combinatorics. Let us discuss the combinatorics of (Σ, β). By Σ we mean a finite collection of connected closed oriented surfaces together with nodal data — a finite subset of Σ (up to isotopy) with a fixed point free involution so that gluing related points yields a connected "nodal surface". By β we mean an integral homology class assigned to each component of Σ. It is important that the set of homology classes realized by "perturbed J-holomorphic" curves lies in a sharp cone[4] in $H_2(M, \mathbb{R})$. This follows from the hypothesis that M is a closed symplectic manifold, see [13] which contains a more general result characterizing symplectic and contact manifolds. Thus, when representations of a realizable class degenerate into components, there are only finitely many possibilities for their homology classes.

4.3. Stable combinatorics and degeneration. The combinatorics is stable if each component with $\beta = 0$ has Euler characteristic strictly less than the number of nodes on that component. Now let us study the stratum or part of the Gromov chain of perturbed J-holomorphic curves corresponding to a certain combinatorics (Σ, β) of "nodal" curves and homological position.

This part of the Gromov chain has an open part with the above combinatorics, and a codimension-two part where more degeneration takes place. If we impose inequalities on the complex structure and on the local energy of the maps, we can carve out a thin (generically tubular) neighborhood of the codimension two part where degeneracies happen. The homological boundary of what is left after carving out is a union of boundaries of tubes around strata where one degeneration happens. Fix one of those terms in the boundary of the cut-off Gromov chain. Along that stratum one sees new combinatorics.

One component of the nodal surface we started with has either had a handle pinched-off or been pinched into two components. The homology class β in the latter case splits into two parts and the genus splits into two parts. Zero homology class and zero genus can occur but the Euler characteristic condition above still holds (partially by definition of the compactification due to Kontsevich).

4.4. The main result. Now we can present the main result. Let S' denote the cut-off Gromov chain (see Subsection 4.3) associated to all stable combinatorial types of connected surfaces with k nodes. So S' defines an element in NC (see Remark 3.9). Set $NC = N$. We will modify S' to S by adding small collars as corrections in the argument below. k fixed.

THEOREM 6. *In ΛN, the free graded commutative algebra generated by N,*

$$\partial S = \delta S + \frac{1}{2}\{S, S\}.$$

PROOF. The ∂ of the total Gromov chain for connected surfaces with k nodes is made out of the boundaries of tubular neighborhoods of the strata corresponding to one additional degeneration. These are approximately described (if the degeneration does not disconnect) by applying the operation δ to the piece of the Gromov

---

[4]I.e., a cone over a compact convex set.

chain corresponding to the new combinatorics obtained by pulling apart the new node and erasing the points (see the last paragraph of Remark 3.10). If the degeneration disconnects, the boundary of the tube is approximately described by applying $\{\,,\}$ to the two pieces obtained by pulling apart the created node and erasing the points.

To replace approximate by exact we add small collar homologies to fill in the gaps. Note we are assuming the families of perturbed J-holomorphic curves satisfy all the transversality required for our operations δ and $\{\,,\}$ to be defined. Thus, a generic perturbation of the $\bar{\partial}(\text{map}) = 0$ equation is required.

We find the equation $\partial S = \delta S + \frac{1}{2}\{S, S\}$ at the chain level in our background. $\square$

References

[1] Alexandrov, M.; Schwarz, A.; Zaboronsky, O.; Kontsevich, M. *The geometry of the master equation and topological quantum field theory.* Internat. J. Modern Phys. **A 12** (1997), 1405–1429.

[2] I.A. Batalin and G.A. Vilkovisky *Quantization of gauge theories with linearly dependent generators.* Phys. Rev. D (3) **28** (1983), 2567–2582.
I.A.Batalin and G.A. Vilkovisky. *Erratum: "Quantization of gauge theories with linearly dependent generators".* Phys. Rev. D (3) **30** (1984), 508.
I.A. Batalin and G. A. Vilkovisky. *Gauge algebra and quantization.* Quantum gravity (Moscow, 1981), 463–480, Plenum, New York, 1984.
I.A. Batalin and G.A. Vilkovisky. *Closure of the gauge algebra, generalized Lie equations and Feynman rules.* Nuclear Phys. **B 234** (1984), 106–124.

[3] M. Chas and D. Sullivan. *String Topology*, Preprint math.GT/9911159, 1999.

[4] V. Chernov, Y. Rudyak. *Algebraic structures on generalized strings.* Preprint math. GT/0306140, 9 pages, 2003.

[5] Y. Eliashberg, A. Givental and H. Hofer. *Introduction to Symplectic Field Theory.* GAFA 2000 (Tel Aviv, 1999). Geom. Funct. Anal. 2000, Special Volume, Part II, 560–673.

[6] K. Fukaya and K. Ono. *Arnold's conjecture and Gromov-Witten invariants for general symplectic manifolds.* Topology, **65**, 933-1048 (1999), Chapter 1, Section 6. See also: The Arnoldfest (Toronto, ON, 1997). Fields Inst. Commun., **24**, 173–190, AMS, Providence, RI, 1999.

[7] E. Getzler and M. Kapranov. *Modular Operads.* Composito Mathematica **110** (1998), 65-125.

[8] M. Gromov. *Pseudoholomorphic curves in symplectic manifolds.* Invent. Math. **82** (1985), 307–347.

[9] I. Kriz and P. May. *Operads, Algebras, Modules, and Motives.* Astrisque, **233**, 1995, iv+145pp.

[10] J. Li and G. Tian. *Virtual moduli cycles and Gromov-Witten invariants of algebraic varieties.* J. Amer. Math. Soc., **11** (1998), 199-174.
Virtual moduli cycles and Gromov-Witten invariants of general symplectic manifolds, Proceeding of UC Irvine Conference on symplectic geometry and topology, 1996, ed. R. Stern.

[11] D. McDuff and D. Salomon. *J-holomorphic curves and Quantum cohomology.* Univ. Lecture Series, AMS, Providence, RI. **6**, 1994.

[12] A. Sen and B. Zwicbach. *Background Independent algebraic structures in closed string field theory.* Commun. Math. Phys. **177** (1996), 305-326

[13] D. Sullivan. *Cycles for the dynamical study of foliated manifolds and complex manifolds.* (dedicated to J.-P. Serre) Inv. Math. **36**, (1976), 225-255.

[14] D. Sullivan. *Open and closed string field theory interpreted in classical algebraic topology.* Proceedings of 2003 Oxford Symposium in honor of Graeme Segal. Geometry, Topology and Quantum Field Theory 2004, ed. U. Tillmann, Cambridge Univ. Press, 344-357.

[15] Scott Wilson. *Unbounded and partial algebras over operads of complexes.* Ph.D. Thesis, Stony Brook University, 2005.

DEPARTMENT OF MATHEMATICS, STONY BROOK UNIVERSITY, STONY BROOK, NY, 11794-3651, USA AND CITY COLLEGE (CUNY) 365 FIFTH AVENUE, NEW YORK, NY 10016, USA

FEYNMAN DIAGRAMS

Feynman diagrams for pedestrians and mathematicians

Michael Polyak

Contents

1. Introduction
2. Finite-dimensional Feynman diagrams
3. Gauge theories and gauge fixing
4. Infinite dimensional case
5. An example of QFT: Chern-Simons theory

References

1. Introduction

1.1. About these lecture notes. For centuries physics was a potent source providing mathematics with interesting ideas and problems. In the last decades something new started to happen: physicists started to provide mathematicians also with technical tools, methods, and solutions. This process seem to be especially strong in geometry and low-dimensional topology. It is enough to mention the mirror conjecture, Seiberg-Witten invariants, quantum knot invariants, etc.

Mathematicians, however, *en masse* failed to learn modern physics. There seem to be two main obstructions. Firstly, there are few textbooks in modern physics written in terms accessible for mathematicians. Mathematicians and physicists speak two different languages, and a good "physical-mathematical dictionary" is missing[1]. Thus, to learn something from a physical textbook, a mathematician should start from a hard and time-consuming process of learning the physical jargon.

Secondly, mathematicians consider (and often rightly so) many physical methods and results to be non-rigorous and do not consider them seriously. In particular, path integrals still remain quite problematic from a mathematical point of view (due

---

2000 *Mathematics Subject Classification.* Primary: 81T18, 81Q30, Secondary: 57M27, 57R56.

Key words and phrases. Feynman diagrams, gauge-fixing, Chern-Simons theory, knots, configuration spaces.

Partially supported by the ISF grant 86/01 and the Loewengart research fund.

[1] With a notable exception of [9], which is somewhat heavy.

to some usually unclear measure aspects), so mathematicians are reluctant to accept any results obtained by using path integrals. Yet, this technique may be put to good use, if at least as a tool to guess an answer to a mathematical problem.

In these notes I will focus on perturbative expansions of path integrals near a critical point of the action. This can be done by a standard physical technique of Feynman diagrams expansion, which is a useful book-keeping device for keeping track of all terms in such perturbative series. I will give a rigorous mathematical treatment of this technique in a finite dimensional case (when it actually belongs more to a course of multivariable calculus than to physics), and then use a simple "dictionary" to translate these results to a general infinite dimensional case.

As a result, we will obtain a recipe how to write Feynman diagram expansions for various physical theories. While in general an input of such a recipe includes path integrals, and thus is not well-defined mathematically, it may be used purely formally for producing Feynman diagram series with certain expected properties. A usual trick is then to "sweep under the carpet" all references to the underlying physical theory, keeping only the resulting series. Their expected properties often can be proved rigorously, directly from their definition.

I will illustrate these ideas on the interesting example of the Chern-Simons theory, which leads to universal finite type invariants of knots and 3-manifolds.

A word of caution: during the whole treatment I will brush aside all questions of measures, convergence, and such; see the discussion in Section 4.5.

1.2. Basics of classical and quantum field theories.

The remaining part of this section is a brief sketch — on the physical level of rigor — of some basic notions and physical jargon used in the quantum field theory (QFT). Its purpose is to give a basic mathematical dictionary of QFT's and a motivation for our consideration of Gaussian-type integrals in this note. An impatient reader may skip it without much harm and pass directly to Section 2. Good introductions to field theories can be found e.g. in [**12**], [**21**]; mathematical overview can be found in [**9**]; various topological aspects of QFT are well-presented in [**23**]. Very roughly, by a *field theory* one usually means the following.

Given a *space-time* manifold X, one considers a space $\mathcal{F}$ of *fields*, which are functions of some kind on X (or, more generally, sections of bundles on X). A *Lagrangian* $L : \mathcal{F} \to \mathbb{R}$ on $\mathcal{F}$ gives rise to the *action* functional $S : \mathcal{F} \to \mathbb{R}$ defined by

$$S(\phi) = \int_X L(\phi) dx.$$

In classical field theory one studies critical points of the action S ("classical trajectories of particles"). These fields can be found from the variation principle $\delta S = 0$, which is simply an infinite-dimensional version of a standard method for finding the critical points of a smooth function $f : \mathbb{R} \to \mathbb{R}$ by solving $\frac{df}{dx} = 0$.

In the quantum field theory one considers instead a *partition function* given by a path integral

(1) $$Z = \int_{\mathcal{F}} e^{ikS(\phi)} \mathcal{D}\phi.$$

over the space of fields, for a constant $k \in \mathbb{R}$ and some formal measure $\mathcal{D}\phi$ on $\mathcal{F}$. This is the point where mathematicians usually stop, since usually such measures are ill-defined. But let this not disturb us.

In the *quasi-classical limit* $k \to \infty$, the stationary phase method (see e.g. [8] and also Exercise 2.5) states that under some reasonable assumptions about the behavior of S this fast-oscillating integral localizes on the critical points of S, so one recovers the classical case.

The *expectation value* $\langle f \rangle$ of an *observable* $f: \mathcal{F} \to \mathbb{R}$ is

$$\langle f \rangle = \frac{1}{Z} \int_{\mathcal{F}} \mathcal{D}\phi \; e^{ikS(\phi)} f(\phi).$$

For a collection $f_1, \ldots, f_m$ of observables their *correlation function* is

$$\langle f_1, \ldots, f_m \rangle = \frac{1}{Z} \int_{\mathcal{F}} \mathcal{D}\phi \; e^{ikS(\phi)} \prod_{i=1}^{n} f_i(\phi).$$

By solving a theory one usually means a calculation of these integrals or their asymptotics at $k \to \infty$.

Increasingly often, due to a simpler behavior and better convergence properties, one considers instead the Euclidean partition function, equally well encoding physical information (and related to (1) by a certain analytic continuation in the time domain, called Euclidean, or Wick, rotation):

$$(2) \qquad Z = \int_{\mathcal{F}} e^{-kS(\phi)} \mathcal{D}\phi.$$

Since at present a general mathematical treatment of path integrals is lacking, we will first consider a finite dimensional case.

1.3. Finite-dimensional version of QFT. Let us take $\mathcal{F} = \mathbb{R}^d$ as the space of fields. An action S and observables f_i are then just functions $\mathbb{R}^d \to \mathbb{R}$. For a constant $k \in \mathbb{R}$, consider the partition function $Z = \int_{\mathbb{R}^d} dx e^{-kS(x)}$ and the correlation functions $\langle f_1, \ldots, f_m \rangle = Z^{-1} \int_{\mathbb{R}^d} dx \; e^{-kS(x)} \prod_i f_i(x)$. We are interested in the behavior of Z and $\langle f_1, \ldots, f_m \rangle$ in the "quasi-classical limit" $k \to \infty$.

A well-known stationary phase method states that for large k the main contribution to Z and $\langle f_1, \ldots, f_m \rangle$ comes from some small neighborhoods of the points x where $\partial S / \partial x = 0$. Thus it suffices to study a behavior of Z and $\langle f_1, \ldots, f_m \rangle$ near such a point x_0. Considering the Taylor expansions of S and f_i in x_0 (and noticing that the linear terms in the expansion of S vanish), after an appropriate changes of coordinates we arrive to the following problem: study integrals

$$\int_{\mathbb{R}^d} dx \; e^{-\frac{1}{2}\langle x, Ax \rangle + \hbar U(x)} P(x)$$

for some bilinear form A, higher order terms $U(x)$, and monomials $P(x)$ in the coordinates x^i.

Further in these notes we will calculate such integrals explicitly. To keep track of all terms appearing in these calculations, we will use Feynman diagrams as a simple book-keeping device. See the notes of Kazhdan in [9] for a more in-depth treatment.

2. Finite-dimensional Feynman diagrams

2.1. Gauss integrals. Recall a well-known formula for the Gauss integral (obtained by calculating the square of this integral in polar coordinates):

PROPOSITION 2.1.
$$\int_{-\infty}^{\infty} dx\, e^{-\frac{1}{2}ax^2} = \sqrt{\frac{2\pi}{a}}.$$

More generally, let $A = (A_{ij})$ be a real $d \times d$ positive-definite matrix, $x = (x^1, \ldots, x^d)$ the Euclidean coordinates in $V = \mathbb{R}^d$, and $\langle\,,\,\rangle : (\mathbb{R}^d)^* \times \mathbb{R}^d \to \mathbb{R}$ the standard pairing $\langle x_i, x^j \rangle = \delta_i^j$. Then

PROPOSITION 2.2.
$$Z_0 = \int_{\mathbb{R}^d} dx\, e^{-\frac{1}{2}\langle Ax, x \rangle} = \left(\det \frac{A}{2\pi}\right)^{-\frac{1}{2}}. \tag{3}$$

Indeed, by an orthogonal transformation (which does not change the integral) we can diagonalize A and apply the previous formula in each coordinate.

REMARK 2.3. In a more formal setting, this may be considered as an equality for a positive-definite symmetric operator $A : V \to V^*$ from a d-dimensional vector space V to its dual (and $\langle\,,\,\rangle : V^* \times V \to \mathbb{R}$). Indeed, A induces $\det A : \Lambda^d V \to \Lambda^d V^*$, so that $\det A \in (\Lambda^d V^*)^{\otimes 2}$ and $(\det A)^{-\frac{1}{2}} \in |\Lambda^d V|$. Hence equality (3) with $\mathbb{R}^d$ changed to V still makes sense if we consider both sides as elements of $|\Lambda^d V|$. In a similar way, for $\mathbb{C}$-valued symmetric operator $A : V \to V^*$ with a positive-definite $\mathrm{Im} A$ one has
$$\int_V dx\, e^{\frac{i}{2}\langle Ax, x\rangle} = \left(\det \frac{A}{2\pi i}\right)^{-\frac{1}{2}}$$
where now both sides belong to $|\Lambda^d V|_{\mathbb{C}}$.

A more general form of equation (3) is obtained by adding a linear term $-\langle b, x \rangle$ with $b \in (\mathbb{R}^d)^*$ to the exponent: define Z_b by
$$Z_b = \int dx\, e^{-\frac{1}{2}\langle Ax, x\rangle + \langle b, x \rangle}. \tag{4}$$

Then, by a change $x \to x - A^{-1}b$ of coordinates, we obtain

PROPOSITION 2.4.
$$Z_b = \left(\det \frac{A}{2\pi}\right)^{-\frac{1}{2}} e^{\frac{1}{2}\langle b, A^{-1}b \rangle} = Z_0 e^{\frac{1}{2}\langle b, A^{-1}b \rangle}. \tag{5}$$

EXERCISE 2.5. Verify the stationary phase method in the simplest case: use an appropriate change of coordinates to pass
$$\text{from} \quad \int_\alpha^\beta dx\, e^{k(-\frac{ax^2}{2} + bx)} \quad \text{to} \quad \int_{\alpha'}^{\beta'} dx\, e^{-\frac{x^2}{2}}.$$

What happens to a small ε-neighborhood of the critical point $x_0 = b/a$ under this change of coordinates? Conclude that in the limit $k \to \infty$ integration over a small neighborhood of $x_0 = b/a$ gives the same leading term in the expansion of this integral in powers of k, as integration over the whole of $\mathbb{R}$.

2.2. Correlation functions.
The correlators $\langle f_1, \ldots, f_m \rangle$ of m functions $f_i : \mathbb{R}^d \to \mathbb{R}$ (also called *m-point functions*) are defined by plugging the product of these functions in the integrand and normalizing:

$$\langle f_1, f_2, \ldots, f_m \rangle = \frac{1}{Z_0} \int dx\, e^{-\frac{1}{2}\langle Ax, x \rangle} f_1(x) \ldots f_m(x). \tag{6}$$

They may be computed using Z_b. Indeed, notice that

$$\frac{\partial}{\partial b_i} \int dx\, e^{-\frac{1}{2}\langle Ax, x \rangle + \langle b, x \rangle} = \int dx\, e^{-\frac{1}{2}\langle Ax, x \rangle + \langle b, x \rangle} x^i,$$

hence for correlators of any (not necessary distinct) coordinate functions we have

$$\langle x^{i_1}, \ldots, x^{i_m} \rangle = \frac{1}{Z_0} \partial_{i_1} \ldots \partial_{i_m} Z_b \big|_{b=0} = \partial_{i_1} \ldots \partial_{i_m} e^{\frac{1}{2}\langle b, A^{-1} b \rangle} \big|_{b=0} \tag{7}$$

where we denoted $\partial_i = \partial/\partial b_i$.

In particular, 2-point functions are given by the Hessian matrix $\frac{\partial^2}{\partial b^2}(Z_b/Z_0)\big|_{b=0}$ with the matrix elements

$$\langle x^i, x^j \rangle = \partial_i \partial_j e^{\frac{1}{2}\langle b, A^{-1} b \rangle} \big|_{b=0} = (A^{-1})_{ij}. \tag{8}$$

Thus the bilinear pairing $\mathrm{Sym}^2(V^*) \to \mathbb{R}$ given by 2-point functions is just the pairing determined by A^{-1}. This explains the similarity of our notations for the 2-point functions and $\langle\,,\,\rangle : V^* \times V \to \mathbb{R}$.

For polynomials, or more generally, formal power series $f_1, \ldots, f_m$ in the coordinates we may apply (7) (with $i_1 = i_2 = \cdots = i_n = i$ for each monomial $(x^i)^n$) and then put the series back together, noting that each x^i should be substituted by ∂_i. This yields:

PROPOSITION 2.6.

$$\langle f_1, f_2, \ldots, f_m \rangle = f_1\left(\frac{\partial}{\partial b}\right) \ldots f_m\left(\frac{\partial}{\partial b}\right) e^{\frac{1}{2}\langle b, A^{-1} b \rangle} \bigg|_{b=0} \tag{9}$$

2.3. Wick's theorem.
Denote by A^{ij} the matrix elements $(A^{-1})_{ij}$ of A^{-1}. The key ingredient of the Feynman diagrams technique is Wick's theorem (see e.g. [**22**]) which we state in its simplest form:

THEOREM 2.7 (Wick).

$$\partial_{i_1} \ldots \partial_{i_m} e^{\frac{1}{2}\langle b, A^{-1} b \rangle} \big|_{b=0} = \begin{cases} \sum A^{j_1 j_2} \ldots A^{j_{m-1} j_m}, & m = 2n \\ 0, & m = 2n+1 \end{cases} \tag{10}$$

where the sum is over all partitions $(j_1, j_2), \ldots, (j_{m-1}, j_m)$ in pairs of the set $i_1, i_2, \ldots, i_m$ of indices.

PROOF. For each k, the expression $\partial_{i_1} \ldots \partial_{i_k} e^{\frac{1}{2}\langle b, A^{-1} b \rangle}$, considered as a function of b, is always of the form $P_{i_1 \ldots i_k}(b) e^{\frac{1}{2}\langle b, A^{-1} b \rangle}$, where $P_{i_1 \ldots i_k}(b)$ is a polynomial. Each new derivative ∂_j acts either on the polynomial part, or on the exponent, by the rule

$$\partial_j\left(P(b) e^{\frac{1}{2}\langle b, A^{-1} b \rangle}\right) = \partial_j(P(b)) e^{\frac{1}{2}\langle b, A^{-1} b \rangle} + P(b)\left(\sum_i A^{ji} b_i\right) e^{\frac{1}{2}\langle b, A^{-1} b \rangle},$$

so the polynomial part $P_{i_1\ldots i_m}(b)$ may be defined recursively by $P_\emptyset(b) = \mathbf{1}$ and

$$\text{(11)} \quad P_{i_1\ldots i_m}(b) = (\partial_{i_1} + \sum_i A^{i_1 i} b_i) P_{i_2\ldots i_m}(b) = \ldots$$

$$= (\partial_{i_1} + \sum_i A^{i_1 i} b_i) \ldots (\partial_{i_m} + \sum_i A^{i_m i} b_i) \mathbf{1},$$

where $\mathbf{1}$ is the function identically equal to 1. We are interested in the constant term $P_{i_1\ldots i_m}(0)$. Directly from (11) we can make two observations. Firstly, if m is odd, $P_{i_1\ldots i_m}(b)$ contains only terms of odd degrees, in particular $P_{i_1\ldots i_m}(0) = 0$. Secondly, unless each derivative ∂_{i_k}, $k < m$ acts on the term $\sum_i A^{i_l i} b_i$ in some l-th, $l > k$, factor of the (11), the evaluation at $b = 0$ would give zero. Each such pair (i_k, i_l) contributes a factor of $A^{i_l i_k}$ to the constant term of $P_{i_1\ldots i_m}$. These observations prove the theorem. $\square$

It is convenient to extend (10) by linearity to arbitrary linear functions of the coordinates, note that in this case we may define the 2-point functions $\langle f, g \rangle$ by $\langle f, A^{-1} g \rangle$ in view of (8), and finally combine it with (7) into the following version of Wick's theorem:

THEOREM 2.8 (Wick). *Let $f_1(x), \ldots, f_m(x)$ be arbitrary linear functions of the coordinates x_i. Then all m-point functions vanish for odd m. For $m = 2n$ one has*

$$\text{(12)} \quad \langle f_1, \ldots, f_m \rangle = \sum \langle f_{i_1}, f_{i_2} \rangle \ldots \langle f_{i_{m-1}}, f_{i_m} \rangle,$$

where the sum is over all pairings $(i_1, i_2), \ldots, (i_{m-1}, i_m)$ of $1, \ldots, m$ and the 2-point functions $\langle f_j, f_k \rangle$ are given by $\langle f_j, A^{-1} f_k \rangle$.

REMARK 2.9. Another idea for a proof of Theorem 2.8 is the following. Note that both sides of (12) are symmetric functions of $1, \ldots, m$, so they may be considered as functions on m-th symmetric power $S^m(V)$ of $V = \mathbb{R}^d$. Thus it suffices to check (12) only for $f_1 = \cdots = f_m = f$; in this case it is obvious.

EXERCISE 2.10. Check that the number of all pairings of $1, \ldots, 2n$ is $(2n)!/2^n n!$. Calculate

$$\int_{-\infty}^{\infty} dx \; x^m e^{x^2/2}$$

using integration by parts and Proposition 2.1. Calculate

$$\left. \frac{d^m}{dx^m} e^{x^2/2} \right|_{x=0}$$

substituting $x^2/2$ instead of x in the Taylor series expansion of e^x. Compare these expressions and explain how are they related to the above number of pairings.

EXERCISE 2.11. Find formulas for the 4-point functions $\langle x_1, x_1, x_2, x_3 \rangle$ and $\langle x_1, x_1, x_1, x_2 \rangle$.

2.4. First Feynman graphs. It is convenient to represent each term

$$\langle f_{i_1}, f_{i_2} \rangle \ldots \langle f_{i_{m-1}}, f_{i_m} \rangle$$

in Wick's formula (12) by a simple graph. Indeed, consider m points, with the k-th point representing f_k. A pairing of $1, \ldots, 2n$ gives a natural way to connect these points by n edges, with an edge (a *propagator* in the physical jargon) $e = (j,k)$ representing $A_e^{-1} = \langle f_j, A^{-1} f_k \rangle$. Equation (12) becomes then

$$\langle f_1, \ldots, f_m \rangle = \sum_{\Gamma} \prod_{e \in \text{edges}(\Gamma)} A_e^{-1}, \tag{13}$$

where the sum is over all univalent graphs as above.

EXAMPLE 2.12. An application of equation (13) for $n = 2$ (see Figure 1a) gives the following:
$$\langle x_1, x_2, x_3, x_4 \rangle = A^{12} A^{34} + A^{13} A^{24} + A^{14} A^{23},$$
$$\langle x_1, x_1, x_2, x_2 \rangle = A^{11} A^{22} + 2 A^{12} A^{12},$$
$$\langle x_1, x_1, x_1, x_1 \rangle = 3 A^{11} A^{11}.$$

2.5. Adding a potential. The above computations may be further generalized by adding a *potential function* $U(x)$ (with some small parameter $\hbar = k^{-1}$) to $\langle Ax, x \rangle$ in the definition of Z_0. Namely, define[2] Z_U by

$$Z_U = \int dx\, e^{-\frac{1}{2}\langle Ax,x\rangle + \hbar U(x)}. \tag{14}$$

Applying (9) for $f = e^{\hbar U(x)}$ we get:

PROPOSITION 2.13.
$$Z_U = Z_0 e^{\hbar U(\frac{\partial}{\partial b})} e^{\frac{1}{2}\langle b, A^{-1} b\rangle}\Big|_{b=0}. \tag{15}$$

Correlation functions $\langle f_1, \ldots, f_m \rangle_U$ are defined similarly to (6):

$$\langle f_1, f_2, \ldots, f_m \rangle_U = \frac{1}{Z_U} \int dx\, e^{-\frac{1}{2}\langle Ax,x\rangle + \hbar U(x)} f_1(x) \ldots f_k(x). \tag{16}$$

Using (9) once again, we get

PROPOSITION 2.14.
$$\langle f_1, f_2, \ldots, f_m \rangle_U = \frac{Z_0}{Z_U} e^{\hbar U(\frac{\partial}{\partial b})} f_1\left(\frac{\partial}{\partial b}\right) \ldots f_m\left(\frac{\partial}{\partial b}\right) e^{\frac{1}{2}\langle b, A^{-1} b\rangle}\Big|_{b=0}. \tag{17}$$

2.6. A cubic potential. Consider the important example of a cubic potential function $U(x) = \sum U_{ijk} x^i x^j x^k$. Let us compute the expansion of the partition function (14) in power series in $\hbar$. The coefficient of $\hbar^n$ in the expansion of (15) is

$$\frac{Z_0}{n!} \left(\sum_{i,j,k} U_{ijk} \partial_i \partial_j \partial_k \right)^n e^{\frac{1}{2}\langle b, A^{-1} b\rangle}\Big|_{b=0}.$$

Let us start with the lowest degrees. By Wick's theorem, the coefficient of $\hbar$ vanishes and the coefficient of $\hbar^2$ is given by

$$\frac{Z_0}{2!} \sum_{i,j,k} \sum_{i',j',k'} U_{ijk} U_{i'j'k'} \partial_i \partial_j \partial_k \partial_{i'} \partial_{j'} \partial_{k'} e^{\frac{1}{2}\langle b, A^{-1} b\rangle}\Big|_{b=0} = \tag{18}$$

$$\frac{Z_0}{2!} \sum_{i,j,k} \sum_{i',j',k'} U_{ijk} U_{i'j'k'} \sum A^{i_1 i_2} A^{i_3 i_4} A^{i_5 i_6},$$

---

[2] Again, let me remind that we ignore problems of convergence: for most $U(x)$ this integral will be divergent!

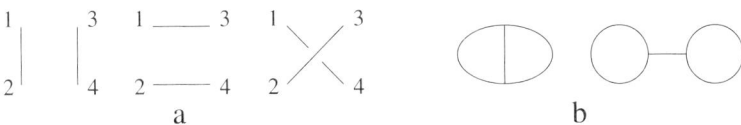

FIGURE 1. Terms of $\langle x_1, x_2, x_3, x_4 \rangle$ and graphs of degree two

where the last sum is over all pairings $(i_1, i_2), \ldots, (i_5, i_6)$ of i, j, k, i', j', k'. We may again encode these pairings by labelled graphs, connecting 6 vertices labelled by i, j, k, i', j', k' by three edges $(i_1, i_2), (i_3, i_4), (i_5, i_6)$ representing $A^{i_1 i_2} A^{i_3 i_4} A^{i_5 i_6}$. This time, however, we have an additional factor $U_{ijk} U_{i'j'k'}$. To represent U_{ijk} graphically, let us glue the triple (i, j, k) of univalent vertices in a trivalent vertex; to preserve the labels, we can write them on the ends of the edges meeting in this new vertex (i.e., on the star of the vertex). Similarly, we represent $U_{i'j'k'}$ by gluing the remaining triple (i', j', k') of univalent vertices into a second trivalent vertex. Thus for each of the $6!/(2^3 3!) = 15$ pairings of i, j, k, i', j', k' we end up with a graph with two trivalent vertices; we get 6 copies of the Θ-graph and 9 copies of the dumbbell graph shown in Figure 1b.

Note, however, that each of these labelled graphs is considered up to its automorphisms, i.e. maps of a graph onto itself, mapping edges to edges and vertices to vertices and preserving the incidence relation. Indeed, while the application of an automorphism changes the labels, it preserves their pairing (edges) and the way they are united in triples (vertices), thus corresponds to the same term in the right hand side of (18). Instead of summing over the automorphism classes of graphs, we may sum over all labelled graphs, but divide the term corresponding to a graph Γ by the number $|\operatorname{Aut}\Gamma|$ of its automorphisms. E.g., for the Θ-graph of Figure 1b $|\operatorname{Aut}\Gamma| = 12$, and twelve copies of this graph (which differ only by transpositions of the labels) all give the same terms $U_{ijk} U_{i'j'k'} A^{ii'} A^{jj'} A^{kk'}$.

Also, when summing the resulting expressions over all indices, note that the terms corresponding to i, j, k, i', j', k' and to i', j', k', i, j, k are the same (which will cancel out with $1/2!$ in front of the sum). Hence, we may write the coefficient of $\hbar^2$ in the following form:

$$Z_0 \sum_\Gamma \frac{1}{|\operatorname{Aut}\Gamma|} \sum_{labels} \prod_v U_v \prod_e A_e^{-1},$$

Here the sum is over all trivalent graphs with two vertices and labellings of their edges, $U_v = U_{ijk}$ for a vertex v with the labels i, j, k of the adjacent edges, and $A_e^{-1} = A^{ij}$ for an edge e with labels i, j.

EXERCISE 2.15. Calculate the number of automorphisms of the dumbbell graph of Figure 1b.

In general, for the coefficient of $\hbar^n$ we get the same formula, but with the summation being over all labelled trivalent graphs with n vertices.

2.7. Correlators for a cubic potential. We may treat m-point functions in a similar way. Let us first consider the power series expansion in $\hbar$ of $Z_U \langle x^{i_1}, \ldots, x^{i_m} \rangle_U$.

FIGURE 2. Degree two graphs with two legs

The coefficient of $\hbar^n$ is

$$\frac{Z_0}{n!} \left(\sum_{i,j,k} U_{ijk} \partial_i \partial_j \partial_k \right)^n \partial_{i_1} \ldots \partial_{i_m} \left. e^{\frac{1}{2}\langle b, A^{-1} b \rangle} \right|_{b=0}.$$

Thus it may again be presented by a sum over labelled graphs, with the only difference being that now in addition to n trivalent vertices these graphs also have m ordered *legs* (i.e. univalent vertices) labelled by $i_1, \ldots, i_m$. See Figure 2 for graphs representing the coefficient of $\hbar^2$ in $Z_U \langle x^1, x^2 \rangle_U$.

However, not all of these graphs will enter in the expression for $\langle x^{i_1}, \ldots, x^{i_m} \rangle_U$, since we should now divide this sum over graphs by Z_U (represented by a similar sum, but over graphs with no legs). This will remove all *vacuum diagrams*, i.e. all graphs which contain some component with no legs. Indeed, the term corresponding to a non-connected graph is a product of terms corresponding to each connected component. Each component with no legs appears also in the expansion of Z_U and thus will cancel out after we divide by Z_U. For example, the first graph of Figure 2 contains a vacuum Θ-graph component. But it also appears in the expansion of Z_U (see Figure 1b). Thus the corresponding factor cancels out after division by Z_U. The same happens with the second graph of Figure 2. As a result, only the last two graphs of Figure 2 will contribute to the coefficient of $\hbar^2$ in the expansion of $\langle x^1, x^2 \rangle_U$.

EXAMPLE 2.16 ("A finite dimensional ϕ^3-theory"). Take $U_{ijk} = \delta_{ij} \delta_{jk}$, i.e. $U = \sum_i (x^i)^3$. Note that since all ends of edges meeting in a vertex are labelled by the same index, we may instead label the vertices. Thus the calculation rules are quite simple: we count uni-trivalent graphs; a vertex represents a sum $\sum_i$ over its labels; an edge with the ends labelled by i, j represents A^{ij}. The coefficients of $\hbar^2$ in Z_U and in $\langle x^1, x^2 \rangle_U$ are given by

$$Z_0 \sum_{i,j} 6(A^{ij})^3 + 9 A^{ij} A^{ii} A^{jj},$$

$$\sum_{i,j} 9 A^{1i} A^{2j} A^{ii} A^{jj} + 6 A^{1i} A^{2j} (A^{ij})^2$$

respectively. We can identify these terms with two graphs of Figure 1b and two last graphs of Figure 2, respectively. The first two graphs of Figure 2 represent

$$\sum_{i,j} 6 A^{12} (A^{ij})^3 + 9 A^{12} A^{ij} A^{ii} A^{jj}.$$

These terms do appear in the $\hbar^2$ coefficient of $Z_U \langle x^1, x^2 \rangle$, but cancel out after we divide it by $Z_U = Z_0 \bigl(1 + \hbar^2 \sum_{i,j} (6(A^{ij})^3 + 9 A^{ij} A^{ii} A^{jj}) + \ldots \bigr)$.

2.8. General Feynman graphs. It is now clear how to generalize the above results to the case of a general potential $U(x)$: the k-th degree term $U_{i_1 \ldots i_k} x^{i_1} \ldots x^{i_k}$ of U will lead to an appearance of k-valent vertices representing factors $U_{i_1 \ldots i_k}$. We will call such a vertex an *internal vertex*. We assume that there are no linear and quadratic terms in the potential, so further we will always assume that all internal vertices of any Feynman graph Γ are of valence ≥ 3; denote their number by $|\Gamma|$. Denote by Γ^0 the set of all graphs with no legs. Also, for $m \geq 1$, denote by Γ^m the set of all non-vacuum (i.e. such that each connected component has at least one leg) graphs with m ordered legs.

Denoting $U_v = U_{i_1 \ldots i_k}$ for an internal vertex v with the labels $i_1, \ldots, i_k$ of adjacent edges, and $A_e^{-1} = A^{ij}$ for an edge e with its ends labelled by i, j, we get

PROPOSITION 2.17.

$$\tag{19} Z_U = Z_0 \sum_{\Gamma \in \Gamma^0} \frac{\hbar^{|\Gamma|}}{|\operatorname{Aut} \Gamma|} \sum_{labels} \prod_v U_v \prod_e A_e^{-1}.$$

Note that instead of performing the internal summation over all labellings, one may include the summation over labels of the star of a vertex into the weight of this vertex.

In a similar way, for m-point functions we get

PROPOSITION 2.18. *For even m,*

$$\tag{20} \langle x^{i_1}, \ldots, x^{i_m} \rangle_U = \sum_{\Gamma \in \Gamma^m} \frac{\hbar^{|\Gamma|}}{|\operatorname{Aut} \Gamma|} \sum_{labels} \prod_v U_v \prod_e A_e^{-1},$$

where the sum is over all labelled graphs Γ with m legs labelled by $i_1, \ldots, i_m$.

Again, we may include the summation over the labels of the star of an internal vertex into the weight of this vertex.

2.9. Weights of graphs. Let us reformulate the above results using a general notion of weights of graphs.

Let V be a vector space. A *weight system* is a collection $(a, \{u_k\}_{k=3}^\infty)$ of $a \in \operatorname{Sym}^2(V)$ and $u_k \in \operatorname{Sym}^k(V^*)$. A weight system W defines a *weight* $W_\Gamma : (V^*)^{\otimes m} \to \mathbb{R}$ of a graph $\Gamma \in \Gamma^m$ in the following way. Assign $u_k \in \operatorname{Sym}^k(V^*)$ to each internal vertex v of valence k, associating each copy of V^* with (an end of) an edge. Also, to the i-th leg of Γ, $i = 1, \ldots, m$ assign some $f_i \in V^*$. Now, for each edge contract two copies of V^* associated to its ends using $a \in \operatorname{Sym}^2(V)$. After all copies of V^* get contracted, we obtain a number $W_\Gamma(f_1, \ldots, f_m) \in \mathbb{R}$.

In our case, a bilinear form A^{-1} and a potential $\hbar U(x)$ determine a weight system in an obvious way: set $a = A^{-1}$ and let u_v to be the degree k part of $\hbar U(x)$. These rules of computing the weights corresponding to a physical theory are called *Feynman rules*.

Formulas (19) and (20) above can be reformulated in these terms as

$$\tag{21} \begin{aligned} Z_U &= Z_0 \sum_{\Gamma \in \Gamma^0} \frac{1}{|\operatorname{Aut} \Gamma|} W_\Gamma, \\ \langle f_1, \ldots, f_m \rangle_U &= \sum_{\Gamma \in \Gamma^m} \frac{1}{|\operatorname{Aut} \Gamma|} W_\Gamma(f_1, \ldots, f_m). \end{aligned}$$

EXERCISE 2.19 (Finite dimensional ϕ^4-theory). Consider a potential $U = \sum_i (x^i)^4$. Formulate the Feynman rules. Find the graphs which contribute to the coefficient of $\hbar^2$ of Z_U and compute their coefficients. Do the same for $\langle x^1, x^2 \rangle_U$.

Draw the graph representing $\sum_i A^{12}(A^{ii})^2$; does it appear in the expansion of $\langle x^1, x^2\rangle_U$ and why?

2.10. Free energy: taking the logarithm. The summation in equation (21) is over all graphs in Γ^0, which are plenty. Denote by Γ^0_{conn} the subset of all connected graphs in Γ^0. There is a simple way to leave only a sum over graphs in Γ^0_{conn}, namely to take the logarithm of the partition function (called the *free energy* in the physical literature):

PROPOSITION 2.20. *Let W be a weight system. Then*

$$\log\left(\sum_{\Gamma\in\Gamma^0}\frac{1}{|\operatorname{Aut}\Gamma|}W_\Gamma\right) = \sum_{\Gamma\in\Gamma^0_{conn}}\frac{1}{|\operatorname{Aut}\Gamma|}W_\Gamma.$$

PROOF. Let us compare the terms of the power series expansion for the right hand side with the terms in the left hand side:

$$\exp\left(\sum_{\Gamma\in\Gamma^0_{conn}}\frac{1}{|\operatorname{Aut}\Gamma|}W_\Gamma\right) = \sum\frac{1}{n_1!\ldots n_k!}W_{\Gamma_1}^{n_1}\ldots W_{\Gamma_k}^{n_k},$$

where the sum is over all k, n_i, and distinct $\Gamma_i \in \Gamma^0_{conn}$, $i=1,\ldots,k$. Consider $\Gamma = (\Gamma_1)^{n_1}\ldots(\Gamma_k)^{n_k} \in \Gamma^0$. Since in addition to automorphisms of each Γ_i there are also automorphisms of Γ interchanging the n_i copies of Γ_i, we have $|\operatorname{Aut}\Gamma| = n_1!\ldots n_k!|\operatorname{Aut}\Gamma_1|\ldots|\operatorname{Aut}\Gamma_k|$. Also, any weight system satisfies $W_{\Gamma'\Gamma''} = W_{\Gamma'}W_{\Gamma''}$, hence $W_\Gamma = W_{\Gamma_1}^{n_1}\ldots W_{\Gamma_k}^{n_k}$. The proposition follows. □

EXERCISE 2.21. Formulate and prove a similar statement for graphs with legs.

REMARK 2.22. It is possible to restrict the class of graphs to 1-connected (in the physical literature usually called 1-*point irreducible*, or 1PI for short) graphs. A graph is 1-connected, if it remains connected after a removal of any one of its edges. This involves a passage to a so-called effective action, which I will not discuss here in details. Mathematically, it simply means an application of a Legendrian transform (a discrete version of a Fourier transform): if $z(b) = \log(Z_b)$ is given by the sum over all connected graphs as in Proposition 2.20, then $\hat{z}(x) = \langle b,x\rangle - z(b)$ is given by a similar sum over all 1PI graphs (and $b(x)$ may be recovered as $\partial\hat{z}/\partial x$).

3. Gauge theories and gauge fixing

3.1. Gauge fixing. All calculations of the previous section dealt only with the case of a non-degenerate bilinear form A; in particular, the critical points of the action $S(x)$ had to be isolated (see Section 1.3). However, gauge theories present a large class of examples when it is not so. Suppose that we have an l-dimensional group of symmetries, i.e. the Lagrangian is invariant under a (free, proper, isometric) action of an l-dimensional Lie group G. Then instead of isolated critical points we have critical orbits, so A has l degenerate directions and the technique of Gauss integration can not be applied.

Let us try to calculate the partition and correlation functions without a superfluous integration over the orbits of G. In other words, we wish to reduce integrals of G-invariant functions on X to integrals on the quotient space $\widetilde{X} = X/G$ of G-orbits. For this purpose, starting from a G-invariant measure on X we should desintegrate it as the Haar measure on the orbits over some "quotient measure" $\tilde{\mu}$ on the base $\tilde{X}$.

If G is compact then $\tilde{\mu}$ is the standard push-forward of μ. For example, if f is a rotationally invariant function on $\mathbb{R}^2$, we can take the pair of polar coordinates (r, ϕ) as coordinates in the quotient space $\tilde{X}$ and the orbit, respectively. The measure on $\tilde{X}$ in this case is $2\pi r \, dr$ and we get the following elementary formula:
$$\int_{\mathbb{R}^2} f(|x|) \, d^2x = 2\pi \int_0^\infty f(r) r \, dr.$$

If G is a locally compact group acting properly on X then $\tilde{\mu}$ can be defined by the property
$$\mu(Y) = \int_{\tilde{X}} |Y \cap G(\tilde{x})| \, d\tilde{\mu}(\tilde{x}),$$
where $G(\tilde{x})$ is the fiber over $\tilde{x} \in \tilde{X}$ and $|\cdot|$ is the Haar measure on it. In this case the integral $\int_X f \, dx$ in question is infinite, but it can be formally defined ("regularized") as $\int_{\tilde{X}} f d\tilde{x}$.

A standard physical procedure for the desintegration that can be applied also to a non-locally compact gauge group is called a *gauge fixing* (see e.g., [**23**]); it goes as follows. Suppose that $f : X \to \mathbb{R}$ is G-invariant, i.e. $f(gx) = f(x)$ for all $x \in X$, $g \in G$. Choose a (local) section $s : \tilde{X} \to X$ which intersects each orbit of G exactly once. Suppose that it is defined by l independent equations $F^1(x) = \cdots = F^l(x) = 0$ for some $F : X \to \mathbb{R}^l$. Firstly, we want to count each G-orbit only once. This is simple to arrange by inserting an l-dimensional δ-function $\delta^l(F(x))$ in the integrand. Secondly, we want to take into account the volume of a G-orbit passing through x, so we should count each orbit with a certain Jacobian factor $J(x)$ (called the *Faddeev-Popov determinant*). How should one define such a factor? We wish to have
$$\int_X f(x) \, dx = \int_X f(x) J(x) \delta^l(F(x)) \, dx.$$

Rewriting the right hand side to include an additional integration over G and noticing that both $f(x)$ and $J(x)$ are G-invariant, we get
$$\int_X f(x) \, dx = \int_X f(x) J(x) \delta^l(F(x) \, dx =$$
$$= \int_X dx \int_G dg \, f(x) J(x) \delta^l(F(gx)) = \int_X dx \, f(x) J(x) \int_G dg \, \delta^l(F(gx)).$$

Thus we see that we should define $J(x)$ by
$$J(x) \int_G dg \, \delta^l(F(gx)) = 1,$$
where dg is the left G-invariant measure on G. Thus, the Faddeev-Popov determinant plays the role of Jacobian for a change of coordinates from x to $(s(\tilde{x}), g)$. Example in §3.3 below provides a good illustration.

REMARK 3.1. A formal coordinate-free way to define $J(x)$ is as follows. The section $s : \tilde{X} \to X$ determines a push-forward $s_* : T_{\tilde{x}}\tilde{X} \to T_x X$ of the tangent spaces. The tangent space $T_x X$ thus decomposes as $s_* \oplus i : T_{\tilde{x}}\tilde{X} \oplus \mathfrak{g} \to T_x X$, where $i : \mathfrak{g} \to T_x X$ is the tangent space to the orbit, generated by the Lie algebra $\mathfrak{g}$ of G. The Jacobian $J(x)$ may be then defined as $J(x) = \det(s_* \oplus i)$.

REMARK 3.2. Equivalently, one may note that the tangent space to the fiber at $x \in s$ may be identified with $\mathfrak{g}$, to directly set $J(x) = \det \Lambda$, where $\Lambda = (\frac{\partial F^i}{\partial \mathfrak{g}^j})$ and $\{\mathfrak{g}^j\}_{j=1}^l$ is a set of generators of the Lie algebra $\mathfrak{g}$ of G, see e.g. [**3**]. I.e., $J(x)$ is the inverse ratio of the volume element of $\mathfrak{g}$ and its image in $\mathbb{R}^l$ under the action of G composed with F.

Indeed, since F has a unique zero on each orbit and since (due to the presence of the delta-function) we integrate only near the section s, we can use F as a local coordinate in the fiber over x. Making a formal change of variables from g to F we get
$$J(x)^{-1} = \int_G dg \, \delta^l(F(gx)) = \int_G dF \, \delta^l(F(gx)) \, \det\left(\frac{\partial g}{\partial F}\right) = \det\left(\frac{\partial g}{\partial F}\right)\Big|_{F=0}.$$
Calculating $(\frac{\partial F(gx)}{\partial g})|_{F=0}$ at a point $x \in s$ and identifying the tangent space to the fiber with $\mathfrak{g}$, we obtain $(\frac{\partial F}{\partial g})|_{F=0} = (\frac{\partial F^i}{\partial \mathfrak{g}^j})$.

EXERCISE 3.3. Let us return to the simple example of a rotationally invariant function $f(x_1, x_2) = f(|x|)$ on $\mathbb{R}^2$, using this time the gauge-fixing procedure. The group $G = S^1$ acts by rotations: $\phi x = e^{i\phi}x$ and the (normalized) measure on G is $\frac{1}{2\pi}d\phi$. We should use the positive x_1-axis for a section s, so we may take e.g. $F = x_2$. A slight complication is that the equation $x_2 = 0$ defines the whole x_1-axis and not only its positive half, so each fiber of G intersects it twice and not once. This can be taken care of, either by dividing the resulting gauge-fixed integral by two, or by restricting its domain of integration to the right half-plane $\mathbb{R}^2_+$ in $\mathbb{R}^2$. In any case, using x_2 instead of ϕ as a local coordinate in the fiber Gx near $x \in s$ we get $d\phi = d(\arctan(x_2/x_1)) = x_1|x|^{-2}dx_2$ Thus for $x \in s$ we have
$$J(x)^{-1} = \int \delta(F(\phi x)) \frac{1}{2\pi} d\phi = \frac{1}{2\pi} \int \delta(x_2) x_1 |x|^{-2} dx_2 = \frac{1}{2\pi x_1},$$
so $J(|x|) = 2\pi |x|$ as expected and
$$\int_{\mathbb{R}^2} f(|x|) \, d^2x = \int_{\mathbb{R}^2_+} f(|x|) 2\pi |x| \delta(x_2) \, dx_1 dx_2 = 2\pi \int_0^\infty f(r) r \, dr.$$

3.2. Faddeev-Popov ghosts. After performing the gauge-fixing, we are left with the gauged-fixed partition function
$$Z_{GF} = \int_{\mathbb{R}^d} dx \, e^{-\frac{1}{2}\langle Ax, x \rangle} \delta^l(F(x)) \det \Lambda.$$
We would like to make it into an integral of the type we have been studying before. We have two problems: to include $\delta(F(x)) \det \Lambda$ in the exponent (i.e., in the Lagrangian) and— more importantly— to make A into a non-degenerate bilinear form.

The δ-function is easy to write as an exponent using the Fourier transform:
$$\delta^l(F(x)) = (2\pi)^{-l} \int_{\mathbb{R}^l} d\xi \, e^{i\langle \xi, F(x) \rangle}.$$
The gauge variables ξ (called *Lagrange multipliers*) supplement the variables x, and the quadratic part of $\langle \xi, F(x) \rangle$ supplements $\langle Ax, x \rangle$ so that the quadratic part A_F of the gauge-fixed Lagrangian is non-degenerate.

The $\det \Lambda$ term is somewhat more complicated; it can be also represented as a Gaussian integral, but over *anti-commuting* variables $c = (c^1, \ldots, c^l)$ and $\bar{c} = (\bar{c}^1, \ldots, \bar{c}^l)$, called *Faddeev-Popov ghosts*. Thus
$$c^i c^j + c^j c^i = \bar{c}^i c^j + c^j \bar{c}^i = \bar{c}^i \bar{c}^j + \bar{c}^j \bar{c}^i = 0 \tag{22}$$
There are standard rules of integration over anti-commuting variables (known to mathematicians as the *Berezin integral*, see e.g. [**23**, Chapter 33] and [**13**, **16**]). The ones relevant for us are
$$\int c^i \, dc^j = \int \bar{c}^i \, d\bar{c}^j = \delta^{ij} \text{ and } \int 1 \, dc^j = \int 1 \, d\bar{c}^j = 0.$$
The multiple integration (over e.g., $dc = dc^l \ldots dc^1$) is defined by iteration. One may show that this implies (see the Exercise below) that for any matrix Λ
$$\int e^{\langle \bar{c}, \Lambda c \rangle} \, dc \, d\bar{c} = \det \Lambda.$$

EXERCISE 3.4. Let $l = 1$ and define the exponent $e^{\lambda \bar{c} c}$ by the corresponding power series. Use the commutation relations (22) to verify that only the two first terms of this expansion do not vanish. Now, use the integration rules to deduce that $\int e^{\lambda \bar{c} c} \, dc \, d\bar{c} = \int (1 + \lambda \bar{c} c) \, dc \, d\bar{c} = \lambda$.

Thus we may rewrite Z_{GF} by adding to the Lagrangian the gauge-fixing term and the ghost term:
$$Z_{GF} = \int dx \, d\xi \, dc \, d\bar{c} \; e^{-\frac{1}{2}\langle Ax, x \rangle + \langle \bar{c}, \Lambda c \rangle + i\langle \xi, F(x) \rangle}.$$

At this stage we may again apply the Feynman diagram expansion to the gauge-fixed Lagrangian. The Feynman rules change in an obvious fashion. The quadratic form now consists of two parts: A_F and Λ, so there are two types of edges. The first type presents A_F, with the labels x^i and ξ^i at the ends. The second type presents Λ, with the labels c^i and $\bar{c}^i$ at the ends. Note that since Λ is not symmetric, these edges are *directed*. Also, there are new vertices, presenting all higher degree terms of the Lagrangian (in particular some where edges of both types meet). An example of the Chern-Simons theory will be provided in Section 5.

3.3. An example of gauge-fixing.
Let us illustrate the idea of gauge-fixing on an example of the standard $\mathbb{C}^*$-action on $\mathbb{C}^2$. In the coordinates $(x_1, \bar{x}_1, x_2, \bar{x}_2)$ on $\mathbb{C}^2$ the gauge group acts by $x_i \to \lambda x_i$, $\bar{x}_i \to \bar{\lambda} \bar{x}_i$. Let us take $A = \frac{x_1}{x_2} \frac{\bar{x}_1}{\bar{x}_2}$ as an invariant function.

Of course, the orbit space $\mathbb{C}P^1$ is quite simple and an appropriate measure on $\mathbb{C}P^1$ is well known; in the coordinates $z = x_1/x_2$, $\bar{z} = \bar{x}_1/\bar{x}_2$ it is given by $dz d\bar{z}/(1 + z\bar{z})^2$ We are thus interested in
$$Z_{GF} = \int \frac{dz d\bar{z}}{(1+z\bar{z})^2} e^{-\frac{1}{2} z \bar{z}}. \tag{23}$$
Let us pretend, however, that we do not know this and proceed with the gauge-fixing method instead.

The invariant measure on $\mathbb{C}^2$ is $dx_1 d\bar{x}_1 dx_2 d\bar{x}_2 / (x_1 \bar{x}_1 + x_2 \bar{x}_2)^2$. In a gauge $F = 0$ we have
$$Z_{GF} = \int \frac{dx_1 dx_2 d\bar{x}_1 d\bar{x}_2}{(x_1 \bar{x}_1 + x_2 \bar{x}_2)^2} e^{-\frac{1}{2} \frac{x_1}{x_2} \frac{\bar{x}_1}{\bar{x}_2}} \delta^2(F(x, \bar{x})) \det \Lambda,$$

where $\Lambda = \begin{vmatrix} x_1 F_{x_1} + x_2 F_{x_2} & \bar{x}_1 F_{\bar{x}_1} + \bar{x}_2 F_{\bar{x}_2} \\ x_1 \bar{F}_{x_1} + x_2 \bar{F}_{x_2} & \bar{x}_1 \bar{F}_{\bar{x}_1} + \bar{x}_2 \bar{F}_{\bar{x}_2} \end{vmatrix}$.

E.g., for $F = x_2 - 1$ we get $\delta^2(|x_2 - 1|)$ and $\det \Lambda = x_2 \bar{x}_2$.

EXERCISE 3.5 (Different gauges give the same result). Consider $F = x_2^\alpha - 1$. Show that $\delta^2(|x_2^\alpha - 1|) = |\alpha x_2^{\alpha-1}|^{-2} \delta^2(|x_2 - 1|)$ and $\det \Lambda = \alpha \bar{\alpha}(x_2 \bar{x}_2)^\alpha$. Check that the dependence on α in Z_{GF} cancels out, thus gives the same result as $F = x_2 - 1$. Show that it coincides with formula (23).

Finally, let us check that while the initial quadratic form A is degenerate, the supplemented quadratic form A_F is indeed non-degenerate. It is convenient to make a coordinate change $x_1' = x_1$, $x_2' = x_2 - 1$. Using a Fourier transform we get

$$\delta^2(x_2 - 1) = (2\pi)^{-2} \int d\xi d\bar{\xi} e^{i(\xi x_2' - \bar{\xi} \bar{x}_2')}.$$

Also, we have $\frac{x_1}{x_2} = x_1' + x_1' \sum_{n=1}^{\infty} (-1)^n x_2'$. We can now compute A and A_F; in the coordinates $(x_1', \bar{x}_1', x_2', \bar{x}_2')$ and $(x_1', \bar{x}_1', x_2', \bar{x}_2', \xi, \bar{\xi})$, respectively, we have:

$$A = \begin{vmatrix} 0 & 1 & 0 & 0 \\ 1 & 0 & 0 & 0 \\ 0 & 0 & 0 & 0 \\ 0 & 0 & 0 & 0 \end{vmatrix}, \quad A_F = \begin{vmatrix} 0 & 1 & 0 & 0 & 0 & 0 \\ 1 & 0 & 0 & 0 & 0 & 0 \\ 0 & 0 & 0 & 0 & i & 0 \\ 0 & 0 & 0 & 0 & 0 & -i \\ 0 & 0 & i & 0 & 0 & 0 \\ 0 & 0 & 0 & -i & 0 & 0 \end{vmatrix}.$$

4. Infinite dimensional case

4.1. The dictionary. Path integrals are generally badly defined, so instead of trying to deduce the relevant results rigorously, we will just provide a basic dictionary to translate the finite dimensional results to the infinite dimensional case.

The main change is that instead of the discrete set $i \in \{1, \ldots, d\}$ of indices we now have a continuous variable $x \in M^n$ (say, in $\mathbb{R}^n$), so we have to change all related notions accordingly. The sum over i becomes an integral over x. Vectors $x = x(i) = (x^1, \ldots, x^d)$ and $b = b(i)$ become fields $\phi = \phi(x)$ and $J(x)$. A quadratic form $A = A(i, j)$ becomes an integral kernel $K = K(x, y)$. Pairings $\langle Ax, x \rangle = \sum_{i,j} x^i A_{ij} x^j$ and $\langle b, x \rangle = \sum_i b^i x^i$ become $\langle K\phi, \phi \rangle = \int dxdy \, \phi(x) K(x, y) \phi(y)$ and $\langle J, \phi \rangle = \int dx \, J(x) \phi(x)$ respectively. The partition function Z_b defined by (4) becomes a path integral Z_J over the space $\mathcal{F}$ of fields

$$Z_J = \int \mathcal{D}\phi \, e^{-\frac{1}{2}\langle K\phi, \phi \rangle + \langle J, \phi \rangle}.$$

The inverse A^{-1} of A defined by $\sum_k A_{ik} A^{kj} = \delta_i^j$ corresponds now to the inverse $G = K^{-1}$ of K defined by

$$\int dz \, K(x, z) G(z, y) = \delta(x - y).$$

Formula (5) for Z_b then translates into

$$Z_J = Z_0 e^{\frac{1}{2}\langle J, GJ \rangle}.$$

Correlators $\langle x^{i_1}, \ldots, x^{i_m} \rangle$ defined by (6) become now m-point functions

$$\langle \phi(x_1), \ldots, \phi(x_m) \rangle = \frac{1}{Z_0} \int \mathcal{D}\phi \, e^{-1/2\langle K\phi, \phi \rangle} \phi(x_1) \ldots \phi(x_m).$$

4.2. Functional derivation. A counterpart of the derivatives $\partial/\partial x^i$ is given by the *functional derivatives* $\delta/\delta\phi(x)$. The theory of functional derivation is well-presented in many places (see e.g. [**10**]), so I will just briefly recall the main notions. Let $F(\phi)$ be a functional. If the differential

$$DF(\phi)(\rho) = \lim_{\varepsilon \to 0} \frac{F(\phi + \varepsilon\rho) - F(\phi)}{\varepsilon}$$

can be represented as $\int \rho(x) h(x) dx$ for some function $h(x)$, then we define $\frac{\delta F}{\delta \phi(x)} = h(x)$. In general, the functional derivative $\delta F/\delta \phi(x)$ is the distribution representing the differential of F at ϕ. The reader can entertain himself by making sense of the following formulas, which show that its properties are similar to usual derivatives:

$$\frac{\delta}{\delta\phi(x)} \phi(y) = \delta(x-y),$$

$$\frac{\delta}{\delta\phi(x)} (F(\phi) H(\phi)) = \frac{\delta}{\delta\phi(x)} (F(\phi)) \cdot H(\phi) + F(\phi) \cdot \frac{\delta}{\delta\phi(x)} (H(\phi)).$$

EXAMPLE 4.1.

$$\frac{\delta}{\delta J(y)} e^{\langle J, \phi \rangle} = \frac{\delta}{\delta J(y)} e^{\int dx J(x) \phi(x)} = \phi(y) e^{\int dx J(x) \phi(x)} = \phi(y) e^{\langle J, \phi \rangle}.$$

EXERCISE 4.2. Consider a (symmetric) potential function

(24) $$U(\phi) = \sum_n \frac{1}{n!} \int dx_1 \ldots dx_n \, U_n(x_1, \ldots, x_n) \phi(x_1) \ldots \phi(x_n).$$

Prove that

$$\frac{\delta}{\delta\phi(y)} U(\phi) = \sum_n \frac{1}{n!} \int dx_1 \ldots dx_n \, U_{n+1}(y, x_1, \ldots, x_n) \phi(x_1) \ldots \phi(x_n).$$

The inverse $G(x, y)$ can be written as a Hessian, similarly to equation (8) for A^{-1}:

$$G(x, y) = \frac{1}{Z_0} \frac{\delta}{\delta J(x)} \frac{\delta}{\delta J(y)} Z_J \bigg|_{J=0}.$$

More generally, for m-point functions we have, similarly to (9),

$$\langle \phi(x_1), \ldots, \phi(x_m) \rangle = \frac{1}{Z_0} \frac{\delta}{\delta J(x_1)} \cdots \frac{\delta}{\delta J(x_m)} Z_J \bigg|_{J=0}.$$

4.3. Wick's theorem and Feynman graphs. Wick's theorem now states that, similarly to (10),

$$\frac{\delta}{\delta J(x_1)} \cdots \frac{\delta}{\delta J(x_m)} e^{\frac{1}{2}\langle J, GJ \rangle} \bigg|_{J=0} = \sum G(x_{i_1}, x_{i_2}) \ldots G(x_{i_{m-1}}, x_{i_m}),$$

where the sum is over all pairings $(i_1, i_2) \ldots (i_{m-1}, i_m)$ of $1, \ldots, m$. Just as in the finite dimensional case, we may encode each pairing by a graph with m univalent vertices labelled by $1, \ldots, m$, and edges connecting vertices i_1 with i_2, ..., and i_{m-1} with i_m presenting the factors of G.

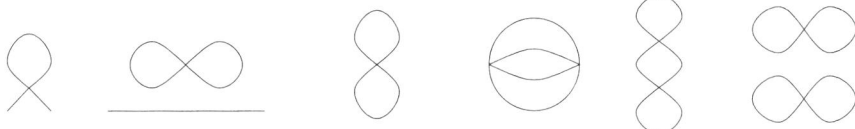

FIGURE 3. Graphs appearing in the ϕ^4-theory

Let us add a potential (24) to the action and define
$$Z_U = \int \mathcal{D}\phi \ e^{-1/2\langle K\phi,\phi\rangle + \hbar U(\phi)}.$$
Then, similarly to (15), we have
$$Z_U = Z_0 e^{\hbar U(\frac{\delta}{\delta J})} e^{\frac{1}{2}\langle J,GJ\rangle}\Big|_{J=0}.$$
Using again the Wick's theorem, we can rewrite the latter expression in terms of Feynman graphs to get
$$(25) \qquad Z_U = \sum_\Gamma \frac{\hbar^{|\Gamma|}}{\operatorname{Aut}\Gamma} \int_{labels} \prod_v U_v \prod_e G_e,$$
where the integral is over all labellings of the ends of edges, $U_v = U(x_1,\ldots,x_k)$ for a k-valent vertex with the labels $x_1,\ldots,x_k$ of the adjacent edges, and $G_e = G(x_i,x_j)$ for an edge with labels x_i, x_j. Sometimes it is convenient to include the integration over the labels of the star of a vertex into the weight of this vertex.

4.4. An example: ϕ^4-theory. Let us write down the Feynman rules for a potential $U(\phi) = \int dx\ \phi^4(x)$. Firstly, the relevant graphs have vertices of valence one or four. Secondly, all edges adjacent to a vertex should be labelled by the same x, so we may instead label the vertices. An edge with labels x, y represents $G(x, y)$ and (including the integration over the vertex labels into the weights of vertices) an x-labelled vertex represents $\int dx$. The linear term in the power series expansion of $\langle x_1, x_2 \rangle_U$ should correspond to non-vacuum graphs with two legs, labelled by x_1 and x_2, and one 4-valent vertex. There is only one such graph, see Figure 3. It represents $\int dx\ G(x_1,x)G(x,x)G(x,x_2)$ and should enter with the multiplicity 12 (the number of all pairings of 6 vertices x_1, x_2, x, x, x, x in which x_1 is not connected to x_2). Let us now check this directly. Indeed, the coefficient of $\hbar$ in $\langle x_1, x_2 \rangle$ is
$$\frac{\delta}{\delta J(x_1)}\frac{\delta}{\delta J(x_2)}\frac{\delta^4}{\delta J(x)^4} e^{\frac{1}{2}\langle J,GJ\rangle}\Big|_{J=0} = 12\int dx\ G(x_1,x)G(x,x)G(x,x_2),$$
where we applied Wick's theorem to obtain the desired equality.

In a similar way, the linear term in the expansion of Z_U should correspond to the graph with no legs and one vertex of valence four (see Figure 3), representing $\int dx\ G^2(x,x)$ (and entering with the multiplicity 3).

EXERCISE 4.3 (ϕ^3-theory). Let $U(\phi) = \int dx\ \phi^3(x)$. Find the Feynman rules for this theory. Which graphs will contribute to the coefficient of $\hbar^2$ in the power series expansion of the 2-point function $\langle x_1, x_2 \rangle_U$? Write down these coefficients explicitly.

4.5. Convergence. Usually the integrals which we get by a perturbative Feynman expansion are divergent and ill-defined in many ways. Often one has to *renormalize* (i.e. to find some way to remove divergencies in a unified and consistent manner) the theory to improve its behavior. Until recently renormalization was considered by mathematicians more like a physical art than a technique; lately Connes and Kreimer [7] have done some serious work to explain renormalization in purely mathematical terms (see a paper by Kreimer in this volume).

But even in the best cases, the Green function $G(x,y)$ usually blows up near the diagonal $x = y$, which brings two problems: Firstly, the weights of graphs with looped edges, starting and ending at the same point (so-called *tadpoles*) are ill-defined and one has to get rid of them in one or another way. Secondly, all diagonals have to be cut out from the spaces over which the integration is performed, so the resulting configuration spaces are open and the convergence of all integrals defining the weights has to be proved. Mathematically these convergence questions usually boil down to the existence of a Fulton-MacPherson-type (see [14]) compactification of configuration spaces, to which the integrand extends.

There is also a challenging problem to interpret the Feynman diagrams series in some classical mathematical terms and to understand the way to produce them without a detour to physics and back. In many examples this may be done in terms of a homology theory of some grand configuration spaces glued from configuration spaces of different graphs along common boundary strata.

We will see all this on an example of the Chern-Simons theory in the next section.

5. An example of QFT: Chern-Simons theory

The Chern-Simons theory has an almost topological character and as such presents an interesting object for low-dimensional topologists. For a connection α in a trivial $SU(N)$-bundle over a 3-manifold M one may define ([6]; see also [11]) the Chern-Simons invariant $CS(\alpha)$ as described in Section 5.1 below. It is the action functional of the classical Chern-Simons theory and was extensively used in mathematics for many years to study properties of 3-manifolds (mostly due to the fact that the classical solutions, i.e. the critical points of $CS(\alpha)$, are flat connections). But it is the corresponding quantum theory which is of interest for us. Its mathematical treatment started only about a decade ago, following Witten's suggestion [25] that it leads to some interesting invariants of links and 3-manifolds, in particular, to the Jones polynomial. While Witten's idea was based on the validity of the path integral formulation of the quantum Chern-Simons theory, his work catalyzed much mathematical activity. By now mathematicians more or less managed to formalize the relevant perturbative series and exorcize from them all physical spirit, leaving a (surprisingly rich) rigorous mathematical extract. In this section I will describe this process in a number of iterations, starting from an intuitive and roughest description and slowly increasing the level of rigor and details. Finally, I will try to reinterpret these Feynman series in some classical topological terms and formulate some corollaries.

5.1. Chern-Simons theory. Further we will use the following data:
- A closed orientable 3-manifold M with an oriented framed link L in M.

- A compact connected Lie group G with an Ad-invariant trace $\mathrm{Tr}: \mathfrak{g} \to \mathbb{R}$ on the Lie algebra $\mathfrak{g}$ of G.
- A principal G-bundle $\mathcal{P} \to M$.

To simplify the situation, we will additionally assume that G is simply connected, since for such groups any principal G-bundle over a manifold M of dimension ≤ 3 (which is our case) is trivializable, see e.g. [**11**].

The appropriate notions of the Chern-Simons theory, considered as a field theory, are as follows. The manifold M plays the role of the space-time manifold X. Denote by $\mathcal{A}$ the space of G-connections on $\mathcal{P}$ and let $\mathcal{G} = \mathrm{Aut}(\mathcal{P})$ be the gauge group. Fields ϕ on M are G-connections on $\mathcal{P}$, i.e. $\mathcal{F} = \mathcal{A}$. The Lagrangian is a functional $L: \mathcal{A} \to \Omega^3(M)$ defined by

$$L(\alpha) = \mathrm{Tr}(\alpha \wedge d\alpha + \frac{2}{3}\alpha \wedge \alpha \wedge \alpha).$$

REMARK 5.1. This choice can be motivated as follows. Let $\theta = d\alpha + \alpha \wedge \alpha$ be the curvature of α. Then $\mathrm{Tr}(\theta \wedge \theta)$ is the Chern-Weil 4-form[3] on $\mathcal{P}$, associated with Tr; this form is gauge invariant and closed. The Chern-Simons Lagrangian $CS(\alpha) = \mathrm{Tr}(\alpha \wedge \theta + \frac{2}{3}\alpha \wedge \alpha \wedge \alpha)$ is an antiderivative of $\mathrm{Tr}(\theta \wedge \theta)$ on $\mathcal{P}$: it is a nice exercise to check that $d(CS(\alpha)) = \mathrm{Tr}(\theta \wedge \theta)$.

The corresponding Chern-Simons action is a function $CS: \mathcal{A} \to \mathbb{R}$ given by

$$CS(\alpha) = \frac{1}{4\pi}\int_M dx\, \mathrm{Tr}(\alpha \wedge d\alpha + \frac{2}{3}\alpha \wedge \alpha \wedge \alpha).$$

It is known that the critical points of this action correspond to flat connections and (assuming that Tr satisfies a certain integrality property[4], which holds in particular for the trace in the fundamental representation of G) it is gauge invariant modulo $2\pi\mathbb{Z}$.

The partition function is given by the following path integral:

$$(26) \qquad Z = \int_\mathcal{A} e^{ikCS(\alpha)} \mathcal{D}\alpha.$$

Here the constant $k \in \mathbb{N}$ is called *level* of the theory; its integrality is needed for the gauge invariance of Z.

Now, let $L = \cup_{j=1}^m L_j$, $j = 1, \ldots m$ be an oriented framed m-component link in M such that each L_j is equipped with a representation R_j of G. Given a connection $\alpha \in \mathcal{A}$, let $\mathrm{hol}_{L_j}(\alpha)$ be the holonomy

$$(27) \qquad \mathrm{hol}_{L_j}(\alpha) = \exp \oint_{L_j} \alpha$$

of α around L_j. Observables in the Chern-Simons theory are so-called Wilson loops. The *Wilson loop* associated with L_j is the functional

$$\mathcal{W}(L_j, R_j) = \mathrm{Tr}_{R_j}(\mathrm{hol}_{L_j}(\alpha)).$$

The m-point correlation function $\langle L \rangle = \langle L_1, L_2, \ldots L_m \rangle$ is defined by

---

[3] Chern-Weil theory states that the de Rham cohomology class of this form is a certain characteristic class of $\mathcal{P}$

[4] Namely that the closed form $\frac{1}{6\pi}\mathrm{Tr}(\alpha \wedge \alpha \wedge \alpha)$ represents an integral class in $H^3(G, \mathbb{R})$

(28) $$\langle L \rangle = Z^{-1} \int_{\mathcal{A}} e^{ikCS(\alpha)} \prod_{j=1}^{m} \mathcal{W}(L_j, R_j) \mathcal{D}\alpha.$$

Since the action is gauge invariant, extrema of the action correspond to points on the moduli space of flat connections. Near such a point the action has a quadratic term (arising from $\alpha \wedge d\alpha$) and a cubic term (arising from $\alpha \wedge \alpha \wedge \alpha$). We would like to consider a perturbative expansion of this theory.

5.2. What do we expect.
Which Feynman graphs do we expect to appear in the perturbative Chern-Simons theory?

Firstly, a gauge-fixing has to be performed, so the ghosts have to be introduced. As a result, we should have two types of edges: the usual non-directed edges (corresponding to the inverse of the quadratic part) and the directed ghost edges.

Secondly, in addition to the quadratic term the action contains a cubic term, so the internal vertices should be trivalent. Also, this time the cubic term is given by an antisymmetric tensor instead of a symmetric one, so one should fix a cyclic order at each trivalent vertex, with its reversal negating the weight of a graph. Two types of edges should lead to two types of internal vertices: usual vertices where three usual edges meet, and ghost vertices where one usual edge meets one incoming and one outgoing ghost edge.

Thirdly, note that the situation with legs is somewhat different from our earlier considerations. Indeed, the legs (i.e. univalent ends of usual edges) of Feynman graphs, instead of being fixed at some points, should be allowed to run over the link L, with each link component entering in $\langle L \rangle$ via its holonomy (27). To reduce this to our previous setting, we can use Chen's iterated integrals to expand the holonomy in a power series where each term is a polynomial in α. In terms of a parametrization $L_j : [0,1] \to \mathbb{R}^3$, this expansion can be written explicitly using the pullback $L_j^* \alpha$ of α to $[0,1]$ via L_j:

$$\mathrm{hol}_{L_j}(\alpha) = 1 + \int_{0<t<1} (L_j^*\alpha)(t) + \int_{0<t_1<t_2<1} (L_j^*\alpha)(t_2) \wedge (L_j^*\alpha)(t_1) +$$
$$\cdots + \int_{0<t_1<\cdots<t_k<1} (L_j^*\alpha)(t_k) \wedge \cdots \wedge (L_j^*\alpha)(t_1) + \ldots$$

where the products are understood in the universal enveloping algebra $U(\mathfrak{g})$ of $\mathfrak{g}$. Thus we should sum over all graphs with any number k_j of cyclically ordered legs on each L_j, and integrate over the positions

$$(x(t_1), \ldots, x(t_{k_j})) \in L_j^{k_j}, \quad 0 < t_1 < \cdots < t_{k_j} < 1$$

of these legs.

These simple considerations turn out to be quite correct. Of course, one should still find an explicit formulas for the weights of such graphs. An explicit deduction of the Feynman rules for the perturbative Chern-Simons theory is described in details in [**3, 15**]. Let me skip these lengthy calculations and formulate only the final results. For simplicity I will consider only an expansion around the trivial connection in $M = \mathbb{R}^3$.

5.3. Feynman rules. It turns out (see [**3**]) that the weight system W^{CS} of the perturbative Chern-Simons theory splits as $W^{CS} = W^G W$, where W^G contains all the relevant Lie-algebraic data of the theory (but does not depend on the location of the vertices of a graph), and W contains only the space-time integration. Since the whole construction should work for any Lie algebra, one may encode the antisymmetry and Jacobi relations already on the level of graphs, changing the weight W_Γ^G of a graph Γ to a "universal weight" $[\Gamma]$, which is an equivalence class of Γ in the vector space over $\mathbb{Q}$ generated by abstract (since we do not care about the location in $\mathbb{R}^3$ of their vertices) graphs, modulo some simple *diagrammatic antisymmetry and Jacobi relations*, shown on Figure 4. The same relations hold for graphs with either usual or ghost edges, so we may think that the relations include the projection making all edges of one type.

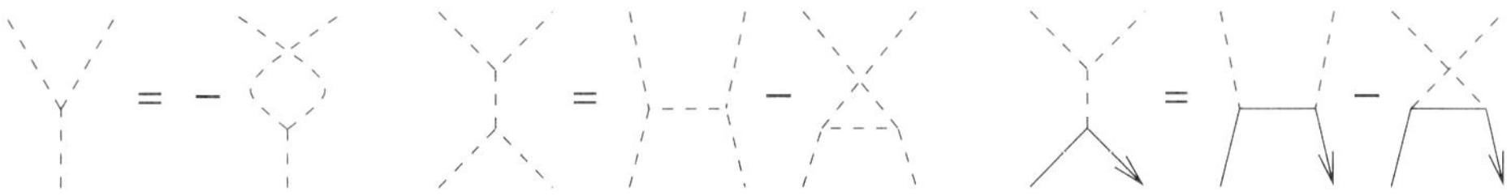

FIGURE 4. Antisymmetry and Jacobi relations

The drawing conventions merit some explanation. It is assumed that the graphs appearing in the same relation are identical outside the shown fragment. In each trivalent vertex we fix a cyclic order of edges meeting there; unless specified otherwise, it is assumed to be counter-clockwise. The edges are shown by dashed lines, and the link component L_j (fixing the cyclic order of the legs) by a solid line. An important consequence of the antisymmetry relation is that for any graph Γ with a tadpole (a looped edge) we have $[\Gamma] = 0$ due to the existence of a "handle twisting" automorphism, rotating the looped edge. Thus from the beginning we can restrict the class of graphs to graphs without tadpoles.

It remains to describe the weight W_Γ of a graph Γ. Roughly speaking, for each internal vertex of Γ we are to perform integration over its position in $\mathbb{R}^3$ (for a ghost vertex we should also take a certain derivative acting on the term corresponding to the outgoing ghost edge); for each leg we are to perform integration over its position in L_j (respecting the cyclic order of legs on the same component). As for the edges, we are to assign to each usual and ghost edge inverses of the operator *curl* and of the Laplacian, respectively.

Somewhat surprisingly (see e.g. [**15**]) two types of edges may be neatly joined into one "combined" edge, thus reducing the graphs in question to graphs with only one type of edges (and just one type of uni- and trivalent vertices). The weight $G(x(e), y(e))$ of such an edge e with the ends in $(x(e), y(e)) \in \mathbb{R}^3 \times \mathbb{R}^3$ has a nice geometrical meaning: it is given by $G(x, y) = \omega(x - y)$, where

$$\omega(x) = \frac{x^1 dx^2 \wedge dx^3}{2\pi ||x||^3} + \text{cyclic permutations of } (1,2,3)$$

is the uniformly distributed area form on the unit 2-sphere $|x| = 1$ in the standard coordinates in R^3. In fact, the usual and the ghost edges (with two possible orientations) give respectively the $(1,1)$, $(2,0)$, and $(0,2)$ parts of $\omega(x-y)$ in terms of its dependence on dx and dy. Abusing notation, I will depict the combined edge again by a dashed line.

REMARK 5.2. A simple explanation for an existence of such a simple unified propagator escapes me. The only explanation which I know is way too complicated: it is the existence (see [**2**]) of the "superformulation" of the gauge-fixed theory, i.e. the fact that the connection together with the ghosts may be united in a "superconnection" of a supertheory, which leads to an existence of a "superpropagator", uniting the usual and the ghost propagator. I believe that there is a simple explanation, probably emanating from the scaling properties and the topological invariance of the Chern-Simons theory, by which one should be able to predict that the combined propagator should be dilatation- and rotation-invariant.

REMARK 5.3. Note that the weight $G(x,x)$ of a tadpole is not well-defined, so it is quite fortunate that we got $[\Gamma] = 0$ for any such graph.

To sum it up, we are interested in the value

$$\langle L \rangle = \sum_{\Gamma} \frac{W_{\Gamma} \hbar^{|\Gamma|}}{\operatorname{Aut} \Gamma} [\Gamma] \tag{29}$$

where $|\Gamma|$ is half of the total number of vertices (univalent and trivalent) of Γ, and the weight of Γ is given by the integral

$$W_{\Gamma} = \int_{C_{\Gamma}} \prod_e G(x(e), y(e)) \tag{30}$$

over the space C_{Γ} of all possible positions of vertices of Γ, such that all vertices remain distinct. Here $\hbar = (k + h^{\vee})^{-1}$, where $h^{\vee}$ is the dual Coxeter number of G (see [**25**]).

I shall describe in more details the type of graphs which appear in this formula and their weights W_{Γ} (both the configuration spaces C_{Γ}, and the integrand).

5.4. Jacobi graphs. Let us start with the graphs. Instead of thinking about graphs embedded in $\mathbb{R}^3$, consider abstract graphs (with just one type of edges), such that

- all vertices have valence one (legs) or three;
- there are no looped edges;
- all legs are partitioned into m subsets $l_1, \ldots, l_m$;
- legs of each subset l_j are cyclically ordered;
- each trivalent vertex is equipped with a cyclic order of three half-edges meeting there;

for technical reasons it will be convenient to think that, in addition to the above,

- all edges are ordered and directed.

We will further address the last three items simply as an orientation of a graph.

For such a graph Γ with a total of $2n$ (univalent and trivalent) vertices define the *degree* of Γ by $|\Gamma| = n$, and denote the set of all such graphs by $\widetilde{\mathfrak{J}}_n$. Set $\widetilde{\mathfrak{J}} = \cup \widetilde{\mathfrak{J}}_n$. The ordering and directions of edges of graphs in $\widetilde{\mathfrak{J}}$ may be dropped by an application of an obvious forgetful map. See Figure 5 for graphs of degree one with $m = 2$ and $m = 1$, and graphs of degree two with $m = 1$. Both antisymmetry and Jacobi relations of Figure 4 preserve the degree of a graph, thus we may consider a vector space over $\mathbb{Q}$ generated by graphs in $\widetilde{\mathfrak{J}}_n$ modulo forgetful, antisymmetry and Jacobi relations. We will call it the *space of Jacobi graphs of degree n* and denote it by $\mathfrak{J}_n$; denote also $\mathfrak{J} = \oplus_n \mathfrak{J}_n$, and let as before $[\Gamma]$ be the class of $\Gamma \in \widetilde{\mathfrak{J}}$ in $\mathfrak{J}$.

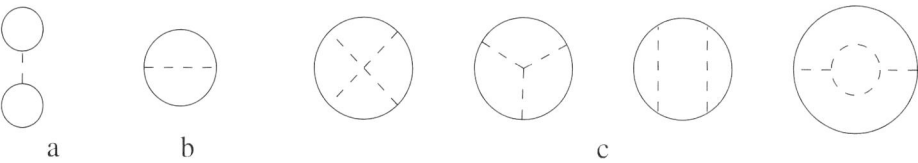

FIGURE 5. Graphs of degree one and two

EXERCISE 5.4. Let $m = 2$. Write the relations between the equivalence classes of degree two graphs shown in Figure 5c. What is the dimension of $\mathfrak{J}_2$?

This settles the type of graphs appearing in formula (29): the summation is over all graphs in $\widetilde{\mathfrak{J}}$, while $\langle L \rangle \in \mathfrak{J}[[\hbar]]$. It is somewhat simpler to study separately the components of different degrees; define

$$(31) \qquad \langle L \rangle_n = \sum_{\Gamma \in \widetilde{\mathfrak{J}}_n} \frac{W_\Gamma}{|\operatorname{Aut}(\Gamma)|} [\Gamma]$$

5.5. Configuration spaces. Let us deal now with the weights (30) of graphs (see [5, 18, 24] for details). The domain of integration in (30) is the configuration space C_Γ of embeddings of the set of vertices of Γ to $\mathbb{R}^3$, such that the legs of each subset l_j lie on the corresponding component L_j of the link L in the correct cyclic order. It is easy to see that for a graph Γ with k trivalent vertices and k_j legs ending on L_j, $j = 1, \ldots, m$ we have $C_\Gamma \cong (\mathbb{R}^3)^k \times \prod_j (S^1 \times \sigma^{k_j - 1}) \setminus \Delta$, where σ^k is a k-dimensional simplex, and Δ is the union of all diagonals where two or more points coincide. Indeed (forgetting for a moment about coincidences of vertices), each trivalent vertex is free to run over $\mathbb{R}^3$, while k_j legs ending on L_j run over $S^1 \times \sigma^{k-1}$, where S^1 encodes the position of the first leg, and the following legs are encoded by their distance from the previous one.

EXERCISE 5.5. Show that the dimension of C_Γ is twice the number of the edges of Γ.

Now, an orientation of a graph Γ determines an orientation of C_Γ; its idea is in fact based on Exercise 5.5. Let me describe this construction in some local coordinates. Near each trivalent vertex of Γ there are three local coordinates (describing its movement in $\mathbb{R}^3$); assign one of them to each of the three ends edges meeting in this vertex using their cyclic order. Near each leg of Γ there is only one local coordinate (describing its movement along the link); assign it to the corresponding end of the edge. By now the end of any edge has one coordinate assigned to it. It remains to order them using the given ordering of all edges of Γ and their directions. Let us order them as $(x_1, y_1, x_2, y_2, \ldots, x_n, y_n)$ where (x_i, y_i) are the coordinates assigned to the beginning and the end of i-th edge. This defines an orientation of C_Γ.

EXERCISE 5.6. The above construction involves a choice in each trivalent vertex since we had only a *cyclic* order of the edges meeting there, while we used a total order of these three edges. Show that a cyclic permutation of the three local coordinates used there preserves the orientation of C_Γ. Also, we used the orientation of Γ; what happens with the orientation of C_Γ if:

(1) The cyclic order of three half-edges in one vertex is reversed?
(2) A pair of edges is transposed in the total ordering of all edges?
(3) The direction of an edge is reversed?

5.6. Gauss-type maps of configuration spaces. To understand the integrand in (30), consider a directed edge e. Its ends (x, y) represent a point in the square $\mathbb{R}^3 \times \mathbb{R}^3$ with the diagonal $\Delta = \{(x,y)|\ x = y\}$ cut out. This cut square $C = \mathbb{R}^3 \times \mathbb{R}^3 \smallsetminus \Delta$ has the homotopy type of S^2, with the Gauss map

$$\phi : (x, y) \mapsto \frac{y - x}{||y - x||}$$

providing the equivalence. The form $\omega(y - x)$ assigned to this edge is nothing more than a pullback of the area form ω on S^2 to C via the Gauss map:

$$\omega(y - x) = \phi^* \omega$$

Each edge e of a graph Γ defines an evaluation map $\mathrm{ev}_e : C_\Gamma \to C$, by erasing all vertices of Γ but for the ends of e. The composition $\phi_e = \phi \circ \mathrm{ev}_e$ defines the Gauss map corresponding to (the ends of) an edge e. The graph Γ with an ordering $e_1, \ldots, e_n$ of edges defines the product $\phi_\Gamma = \prod_{i=1}^n \phi_{e_i} : C_\Gamma \to (S^2)^n$ of Gauss maps. Finally, the weight W_Γ is given by integrating the pullback of the volume form $dvol = \wedge_{i=1}^n \omega$ on $(S^2)^n$ to C_Γ by the product Gauss map ϕ_Γ:

$$(32) \qquad W_\Gamma = \int_{C_\Gamma} \phi_\Gamma^* \, dvol$$

EXERCISE 5.7. Suppose that a graph Γ has a double edge (i.e., a pair of edges both endpoints of which coincide). Show that $\dim(\phi_\Gamma(C_\Gamma)) \leq \dim(C_\Gamma) - 1$. Deduce that $W_\Gamma = 0$.

The following important example shows that at least in some simple cases W_Γ has an interesting topological meaning:

EXAMPLE 5.8. Let $\Gamma = e$ be a graph with one edge with the ends on two link components L, see Figure 5a. The configuration space $C_e \cong S^1 \times S^1 \subset C$ is a torus. It is mapped to S^2 by the Gauss map $\phi = \phi_e$. The weight $W_e = \int_{C_e} \phi^* \omega = \deg(\phi)$ is in this case just the degree of the map ϕ. This fact has many important consequences. In particular W_e takes only integer values and is preserved if we change the uniformly distributed area form ω to any other volume form $dvol$ on S^2 normalized by $\int_{S^2} dvol = 1$. It is also preserved if we deform the link by isotopy (since then the configuration space changes smoothly and the degree can not jump), so is a link invariant. This invariant is easy to identify: $\int_{C_e} \phi^* \omega = \mathrm{lk}(L_1, L_2)$ is the famous Gauss integral formula for the linking number $\mathrm{lk}(L_1, L_2)$ of L_1 with L_2. Thus we get

PROPOSITION 5.9. *Let $\Gamma = e$ be a graph with one edge with the ends on two link components L. Then the weight W_e is the linking number $\mathrm{lk}(L_1, L_2)$.*

EXERCISE 5.10. There is a simple combinatorial way to compute $\mathrm{lk}(L_1, L_2)$ from any link diagram: count all crossings where L_1 passes over L_2, with signs shown in Figure 6a. Interpret this formula as a calculation of $\deg(\phi)$ by counting

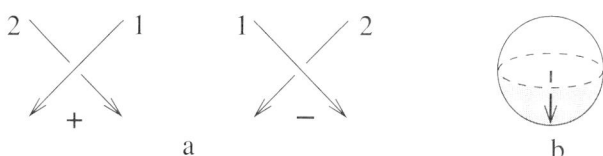

FIGURE 6. Signs of crossings and the south pole on S^2

(with signs) the number of preimages of a certain regular value of ϕ (hint: look at Figure 6b). What formula would we get if we counted the preimages of the north pole?

For other graphs the situation is more complicated. For example, let $\Gamma = e$ be the graph with one edge, both ends of which end on *the same* link component, see Figure 5b. Then the configuration space C_e is an *open* annulus $(R^3)^0 \times S^1 \times \sigma^1 \smallsetminus \Delta = S^1 \times (0, 1)$ (torus cut along the diagonal). The Gauss map ϕ is badly behaved near the diagonal, so the integrand blows up near the diagonal and we can not extend it to the closed torus. The integral nevertheless converges; one way to see it is to compactify C_e, cutting out of it some small neighborhood of the diagonal. This makes C_e into a closed annulus $C_e^\varepsilon = S^1 \times [\varepsilon, 1 - \varepsilon]$ (thus making the integral convergent) and we can recover the initial integral by taking $\varepsilon \to 0$. But the Gauss integral W_e is no more a knot invariant: it may take any real value under a knot isotopy. A detailed discussion on this subject may be found in [**5**]. Why does this happen? The reason is that the compactified space C_e is not a torus, but an annulus, so has a boundary and the degree of the Gauss map is not well-defined. When both ends of the edge start to collide together, the direction of the vector connecting them (which appears in the Gauss map) tends to the (positive or negative) tangent direction to the knot. The image of the unit tangent to the knot under the Gauss map is a certain curve γ on S^2. One of the boundary circles $S^1 \times \varepsilon$ and $S^1 \times (1 - \varepsilon)$ of C_e^ε is mapped into γ, while the other is mapped into $-\gamma$, and the weight W_e is part of the area of S^2 covered by the annulus $\phi(C_e^\varepsilon)$ between these curves. Unfortunately, γ may move on S^2 under an isotopy of L, so this area may change.

In this particular case there is a neat way to solve this problem: let L be *framed* (i.e. fix a section of its normal bundle). We may think about the framing as about a unit normal vector $n(x)$ in each point x of a knot. This allows us to slightly deform the Gauss map: $\phi(x, y) \to \phi(x, y) + \varepsilon n(y)$. Now both boundary circles of the annulus C_e map into the same curve on S^2 (why?) and we may glue the annulus into the torus so that the map ϕ_ε extends to it. It makes W_e into an invariant of framed knots, called the self-linking number (the same result may be obtained by slightly pushing L off itself along the framing and considering the linking number of the knot with its pushed-off copy).

It turns out that for other graphs there are also no divergence problems, so all integrals W_Γ converge, and that a collision of all vertices of a graph to one point (so-called anomaly, see [**18, 24**]) is the only source of non-invariance, exactly as for W_e above. Thus there is a suitable normalization of the expression (31) for $\langle L \rangle = \sum_n \langle L \rangle_n$ which gives a link invariant. To avoid a complicated explicit description of this normalization, let me formulate this result as follows:

THEOREM 5.11 ([**1**], [**18**], [**24**]). *Let $L = \cup_{i=1}^{m} L_i$ be a link. Then $\langle L \rangle$ depends only on the isotopy class of L and on the Gauss integrals $W_e(L_i)$ of each component L_i. In particular, an evaluation of $\langle L \rangle$ at representatives of L for which $W_e(L_1) = \cdots = W_e(L_m) = 0$ is a link invariant.*

REMARK 5.12. It is known that this is a universal invariant of finite type. In particular this means that it is stronger than both the Alexander and the Jones polynomials (it contains the two-variable HOMFLY polynomial) and all other quantum invariants. Conjecturally the anomaly vanishes and this invariant coincides with the Kontsevich integral, see [**18**].

EXAMPLE 5.13. Let L be a knot, and take $n = 2$. There are four graphs of degree two, shown in Figure 5c. We will denote the first of them X, and the second by Y. By Exercise 5.7 the weight of the third graph vanishes. Also, choose a framing of L so that the self-linking is 0; then the contribution of the last graph vanishes (another way to achieve the same result is to add to $\langle L \rangle_2$ a certain multiple of the self-linking number squared); we can set then $[X] = [Y]$. Thus we will consider simply

$$v_2 = \frac{1}{4} \int_{C_X} \phi_X^*(\omega \wedge \omega) + \frac{1}{3} \int_{C_Y} \phi_Y^*(\omega \wedge \omega \wedge \omega).$$

The first integral is 4-dimensional, while the second is 6-dimensional; none of them separately is a knot invariant (see [**20**] for a discussion); however, their sum v_2 is (see [**3**], [**20**])! This invariant is, up to a constant, the second coefficient of the Alexander-Conway polynomial. See [**20**] for its detailed treatment as the degree of a Gauss-type map.

5.7. Degrees of maps. How can we explain the result of Theorem 5.11? We may try to repeat the reasoning of Example 5.8 in the general case. Recall that by Exercise 5.5 the dimensions of C_Γ and $(S^2)^n$ match, so if C_Γ would be a closed manifold, then equation (32) would be a formula for a calculation of the degree of ϕ_Γ. In other words, if C_Γ would have a fundamental class, (32) would be its pairing with the pullback $\phi_\Gamma^* dvol$. That would be great: we would know that W_Γ takes only integer values, and would be able to compute it in many ways, including a simple counting of preimages of any regular value of ϕ.

Unfortunately, the reality is much worse: C_Γ is an open space, so the degree is not well-defined and even the convergence is unclear. To guarantee the convergence, we should construct a compactification $\bar{C}_\Gamma$ of C_Γ to which ϕ_Γ extends. This however will cause new problems: the space $\bar{C}_\Gamma$ will have many boundary strata. There are various way to deal with them: we can relativize some of them (i.e. consider a relative version of the theory), cap-off some others (i.e. glue to them some new auxiliary configuration spaces), or zip them up (gluing a stratum to itself by an involution). But in general, some boundary strata will remain; indeed, in Example 5.13 we have seen that none of W_X or W_Y separately can be made into a knot invariant. The remedy would be to glue together the configuration spaces for different graphs in $\tilde{\mathfrak{J}}_n$ along the common boundary strata. This tedious work can be done indeed [**18**, **24**] and (up to a certain anomaly correction) one can interpret $\langle L \rangle$ as the degree of a certain map Φ_n from a grand configuration space $\mathcal{C}_n$ to $(S^2)^n$. Too many technicalities are involved to describe this construction in necessary details, so I refer the interested reader to [**18**, **24**] and will present only a brief sketch of this construction.

The first problem is that initially the dimensions of C_Γ for various $\Gamma \in \widetilde{\mathfrak{J}}_n$ do not match. E.g., in Example 5.13, for $n=2$ the spaces C_X and C_Y have dimensions 4 and 6 respectively. This can be fixed by considering a product $C_\Gamma \times (S^2)^k$ of C_Γ with enough spheres to make the maps $\phi_\Gamma \times (\mathrm{id})^k$ to have the same target space $(S^2)^N$ for all Γ. Now one should do the gluings. When two endpoints of an edge e of a graph Γ collide, the corresponding boundary stratum of C_Γ looks like $C_G \times S^2$ for $G = \Gamma/e$. Thus we can glue together such strata for all pairs (Γ, e) with isomorphic $G = \Gamma/e$. Some more, so-called hidden, strata remain after these main gluings. Fortunately, each of them can be zipped-up (i.e. glued to itself by a certain involution). The only codimension one boundary strata which remains after all these gluings are the anomaly strata, where all vertices of a graph $\Gamma \in \widetilde{\mathfrak{J}}$ collide together. These problematic anomaly strata can be glued [**18**] to a new auxiliary space. One ends up with a grand configuration space $\mathcal{C}_n$ endowed with a map $\Phi : \mathcal{C}_n \to (S^2)^N$ (glued from the corresponding maps $\phi_\Gamma : \bar{C}_\Gamma \to (S^2)^N$). One may show that the cohomology $H^{2N}(\mathcal{C}_n)$ of this space projects surjectively to $\mathfrak{J}_n$ (see [**17**] for a similar case of 3-manifold invariants). Then $\langle L \rangle_n^0 = \langle L \rangle_n +$ anomaly correction can be interpreted as the degree of Φ_n, or more exactly, the image in $\mathfrak{J}_n$ of the fundamental class $[(S^2)^N]$ under the induced composite map $\pi \circ \Phi^* : H^{2N}(S^2)^N \to H^{2N}(\mathcal{C}_n) \to \mathfrak{J}_n$.

5.8. Final remarks. There remain many questions: which compactification should we take, why do the antisymmetry and Jacobi relations appear in the cohomology of the grand configuration space, etc. Each of them is quite lengthy and is out of the scope of this note. We refer the interested reader to [**5, 18, 24**]. A mathematical treatment of invariants of 3-manifolds arising from the Chern-Simons theory was done in [**2, 4**]; I especially recommend [**17**]. While I do not know whether similar Feynman series arising in other topological problems always have a reformulation in terms of degrees of maps of some grand configuration space, it seems quite plausible. There are at least some other notable examples, see e.g. [**19**] for a similar interpretation of Kontsevich's quantization of Poisson structures.

References

[1] D. Altschler, L. Freidel, *On universal Vassiliev invariants*, Comm. Math. Phys. **170** (1995) 41–62.

[2] S. Axelrod, I. M. Singer, *Chern-Simons perturbation theory*, Proc. XX DGM Conf. (New-York, 1991) (S. Catto and A. Rocha, eds.) World Scientific, 1992, 3–45; *Chern-Simons perturbation theory II*, J. Diff. Geom. **39** (1994) 173–213.

[3] D. Bar-Nathan, *Perturbative aspects of the Chern-Simons topological quantum field theory*, Ph.D. thesis, Princeton Univ. 1991; *Perturbative Chern-Simons theory*, J. Knot Theory and Ramif. **4** (1995) 503–548.

[4] R. Bott, A. Cattaneo, *Integral invariants of 3-manifolds I, II*, J. Diff. Geom. **48** (1998), 91–133, and J. Diff. Geom. 53 (1999), no. 1, 1–13.

[5] R. Bott, C. Taubes, *On the self-linking of knots*, J. Math. Phys. 35 (1994) 5247–5287.

[6] S. S. Chern, J. Simons, *Some cohomology classes in principal fiber bundles and their application to riemannian geometry*, Proc. Nat. Acad. Sci. U.S.A. **68** (1971) 791–794.

[7] A. Connes, D. Kreimer, *Renormalization in quantum field theory and the Riemann-Hilbert problem I, II*, Commun.Math.Phys. **210** (2000) 249–273, Commun.Math.Phys. **216** (2001) 215–241.

[8] E. T. Copson, *Asymptotic expansions*, Cambridge Tracts in Math. and Math. Phys. **55**, 1967.

[9] *Quantum fields and strings: a course for mathematicians*, vol 1-2, (P. Deligne et al. eds.) AMS 1999.

[10] A. Dubrovin, A. Fomenko, S. Novikov, *Modern geometry— methods and applications. Part II. The geometry and topology of manifolds*, Graduate Texts in Mathematics, **104**, Springer, 1985.
[11] D. Freed, *Classical Chern-Simons theory I, II*, Adv. Math. **113** (1995) 237–303, Houston J. Math. **28** (2002) 293–310.
[12] J.-M. Drouffe, C. Itzykson, *Statistical field theory*, Cambridge Univ.Press, 1989.
[13] L. D. Faddeev, V. N. Slavnov, *Gauge fields, introduction to quantum theory*, Benjamin/Cummings, Reading, 1980.
[14] W. Fulton, R. D. MacPherson, *A compactification of configuration spaces.*, Annals of Math. (2) **139** (1994), 183–225.
[15] E. Guadagnini, M. Martinelli, M. Mintchev, *Perturbative aspects of the Chern-Simons field theory*, Phys. Let. **B277** (1989) 111; *Chern-Simons field theory and link invariants*, Nucl. Phys. **B330** (1990) 575–607.
[16] C. Itzykson, J. Zuber, *Quantum field theory*, McGraw-Hill, New-York, 1985.
[17] G. Kuperberg, D. Thurston, *Perturbative 3-manifold invariants by cut-and-paste topology*, math.GT/9912167.
[18] S. Poirier, *The Configuration space integral for links in R^3*, Algebr. Geom. Topol. **2** (2002) 1001–1050.
[19] M. Polyak, *Quantization of linear Poisson structures and degrees of maps*, Let. Math. Phys. **60** (2003) 15–35.
[20] M. Polyak, O. Viro, *On the Casson knot invariant*, J. Knot Theory and Ramif. **10** (2001) 711–738.
[21] A. Polyakov, *Gauge fields and strings*, Harwood academic publishers, 1987.
[22] S. Schweber, *An introduction to relativistic quantum field theory*, Row, Peterson 1961.
[23] A. S. Schwarz, *Quantum field theory and topology*, Springer, 1993.
[24] D. Thurston, *Integral expressions for the Vassiliev knot invariants*, M.A. thesis, Harvard Univ. 1995, math.QA/9901110.
[25] E. Witten, *Quantum field theory and the Jones polynomial*, Comm. Math. Phys. **121** (1989) 351–399.

DEPARTMENT OF MATHEMATICS, THE TECHNION, 32000 HAIFA, ISRAEL
E-mail address: `polyak@math.technion.ac.il`

Structures in Feynman graphs: Hopf algebras and symmetries

Dirk Kreimer

ABSTRACT. We review the combinatorial structure of perturbative quantum field theory with emphasis given to the decomposition of graphs into primitive ones. The consequences in terms of unique factorization of Dyson–Schwinger equations into Euler products are discussed.

CONTENTS

1. Introduction
2. Motivation
3. Lie and Hopf algebra structures in a perturbative expansion
4. Renormalization and the Riemann–Hilbert problem
5. Multiple rescalings
6. Derivations on the Hopf algebra
7. Primitivity
8. Renormalization and Hochschild cohomology
9. Unique factorization and Dyson–Schwinger equations
10. Conclusion and acknowledgments
References

1. Introduction

The reputation of quantum field theory has always been mixed. As a predictive theory, it is the best theory ever formulated. It has been plagued by inconsistencies and conceptual flaws though, ever since it was first spelled out. Roughly speaking, these shortcomings come in two forms: order by order in the perturbation theory short-distance singularities seemingly destroy the meaning of the Feynman rules, and the predictive power of a perturbative calculation. Elimination of this flaw in perturbative renormalization works self-consistently, but this does not satisfy a

---

1991 *Mathematics Subject Classification.* 81T15, 81T18, 81Q30.
Key words and phrases. Renormalization, Feynman Graphs, Hopf Algebras, Quantum Field Theory.

mathematician: without any guiding structure, renormalization remained in disrepute, as only a means to hide the infinities under the carpet. This is surely not a pillar on which the foundations of the theory could rest. Eventually, recourse to extended objects, avoiding the presence of point-like short-distance singularities, seemed unavoidable. So far, this has not led to the advent of a predictive theory replacing local quantum field theory.

The second shortcoming of perturbative quantum field theory was its inability to make contact with non-perturbative approaches: a demon, enabled to renormalize any loop order in arbitrary short time indeed seems to be monstrous: we are confronted typically with a series of finite numbers whose asymptotic behaviour so far defies understanding, i.e. resummation. Singularities in the Borel plane on the positive axis can be generated by renormalons, and by instanton singularities [1]. The former typically result from chains of one-loop subgraphs which lead to an exponential growth of the perturbation series. Instantons meanwhile incorporate non-perturbative behaviour parametrized by solutions of the classical field equations.

Some of these flaws gave way recently: the ugly duckling of short-distance singularities and their elimination in perturbative renormalization turned out to be a conceptual asset of the theory.

We will review these developments and put them into context, with emphasis given to comment on future potential for progress beyond perturbation theory. Much of what is reported here in the first six sections has been published elsewhere, or was, in much greater detail, the content of a course in renormalization theory recently given [2]. The final section is devoted to some new ideas.

2. Motivation

The structure of the perturbative expansion of a Quantum Field Theory (QFT) is in many ways determined by Hopf algebra and Lie algebra structures on Feynman graphs (see §3.3, in particular the four lines after Proposition 7). Forest formulae originate from the Hopf algebra structure, while notions like anomalous dimensions and β-functions relate to the Lie algebra structure. This allows for considerable simplifications in the conceptual interpretation of renormalization theory. Indeed, the identification with the Riemann–Hilbert problem allows to summarize renormalization theory in a single line: find the Birkhoff decomposition of a regularized but unrenormalized physical parameter of interest. The positive part will be its renormalized contribution (in a MS scheme), the negative part will be the corresponding *counterterm* [3, 4]. Here, MS refers to minimal subtraction: the negative part in the Birkhoff decomposition eliminates the poleterms, while leaving the finite part untouched.

Nevertheless, this result, a direct consequence of the Hopf and Lie algebra structure and of the existence of a group homomorphism to a certain diffeomorphism group, does not exhaust the tools given at our hand by these algebraic structures.

Indeed it seems wise to begin a consideration of quantum field theory from the viewpoint of combinatorics and graph theory, a viewpoint already mandated by 't Hooft and Veltman's famous "diagrammar" [5]. By its very definition, QFT will ultimately reflect, in its short distance singularities and its most notorious properties reflecting those, the structure of spacetime at the infinitesimal small. The lack of the ability to perform experiments at essentially infinitely high energies means the observables which arise from the presence of short distance singularities are the only window we have towards that structure.

But then, it is desirable to rest the pillars of the foundations of QFT on structures which are robust enough to accommodate the unknown structure of the very small, and hence combinatorics is certainly a good candidate.

From this viewpoint, it seems favorable to start from Feynman diagrams, and try to derive the features which we hope to see in a QFT from their combinatorial properties. The structure of the very small might still be queerer than we think, and maybe even queerer than we can think, and so any attempt to axiomatize or construct QFT from principles gained from experience with the not so very small might ultimately turn out to be demanding more than Nature is prepared to deliver. So we will set out to explore the combinatorial structures behind a perturbative expansion, which, as we will see, in itself provides the means to handle short-distance singularities, and offers much in terms of a conceptual analysis of QFT. The development of this combinatorial viewpoints owes much to the efforts of practitioners of QFT, who exposed it to the most cruel tests in radiative correction calculations. One is left with awe when one studies in detail how well perturbative QFT fares in such tests. None of the rigorous approaches to QFT ever produced tools which contributed to the art of radiative correction calculations significantly while the combinatorial notions reported here build a rigorous mathematical background for the practice of QFT, and hopefully start to close a gap between such practice of QFT and its mathematical foundations which has grown far too large in the last decades.

There are two basic operations on Feynman graphs which govern their combinatorial structure, organize their contributions to a chosen Green function as well as organize the process of renormalization.

These two basic operations are the decomposition of a graph into subgraphs, and the opposite operation, insertion of subgraphs into a graph. While insertion of subgraphs is needed to generate the formal series over graphs which provide a fixed point for the Dyson–Schwinger equation of a given Green function, decomposition of graphs is necessary to achieve renormalization by counterterms which are local expressions, polynomial in (derivatives of) fields in the Lagrangian. Such a Lagrangian $\mathcal{L}$ is typically a finite sum of monomials

$$\mathcal{L} = \sum_i \mathcal{M}_i,$$

where, for example in a massive scalar theory with cubic interaction, we have monomials $\mathcal{M}_1 = 1/2 \ \partial_\mu \phi \partial^\mu \phi$, $\mathcal{M}_2 = 1/2 \ m^2 \phi^2$, $\mathcal{M}_3 = g/6 \ \phi^3$. Each such monomial $\mathcal{M}_i$ yields a Z-factor $\mathcal{M}_i \to Z_i \mathcal{M}_i$ to absorb short-distance singularities. Feynman graphs arise when we expand in terms of a "weak coupling" g (i.e., g is small). The Z-factors provide invertible series in g, their constant term is unity. The theory is typically calculated using some regulator. Z-factors are arranged such that they eliminate all divergences so that the regulator can be switched off eventually. As always, absorbing singularities allows for choice of the remaining finite part, which gives rise to the various renormalization schemes used in practice.

Let us take a first look at a Feynman graph and the roles these operations play. Consider a "three-loop vertex-correction" Γ, this time in QED in four dimensions, with the usual identification of wavy lines with photons and straight lines with fermions

$$\Gamma = \;\raisebox{-0.5em}{\includegraphics[height=2em]{vertex}}\;.$$

This graph Γ consists of twelve edges and seven vertices. We denote by $\Gamma^{[0]}$ the set of vertices, and by $\Gamma^{[1]}$ the set of edges. There are three external edges which have an open end. They are just a reminder of the meaning a physicist gives to such a graph: it is a contribution to the probability amplitude of a scattering process involving, in this case, a fermion anti-fermion pair and a photon, so a decay $1 \to 2$ or recombination $2 \to 1$. To these external edges we can assign quantum numbers, specifying the spin, mass, momenta and other characteristics of the particles involved in the scattering process.

The set of edges decomposes in this obvious manner into internal and external ones $\Gamma^{[1]} = \Gamma^{[1]}_{int} \cup \Gamma^{[1]}_{ext}$. To calculate the actual contribution of a graph Γ, one needs *Feynman rules* [1] that can be heuristically derived from the Lagrangian of the theory in a straightforward way. They come with a surprise though: typically, in sensible quantum field theories they do not seem to make sense, at first sight.

Obviously, we use two different meanings of sense. What goes on here is that the theories most sensible from a particle physicists viewpoint are those which agree best with observations. Nature singles out by this criterion renormalizable quantum field theories in four dimensions. But then, their Feynman rules seem to violate common sense: evaluating the Feynman graphs in such theories by the Feynman rules produces ill-defined quantities galore. It is a relief then that these senseless quantities actually make good mathematical sense when one looks at the structure of graphs much more closely.

Let us go back to the example of the graph Γ, regarded as a QED graph in four dimensions. Let us describe the structure of the ill-defined quantities we get from this graph. First of all, we assign a variable k_e to each edge $e \in \Gamma^{[1]}$. Variables attached to internal edges we call *internal momenta*, while variables attached to external edges we call *external momenta*, which we assume to be fixed and given as part of the quantum numbers of external particles.

Each vertex in the set $\Gamma^{[0]}$ of vertices of Γ imposes a constraint on these variables, such that the momenta attached to a vertex add to zero. One easily recognizes that the number of free variables left is then equal to the number of loops in the graph. Those free variables, corresponding to internal unobserved momenta, have to be integrated out. The Feynman rules attach "propagators" $P^{-1}(k_e)$ to each edge e, and the edge variables k_e have to be integrated over a D-dimensional Euclidean space (as far as short-distance singularities go we can indeed avoid the complications provided by other signatures of the metric, or by some non-vanishing curvature). Depending on the scaling degree ω_P of the inverse propagators $P(k_e)$, $P(\lambda k_e) = \lambda^{\omega_P} P(k_e)$ for large k_e, this might or might not be a well-defined integral. This can be easily decided by power counting, and leads us to the notion of a degree of divergence: assigning weights ω_P to edges (and, in general, also to vertices), allows, by sole consideration of these weights and the number of loops in a graph, to decide in advance if the integrals attached to a graph will have short-distance (UV $\equiv$ ultra-violet) divergences. Such an integral is typically of the

---

[1] See the article by M. Polyak in this volume.

form

$$\phi(\Gamma) = \int \prod_{e \in \Gamma^{[1]}_{int}} d^D k_e \, P^{-1}(k_e) \prod_{v \in \Gamma^{[0]}} \delta\left(\sum_{j \in f_v} k_j\right) g(v). \quad (1)$$

Here δ is the Dirac δ-function, f_v is the set of edges attached to v, and $g(v)$ is the factor which the Feynman rules assign to the vertex v (with an appropriate ordering of the factors along fermionic lines). Note that this integral representation implies momentum conservation for the external momenta.

Understanding the singularity structure of such an expression amounts to an identification of singular subintegrals, which can only be provided by subgraphs which contain closed loops, and it thus suffices to consider "1-particle irreducible" (1PI) graphs (where every edge is part of an embedded cycle) and their disjoint unions to identify all singular subsectors.

So then, what is the message for our example? It turns out that there is one divergent subgraph for QED in $D = 4$ dimensions. So what we get is an ill-defined quantity containing another ill-defined quantity as a subintegral.

How do we make sense of this? There are two steps in this process. The first is to understand how to make sense out of the case where the graphs have no divergent subgraphs. The second and harder step is to understand how to do it when subproblems are present. In between lies the step to understand why divergent subgraphs make life so much harder.

Consider

$$\Gamma_0 := \quad .$$

This is a divergent QED graph which for $D = 4$ has no proper divergent subgraphs. By the above such a graph can be written in the form

$$\phi(\Gamma)(m; p_i) = \int_0^\infty \frac{F_\Gamma(r; m; p_i)}{r} dr, \quad (2)$$

with

$$F_\Gamma(0; m; p_i) = 0, \quad F_\Gamma(r; m; p_i) \sim Q(\log r), \quad (3)$$

where $Q(\log r)$ is a polynomial in $\log r$ with coefficients independent of $m; p_i$ for large positive $\log r$, and where $r = \sum_i |k_{e_i}|$, see [13]. Furthermore, in the above p_i are the momenta attached to external edges which remain after integrating out the momenta k_e where e is an internal edge, and m refers to possible masses on which the propagators $P^{-1}(k_e)$ might depend.

Hence, what is sick about this graph calculation remains invariant when we vary these external parameters – the disease is localized, hence curable: the difference

$$\phi(\Gamma)(m; p_i) - \phi(\Gamma)(m; \tilde{p}_i) \quad (4)$$

exists for any modified external momenta $\tilde{p}_i$. Actually, in a log divergent graph free of subdivergences the divergence remains invariant under any diffeomorphism ψ of external parameters $(m; p_i) \to \psi((m; p_i))$.

So we can give no absolute meaning to the value of a Feynman graph, but the relative value defined by comparison with another graph with modified continous quantum numbers exists. An obvious example of a graph without subdivergences is a one-loop graph (where every subgraph with the same external vertices has no loops). These were the early successes of QFT indeed: the comparison of observables distinguished by different external parameters.

One point is worth emphasizing here: it is not just the loop number which makes a simple subtraction sufficient, but the fact that there are no subdivergences. That is one of the crucial advantages of the Hopf algebraic description of short distance singularities: the number of divergent sectors provides a well-defined grading on that Hopf algebra, and induction over that grading provides a much clearer understanding of how finite results are achieved. With respect to this grading, the bi-degree or augmentation degree as we will call it, divergent graphs free of subdivergences are of bidegree one, and correspond to the primitive elements in the Hopf algebra. We will often call them primitive graphs. Ultimately, they are the building blocks out of which we can assemble the full perturbative expansion, once we learn how to insert them one into the other.

This story has a Lagrangian version: the reference to a chosen scheme is established by plugging counterterms into the Lagrangian, such that all Green functions vanish at this reference "point", from now on called renormalization point. The choice of this point corresponds to a choice of a subtraction scheme R. Linguistically, we are rather lax: the choice of any scheme like minimal subtraction, momentum scheme, on-shell scheme and so on will be allowed. Any such choice, as we will see, corresponds to the choice of a certain element in the group of characters of the Hopf algebra, and hence indeed to a point in the corresponding group.

What goes wrong when subdivergences are present is obvious - simple differences like the above will fail. Indeed, the presence of divergences generates a dependence of the illness on external parameters. Is this the end of the theory?

Fortunately not. Let us consider what happens when we take two primitively divergent graphs and insert them one into the other, say we insert

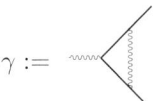

into Γ_0 so that Γ is obtained. Evaluating by the Feynman rules, the integral $\phi(\gamma)$ will appear as a subintegral of $\phi(\Gamma)$. Typically, the continous parameters (momenta) attached to the external edges of γ will be integrated over in the larger integral $\phi(\Gamma)$. But really, what we should insert in that larger integral is $\phi(\gamma)$ minus its value at the renormalization point. This is the actual trick: the elimination of subdivergences goes first, before the cure is available for the larger problem posed by the larger graph Γ. This is consistent with the Lagrangian story: curing the sickness of γ required the insertion of its counterterm into the Lagrangian. Thus, this modified Lagrangian provides for each γ its counterterm, hence provides the cured version of γ.

Summarizing, Γ in our example contains one interesting subgraph, the one-loop "vertex correction" γ. It is the only subgraph which provides a divergence, and the whole UV-singular structure comes from this subdivergence and from the overall divergence of Γ itself. From the analytic expressions corresponding to Γ, to Γ_0 and to γ we can form the analytic expression corresponding to the renormalization of the graph Γ. It is given by

$$(5) \qquad \phi(\Gamma) - R(\phi(\Gamma)) - R(\phi(\gamma))\phi(\Gamma_0) + R\left(R(\phi(\gamma))\phi(\Gamma_0)\right).$$

We emphasize that the crucial step in obtaining this expression is the use of the graph Γ and its disentangled pieces, γ and $\Gamma_0 = \Gamma/\gamma$. Here Γ_0 is obtained from Γ by collapsing the subgraph γ to a point. Diagrammatically, the above expression reads (omitting ϕ)

$$\left\langle\!\!\!\!\diagup - R\left(\diagup\right) - R\left(\diagup\right)\diagup \right. $$
$$\left. + R\left(R\left(\diagup\right)\diagup\right)\right).$$

The unavoidable arbitrariness in the so-obtained expression lies in the choice of the map R which we suppose to be such that it does not modify the short-distance singularities (UV divergences) in the analytic expressions corresponding to the graphs. It just evaluates graphs at the chosen renormalization point, so it employs the chosen scheme. Certain requirements on R have to be demanded [6, 3, 7]: it has to be faithful to short-distance singularities so that elements in the image of $\mathrm{id} - R$ are finite, and it has to establish a Baxter algebra on the target space of the Feynman rules $\phi: H \to V$:

(6) $\quad R(ab) + R(a)R(b) = R(aR(b)) + R(R(a)b), \ R: V \to V, \ a, b \in V.$

Let us summarize the overall structure envisaged at this moment: Feynman rules provide a character on the Hopf algebra of graphs with values in some suitable space V, which itself can be a ring or an algebra (see below). Often, it is for example the ring of Laurent series in some regularization parameter, with poles of finite order. This ring has a multiplication, and a further structure map $R: V \to V$ which fulfills (6). Such an algebra or ring V is then called a Baxter algebra or ring, allowing to connect renormalization theory to the study of integrable systems, see [**6, 30, 31**][2].

This then renders the above combination of four terms finite. If there were no subgraphs, a simple subtraction $\phi(\Gamma) - R(\phi(\Gamma))$ would suffice to eliminate the short-distance singularities, but the necessity to obtain local counterterms forces us to first subtract subdivergences. This is Bogoliubov's famous $\bar{R}$ operation [**8**], which delivers here:

(7) $\qquad \phi(\Gamma) \to \bar{R}(\phi(\Gamma)) = \phi(\Gamma) - R(\phi(\gamma))\phi(\Gamma_0).$

This provides two of the four terms above. Among them, these two are free of subdivergences and hence provide only a local overall divergence. The projection of these two terms into the range of R provides the other two terms, which combine to the counterterm

(8) $\qquad Z_\Gamma = -R(\phi(\Gamma)) + R(R(\phi(\gamma))\phi(\Gamma_0))$

of Γ, and addition of this counterterm delivers the finite result above, in the kernel of R, by the fact that the UV divergences are not changed by the renormalization map R.

---

[2]This refers not the the physicist R.Baxter of the Yang–Baxter map, but to the mathematician G.Baxter, though their work is intimately related, see section two of [**31**].

Locality is indeed connected to the absence of subdivergences: if a graph has only an overall divergence, UV singularities only appear when all loop momenta tend to infinity jointly. Regarding the analytic expressions corresponding to a graph as a Taylor series in external parameters like masses or momenta, power counting establishes that only the coefficients of the first few polynomials in these parameters are UV singular. Hence they can be subtracted by a counterterm which is a polynomial in fields and their derivatives. The argument fails as long as one has not eliminated all subdivergences: their presence can force each term in the Taylor series to be divergent. For example, if none of the edges or vertices of the subgraph involves the external momenta (by routing external momenta so that they avoid the subgraph under consideration), then no derivative with respect to those parameters can possibly eliminate the divergence generated by this subgraph. Hence, the preparation of a graph for a local subtraction by Bogoliubov's operation is unavoidable. The independence of the singularities of a prepared graph on the variation (diffeomorphism) of external parameters is a strong hint to regard the remaining singularity as a residue, an analogy with far-reaching consequences [**4**] to which we will come back below.

The basic operation so far was the disentanglement of the graph Γ into subgraphs γ and cographs $\Gamma_0 = \Gamma/\gamma$, and this very disentanglement gives rise to a Hopf algebra structure, as was first observed in [**9**], which we will describe shortly.

It is useful to study the invariants of a permutation of places where a subgraph γ is inserted in a graph Γ_0. What obviously remains invariant is the hierarchical structure of subdivergences, what varies is the topology of the graph. Indeed, the counterterm for γ provided by the Lagrangian is the same wherever we insert γ, and the difference of two such insertions will need no counterterm for γ. This has immediate consequences for number-theory [**10**] to be commented at the end of section seven, when we connect such invariance to a Galois symmetry.

For now, as an example, consider two graphs

$$(9) \qquad \Gamma_1 = \quad , \quad \Gamma_2 = \quad .$$

They have one common property: both of them can be regarded as the graph

$$(10) \qquad \Gamma_0 = \Gamma_1/\gamma = \Gamma_2/\gamma = \quad ,$$

into which the subgraph

$$\gamma = \quad$$

is inserted. Such graphs are equivalent, in the sense that the combinatorial process of renormalization produces exactly the same subtraction terms for both of them [**9**]. This equivalence can be most meaningfully stated using the language of operads [**11**][3]: inserting a subgraph at different places is an operad composition, with a labelled composition to denote the places where to insert subgraphs. Vanishing of

---

[3]See also the article by Voronov in this volume

the leading singularity for that difference then means that the permutation group for that operad composition is trivially represented on that leading short-distance singularity.

The combinatorics of renormalization is essentially governed by this bookkeeping process of the hierarchies of subdivergences, and this bookkeeping is what is delivered by rooted trees. They are just the appropriate tool to store the hierarchy of disjoint and nested subdivergences, and ultimately, overlapping subdivergences [**9, 12, 6**].

Hence the Hopf algebra of Feynman graphs indeed has a "role model": the Hopf algebra of rooted trees. This Hopf algebra has been first described in [**9, 22**], and is now textbook material [**32**]. It can be quickly described as the span of monomials in images of a closed Hochschild one cocycle B_+:

$$1, B_+(1), B_+(B_+(1)), B_+(B_+(B_+(1))), B_+(B_+(1)^2), \ldots$$

are the first few linear generators, and the relation to trees is obvious. Here,

$$bB_+ = 0 \Leftrightarrow \Delta B_+ = B_+ \otimes 1 + [\mathrm{id} \otimes B_+]\Delta$$

the coproduct structure is completely determined by demanding Hochschild closedness $bB_+ = 0$, and the usual requirement that the coproduct Δ is an algebra map.

Rooted trees can be assigned in two natural ways to a Feynman graphs:

(i) Decomposing momentum space Feynman integrals into divergent sectors [**12**]. This amounts to a resolution of overlapping divergences into disjoint or nested sectors in the integral representation of graphs provided by the momentum space Feynman rules. This can be done, and will be exhibited later on when we comment on how to use Hochschild cohomology[4] to provide a proof for renormalizability.

(ii) On the other hand, starting from coordinate space Feynman rules, the singularities stratify in rooted trees directly, upon the fact that they are supported along diagonals in the configuration space of the location of vertices.

Combining both viewpoints, the Fourier transform becomes a map between two Hopf algebras of rooted trees. Before we define the Hopf algebra of Feynman graphs more formally, we briefly describe Feynman graphs from the configuration space viewpoint.

First, let us ask what to make out of graphs which have overlapping divergences. For graphs without them like Γ_1, Γ_2 above there is a unique way to obtain them from

$$\Gamma_0 = \Gamma_1/\gamma = \Gamma_2/\gamma =$$

and the one-loop vertex correction γ. We plug γ into the other vertex-correction at an appropriate internal vertex to obtain the desired graphs. In such graphs divergent subgraphs can be identified in a unique manner.

This has to be contrasted with graphs which have overlapping divergences. We have typically no unique manner but several ways instead to identify subdivergent

---

[4]See, e.g., the article by Tamarkin and Tsygan in this volume.

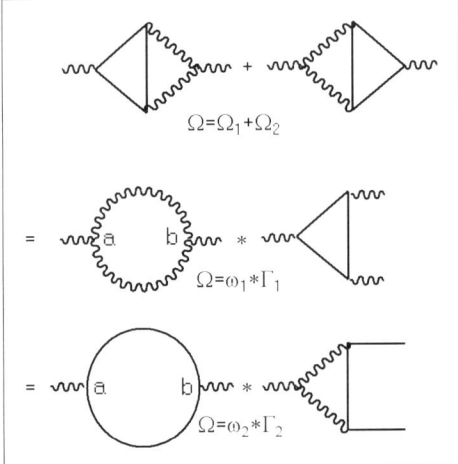

FIGURE 1. This sum Ω of two graphs Ω_1, Ω_2 can be obtained in two different ways. We can either glue Γ_1 into ω_1 by identifying the external edges of Γ_1 with the edges adjacent to the vertices a or b of ω_1, or similarly, do so for the graphs ω_2, Γ_2, with the same result Ω.

graphs: such graphs contain more than one maximal forest, more than one way how to shrink all divergent subgraphs to a point such that a primitive cograph remains.

It is in this loss of uniqueness of decomposition, or factorization if you like, where some of the most fascinating aspects of QFT reside: indeed, we will sketch at the end of this paper that we are fighting here with the famous problem of unique factorization, and will argue that the cure is quite similar to what one does for algebraic number fields: factorize into prime ideals.

Consider Figure 1. It shows two ways of obtaining a sum Ω of two graphs, by inserting a vertex graph Γ_1 into the two internal vertices a, b of a self-energy ω_1, or a vertex graph Γ_2 into ω_2,

(11) $$\Omega = \omega_1 \star \Gamma_1 = \omega_2 \star \Gamma_2.$$

Note that each of the two graphs Ω_1, Ω_2 in Ω has four internal vertices. There are two subsets of three vertices in each graph that belong to two divergent vertex subgraphs. In the coordinate space, these subsets provide a singular strata located along the corresponding diagonals where these subsets collapse to a single point. The Feynman rules in the coordinate representation do not make sense along these diagonals, and their continuation to these diagonals is a problem dual to the compactification of the configuration space along the diagonals. The choice of compactification corresponds to the choice of a renormalization scheme, and the fact that diagonals in the configuration space come stratified by rooted trees [14] invites one to establish the Hopf algebra techniques of renormalization theory in the study of configuration spaces, and vice versa. The short distance singularities of Feynman graphs then come solely from regions where all vertices are located at coinciding points. One has no problem to define the Feynman integrand in the configuration space of vertices at distinct locations, while a proper extension to diagonals is what is required [15, 16, 17].

Due to the Hopf algebra structure of Feynman graphs we can define the renormalization of all such sectors without making recourse to any specific analytic properties of the expressions (Feynman integrals) representing those sectors. The only assumption we make is that in a sufficiently small neighborhood of such an ultralocal region (the neighborhood of a diagonal) we can define the scaling degree, that is, the combinatorial power counting. Analytic detail arising from quests for causality for example can be imposed later. Having determined these power counting degrees and having chosen a renormalization scheme, we will soon formally define the principle of multiplicative subtraction which will tell us how we get local counterterms and finite renormalized quantities whatever the finer detail is of the Feynman rules. This is the strong combinatorial backbone of QFT, which ultimately has its source in the notion of a residue, and in the fundamental invariance properties of residues.

This is indeed a very useful application of the Hopf algebra: the decomposition into its primitives reveals the range over which the combinatorial structures of renormalization are stable to include, for example, fluctuations of the metric as long as the scaling degree of propagators is microlocally unchanged. Such fluctuation do not alter the residue of a bidegree-one graph. This allows us to understand the recent results of Brunetti and Fredenhagen [18] from a combinatorial viewpoint.

Its time now to define the Hopf algebra of Feynman graphs, which expresses this combinatorial backbone of renormalization theory.

3. Lie and Hopf algebra structures in a perturbative expansion

3.1. Lie and Hopf algebras of Feynman graphs.
Here we give some formal definitions following [20].

For any graph Γ, denote its set of edges by $\Gamma^{[1]}$ and its set of vertices by $\Gamma^{[0]}$. The graphs we consider are labelled. This means that there is a label attached to each edge which determines its "type". Diagramatically, the type of an edge is often indicated by the way it is drawn: (oriented) straight lines, curly lines, dashed lines, and so on. These edges (representing propagators in physicists parlance), are chosen with reference to Lorentz covariant wave equations: the propagator as the analytic expression assigned to an edge is an inverse wave operator with boundary conditions typically chosen in accordance with causality. We can, if desired, ignore such consideration by the choice of an Euclidean metric.

We exclude graphs with self-loops: no edge connects a vertex to itself. But we allow two different vertices to be connected by several edges. Note that self-loops are naturally excluded in a massless theory.

DEFINITION 1. *The star of a vertex* $v \in \Gamma^{[0]}$ *is the set* $f_v := \{f \in \Gamma^{[1]} \mid v \cap f \neq \emptyset\}$.

The type of a vertex is determined by the type of its star.

DEFINITION 2. *An edge* $f \in \Gamma^{[1]}$ *is internal, if* $\{v_f\} := f \cap \Gamma^{[0]}$ *consists of two vertices.*

Let us define Γ_{int} as the subgraph of Γ consisting of all internal edges of Γ.

DEFINITION 3. *An edge* $f \in \Gamma^{[1]}$ *is external, if* $f \cap \Gamma^{[0]}$ *is a single vertex.*

As we exclude self-loops, this means that an external edge has an open end. These edges correspond to external particles interacting in the way prescribed by the graph.

Let us now define n-particle irreducible (n-PI) graphs.

DEFINITION 4. *A n-particle irreducible graph (n-PI graph) Γ is a graph without self-loops such that it remains connected after removal of any n edges.*

Of particular importance are the 1PI graphs. They do not decompose into disjoint graphs upon removal of an edge. Note that any n-PI graphs is also $(n-1)$-PI, $\forall n \geq 2$. A graph which is not 1-PI is called reducible. Also, any connected graph is considered as 0-PI.

We now turn to the possibilities of inserting graphs into each other. Our first requirement is to establish bijections between sets of edges so that we can define gluing operations.

DEFINITION 5. *We call two sets of edges I_1, I_2 compatible, $I_1 \sim I_2$, if they contain the same number of edges of the same type.*

We say that two vertices v_1, v_2 are of the same type if their stars are compatible.

Quite often, we will collapse the internal subgraph to a point. The only useful information still available after that process is about its set of external edges:

DEFINITION 6. *We define $\mathbf{res}(\Gamma)$ to be the result of collapsing $\Gamma_{\text{int}}^{[1]}$ to a point.*

An example is

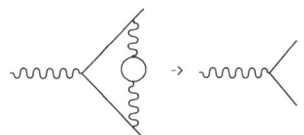

Note that $\mathbf{res}(\Gamma)^{[1]} \equiv \mathbf{res}(\Gamma)_{\text{ext}}^{[1]} \sim \Gamma_{\text{ext}}^{[1]}$. By construction all graphs which have compatible sets of external edges have the same residue.

If $\Gamma_{\text{ext}}^{[1]}$ is empty, we call Γ a *vacuum graph*; if $\Gamma_{\text{ext}}^{[1]}$ consists of a single edge, we call Γ a *tadpole graph*. Vacuum graphs and tadpole graphs will be discarded in what follows. If $\Gamma_{\text{ext}}^{[1]}$ contains two elements, we call Γ a *self-energy graph*; if $\Gamma_{\text{ext}}^{[1]}$ contains more than two edges, we call Γ an *interaction* or *vertex graph*. Further we restrict ourselves to graphs which have vertices such that the cardinality of their types is ≥ 2. We thus do not consider theories involving external field backgrounds mandating vertices having only a single edge attached to them.

At this stage, we are provided with a list of edges and vertices obtained typically, but not necessarily, from some QFT Lagrangian. Weights assigned to these elements allow us to discriminate graphs formed from these objects by the means of power counting. We can thus meaningfully speak about 1PI graphs which are superficially divergent or convergent, that is, graphs Γ such that the integral $\phi(\Gamma)$ diverges or converges.

We are after a mechanism which eliminates all possible divergences coming from subintegrations. These can be detected by power counting on the 1PI subgraphs, and we will soon introduce a Hopf algebra structure based on superficially divergent 1PI graphs.

3.2. External structures. Let us mention here one more useful notational device: the external structures of [**3**]. To each external edge $f \in \Gamma_{\text{ext}}^{[1]}$ we assign an external momentum p_f, a vector in some (to be appropriately specified) vectorspace and impose the condition $\sum_{f\in\Gamma^{[1]}_{ext}} p_f = 0$. We let $E_{\mathbf{res}(\Gamma)}$ be the linear space of functions on the subset of assignments $f \mapsto p_f$ fulfilling this condition.

Following [3] we use a notation familiar from distribution theory to describe the evaluation of a graph Γ by some Feynman rules $\phi : \mathcal{H} \to E_{\mathbf{res}(\Gamma)}$: with $\phi(\Gamma) \in E_{\mathbf{res}(\Gamma)}$, we denote by $\langle \sigma, \phi(\Gamma) \rangle$ the evaluation of $\phi(\Gamma)$ with respect to the distribution σ on $E_{\mathbf{res}(\Gamma)}$, in the same way as $\langle \delta_a, f \rangle = f(a)$ defines the evaluation of a function at a by the pairing with a Dirac δ-distribution supported at a.

A graph Γ together with a distribution σ we call a specified graph, and write (Γ, σ) for such a pair. We can regard it as a graph with a further prescription how to evaluate it.

We will need these notions to keep track of our perturbative expansion: let us assume that the Lagrangian L which is the source for our perturbative expansion is a sum over k field monomials. Assume that L contains n different fields $\phi_1, \ldots \phi_n$. Then to each monomial $P(\{\phi_i\})$ in L we can assign the sequence $(i_1, \ldots, i_n)_P$, which tells us the degree of P in each field. We call this sequence the field degree of P. If two monomials P, P' deliver equal field degrees, they both can be described by graphs with an equal residue in the above sense: their corresponding Feynman graphs Γ will have identical external legs, hence identical residues $\mathbf{res}(\Gamma)$. Counterterms for such monomials are then calculated by suitable projections in $E_{\mathbf{res}(\Gamma)}$, implemented by distributions $\sigma_P, \sigma_{P'}$.

A typical example are the mass and wave function renormalization of a scalar propagator, with monomials $P = -m^2\phi^2/2$ and $P' = \partial_\mu\phi\partial^\mu\phi/2$, say. Both monomials are quadratic in a scalar field ϕ and the contributions of two-point graphs to P, P' are obtained from

$$\langle \sigma_P, \phi(\Gamma) \rangle$$

and

$$\langle \sigma_{P'}, \phi(\Gamma) \rangle,$$

with $\langle \sigma_P, f(m^2, q^2) \rangle = \partial_{m^2} f \mid_{m^2 = q^2}$ and $\langle \sigma_{P'}, f(m^2, q^2) \rangle = \partial_{q^2} f \mid_{m^2 = q^2}$, for any function $f(m^2, q^2)$. Similarly, one extends to other cases for monomials of equal field degree, and one can conveniently express the formfactor decomposition of Green functions in this language.

In such a notation, all field monomials P in the Lagrangian correspond to expressions $\langle \sigma_P, \phi(\mathbf{res}(\Gamma)) \rangle$, where Γ is a graph whose external legs are in accordance with the field degree of P, and one always has that

(12) $$\langle \sigma_P, \phi(\Gamma) \rangle = \rho_P(\Gamma) \langle \sigma_P, \phi(\mathbf{res}(\Gamma)) \rangle.$$

Here $\rho_P(\Gamma)$ is a scalar function (the form-factor) which satisfies $\rho_P(\Gamma_1\Gamma_2) \equiv \rho_P(\Gamma_2\Gamma_1)$ $= \rho_P(\Gamma_1)\rho_P(\Gamma_2)$ $\forall P$ for disjoint diagrams, regardless of the fact that $\langle \sigma_P, \phi(\Gamma_P) \rangle$ can be matrix-valued. The simple fact that the action is a Lorentz scalar indeed mandates that a formfactor decomposition into scalar coefficients is always possible, and is in accordance with the product structure of graphs: the evaluation of a product is the commutative product of the evaluations.

It is those scalar coefficients which appear as characters on the Hopf algebra of graphs which we describe below. Our route will be to first define a pre-Lie product on graphs, which at the same time establishes a Lie bracket upon its antisymmetrization, and then take the dual of the universal enveloping algebra of the so-obtained Lie algebra as the Hopf algebra under consideration. This indeed provides a Hopf algebra which is always commutative. It underlies the disentanglement of graphs into their divergent parts, the combinatorial backbone of local quantum field theory.

3.3. The pre-Lie structure. Consider two graphs Γ_1, Γ_2. Assume that Γ_2 is an interaction graph, i.e., it has at least 3 external edges. For a chosen vertex $v \in \Gamma_1^{[0]}$ such that $f_v \sim \Gamma_{2,\text{ext}}^{[1]}$, we define

$$(13) \qquad \Gamma_1 \star_v \Gamma_2 = (\Gamma_1 \backslash \{v\}) \bigsqcup_{f_v \sim \Gamma_{2,\text{ext}}^{[1]}} \Gamma_2,$$

where $\bigsqcup$ stands for disjoint union. In other words, we blow-up the vertex v to the internal graph $\Gamma_{2,\text{int}}$. Then we sum over all possible bijections between f_v and $\Gamma_{2,\text{ext}}^1$, and normalize such that topologically different graphs are generated precisely once: if k bijections lead to the same graph, we divide each term in the sum by k.

We now define

$$(14) \qquad \Gamma_1 \star \Gamma_2 = \sum_{v \in \Gamma_1^{[0]}, f_v \sim \Gamma_{2,\text{ext}}^{[1]}} \Gamma_1 \star_v \Gamma_2.$$

The $\star$ operation can be extended to the insertion of self-energy graphs (see §3.1), replacing internal edges by self-energy graphs which have the corresponding external edges.

We also define the insertion of specified graphs in the similar vein, by requiring that

$$(15) \qquad (\Gamma_1, \sigma_1) \star (\Gamma_2, \sigma_2) = (\Gamma_1 \star \Gamma_2, \sigma_1).$$

By inserting a graph, we forget about conditions imposed on its external legs. Dually, for the Hopf algebra below, this corresponds to the fact that upon disentangling a graph, we have the freedom to impose constraints -renormalization conditions on its subgraphs.

PROPOSITION 7. *The operation $\star$ is pre-Lie, namely*

$$[(\Gamma_1, \sigma_1) \star (\Gamma_2, \sigma_2)] \star (\Gamma_3, \sigma_3) - (\Gamma_1, \sigma_1) \star [(\Gamma_2, \sigma_2) \star (\Gamma_3, \sigma_3)]$$
$$= [(\Gamma_1, \sigma_1) \star (\Gamma_3, \sigma_3)] \star (\Gamma_2, \sigma_2) - (\Gamma_1, \sigma_1) \star [(\Gamma_3, \sigma_3) \star (\Gamma_2, \sigma_2)]$$

To sketch the proof, which is elementary for unspecified graphs, we note that the insertion of subgraphs is a local operation, and that on both sides, the difference amounts to plugging the subgraphs Γ_2, Γ_3 into Γ_1 at disjoint places, which is evidently symmetric under the exchange of Γ_2 with Γ_3.

Taking external structures into account, we can choose a color for each evaluation σ_i under consideration, and represent the pair (σ, Γ) by a colored graph. We then realize that coloring does not spoil the locality of the insertions.

If needed, this can be generalized in a way as to maintain constraints on external legs when inserting a graph into another, still maintaining the pre-Lie structure. Integration over the edges according to (1) is then done in a way such that it obeys the constraints valid for each subgraph. Such extensions are useful in practice when one desires to decompose graphs into various analytic subfactors [21], and underly the decomposition into primitives discussed in section five.

In most of what follows we only use unspecified graphs, with the generalization to specified graphs being evident.

We get a Lie algebra $\mathcal{L}$ by antisymmetrizing this operation,

$$(16) \qquad [\Gamma_1, \Gamma_2] = \Gamma_1 \star \Gamma_2 - \Gamma_2 \star \Gamma_1$$

and a Hopf algebra $\mathcal{H}$ as the dual of the universal enveloping algebra of this Lie algebra. Typically, one restricts attention to graphs which are superficially divergent, with residues corresponding to field monomials in the Lagrangian.

Superficially convergent graphs can be incorporated using the trivial abelian Lie algebra which they span when one regards them as specified graphs. The fact that they do not contribute to counterterms in the Lagrangian means that they are annihilated by external structures which project onto the contributions to superficially divergent field monomials P. As a result they have a vanishing Lie bracket among themselves and furthermore form a semi-direct product with their superficially divergent cousins [**3**].

We close this section by giving an example of the Lie bracket

3.4. The principle of multiplicative subtraction.

Having defined Lie algebra structures on graphs, it is now easy to harvest them to give a clear conceptual meaning to the renormalization process. As announced, we just have to dualize the universal enveloping algebra $\mathcal{U}(\mathcal{L})$ of $\mathcal{L}$ and obtain a commutative, but not cocommutative Hopf algebra $\mathcal{H}$ [**3**].

From now on, when we want to distinguish carefully between the Hopf and Lie algebras of Feynman graphs we write δ_Γ for the multiplicative generators of the Hopf algebra and write Z_Γ for the dual maltiplicative generators of $\mathcal{U}(\mathcal{L})$ with pairing

$$(17) \qquad \langle Z_\Gamma, \delta_{\Gamma'} \rangle = \delta^K_{\Gamma, \Gamma'},$$

where on the rhs we have the Kronecker δ^K, and extend the pairing by means of the coproduct

$$(18) \qquad \langle Z_{\Gamma_1} Z_{\Gamma_2}, X \rangle = \langle Z_{\Gamma_1} \otimes Z_{\Gamma_2}, \Delta(X) \rangle.$$

Quite often, we want to refer to the graph(s) which index an element in $\mathcal{H}$ or $\mathcal{L}$. For that purpose, for each element in $\mathcal{H}$ and each element in $\mathcal{L}$ we introduce a map to graphs, indicated by an overbar:

$$(19) \qquad \overline{Z_X} = X, \overline{\delta_X} = X.$$

As we already have emphasized the Hopf algebra of rooted trees is the role model for the Hopf algebras of Feynman graphs. It underlies the process of renormalization when formulated perturbatively at the level of Feynman graphs. The following formulas should be of no surprise for the reader acquainted with the universal Hopf algebra of rooted trees and are a straightforward generalization of similar formulas for rooted tree Hopf algebras [**22**].

First of all, we start considering one-particle irreducible graphs as the linear generators of the Hopf algebra, with their disjoint union as product. We then identify the Hopf algebra as described above by a coproduct $\Delta : \mathcal{H} \to \mathcal{H} \otimes \mathcal{H}$:

$$(20) \qquad \Delta(\Gamma) = \Gamma \otimes 1 + 1 \otimes \Gamma + \sum_{\gamma \subset \Gamma} \gamma \otimes \Gamma/\gamma,$$

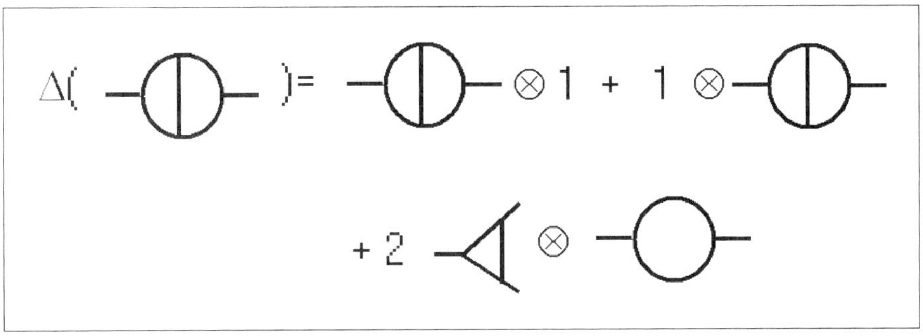

FIGURE 2. The coproduct $\Delta(\Gamma)$.

where the sum is over all unions of one-particle irreducible (1PI) superficially divergent proper subgraphs and we extend this definition to products of graphs so that we get a bialgebra. The above sum should, when needed, also run over appropriate external structures to specify the appropriate type of local insertion [3] which appear in local counterterms, which we omitted in the above sum for simplicity. In general, it is the Milnor–Moore Theorem [19] which says that any such Hopf algebra (commutative, coassociative, and equal to $\mathbb{C}$ in degree zero) is constructed as the dual of the universal enveloping algebra of a Lie algebra (the primitives). Fig.(2) gives an example of a coproduct.

For any Hopf algebra element X we often write a shorthand for its coproduct

$$\Delta(X) = \widetilde{\Delta}(X) + X \otimes 1 + 1 \otimes X = X \otimes 1 + 1 \otimes X + X' \otimes X''.$$

Let now X be a 1PI graph. For each term in the sum $\widetilde{\Delta}(X) = \sum_i X' \otimes X''$ we have unique gluing data G_i such that

$$(21) \qquad X = X'' \star_{G_i} X', \ \forall i.$$

These gluing date describe the necessary bijections to glue the components X' back into X'' so as to obtain X.

The counit $\bar{e}$ vanishes on any non-trivial Hopf algebra element, $\bar{e}(1) = 1$, $\bar{e}(X) = 0$. At this stage we have a commutative, but typically not cocommutative bialgebra. It actually is a Hopf algebra as the antipode in such circumstances comes for free as

$$(22) \qquad S(\Gamma) = -\Gamma - \sum_{\gamma \subset \Gamma} S(\gamma)\Gamma/\gamma.$$

The next thing we need are Feynman rules, which we regard as maps $\phi : \mathcal{H} \to V$ from the Hopf algebra of graphs $\mathcal{H}$ into an appropriate space V.

Over the years, physicists have invented many calculational schemes in perturbative quantum field theory, and hence it is of no surprise that there are many choices for this space.

For example, if we want to work on the level of Feynman integrands in a BPHZ scheme, we could take as this space a suitable space of Feynman integrands (realized either in momentum space or configuration space, whatever suits). An alternative scheme would be the study of regularized Feynman integrals, for example the use of dimensional regularization would assign to each graph a Laurent-series with poles of finite order in a variable ε near $\varepsilon = 0$, and we would obtain characters evaluating in this ring, an approach leading to the Riemann-Hilbert decomposition described below. In any case, we will have $\phi(\Gamma_1\Gamma_2) \equiv \phi(\Gamma_2\Gamma_1) = \phi(\Gamma_1)\phi(\Gamma_2)$, $\forall \phi: \mathcal{H} \to V$.

Then, with the Feynman rules providing a canonical character ϕ, we will have to make one further choice: a renormalization scheme. This is is a map $R: V \to V$, and we demand that is does not modify the UV-singular structure: in BPHZ language, it should not modify the Taylor expansion of the integrand for the first couple of terms divergent by power counting. In dimensional regularization, then, we demand that it does not modify the pole terms in ε. Furthermore, we require it to make the pair (V, R) into a Baxter algebra, as in (6).

Finally, the principle of multiplicative subtraction emerges: we define a further character S_R^ϕ which deforms $\phi \circ S$ slightly and delivers the counterterm for Γ in the renormalization scheme R:

$$(23) \qquad S_R^\phi(\Gamma) = -R[\phi(\Gamma)] - R\left[\sum_{\gamma \subset \Gamma} S_R^\phi(\gamma)\phi(\Gamma/\gamma)\right]$$

which should be compared with the undeformed

$$(24) \qquad \phi \circ S = -\phi(\Gamma) - \sum_{\gamma \subset \Gamma} \phi \circ S(\gamma)\phi(\Gamma/\gamma).$$

Then, the classical results of renormalization theory follow immediately [9, 12, 22]. We obtain the renormalization of Γ by the application of a renormalized character

$$\Gamma \to S_R^\phi \star \phi(\Gamma)$$

and the $\bar{R}$ operation as

$$(25) \qquad \bar{R}(\Gamma) = \phi(\Gamma) + \sum_{\gamma \subset \Gamma} S_R^\phi(\gamma)\phi(\Gamma/\gamma),$$

so that we have

$$(26) \qquad S_R^\phi \star \phi(\Gamma) = \bar{R}(\Gamma) + S_R^\phi(\Gamma).$$

In the above, we have given all formulas in their recursive form. Zimmermann's original forest formula solving this recursion is obtained when we trace our considerations back to the fact that the coproduct of rooted trees can be written in non-recursive form, and similarly the antipode [12]. We will come back below to this transition between graphs and trees. Also, we note that the principle of multiplicative subtraction can be formulated in much larger generality, as it is a basic combinatorial principle, see for example [23] for another appearance of this principle.

3.5. The bidegree. A fundamental notion is the bidegree of a 1PI graph (see [24] for a convenient review of notions needed here), resulting from iterated projections into the augmentation ideal. Usually, induction in perturbative QFT, aiming to prove a desired result is carried out using induction over the loop number, an obvious grading for 1PI graphs. On quite general grounds, for our Hopf algebras there exists another grading, which is actually much more useful. We call it the bidegree, bid(Γ). To motivate it, consider a superficially divergent n-loop graph Γ which has no divergent subgraph. It is evident that its short-distance singularities can be treated by a single subtraction, for any n. It is not the loop number, but the number of divergent subgraphs which is the most crucial notion here. Fortunately, this notion has a precise meaning in the Hopf algebra of superficially divergent graphs using the projection into the augmentation ideal. This indeed counts the degree in renormalization parts of a graph: an overall superficially convergent graph has bidegree zero by definition, a primitive Hopf algebra element has bidegree one, and so on.

So we have $\mathcal{H} = \bigoplus_{i=0}^{\infty} \mathcal{H}^{(i)}$, with $\mathrm{bid}(\mathcal{H}^{(i)}) = i$. To define this decomposition, let $\mathcal{H}_{\mathrm{Aug}}$ be the augmentation ideal of the Hopf algebra, and let $P : \mathcal{H} \to \mathcal{H}_{\mathrm{Aug}}$ be the corresponding projection $P = \mathrm{id} - E \circ \bar{e}$, with $E(q) = qe$. Let $\widetilde{\Delta}(X) = \Delta(X) - e \otimes X - X \otimes e$, as before. $\widetilde{\Delta}$ is still coassociative, and for any $X \in \mathcal{H}_{\mathrm{Aug}}$ there exists a unique maximal k such that $\widetilde{\Delta}^{k-1}(X) \in [\mathcal{H}^{(1)}]^{\otimes k}$. Here, $\mathcal{H}^{(1)}$ is the linear span of primitive elements z: $\Delta(z) = z \otimes e + e \otimes z$.

DEFINITION 8. *The map $X \to k$ obtained as described above is called the bidegree $\deg_p$ of X.*

The bidegree is conserved under the coproduct and under the product (disjoint union). Typically, all properties connected to questions of renormalization theory can be proven more efficiently using the grading by the bidegree instead of the loop number. In particular, we will discuss below the interplay between Hochschild cohomology and the bidegree, which explains locality of counterterms and finiteness of renormalized graphs rather succinctly. But let us first come to an even more succinct formulation of renormalization theory, making use of the existence of complex regularizations.

4. Renormalization and the Riemann–Hilbert problem

4.1. The Birkhoff decomposition. The Feynman rules in dimensional or analytic regularization determine a character ϕ on the Hopf algebra which evaluates as a Laurent series in a complex regularization parameter ε, with poles of finite order, this order being bounded by the bidegree of the Hopf algebra element to which ϕ is applied. In minimal subtraction, $\phi_- := S^{\phi}_{R=MS}$ has similar properties: it is a character on the Hopf algebra which evaluates as a Laurent series in a complex regularization parameter ε, with poles of finite order, this order being bounded by the bidegree of the Hopf algebra element to which $S^{\phi}_{R=MS}$ is applied, only that there will be no powers of ε which are ≥ 0. Then, $\phi_+ := S^{\phi}_{R=MS} \star \phi$ is a character which evaluates in a Taylor series in ε, all poles are eliminated. We have the Birkhoff decomposition

(27) $$\phi = \phi_-^{-1} \star \phi_+.$$

This establishes an amazing connection between the Riemann–Hilbert problem and renormalization [**3, 4**]. It uses in a crucial manner once more that the multiplicativity constraints (6),

$$R[xy] + R[x]R[y] = R[R[x]y] + R[xR[y]],$$

ensure that the corresponding counterterm map S_R is a character as well,

(28) $$S_R[xy] = S_R[x]S_R[y], \ \forall x, y \in H,$$

by making the target space of the Feynman rules into a Baxter algebra, characterized by this multiplicativity constraint. The connection between Baxter algebras and the Riemann–Hilbert problem, which lurks in the background here, remains largely unexplored, as of today.

Renormalization in the MS scheme can now be summarized in one sentence: with the character ϕ given by the Feynman rules in a suitable regularization scheme and well-defined on any small curve around $\varepsilon = 0$, find the Birkhoff decomposition $\phi_+(\varepsilon) = \phi_- \star \phi$.

The unrenormalized analytic expression for a graph Γ is then $\phi[\Gamma](\varepsilon)$, the MS-counterterm is $S_{MS}(\Gamma) \equiv \phi_-[\Gamma](\varepsilon)$ and the renormalized expression is the evaluation $\phi_+[\Gamma](0)$. Once more, note that the whole Hopf algebra structure of Feynman graphs is present in this group: the group law demands the application of the coproduct, $\phi_+ = \phi_- \star \phi \equiv S_{MS}^{\phi} \star \phi$.

But still, one might wonder what a huge group this group of characters really is. What one confronts in QFT is the group of diffeomorphisms of physical parameter: low and behold, changes of scales and renormalization schemes are just such (formal) diffeomorphisms. So, for the case of a massless theory with one coupling constant g, for example, this just boils down to formal diffeomorphisms of the form

$$g \to \psi(g) = g + c_2 g^2 + \ldots.$$

The group of one-dimensional diffeomorphisms of this form looks much more manageable than the group of characters of the Hopf algebras of Feynman graphs of such a theory.

4.2. Diffeomorphisms of physical parameters. Thus, it would be very nice if the whole Birkhoff decomposition could be obtained at the level of diffeomorphisms of the coupling constants. The crucial ingredient is to realize the role of a standard QFT formula of the form

$$(29) \qquad g_{\text{new}} = g_{\text{old}}\, Z_1 Z_2^{-3/2},$$

which expresses how to obtain the new coupling in terms of a diffeomorphism of the old. This was achieved in [4], recognizing this formula as a Hopf algebra homomorphism from the Hopf algebra of diffeomorphism to the Hopf algebra of Feynman graphs, regarding $Z_g = Z_1/Z_2^{3/2}$, a series over counterterms for all 0PI graphs with the external leg structure corresponding to the coupling g, in two different ways. It is at the same time a formal diffeomorphism in the coupling constant g_{old} and a formal series in Feynman graphs. As a consequence, there are two competing coproducts acting on Z_g. That both give the same result defines the required homomorphism, which transposes to a homomorphism from the largely unknown group of characters of $\mathcal{H}$ to the one-dimensional diffeomorphisms of this coupling.

The crucial fact in this is the recognition of the Hopf algebra structure of diffeomorphisms by Connes and Moscovici [25]: Assume you have formal diffeomorphisms ϕ, ψ in a single variable

$$(30) \qquad x \to \phi(x) = x + \sum_{k>1} c_k^{\phi} x^k,$$

and similarly for ψ. How do you compute the Taylor coefficients $c_k^{\phi \circ \psi}$ for the composition $\phi \circ \psi$ from the knowledge of the Taylor coefficients c_k^{ϕ}, c_k^{ψ}? It turns out that it is best to consider the Taylor coefficients

$$(31) \qquad \delta_k^{\phi} = \log(\phi'(x))^{(k)}(0)$$

instead, which are as good to recover ϕ as the usual Taylor coefficients. The answer lies then in a Hopf algebra structure:

$$\delta_k^{\phi\circ\psi} = m \circ (\tilde{\psi} \otimes \tilde{\phi}) \circ \Delta_{CM}(\delta_k), \tag{32}$$

where $\tilde{\phi}, \tilde{\psi}$ are characters on a certain Hopf algebra $\mathcal{H}_{CM}$ (with coproduct Δ_{CM}) so that $\tilde{\phi}(\delta_i) = \delta_i^\phi$, and similarly for $\tilde{\psi}$. Thus one finds a Hopf algebra with abstract generators δ_n such that it introduces a convolution product on characters evaluating to the Taylor coefficients $\delta_n^\phi, \delta_n^\psi$, such that the natural group structure of these characters agrees with the diffeomorphism group.

It turns out that this Hopf algebra of Connes and Moscovici is intimately related to rooted trees in its own right [22], signalled by the fact that it is linear in generators on the rhs, as are the coproducts of rooted trees and graphs.[5]

There are a couple of basic facts which enable to make in general the transition from this rather foreign territory of the abstract group of characters of a Hopf algebra of Feynman graphs (which, by the way, equals the Lie group assigned to the Lie algebra with universal enveloping algebra the dual of this Hopf algebra) to the rather concrete group of diffeomorphisms of physical observables. These steps are

- Recognize that Z factors are given as counterterms over formal series of graphs starting with 1, graded by powers of the coupling, hence invertible.
- Recognize the series Z_g as a formal diffeomorphism, with Hopf algebra coefficients.
- Establish that the two competing Hopf algebra structures of diffeomorphisms and graphs are consistent in the sense of a Hopf algebra homomorphism.
- Show that this homomorphism transposes to a Lie algebra and hence Lie group homomorphism.

This works out extremely nicely, with details given in [4]. In particular, the effective coupling $g_{\text{eff}}(\varepsilon)$ now allows for a Birkhoff decomposition in the space of formal diffeomorphisms

$$g_{\text{eff}}(\varepsilon) = g_{\text{eff}-}(\varepsilon)^{-1} \circ g_{\text{eff}+}(\varepsilon), \tag{33}$$

where $g_{\text{eff}-}(\varepsilon)$ is the bare coupling and $g_{\text{eff}+}(0)$ the renormalized effective coupling while the coupling can be regarded as the coupling in any theory which has a single interaction term. If there are multiple interaction terms in the Lagrangian, one finds similar results relating the group of characters of the corresponding Hopf algebra to the group of formal diffeomorphisms in the multidimensional space of coupling constants.

There are some false but pertinent claims in the literature with regards to the extension of this result to massless QED and other theories with spin [26]. The above results hold as they stand for massless QED, with the relevant Hopf algebra homomorphism given by $e_{\text{new}} = Z_3^{-1/2} e_{\text{old}}$.

---

[5]Taking the δ_n as naturally grown linear combination of rooted trees imbeds the commutative part of the Connes-Moscovici Hopf algebra in the Hopf algebra of rooted trees, which on the other hand allows for extensions similar to the ones needed by Connes and Moscovici. Details are in [22], with an extended discussion of generalized natural growth and bicrossed product structures in [20].

The confusion arises from the assumption that in theories with spin, the non-commutativity of Green functions demands the use of a noncommutative noncocommutative Hopf algebra, ignoring the commutative Hopf algebra obtained from the pre-Lie algebra of graphs insertions above. This is patently wrong and ignores the fact that the relevant characters on the Hopf algebra are given by the coefficient functions of the tree-level form factors, and these coefficient functions are scalar characters

$$\rho(\Gamma_1\Gamma_2) = \rho(\Gamma_2\Gamma_1) = \rho(\Gamma_1)\rho(\Gamma_2), \tag{34}$$

on a commutative Hopf algebra, with all noncommutativity residing in the matrix structures multiplying those form-factors. A typical such character ρ is provided by the Feynman rules ϕ of massless QED applied to a fermion self-energy graph Γ with external momentum $\slashed{p}$ say, where we have

$$\phi(\Gamma) = \rho(\Gamma)\slashed{p}, \tag{35}$$

with $\rho(\Gamma) = Tr(\slashed{p}\phi(\Gamma))/p^2$, and $\slashed{p} = \phi(\mathbf{res}(\Gamma))$, and indeed, the matrix structure of $\slashed{p}$ plays no further role with respect to the commutativity of $\mathcal{H}$.

5. Multiple rescalings

So we have singled out MS as a special character, providing us with a Birkhoff decomposition. So what about other renormalization schemes: are they less meaningful? Not quite.

Let us come back to unrenormalized Feynman graphs, and their evaluation by some chosen character ϕ, and let us also choose a renormalization scheme R. The group structure of such characters on the Hopf algebra can be used in an obvious manner to describe the change of renormalization schemes. This has very much the structure of a generalization of Chen's Lemma [6].

5.1. Chen's Lemma. Consider $S_R \star \phi$. Let us change the renormalization scheme from R to R'. How is the renormalized character $S_{R'} \star \phi$ related to the renormalized character $S_R \star \phi$? The answer lies in the group structure of characters:

$$S_{R'} \star \phi = [S_{R'} \star S_R \circ S] \star [S_R \star \phi], \tag{36}$$

which generalizes Chen's Lemma on iterated integrals [6]. We inserted a unit η with respect to the $\star$-product in form of $\eta = S_R \circ S \star S_R \equiv S_R^{-1} \star S_R$, and can now read the rerenormalization, switching between the two renormalization schemes, as composition with the renormalized character $S_{R'} \star S_R^{-1}$. Note that $S_{R'} \star S_R \circ S$ is a renormalized character indeed: if R, R' are both self-maps of V which do not alter the short-distance singularities as discussed before, then in the ratio $S_{R'} \star S_R \circ S$ those singularities drop out.

Similar considerations apply to a change of scales which determine a character [6]. If μ is a dimensionful parameter which dominates the process under consideration and which appears in a character $\phi = \phi(\mu)$, then the transition $\mu \to \mu'$ is implemented in the group by acting on the right with the renormalized character $\psi^\phi_{\mu,\mu'} := \phi(\mu) \circ S \star \phi(\mu')$ on $\phi(\mu)$,

$$\phi(\mu') = \phi(\mu) \star \psi^\phi_{\mu,\mu'}. \tag{37}$$

Let us note that this Hopf algebra structure can be efficiently automated as an algorithm for practical calculations exhibiting the full power of this combinatorics [**27**].

Now, assume we compute Feynman graphs by some Feynman rules in a given theory and decide to subtract UV singularities at a chosen renormalization point μ. This amounts, in our language, to saying that the map S_R is parametrized by this renormalization point: $S_R = S_{R_\mu}$. Typically, one has for $R_\mu : V \to V$,

$$(38) \qquad R_\mu(ab) = R_\mu(a)\, R_\mu(b),$$

in accordance with but stronger than the multiplicativity constraints.

Then, let $\Phi(\mu, \mu')$ be the ratio $\Phi(\mu, \mu') = S_{R_\mu} \star \phi(\mu')$. We then have the groupoid law as part of the before-mentioned Chen's lemma [**6, 7**]

$$(39) \qquad \Phi(\mu, \eta) \star \Phi(\eta, \mu') = \Phi(\mu, \mu'),$$

thanks to (38).

5.2. Automorphisms of $\mathcal{H}$.

We further note that the use of external structures always allows to pullback renormalization schemes to automorphisms of the Hopf algebra by solving the equation

$$(40) \qquad S_R^\phi \circ S = \phi \circ \Theta_R,$$

for the automorphism $\Theta_R : \mathcal{H} \to \mathcal{H}$. Starting from primitive elements Γ, this can be solved to determine appropriate external structures, following the guidance of [**6**], solving by a recursion over the bidegree.

The full group structure of the group of characters of the Hopf algebra has barely been used in practice yet, with some notable exceptions in [**27, 28**], but it provides an enormously rich set of new tools for the investigation of QFT. One striking aspect is that it allows insight into the hardest problem of QFT: understanding the analytic aspects of the perturbative expansion, by laying bare the way in which analytic input enters the combinatorics of renormalization, and by allowing nonperturbative results completely based on the Hopf and Lie algebra structures, for example the beautiful duality between the scaling variable and the coupling discovered in [**29**].

Harvesting the combinatorial structure of QFT emphasizes the crucial role played by bidegree one graphs and their residues. Relations between such diagrams are often consequences of symmetries in the Lagrangian -BRST symmetry after the quantization of a local gauge symmetry, supersymmetry, or both. Even more striking though are relations which can not be traced back to such an origin. Typically they come as analytic surprises after the calculation of different diagrams. To get an idea of an underlying conceptual source for such relations, we have to look at our main theme, insertion and elimination of subgraphs, more closely.

6. Derivations on the Hopf algebra

Having defined the Hopf- and Lie algebras of Feynman graphs, it is profitable to look into representations of the Lie algebra as derivations on the Hopf algebra, following [**20**].

6.1. Representations of $\mathcal{L}$.

The Lie algebra $\mathcal{L}$ gives rise to two representations acting as derivations on the Hopf algebra $\mathcal{H}$:

(41) $$Z_\Gamma^+ \times \delta_X = \delta_{X \star \Gamma}$$

and

(42) $$Z_\Gamma^- \times \delta_X = \sum_i \langle Z_\Gamma^+, \delta_{X'_{(i)}} \rangle \delta_{X''_{(i)}}.$$

Furthermore, we stress again that any term in the coproduct of a 1PI graph Γ determines gluing data G_i such that

$$\Gamma = \Gamma''_{(i)} \star_{G_i} \Gamma'_{(i)}, \forall i.$$

Here, G_i specifies vertices in $\Gamma''_{(i)}$ and bijections of their types with the elements of $\Gamma'_{(i)}$ such that Γ is regained from its parts:

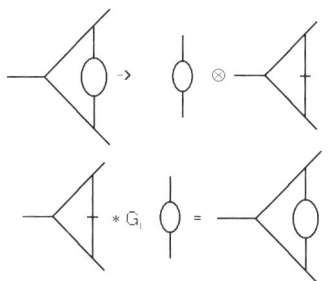

The first line gives a term (i) in the coproduct, decomposing this graph into its only divergent subgraph (assuming we have chosen ϕ^3 in six dimensions, say) and the corresponding cograph, the second line shows the gluing G_i for this term, in this example.

We want to understand the commutator

(43) $$[Z_{\Gamma_1}^+, Z_{\Gamma_2}^-],$$

acting as a derivation on the Hopf algebra element δ_X. To this end introduce

(44) $$Z_{[\Gamma_1, \Gamma_2]} \times \delta_X = \sum_i \langle Z_{\Gamma_2}^+, \delta_{X'_{(i)}} \rangle \delta_{X''_{(i)} \star_{G_i} \Gamma_1}.$$

Here, the gluing operation G_i still acts such that each topologically different graph is generated with unit multiplicity:

6.2. Insertion and elimination.

So finally we are free to exchange graphs, by eliminating one and inserting the other. This is an immensely structured operation, even an one restricts these operations to the cocommutative Hopf algebra of ladder graphs [33].

THEOREM 9. [20]. *For all 1PI graphs Γ_i, s.t. $\mathbf{res}(\Gamma_1) = \mathbf{res}(\Gamma_2)$ and $\mathbf{res}(\Gamma_3) = \mathbf{res}(\Gamma_4)$, the bracket*

$$\begin{aligned}[Z_{[\Gamma_1,\Gamma_2]}, Z_{[\Gamma_3,\Gamma_4]}] &= +Z_{[\overline{Z_{[\Gamma_1,\Gamma_2]} \times \delta_{\Gamma_3}},\Gamma_4]} - Z_{[\Gamma_3,\overline{Z_{[\Gamma_2,\Gamma_1]} \times \delta_{\Gamma_4}}]} \\ &\quad - Z_{[\overline{Z_{[\Gamma_3,\Gamma_4]} \times \delta_{\Gamma_1}},\Gamma_2]} + Z_{[\Gamma_1,\overline{Z_{[\Gamma_4,\Gamma_3]} \times \delta_{\Gamma_2}}]} \\ &\quad - \delta^K_{\Gamma_2,\Gamma_3} Z_{[\Gamma_1,\Gamma_4]} + \delta^K_{\Gamma_1,\Gamma_4} Z_{[\Gamma_3,\Gamma_2]},\end{aligned}$$

defines a Lie algebra of derivations acting on the Hopf algebra $\mathcal{H}$ via

$$Z_{[\Gamma_i,\Gamma_j]} \times \delta_X = \sum_I \langle Z^+_{\Gamma_2}, \delta_{X'_{(i)}} \rangle \delta_{X''_{(i)} \star_{G_i} \Gamma_1}, \tag{45}$$

where the gluing data G_i are normalized as before.

The Kronecker δ^K terms just eliminate the overcounting when combining all cases in a single equation.

We note that $Z_{[\Gamma,\Gamma]} \times \delta_X = k_\Gamma \delta_X$, where k_Γ is the number of appearances of Γ in X and where we say that a graph Γ appears k times in X if k is the largest integer such that

$$\langle \Gamma^k \otimes \text{id}, \Delta(\delta_X) \rangle \tag{46}$$

is non-vanishing. Also $I : Z_{[\Gamma_1,\Gamma_2]} \to Z_{[\Gamma_2,\Gamma_1]}$ is an anti-involution such that

$$I([Z_{[\Gamma_1,\Gamma_2]}, Z_{[\Gamma_3,\Gamma_4]}]) = -[I(Z_{[\Gamma_1,\Gamma_2]}), I(Z_{[\Gamma_3,\Gamma_4]})]. \tag{47}$$

Furthermore

$$[Z_{[\Gamma_1,\Gamma_2]}, Z_{[\Gamma_2,\Gamma_1]}] = Z_{[\Gamma_1,\Gamma_1]} - Z_{[\Gamma_2,\Gamma_2]}.$$

By construction, we have

PROPOSITION 10.

$$Z^+_\Gamma \equiv Z_{[\Gamma,\mathbf{res}(\Gamma)]},$$

$$Z^-_\Gamma \equiv Z_{[\mathbf{res}(\Gamma),\Gamma]},$$

and $[Z^-_X, Z^-_Y] = -Z^-_{[Z^+_X, Z^+_Y]}$.

Finally,

PROPOSITION 11.

$$\begin{aligned}[Z_{[\Gamma_1,\mathbf{res}(\Gamma_1)]}, Z_{[\mathbf{res}(\Gamma_2),\Gamma_2]}] &= \delta^K_{\mathbf{res}(\Gamma_1),\mathbf{res}(\Gamma_2)} Z_{[\Gamma_1,\Gamma_2]} + \delta^K_{\Gamma_1,\Gamma_2} Z_{[\mathbf{res}(\Gamma_2),\mathbf{res}(\Gamma_1)]} \\ &\quad - Z^-_{\overline{Z_{[\mathbf{res}(\Gamma_1),\Gamma_1]} \times \delta_{\Gamma_2}}} - Z^+_{\overline{Z_{[\mathbf{res}(\Gamma_2),\Gamma_2]} \times \delta_{\Gamma_1}}}.\end{aligned}$$

These derivations provide a convenient means to relate the insertion of subgraphs at different places to Galois symmetries in Feynman graphs, to be discussed below.

7. Primitivity

The letters in which QFT wants to be formulated are primitive 1PI graphs. But QFT speaks in manners more subtle than words: the letters do not come in linear order, but are inserted into each other in a highly structured way as we saw. How fluent are we in that language? Well, the combinatorial structures revealed so far allow us to decipher the content of QFT -the general Feynman graphs- completely in terms of these underlying letters. This features the residue as the central notion in quantum field theory, with each letter providing its own unique residue, a renormalization group invariant which connects the topology of the graph to number theory [**21, 35, 36, 37, 38**]. Let us discuss the disentanglement of Feynman graphs into these residues now.

Consider $\Gamma_1 \star_i \Gamma_2$, for primitive graphs Γ_1, Γ_2 and insertion at a compatible vertex i. Both graphs provide a first order pole, how do these two first order poles determine the second order pole in $\Gamma_1 \star_i \Gamma_2$? And how are the higher order poles determined in general? Such questions can now be answered completely thanks to the knowledge of the Hopf algebra structure.

7.1. Higher poles. The explicit formulas in [**4**] allow to find the combinations of primitive graphs into which higher order poles resolve. The weights are essentially given by iterated integrals which produce coefficients which generalize the tree-factorials obtained for the undecorated Hopf algebra in [**6, 39, 27**]. Iterated application of this formula allows to express inversely the first-order poles contributing to the β-function as polynomials in Feynman graphs free of higher-order poles.

This decomposition into pole terms is accompanied by a decomposition according to the bidegree [**40, 24**]. In practice, this decomposition is evident for subgraphs with two legs in a massless theory, using that the only effect of self-energies is to raise edge variables to non-integer scaling degrees in (1) [**21**]. We thus concentrate on the decomposition of vertex subgraphs.

It is very instructive to use automorphisms $\Theta : \mathcal{H} \to \mathcal{H}$ of the Hopf algebra of specified graphs which vary external structures. We will assume that Θ is chosen in a way such that $R_\Theta \circ \phi := \phi \circ \Theta$ defines a map $R_\Theta : V \to V$ in accordance with the usual requirements we impose on renormalization maps. Typically, Θ will pose further conditions on momenta attached to external edges of vertex correction subgraphs and thus modifies external structures so that an appropriate representation of Feynman graphs will be by suitably colored graphs. Let then

$$(48) \qquad S_\Theta[(\Gamma, \sigma)] = -\Theta \left[(\Gamma, \sigma) + \sum_{\gamma \subset \Gamma} (\Gamma/\gamma \star_{G_i} S_\Theta[(\gamma, \sigma)]) \right],$$

with gluing data G_i as before and where we let $\Theta[(\Gamma, \sigma)] := (\Gamma, \Theta(\sigma))$ be such a change of external structure. If we choose Θ to be trivial on graphs with two external legs, but nontrivial on interaction graphs (setting, for example, all but two chosen external momenta to zero), one gets

THEOREM 12. $\mathrm{bid}(S_\Theta \star \mathrm{id}((\Gamma, \sigma))) = 0$ and $\mathrm{bid}(S_\Theta((\Gamma, \sigma))) = 1$.

This is indeed obvious when we compare with the principle of multiplicative subtraction: in the above: recursively, each (sub)graph is decomposed into the image of $\mathrm{id} - \Theta$ or Θ. The usual recursion over subgraphs then gives a complete decomposition over the bidegree such that each superficial divergent subgraph is in the image of Θ. Hence:

COROLLARY 13. *For each Θ as above, $\Gamma = \sum_{i=0}^{\mathrm{bid}(\Gamma)} \Gamma_i$, with $\mathrm{bid}(\Gamma_i) = i$ and $\Gamma_i = \prod_{j=1}^{i} \gamma_j$, with $\mathrm{bid}(\gamma_j) = 1$.*

This provides a complete factorization of higher bidegree graphs into primitive ones: we get indeed a decomposition of Γ in terms of increasing bidegree such that all subgraphs in the term of highest bidegree are in the image of Θ. A simple analytic argument based on the universal integral representation (1) then gives the higher order pole terms in terms of residues. As an example, we get for the graph $\Gamma = \Gamma_0 \star_i \gamma$ of the introduction, with $\mathrm{bid}(\Gamma) = 2$:

$$S_\Theta \star \mathrm{id}(\Gamma) = \left(\raisebox{-0.5em}{\includegraphics[height=2em]{g1}}, \sigma \right)(p_1, p_2) - \left(\raisebox{-0.5em}{\includegraphics[height=2em]{g2}}, \Theta(\sigma) \right)(p_1)$$

$$(49) \qquad - \left(\raisebox{-0.5em}{\includegraphics[height=2em]{g3}}, \sigma \right)(p_1, p_2) + \left(\raisebox{-0.5em}{\includegraphics[height=2em]{g4}}, \Theta(\sigma) \right)(p_1),$$

where we let σ be an evaluation at external momenta $p_1, p_1 + p_2, p_2$ and $\Theta(\sigma)$ be an evaluation at zero momentum transfer, keeping p_1. Upon evaluation by ϕ, in the second line, the subgraph in a circle is inserted as a subintegral into $\phi(\Gamma_0)$ such that it also has zero momentum transfer at the appropriate leg. It modifies the integral corresponding to $\phi(\Gamma_0)$ only in the dependence on a single edge variable corresponds to the distinguished momentum upon which the subgraph still depends when evaluated with the external structure $\Theta(\sigma)$. This results then in the same factorization in calculations as one gets from self-energy graphs.

It is easy to see that the four terms above combine to an Feynman integral

$$\phi(S_\Theta \star \mathrm{id}(\Gamma))$$

which is convergent, and that the first and the third, as well as the second and the fourth term in (49), combine to a Feynman integral free of subdivergences, while the fourth term alone is of bidegree two, but decomposes in the requested manner, due to the fact that the insertion of the subintegral only modifies the dependence on a single edge variable in $\phi(\Gamma_0)$. Note that this decomposition amounts to adding zero in a way such that each subdivergence has the desired external structure.

7.2. The scattering type formula.

All this combines nicely to a scattering type formula, explicitly worked out for the case of ϕ^3 in [4]. Underlying are some asymptotic scaling properties of graded complex Lie groups worked out in [4] as well. It automatically reproduces the right weights with which residues combine to coefficients of higher pole terms, taking into account both the grading by loop number and the bidegree.

In this context, locality of counterterms ensures that a modification of scales will not change the negative part of the Birkhoff decomposition of the character γ of $\mathcal{H}$ under consideration (following the notation of [3], we let γ be the evaluation

of the unrenormalized Feynman rules ϕ on an infinitesimal circle around $\varepsilon = 0$). Hence,
$$\gamma_-(\varepsilon)\,\theta_{t\varepsilon}(\gamma_-(\varepsilon)^{-1}) \text{ is convergent for } \varepsilon \to 0\,,$$
where $\theta_{t\varepsilon}$ implements a one-parameter group of rescalings [3]. The generator $\beta = \left(\frac{\partial}{\partial t} F_t\right)_{t=0}$ of this one parameter group is related to the *residue* of γ
$$\operatorname*{Res}_{\varepsilon=0} \gamma = -\left(\frac{\partial}{\partial u} \gamma_-\left(\frac{1}{u}\right)\right)_{u=0}$$
by the simple equation,

(50) $$\beta = Y \operatorname{Res} \gamma\,,$$

where $Y = \left(\frac{\partial}{\partial t} \theta_t\right)_{t=0}$ is the grading.

Amazingly, one can give $\gamma_-(\varepsilon)$ in closed form as a function of β [3]. We shall for convenience introduce an additional generator in the Lie algebra of G (i.e. primitive elements of $\mathcal{H}^*$) such that,
$$[Z_0, X] = Y(X) \qquad \forall\, X \in \text{Lie } G\,,$$
where Y implements the grading by the loop number. The scattering type formula for $\gamma_-(\varepsilon)$ is then,

(51) $$\gamma_-(\varepsilon) = \lim_{t \to \infty} e^{-t\left(\frac{\beta}{\varepsilon} + Z_0\right)} e^{t Z_0}\,,$$

exemplifying that the higher pole structure of the divergences is uniquely determined by the residue.

Note that the above decomposition into residues allows for a systematic investigation into the properties of the insertion operad underlying the insertion of subgraphs.

This operad has as indexed composition $\circ_{v,b}$ for the insertion of a subgraph γ, at a chosen place v (an edge or vertex) in another graph Γ, using a chosen bijection b of $\gamma_{\text{int}}^{[1]}$ with f_v. Relabelling rules of internal edges and vertices such that axioms of an operad are fulfilled are straightforward [7]. We can now start asking for the question how a variation of $\circ_{v,b} \to \circ_{v',b'}$ modifies the higher order pole terms, with invariance of the highest degree pole terms easily obtained from the scattering formula, as well as its decomposition into the residues of the underlying bidegree one graphs.

Invariances of the residue of a higher bidegree graph reflect, as we will see, number-theoretic properties of a graph and will be discussed below. But let us first comment on the usefulness of Hochschild cohomology in the renormalization process, which will finish our combinatorial review of standard renormalization theory.

8. Renormalization and Hochschild cohomology

A particularly nice way [34] to proof locality of counterterms and finiteness of renormalized Green functions can be obtained using the Hochschild properties of the operator B_+^x, where x indicates an appropriate primitive graph and its gluing data. Indeed it raises the bidegree by one unit and is therefore a natural candidate to obtain such feasts. To do this nicely, it pays to describe how to map Feynman graphs to decorated rooted trees.

8.1. From graphs to trees.

Let us set out to define a Hopf algebra of non-planar decorated rooted trees (see [24] for a census of such algebras) $\mathcal{H}_{DRT}$ and a homomorphism ρ from the one of graphs $\mathcal{H}_{FG}$ to this one, $\rho: \mathcal{H}_{FG} \to \mathcal{H}_{DRT}$,

$$\rho(\Gamma_1 \Gamma_2) = \rho(\Gamma_1)\rho(\Gamma_2), \tag{52}$$

such that

$$[\rho \otimes \rho] \circ \Delta_{FG}(\Gamma) = \Delta_T(\rho(\Gamma)). \tag{53}$$

Here, Δ_T and Δ_{FG} are the coproducts in the Hopf algebras of rooted trees and Feynman graphs, respectively. The range of ρ defines a sub-Hopf algebra $\mathcal{H}_\rho \subset \mathcal{H}_{DRT}$, and ρ is a bijection between $\mathcal{H}_{FG}$ and this closed sub-Hopf algebra $\mathcal{H}_\rho$. We let 1 be the unit in both Hopf algebras so that $\rho(1) = 1$ and identify scalars.

In both algebras we have a decomposition with respect to the bidegree

$$\mathcal{H}_{FG} = \oplus_{i \geq 0} \mathcal{H}_{FG}^{(i)} \text{ and } \mathcal{H}_{DRT} = \oplus_{i \geq 0} \mathcal{H}_{DRT}^{(i)}. \tag{54}$$

We set, to start an induction over the bidegree, $\mathcal{H}_{FG}^{(i)} = \mathcal{H}_{DRT}^{(i)}$, $i = 0, 1$, where we identify a primitive graph Γ with the decorated rooted tree $\rho(\Gamma) = (*, \Gamma) \in \mathcal{H}_{DRT}^{(1)}$.

Then, for some positive integers r and k_j, $j = 1, \ldots, r$, any non-primitive 1PI graph Γ can be written at most in r different forms (here, r is the number of maximal forests alluded to in the introduction)

$$\Gamma = \prod_{i=1}^{k_j} \Gamma_j \star_{j,i} \gamma_{j,i}, \ \forall j = 1, \ldots, r.$$

We call Γ overlapping if $r > 1$. We saw an example for the case $r = 2$ earlier, in Fig.(1). Then, we set

$$\rho(\Gamma) = \sum_{j=1}^{r} B_+^{\Gamma_j, G_{j,i}} \left[\prod_{i=1}^{k_j} \rho(\gamma_{j,i}) \right], \tag{55}$$

with appropriate gluing data $G_{j,i}$, obtained from Δ_{FG} as before, see (21). One immediately proves by induction the required properties of ρ, using

$$\Delta_T(\rho(\Gamma)) = \rho(\Gamma) \otimes 1 + \sum_{j=1}^{r} (1 \otimes B_+^{\Gamma_j, G_{j,i}}) \Delta_T(\prod_{i=1}^{k_j} \rho(\gamma_{j,i})), \tag{56}$$

which allows for an inductive construction of ρ. The so obtained decorated rooted trees will have decorated vertices where the decorations consist of bidegree one (primitively divergent) graphs and gluing data, which store information how to glue the descendent branches into that decoration. Note that the action of the B_+ operator is well defined under the coproduct: if $X = B_+^{\Gamma, G_Y}(Y)$, then $B_+^{\Gamma, G_Y}(Y'')$ is well-defined, as the gluing data only use the external legs of Y, and $\mathbf{res}(Y) = \mathbf{res}(Y'')$ for all cographs Y'' appearing in the coproduct of Y. All these operators $B_+^{\Gamma_j, G_{j,i}}$ are closed Hochschild one-cocycles.

The map ρ is not uniquely given though: the usual ambiguity in a transition from a filtration to a degree means that in this construction of ρ in terms of increasing bidegree, ρ can always be modfied by terms of lower bidegree. By construction, we fixed ρ at bidegrees 0,1. The remaining freedom is actually an asset in practical calculations [**22, 12, 41**].

8.2. Locality and finiteness.

By the above, we can restrict an inductive proof of locality of counterterms and finiteness of renormalized Green functions to the study of elements of the form $\Gamma = B_+^{\gamma_i, G_i}(X_i) \in \mathcal{H}_{FG}$, where we use the same notation $B_+^{\gamma, G}$ in the Hopf algebra of graphs as in the one of decorated rooted trees.

We will proceed by an induction over the bidegree which is much more natural than the usual induction over the number of loops.

Hence our task is: assume that $S_R \star \phi(\Gamma)$ is finite and and $S_R(\Gamma)$ a local counterterm for all Γ with $\mathrm{bid}(\Gamma) \leq k$. Show these properties for all Γ with $\mathrm{bid}(\Gamma) = k+1$.

The start of the induction is easy: at unit bidegree, $\phi(\Gamma) - R[\phi(\Gamma)]$ is finite and $S_R(\Gamma)$ is local by assumption on R.

Let us assume we have established the desired properties of S_R and $S_R \star \phi$ acting on all Hopf algebra elements up to bidegree k. Assume $\mathrm{bid}(\Gamma) = k+1$. We have

$$\Gamma = B_+^{\gamma, G}(X), \tag{57}$$

where $\mathrm{bid}(\gamma) = 1$, $\mathrm{bid}(X) = k$, X some monomial in the Hopf algebra.

Next,

$$\Delta(\Gamma) = B_+^{\gamma, G}(X) \otimes 1 + [1 \otimes B_+^{\gamma, G}]\Delta(X), \tag{58}$$

which expresses the fact that $B_+^{\gamma, G}$ is a closed Hochschild one-cocycle.

Let us decompose, as usual,

$$\Delta(X) = X \otimes 1 + 1 \otimes X + X' \otimes X''.$$

It is easy to see from the structure of (1) that we can define

$$\mathbf{B}_+(\phi; \phi; \gamma, G; X) \equiv \phi(B_+^{\gamma, G}(X)) \tag{59}$$

and extend this definition to a map $\mathbf{B}_+(\phi; S_R \star \phi; \gamma, G; X)$ which glues the renormalized results $S_R \star \phi$ into the integral $\phi(\gamma)$.

Using the Hochschild closedness of $B_+^{\gamma, G}$ one immediately gets

$$S_R \star \phi(\Gamma) = S_R(\Gamma) + \mathbf{B}_+(\phi; S_R \star \phi; \gamma, G; X) \tag{60}$$

and

$$S_R(\Gamma) = -R[\mathbf{B}_+(\phi; S_R \star \phi; \gamma, G; X)]. \tag{61}$$

From here, the induction step boils down to a simple estimate using the fact that the power counting for asymptotically large internal loop momenta in $\phi(\gamma)$ is modified by the insertion of $S_R \star \phi(X)$ (which is finite by assumption, having bidegree k) only by powers of $\log(|k_e|)$, and that delivers the result easily, from (1).

9. Unique factorization and Dyson–Schwinger equations

So far, we have described the combinatorial structures underlying renormalization theory. As it befits a Hopf algebra, it all comes down to the study of residues, invariants under diffeomorphisms of continous quantum numbers, parametrizing the primitive elements of the Hopf algebra. Locality hides in the fact that every primitive element has a residue proportional to its graphical residue:

$$\mathrm{res}[\phi(\Gamma)] = \rho(\Gamma) \langle \sigma_P, \phi(\mathbf{res}(\Gamma)) \rangle, \quad \forall \Gamma, \mathrm{bid}(\Gamma) = 1. \tag{62}$$

where ρ is a character on $\mathcal{H}$. This gives a fascinating way of actually defining the tree level terms $\langle \sigma_P, \phi(\mathbf{res}(\Gamma)) \rangle$ in the Lagrangian as the residues of primitive graphs.

This, it occurs to me, is a proper way to define locality in noncommutative QFT. For such theories the traditional notion of locality is obscured by the fact that non-vanishing commutators of spacetime coordinates exponentiate to non-local functions even at the tree level [**42, 43**]. Not too surprisingly, counterterms involve terms which are then non-local in the traditional sense of not being powers of fields and their derivatives. Nevertheless, the real question to my mind is if they obey the equation above: are the residues (in the operator-theoretic sense) of the bidegree one-part in the perturbative expansion the generators of the tree level (the bidegree zero part) terms?

This is a well-known phenomenon in the study of gauge theories over ordinary spacetime [**44**], which allows to recover the Yang-Mills action from integrating the one-loop fermion determinant. This has far reaching generalizations incorporating the noncommutative case and connecting gauge theories to Connes' spectral action, via an operator theoretic residue [**45**]. The significance of such an approach certainly lies in the emphasis it gives to the Dirac operator and the residue. But in this, one-loop graphs played a distinguished role, and the most promising way to generalize such results to the whole of QFT is to factorize QFT completely into bidegree one graphs, into residues and traces, that is.

So let us muse, in this final section, about the structure of a perturbative expansion in general, using properties of the Dyson–Schwinger equations. These quantum equations of motion are equations which are formally solved in an infinite series of graphs, providing a fixpoint for these equations, and evaluating these graphs order by order using the renormalized character $S_R \star \phi$. These equations come typically as integral, or integro-differential equations with a firm reputation to be unsolvable: indeed, their solution would solve and establish quantum field theory in four dimensions.

We essentially learned above how to disentangle Feynman graphs into primitive graphs of unit bidegree. How can we utilize this fact in those equations?

A typical example of such an equation is

(63)

This equation as as a fixpoint a formal series over graphs which starts like

(64)

9.1. Unique factorization into residues.
To start a more systematic understanding of this equations along the lines of [**34**] let us sketch here an approach based on the fact that the series of graphs which provide a solution for it can be written with the help of a commutative product. This product generalizes the shuffle product (a rather interesting construct in its own right [**46**]) of ordered sequences to our (partially ordered) Feynman graphs.

To this end, let $w = (a_1, a_2, \ldots, a_n)$ be a sequence of primitive graphs, a word, say, of letters ordered in such a sequence.

We now can define a partial order on Feynman graphs by defining the set $I(w)$ of all Feynman graphs compatible with the ordered sequence w as those 1PI graphs Γ which fulfill

(65) $$\langle Z_{a_1} \otimes \ldots \otimes Z_{a_k}, \Delta^{k-1}(\Gamma) \rangle = 1.$$

Here, Z_{a_j} are the duals of the primitive elements a_j. Let $\Sigma(w) = \sum_{\Gamma \in I(w)} \Gamma$ be the sum over all graphs compatible with w. Note that with the underlying partial order based on 'being a subgraph',

(66) $$\Gamma_1 \leq \Gamma_2 \Leftrightarrow Z^-_{\Gamma_1} \times \delta_{\Gamma_2} \neq 0,$$

we can write for any graph Γ

(67) $$\Delta(\Gamma) = \sum_{\Gamma_1, \Gamma_2 \in \mathcal{H}} \zeta(\Gamma_1, \Gamma_2) \Gamma_1 \otimes \Gamma_2,$$

where $\zeta(\Gamma_1, \Gamma_2)$ is the ζ-function of the incidence algebra assigned to this partial order [**47**].

The permutation group on n elements acts naturally on the word w with n letters. Let $n_\Gamma := \sum_{\sigma \in S_n} \sum_{\Gamma_1 \in I(\sigma(w))} \langle Z_\Gamma, \delta_{\Gamma_1} \rangle$. Then, we define $\overline{\Sigma}(w)$ to be the same sum as $\Gamma(w)$, but each graph with coefficient $1/n_\Gamma$:

(68) $$\overline{\Sigma}(w) = \sum_{\Gamma \in I(w)} \frac{1}{n_\Gamma} \Gamma.$$

On words $w_1 = (a_1, v_1), w_2 = (b_1, v_2)$ (a_1, b_1 letters, v_1, v_2 subwords) one has the usual commutative associative product of rifle shuffles

(69) $$w_1 \perp w_2 = (a_1, v_1 \perp w_2) + (b_1, w_1 \perp v_2).$$

Then, one can consider the commutative associative product

(70) $$\overline{\Sigma}(w_1) \cdot \overline{\Sigma}(w_2) := \overline{\Sigma}(w_1 \perp w_2).$$

The series of graphs giving a formal solution to the Dyson Schwinger equation above can now be obtained as the product of two Euler factors

$$\frac{1}{1-\overline{\Sigma}(\gamma_1)} \cdot \frac{1}{1-\overline{\Sigma}(\gamma_2)}. \tag{71}$$

Note that the fact that the Dyson-Schwinger equation sums democratically over all possible insertions of graphs into each other is crucial here. Consequences of factorization for Dyson–Schwinger equations are considered in [**34**], but much more about this will be said in future work.

Not much stops us to consider actually an Euler product over all primitive graphs to get a formal solution to Dyson–Schwinger equations in general. We should just construct ζ-functions dedicated to a chosen Green function, defined via an Euler product over primitive elements. Or does it?

First of all, we would like to have the structure of an Euler product not only on the level of graphs, but would like it to be compatible with the Feynman rules. That would deliver a genuine factorization into residues, with far reaching consequences for non-perturbative aspects, in particular with respect to the notorious renormalon problem: indeed, if we had for example for general Euler factors

$$\phi\left(\frac{1}{1-\overline{\Sigma}(\gamma_1)} \cdot \frac{1}{1-\overline{\Sigma}(\gamma_2)}\right) = \frac{1}{1-\phi(\overline{\Sigma}(\gamma_1))} \frac{1}{1-\phi(\overline{\Sigma}(\gamma_2))},$$

we would easily verify that no chains of subgraphs γ_2 say would produce the renormalon problem: factorization into Euler products resums the perturbation series in a way eliminating the factorial growth attached to renormalons, decomposing the whole perturbation series into an Euler product over geometric series.

Note that the existence of Feynman rules compatible with the above factorization essentially would establish a shuffle identity on Feynman graphs, in complete generalization of the situation for iterated integrals. Again, much more on that will be reported in future work.

A further difficulty in this approach is that a full D-S equation mixes graphs with a varying number of external legs. In particular, in the above example we avoided the presence of overlapping divergences. In their study, to my mind, are hidden some of the most valuable treasures of quantum field theory.

9.2. Uniqueness of factorization and ideals in gauge theories. In the presence of overlapping divergences, unique factorization is lost. Reconsider Fig.(1), an example taken from non-abelian gauge theory.

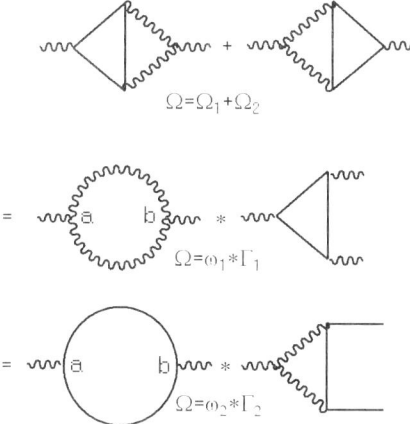

The linear combination Ω of the two graphs Ω_1, Ω_2 in the first line describes either all ways of inserting Γ_1 into ω_1, or all ways of inserting Γ_2 into ω_2. Ω does not factorize in a unique manner, a phenomenon typical for local gauge theories, as will be demonstrated elsewhere. We only very briefly sketch some relevant ideas here.

Comparing the situation with algebraic number theory (see [**48**] for an excellent review) gives the right idea how to restore unique factorization: consider ideals. Apparently, if we let $(a, b, c, \dots)$ the ideal of Feynman graphs containing letters $(a, b, c, \dots)$, then we can restore unique factorization of the ideal (Ω) as

(72) $$(\Omega) = (\omega_1, \omega_2)(\omega_1, \Gamma_2)(\Gamma_1, \omega_2)(\Gamma_1, \Gamma_2).$$

This implies that we can build products of these components - we must have

$$(\omega_1, \omega_2)(\omega_1, \Gamma_2) = (\omega_1 \cdot \omega_1, \omega_1 \cdot \Gamma_2, \omega_2 \cdot \omega_1, \omega_2 \cdot \Gamma_2) = (\omega_1),$$
$$(\Gamma_1, \omega_2)(\Gamma_1, \Gamma_2) = (\Gamma_1 \cdot \Gamma_1, \Gamma_1 \cdot \Gamma_2, \omega_2 \cdot \Gamma_1, \omega_2 \cdot \Gamma_2) = (\Gamma_1)$$

and

(73) $$(\omega_1, \omega_2)(\Gamma_1, \omega_2) = (\omega_1 \cdot \Gamma_1, \omega_1 \cdot \omega_2, \omega_2 \cdot \Gamma_1, \omega_2 \cdot \omega_2) = (\omega_2),$$
$$(\omega_1, \Gamma_2)(\Gamma_1, \Gamma_2) = (\omega_1 \cdot \Gamma_1, \omega_1 \cdot \Gamma_2, \Gamma_2 \cdot \Gamma_1, \Gamma_2 \cdot \Gamma_2) = (\Gamma_2).$$

But this demands that there are relations between those graphs: for example, the product $\omega_2 \cdot \Gamma_1$ is not allowed to vanish, if we want to maintain the structure of an integral domain for our democratic insertions of graphs. How can this be achieved when for example there is no place v in ω_2 such that $\mathbf{res}(\Gamma_1) = f_v$?

The answer lies in the local gauge symmetry, which actually mixes external structures. An easy calculation shows that the one-loop transversal gauge boson propagator and the one-loop fermion propagator are connected:

(74) $$\rho_{F^2}\left(\text{\includegraphics{}} \right) = \xi^{-1} \rho_{\bar\psi D \psi}\left(\text{\includegraphics{}} \right) \frac{D-3}{D-1} Tr(I),$$

where we define the gauge parameter ξ as the deviation from the transversal gauge, that is we define the massless Euclidean gauge boson propagator as

$$P_{\mu\nu}^{-1}(k) = \frac{g_{\mu\nu}}{k^2} + (\xi - 1)\frac{k_\mu k_\nu}{k^4},$$

and use suitable characters ρ_P. Note that both sides of (74) are independent of ξ. This identity alone is sufficient to guarantee that none of the above products is degenerate. Equation (74) allows for a very beautiful reinterpretation of gauge symmetry which will be commented on in much greater detail in future work.

As a final remark, note that along with such a factorization comes a considerable simplification in the study of quantum action principles and renormalization of gauge symmetries, as the Leibniz rules for the relevant (BRST-) differentials ensure that one can restrict the study of such questions to single primitively divergent bidegree one factors, which allows to lift results as [**49**] to all orders once factorization is established.

9.3. Galois symmetries in perturbative QFT.

In the above, we started drifting towards a treatment of Feynman graphs as a ring, with associated field of fractions say, where the role of primes is played by primitive graphs, and an Euler product combined with an appropriate shuffle identity for Feynman rules should guide us towards an appropriate notion of a ζ-function for a given Green function.

It is essentially unique factorization which then summarizes all the wisdom of the forest formulas with the renormalization group playing the role of a group of units in that field, and with gauge symmetries making life interesting in the quest for unique factorization.

But then, the solution of the Dyson-Schwinger equation can be factorized in various different sets of primitives. This brings us back to the behaviour of graphs which are build from similar primitives, inserted into each other at different places though. In the mental picture laid bare here this connects to differential Galois theory: Consider the combination $\Gamma_0(\star_i - \star_j)\gamma$, studied in the introduction (9).

We can consider the "differential equation" ($Z_{[\mathbf{res}(\gamma),\gamma]}$ is a derivation!)

$$Z_{[\mathbf{res}(\gamma),\gamma]} \times \delta_\Gamma = \Gamma_0, \tag{75}$$

which is solved by the bidegree two graph $\Gamma = \Gamma_0 \star_i \gamma$ as well as by the bidegree two $\Gamma = \Gamma_0 \star_j \gamma$, while the bidegree one primitive $X = \Gamma_0(\star_i - \star_j)\gamma$ solves the homogenous equation

$$Z_{[\mathbf{res}(\gamma),\gamma]} \times \delta_X = 0. \tag{76}$$

This justifies to connect the insertion of subgraphs at various different places with Galois symmetries, and was the motivation to indeed look at invariants under such symmetries in Feynman graphs, with a beautiful first result reported in [**10**]: the coefficient of the highest weight transcendental in the residues of two graphs connected by such a symmetry is invariant. While this is obvious, thanks to the scattering type formula, for the coefficient of the highest pole in the regularization parameter, it is a very subtle result for the residue in a graph of large bidegree.

10. Conclusion and acknowledgments

Quantum field theory is to me still the most subtle construct we came across so far in theoretical physics. In particular, in its short distance singularities, it hides mathematical wisdom and beauty which parallel its unmatched achievements as a predictive theory. It is my hope that the recently discovered combinatorial structures in quantum field theory provide powerful tools to unravel these structures, for the benefit of the mathematician and practitioner of quantum field theory alike.

Many thanks to Dennis Sullivan and the organizers of the *Dennisfest* for giving me the opportunity to speak at this most entertaining workshop.

It is a pleasure to thank Ron Donagi, Takashi Kimura and Jim Stasheff for discussions on various algebraic and operadic properties of Feynman graphs, and my friends and collaborators David Broadhurst and Alain Connes for much needed help in sharpening more of the tools necessary to deal with QFT.

Partial support of this work by the Clay Mathematics Institute is gratefully acknowledged.

References

[1] A.S. Wightman, *Some Lessons of Renormalization Theory*, in "The Lessons of Quantum Theory", J. de Boer et.al. Eds., Elsevier 1986.
[2] D. Kreimer, *Lectures on Renormalization Theory*, Boston University 2001, to appear.
[3] A. Connes, D. Kreimer, *Renormalization in quantum field theory and the Riemann-Hilbert problem. I: The Hopf algebra structure of graphs and the main theorem*, Commun. Math. Phys. **210** (2000).
[4] A. Connes, D. Kreimer, *Renormalization in quantum field theory and the Riemann-Hilbert problem. II: The beta-function, diffeomorphisms and the renormalization group*, Commun. Math. Phys. **216**.
[5] G. 't Hooft, M. Veltman, *DIAGRAMMAR*, Cern report 73/9 (1973), reprinted in "Particle interactions at very high energies", NATO Adv. Study Inst. Series, Sect. B, vol. 4B, 177.
[6] D. Kreimer, *Chen's iterated integral represents the operator product expansion*, Adv. Theor. Math. Phys. **3** (2000).
[7] D. Kreimer, *Combinatorics of (perturbative) quantum field theory*, Phys. Reports **363** (2002).
[8] Collins, *Renormalization*, Cambridge Univ. Press 1984.
[9] D. Kreimer, *On the Hopf algebra structure of perturbative quantum field theories*, Adv. Theor. Math. Phys. **2** (1998).
[10] I. Bierenbaum, R. Kreckel, D. Kreimer, *On the invariance of residues of Feynman graphs*, J. Math. Phys. **43** (2002) 4721.
[11] J.L. Loday, *La renaissance des opérades*, (French) [The rebirth of operads] Séminaire Bourbaki, Vol. 1994/95. Astérisque No. 237, (1996), (Exp. No. 792, 3) 47.
[12] D. Kreimer, *On overlapping divergences*, Commun. Math. Phys. **204** (1999).
[13] S. Weinberg, *High-Energy Behavior In Quantum Field Theory*, Phys. Rev. **118** (1960) 838.
[14] W. Fulton, R. MacPherson, *A compactification of configuration spaces*, Ann. Math. **139** (1994) 183.
[15] H. Epstein, V. Glaser, *The role of locality in perturbation theory*, Ann. Inst. H. Poincaré **19** (1973) 211.
[16] C. Bergbauer, D. Kreimer, *The Hopf algebra of rooted trees in Epstein-Glaser renormalization*, Ann. H. Poincaré, to appear, arXiv:hep-th/0403207.
[17] R. Stora, *Renormalized perturbation theory: a theoretical laboratory*, talk given at *Mathematical Physics in Mathematics and Physics*, Siena, June 2000.
[18] R. Brunetti, K. Fredenhagen, *Microlocal analysis and interacting quantum field theories: Renormalization on physical backgrounds*, Commun. Math. Phys. **208** (2000).
[19] J.W. Milnor, J.C. Moore, *On the structure of Hopf algebras*, Ann. Math. (2) 81 (1965) 211.
[20] A. Connes, D. Kreimer, *Insertion and elimination: The doubly infinite Lie algebra of Feynman graphs*, Annales Henri Poincaré **3** (2002).
[21] D. Kreimer, *Knots and Feynman Diagrams*, Cambridge Univ. Press 2000.
[22] A. Connes, D. Kreimer, *Hopf algebras, renormalization and noncommutative geometry*, Commun. Math. Phys. **199** (1998).
[23] F. Markopoulou, *Coarse graining in spin foam models*, Class. Quant. Grav. **20** (2003) 777.
[24] L. Foissy, *Les algèbres des Hopf des arbres enracinés décorées*, Thesis, Univ. Reims, Dept. of Math., available from the author: loic.foissy@univ-reims.fr (2001).
[25] A. Connes, H. Moscovici, *Hopf algebras, cyclic cohomology and the transverse index theorem*, Commun. Math. Phys. **198** (1998) 199.
[26] C. Brouder, A. Frabetti, *Noncommutative renormalization for massless QED*, hep-th/0011161.

[27] D. J. Broadhurst, D. Kreimer, *Renormalization automated by Hopf algebra*, J. Symb. Comput. **27** (1999).

[28] D. J. Broadhurst, D. Kreimer, *Combinatoric explosion of renormalization tamed by Hopf algebra: 30-loop Pade-Borel resummation*, Phys. Lett. **B475** (2000).

[29] D. J. Broadhurst, D. Kreimer, *Exact solutions of Dyson-Schwinger equations for iterated one-loop integrals and propagator-coupling duality*, Nucl. Phys. B **600** (2001).

[30] K. Ebrahimi-Fard, L. Guo and D. Kreimer, *Integrable renormalization. I: The ladder case*, J.Math.Phys., to appear, arXiv:hep-th/0402095.

[31] K. Ebrahimi-Fard, L. Guo and D. Kreimer, *Integrable renormalization. II: The general case*, arXiv:hep-th/0403118.

[32] J.M. Gracia-Bondia, J.C. Varilly and H. Figueroa, *Elements of noncommutative geometry*, Birkhaeuser, Boston 2000.

[33] I. Mencattini and D. Kreimer, *Insertion and Elimination Lie Algebra: the Ladder case*, Lett. Math. Phys. **67** (2004) 61.

[34] D. Kreimer, *Factorization in quantum field theory: An exercise in Hopf algebras and local singularities*, contributed to Les Houches School of Physics: *Frontiers in Number Theory, Physics and Geometry*, Les Houches, France, 9-21 Mar 2003.

[35] D. J. Broadhurst, D. Kreimer, *Knots and numbers in Φ^4 theory to 7 loops and beyond*, Int. J. Mod. Phys. **C6** (1995).

[36] D. J. Broadhurst, D. Kreimer, *Association of multiple zeta values with positive knots via Feynman diagrams up to 9 loops*, Phys. Lett. **B393** (1997).

[37] J.M. Borwein, D.M. Bradley, D.J. Broadhurst, *Evaluation of k-fold Euler–Zagier sums: a compendium of results for arbitrary k*, Elec. J.Comb.**4**(2), R5, (1997);
D.J. Broadhurst, *Conjectured Enumeration of irreducible Multiple Zeta Values, from Knots and Feynman Diagrams*, hep-th/9612012;
D.J. Broadhurst, *On the enumeration of irreducible k-fold Euler sums and their roles in knot theory and field theory*, hep-th/9604128.

[38] D. Kreimer, *Feynman diagrams and polylogarithms: Shuffles and pentagons*, Nucl. Phys. Proc. Suppl. **89** (2000).

[39] D. Kreimer, R. Delbourgo, *Using the Hopf algebra structure of QFT in calculations*, Phys. Rev. **D60** (1999).

[40] D. J. Broadhurst, D. Kreimer, *Towards cohomology of renormalization: Bigrading the combinatorial Hopf algebra of rooted trees*, Commun. Math. Phys. **215** (2000).

[41] D.J. Broadhurst, R. Delbourgo, D. Kreimer, *Unknotting the polarized vacuum of quenched QED*, Phys. Lett. **B366** (1996).

[42] I. Chepelev, R. Roiban, *Convergence theorem for non-commutative Feynman graphs and renormalization*, JHEP **0103** (2001) 001[hep-th/0008090].

[43] M. R. Douglas, N. A. Nekrasov, *Noncommutative field theory*, Rev. Mod. Phys. **73** (2002) 977.

[44] M.A. Shifman et al., *ABC of instantons*, Fortschr. Phys. **32,11** (1984) 585.

[45] A. Connes, *Lectures at Collège de France*, winter 2000/2001.

[46] M.E. Hoffman, *Quasi Shuffle Products*, J. Algebraic Combin. **11** (2000) 49.

[47] G.-C. Rota, *On the Foundations of Combinatorial Theory I: Theory of Möbius Functions*, Zeitsch. f. Wahrscheinlichkeitstheorie **2,4** (1964) 340, reprinted in *Gian-Carlo Rota on Combinatorics : Introductory Papers and Commentaries*, (Contemporary Mathematicians) by Gian-Carlo Rota (Editor), Joseph P. S. Kung (Editor).

[48] H.M. Stark, *Galois Theory, Algebraic Number Theory and Zeta Functions*, in Proc. "From Number Theory to Physics", Les Houches March 1989, M. Waldschmidt et.al., Eds., 2nd corr. printing, Springer 1995.

[49] C.P. Martin, D. Sanchez-Ruiz, *Action principles, restoration of BRS symmetry and the renormalization group equation for chiral non-Abelian gauge theories in dimensional renormalization with a non-anticommuting γ_5*, Nucl. Phys. **B572** (2000).

CNRS-IHES, FRANCE AND CENTER F. MATH.-PHYS., BOSTON U., US.

ALGEBRAIC STRUCTURES

Notes on Universal Algebra

Alexander A. Voronov

Dedicated to Dennis Sullivan on the occasion of his sixtieth birthday.

ABSTRACT. These are notes of a mini-course on "Universal Algebra," given at Dennisfest in June 2001. The goal of these notes is to give a self-contained survey of deformation quantization, operad theory, and graph homology. Some new results related to "String Topology" and cacti are announced in Section 2.7.

CONTENTS

Introduction
1. Graphs and formal algebraic quantization
2. Trees and operads
3. Graph homology
References

Introduction

Either due to the influence of string theory or just because this is what the face of mathematics was supposed to look like before the beginning of the third millennium, graphs have recently stepped forward and overwhelmed many areas of mathematics, including universal algebra. The use of graphs is similar to the Feynman diagram technique in physics: the amazing thing is that its applications to pure mathematics are extremely powerful. In these lectures we are going to discuss three topics: deformation quantization, operad theory, and graph homology, in which significant progress has been made with the help of graphs as a unifying pattern:-) These topics hardly fit the conventional meaning of Universal Algebra, which considers varieties of algebras. However, since the topics deal with universal

---

2000 *Mathematics Subject Classification.* 18D50; Secondary 14H10, 53D55, 55P48, 58D15.
Key words and phrases. deformation quantization, graph homology, operad.
The author was supported in part by NSF grant DMS-0227974 and the James H. Simons Math-Physics Endowment.

patterns in algebra, such as the study of n-ary operations via operads in Section 2, we thought the title would be appropriate. See also a disclaimer below.

Acknowledgments. I would like to thank Dennis Sullivan and the organizers of Dennisfest for inviting me to give this mini-course. These notes are based on several ideas of Maxim Kontsevich, which have largely influenced the subject. I am also grateful to Giovanni Felder, Eric Harrelson, Kolya Ivanov, John Jones, Tom Leinster, Bob Penner, Jim Stasheff, Dennis Sullivan, the anonymous referee, and the lively Dennisfest and the University of Minnesota audiences for many helpful suggestions, incorporated in this version of the notes. I would also like to thank Kyoto University for its hospitality in July 2001, when the paper was written up.

Disclaimer. The characters and figures portrayed and the titles and notions used herein are fictitious and any resemblance to the names, character, or history of any person is coincidental and unintentional.

1. Graphs and formal algebraic quantization

One of the oldest open problems solved with the essential help of graphs is perhaps the problem of deformation quantization.

1.1. Deformations of associative algebras.
Let us start with a review of deformation theory of associative algebras. Let A be an associative algebra over a field k of characteristic zero.

DEFINITION 1.1. A *formal deformation* of A is a $k[[t]]$-bilinear multiplication law $m_t : A[[t]] \otimes_{k[[t]]} A[[t]] \to A[[t]]$ on the space $A[[t]]$ of formal power series in a variable t with coefficients in A, satisfying the following properties:

$$m_t(a,b) = a \cdot b + m_1(a,b)t + m_2(a,b)t^2 + \cdots \qquad \text{for } a,b \in A,$$

where $a \cdot b$ is the original multiplication on A, and m_t is associative, which is equivalent to the equation

$$m_t(m_t(a,b),c) = m_t(a,m_t(b,c)) \qquad \text{for } a,b,c \in A.$$

REMARK 1. Note that for a formal deformation m_t of a *commutative* algebra A, the bracket defined by the first-order part of the commutator,

$$\begin{aligned}\{a,b\} &:= \frac{1}{2t}(m_t(a,b) - m_t(b,a)) \mod t, \\ &= \frac{1}{2}(m_1(a,b) - m_1(b,a))\end{aligned}$$

defines on A the structure of a Poisson algebra, see Section 2.4.4. (All the identities of a Poisson algebra follow from the fact that $A[[t]]$ with the product $m_t(a,b)$ and the bracket $\frac{1}{2}(m_t(a,b) - m_t(b,a))$ is a noncommutative Poisson algebra in an obvious sense, e.g., see [**FGV95**].) In physical terms, one can regard t as a quantum parameter, such as the Planck constant, the Poisson algebra A as the *quasi-classical limit* of the associative algebra $A[[t]]$, and the algebra $A[[t]]$ as a *deformation quantization* of the Poisson algebra A. The *deformation quantization problem* is the inverse problem: given a Poisson algebra A, find a formal deformation returning the original Poisson algebra structure on A in the quasi-classical limit.

The main tool in studying deformation theory of associative algebras is the *Hochschild complex*

$$0 \to C^0(A,A) \xrightarrow{d} \cdots \xrightarrow{d} C^n(A,A) \xrightarrow{d} C^{n+1}(A,A) \xrightarrow{d} \cdots,$$

where $C^n(A,A) := \operatorname{Hom}(A^{\otimes n}, A)$ is the space of *Hochschild n-cochains*, i.e., the n-linear maps $f(a_1,\ldots,a_n)$ on A with values in A, and the *differential d*, $d^2 = 0$, is defined as

$$(df)(a_1,\ldots,a_{n+1}) := a_1 f(a_2,\ldots,a_{n+1})$$
$$+ \sum_{i=1}^{n} (-1)^i f(a_1,\ldots,a_{i-1}, a_i a_{i+1}, a_{i+2},\ldots,a_{n+1})$$
$$- (-1)^n f(a_1,\ldots,a_n) a_{n+1},$$

for $f \in C^n(A,A)$, $a_1,\ldots,a_{n+1} \in A$. The *Hochschild cohomology* is then the cohomology

$$H^\bullet(A,A) := \operatorname{Ker} d / \operatorname{Im} d$$

of this complex. The Hochschild complex admits a bracket

$$[,] : C^m(A,A) \otimes C^n(A,A) \to C^{m+n-1}(A,A),$$

called the *Gerstenhaber bracket, or the G-bracket*. This bracket was defined by M. Gerstenhaber [**Ger63**] by an explicit formula extending the following one, defined for elements $f, g \in C^2(A,A)$:

(1) $\quad [f,g](a,b,c) := f(g(a,b),c) - f(a,g(b,c)) + g(f(a,b),c) - g(a,f(b,c)).$

We will give a conceptual definition, due to J. Stasheff, based on the following idea. The matter is that the Hochschild complex may be identified with the space of graded derivations of the tensor coalgebra: $T^c(A[1]) := \bigoplus_{n \geq 0} A[1]^{\otimes n}$, and the bracket is just the commutator of derivations [**Sta93**]. Here $A[1]$ denotes the graded vector space whose only nonzero graded component is A, placed in degree -1. It is a good exercise to deduce Equation (1) and a more general formula directly from the definition of the G-bracket as the commutator of derivations.

In general, for a graded vector space $V = \bigoplus_n V^n$, $V[k]$ denotes *grading shift*, or what is known to topologists as *k-fold suspension*: it is a graded vector space $V[k]$ whose component of degree n is $V[k]^n := V^{k+n}$. Note that a derivation determined by a map $A[1]^{\otimes n} \to A[1]$ has degree $n-1$. Therefore, the bracket defines a differential graded (DG) Lie algebra structure on the Hochschild complex $C^\bullet(A,A)[1]$ with a shifted grading $\deg f = n-1$ for $f \in C^n(A,A)$. The (appropriately shifted) Hochschild cohomology $H^\bullet(A,A)[1]$ inherits the structure of a graded Lie algebra.

The importance of the bracket comes from the following tautological fact, which however may be regarded as the cornerstone of deformation theory.

PROPOSITION 1.1. *A formal multiplication*

$$m_t(a,b) = m_0(a,b) + m_1(a,b)t + m_2(a,b)t^2 + \cdots, \qquad a,b \in A,$$

is associative, iff $[m_t, m_t] = 0$.

PROOF. Using (1) for the G-bracket of m_t with itself, we get

$$[m_t, m_t](a,b,c) = 2(m_t(m_t(a,b),c) - m_t(a, m_t(b,c))).$$

The right-hand side of the formula contains the associativity equation, and we are done. $\square$

REMARK 2. Because the original multiplication $m_0(a,b) := a \cdot b$ is associative, the G-bracket square of it vanishes: $[m_0, m_0] = 0$. Therefore, the commutator with m_0 defines an inner differential on the Hochschild complex. This differential is in fact the Hochschild differential: another exercise is to verify that $df = [f, m_0]$.

Classical deformation theory was about the following "perturbative" results.

PROPOSITION 1.2. (1) *A formal multiplication*
$$m_t(a,b) = a \cdot b + m_1(a,b)t + m_2(a,b)t^2 + \cdots, \qquad a, b \in A,$$
is associative modulo t^2, iff $dm_1 = 0$. In this case m_1 defines a Hochschild cohomology class $m_1 \in H^2(A, A)$.

(2) *Suppose that a formal multiplication as above is associative modulo t^2. Then the existence of m_2 such that m_t is associative modulo t^3 is equivalent to the vanishing $[m_1, m_1] = 0$ in Hochschild cohomology $H^\bullet(A, A)$.*

PROOF. Expand the equation $\frac{1}{2}[m_t, m_t] = 0$ in powers of t and collect terms by t^n for $n = 0, 1$, and 2. We will get the following.

$$t^0: \qquad \frac{1}{2}[m_0, m_0] = 0,$$
$$t^1: \qquad dm_1 = 0,$$
$$t^2: \qquad dm_2 + \frac{1}{2}[m_1, m_1] = 0.$$

The first equation is always satisfied, because the original multiplication is associative, see Remark 2. The other two equations explain both statements of the corollary. □

1.2. Deformation quantization. Deformation quantization usually refers to a specific deformation quantization problem in a geometric/physical setting. The following theorem, which solves the deformation quantization problem posed by Bayen, Flato, Frønsdal, Lichnerowicz, and Sternheimer [**BFF**[+]**78**], is a remarkable breakthrough in pure mathematics achieved by applying ideas motivated by Feynman diagrams. Recall that a *Poisson manifold* is a pair $(M, \{,\})$, where M is a smooth manifold and $\{f, g\}$ is a local bilinear operation defined for smooth functions f, g on M in a way that, combined with the pointwise multiplication, it defines the structure of a Poisson algebra, see 2.4.4, on the smooth functions. Here *local* means that $\{f, g\}$ is a bidifferential operator, *i.e.*, a differential operator with respect to f and g.

THEOREM 1.3 (M. Kontsevich [**Kon03**]). *Every Poisson manifold $(M, \{,\})$ may be deformation quantized, i.e., there exists a formal deformation quantization, see Remark 1,*
$$f \star g := m_t(f, g) = fg + m_1(f, g)t + m_2(f, g)t^2 + \cdots, \qquad f, g \in C^\infty(M),$$
of the Poisson algebra $A = C^\infty(M)$ of smooth functions, so that all the m_i's are local. According to our definition of deformation quantization, the star product must be associative and also recover the Poisson algebra of functions in the quasi-classical limit, i.e., $(m_1(f, g) - m_1(g, f))/2 = \{f, g\}$.

PROOF. We will only consider the case of $M = \mathbb{R}^d$ with an arbitrary Poisson structure, where the situation is already highly nontrivial. Globalization, which is done using a Fedosov-type connection, see [**Kon03, CFT02**], lies outside the main theme of these notes: no pattern in it has to do with graphs.

First, we will sketch Kontsevich's original proof, giving an explicit formula for the *star product* $f \star g$:

$$（2） \qquad f \star g := \sum_{n=0}^{\infty} \frac{t^n}{n!} \sum_{\Gamma \in \mathcal{G}_{n,2}} W_\Gamma B_\Gamma(f,g).$$

which will be explained in this paragraph. The interior summation runs over the set $\mathcal{G}_{n,2}$ of directed graphs Γ of a certain type with vertices labeled $1, 2, \ldots, n, \bar{1}, \bar{2}$. The set $\mathcal{G}_{n,2}$ of graphs consists of the graphs satisfying the following conditions. Each vertex of the first type, *i.e.*, labeled $1, 2, \ldots$, or n, has exactly two outgoing edges, labeled the first one and the second, and there are no other edges whatsoever. No edge may form a loop, *i.e.*, start and end at one and the same vertex. $B_\Gamma(f,g)$ is a bidifferential operator defined by an explicit formula [**Kon03**], which we will describe using an example.

$$\text{For} \quad \Gamma = \begin{array}{c} 1 \quad j \quad 2 \\ i \quad k \quad \bar{1} \\ \bar{1} \quad \bar{2} \end{array}$$

the corresponding bidifferential operator will be

$$B_\Gamma(f,g) := \alpha^{ij}\partial_j(\alpha^{kl})\partial_k\partial_i(f)\partial_l(g),$$

where α^{ij} denotes the corresponding component of the Poisson tensor in a fixed coordinate system $(x_1, \ldots, x_d)$ in $\mathbb{R}^d$ and we assume summation over the repeating indices. Finally, the coefficient W_Γ is given by another formula:

$$W_\Gamma := \frac{1}{n!(2\pi)^{2n}} \int_{C_{n,2}^+} \bigwedge_{\text{edges } r \to s \text{ of } \Gamma} d\phi(z_r, z_s),$$

where $C_{n,2}^+$ is the configuration space of n distinct points $z_1, \ldots, z_n$ in the upper half-plane and two fixed points $z_{\bar{1}} = 0$ and $z_{\bar{2}} = 1$ on the real line; $\phi(z_r, z_s)$, r and s running over $\{1, 2, \ldots, n, \bar{1}, \bar{2}\}$, is the directed angle at z_r between the hyperbolic line through z_r and ∞ and the hyperbolic line through z_r and z_s. The order in the wedge product is given by the lexicographic order of the vertices $\{1, \ldots, n\}$ and the orders of the set of edges going out of the vertices. Kontsevich proves that this improper integral is absolutely convergent.

The associativity of the star product may be verified explicitly, see [**Kon03**]. However, I will sacrifice the rigor for the moral and give a more conceptual, physical explanation of the associativity, following A. Cattaneo and G. Felder [**CF00**]. Cattaneo and Felder define the star product as

$$（3） \qquad (f \star g)(x) := \int_{\mathcal{P}_x} f(X(0))g(X(1))e^{iS(X,\eta)/t} DX D\eta.$$

This is a Feynman integral over the infinite dimensional "path" space, which is the following space of fields X and η on the upper half-plane H:

$$\mathcal{P}_x := \{X : \bar{H} \to \mathbb{R}^d, \eta \in \Omega^1(\bar{H}) \otimes \mathbb{R}^d \mid X(\infty) = x$$
$$\text{and } \eta \text{ vanishes on tangent vectors to the boundary}\}.$$

The function $S(X, \eta)$ is a certain action functional defining a Poisson sigma model on H, see [**CF00**]:

$$S = \int_H (\eta_{\mu i} \partial_\nu X^i + \frac{1}{2}\alpha^{ij}(X)\eta_{\mu i}\eta_{\nu j}) du^\mu du^\nu.$$

A rigorous definition of the Feynman integral would be the very formula (2). However, physics takes the opposite viewpoint and treats (2) as the saddle-point expansion of the integral (3) in parameter t obtained by formally applying the rules by analogy with the finite-dimensional case. The advantage of this approach is that the mystery of Kontsevich's formula (2) is now replaced by the mystery of Equation (3), which is not so mysterious to a physicist, for whom it represents a standard integral quantization formula. Another advantage is that it offers the following explanation of the associativity.

Consider the integral

$$\langle f, g, h \rangle_p(x) := \int_{\mathcal{P}_x} f(X(0))g(X(1))h(X(p))e^{iS(X,\eta)/t} DX D\eta,$$

where $p \in (1, \infty) \subset \mathbb{R} \subset \bar{H}$ is a fixed point on the real line between 1 and ∞ and $\mathcal{P}_x$ is as above in (3). This integral is independent of the choice of this point p, because the action S is diffeomorphism invariant and, roughly speaking, by integrating over all fields X and η, we take an average over all possible positions of p. Thus, the limits of $\langle f, g, h \rangle_p$ as $p \to 1$ and $p \to \infty$ will be the same. On the other hand, in the moduli space of configurations of four points $0, 1, p$, and ∞ on the boundary of H, these configurations will degenerate as follows:

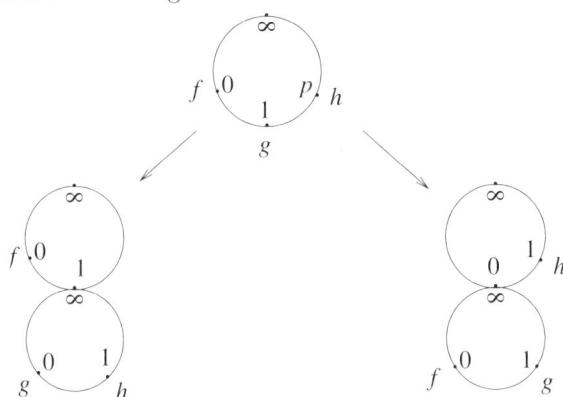

which means that

$$\lim_{p \to 1} \langle f, g, h \rangle_p = f \star (g \star h),$$
$$\lim_{p \to \infty} \langle f, g, h \rangle_p = (f \star g) \star h,$$

yielding the associativity.

In reality, things are more complicated than I have made them appear: the Feynman diagram expansion involves gauge fixing and renormalization, which is achieved by introducing ghosts and antighosts (and what not) and using the BV formalism, see [**CF00**]. The behavior of the Feynman integral with respect to the compactification of the configuration spaces is another issue suppressed in the above. I hope to learn these things before the next Dennisfest:-) □

2. Trees and operads

The combinatorics of trees essentially describes the combinatorics of operads, while more general graphs would be related to PROP's and modular operads. The main idea is that the trees form a free operad (More precisely, the linear span of the set of labeled trees with a certain differential forms the A_∞ operad), and vice versa, any free operad may be described via decorated trees: the tree remembers how to get to its root by applying the operations put at its vertices, given no identities, such as the associativity, whatsoever. Here we will review these notions and related concepts of homotopy algebra.

Operads in general are spaces of operations with certain rules on how to compose operations to get new ones. In this sense operads are directly related to Lawvere's algebraic theories and represent true objects of universal algebra. However, operads as such appeared in topology in the works of J. P. May, J. M. Boardman and R. M. Vogt as a recognition tool for based multiple loop spaces. Stasheff earlier described the first example of an operad, the associahedra, which recognized based loop spaces, while about the same time, Gerstenhaber, studying the algebra of the Hochschild complex, introduced the notion of a composition algebra, which is nothing but the notion of an operad of graded vector spaces.

2.1. PROP's. Operads bear certain superficial resemblance with Deligne-Mumford stacks: both notions have proven to be very useful in mathematics and their formal definitions often leave a novice with nothing more than a "formality" shock — it is as easy to get lost in nested multi-indices in a huge commutative diagram, as it is to drown in the ocean of sheaves of groupoids over the site of schemes on your first trip to the beach. It may be easier to define a PROP (=PROducts and Permutations) and think of an operad as certain part of a PROP. However, later we will give an independent definition of an operad.

DEFINITION 2.1. A *PROP* is a symmetric monoidal (sometimes called tensor) category whose set of objects is identified with the set $\mathbb{Z}_+$ of nonnegative integers, where the tensor law on $\mathbb{Z}_+$ is given by addition and the associativity transformation α is equal to identity. See the founding fathers' sources, such as, J. F. Adams' book [**Ada78**] or S. Mac Lane's paper [**ML65**] for more detail.

Usually, PROP's are enriched over another symmetric monoidal category, that is, the morphisms in the PROP are taken as objects of the other symmetric monoidal category. This gives the notions of a PROP of sets, vector spaces, complexes, topological spaces, manifolds, etc. Examples of PROP's include the following. We will only specify the morphisms, because the objects are already given by the definition.

EXAMPLE 2.1. The *endomorphism PROP* of a vector space V has the space of morphisms $\mathrm{Mor}(m,n) = \mathrm{Hom}(V^{\otimes m}, V^{\otimes n})$. This is a PROP of vector spaces. The composition and tensor product of morphisms are defined as the corresponding operations on linear maps.

EXAMPLE 2.2. The *Segal PROP* is a PROP of infinite dimensional complex manifolds. A morphism is defined as a point in the moduli space $\mathcal{P}_{m,n}$ of isomorphism classes of complex Riemann surfaces bounding $m+n$ labeled nonoverlapping holomorphic holes. The surfaces should be understood as compact smooth complex curves, not necessarily connected, along with $m+n$ biholomorphic maps of the closed unit disk to the surface, thought of as holes. The biholomorphic maps are part of the data, which in particular means that choosing a different biholomorphic map for the same hole is likely to change the point in the moduli space. The more exact nonoverlapping condition is that the closed disks in the inputs do not intersect pairwise and the closed disks in the outputs do not intersect pairwise, however, an input and an output disk may have common boundary, but are still not allowed to intersect at an interior point. This technicality brings in the identity morphisms to the PROP, but does not create singular Riemann surfaces by composition. The composition of morphisms in this PROP is given by sewing the Riemann surfaces along the boundaries, using the equation $zw = 1$ in the holomorphic parameters coming from the standard one on the unit disk. The tensor product of morphisms is the disjoint union. This PROP plays a crucial role in Conformal Field Theory, as we will see now.

2.2. Algebras over a PROP. We need to define another important notion before we proceed.

DEFINITION 2.2. We say that a vector space V is an *algebra over a PROP P*, if a morphism of PROP's from P to the endomorphism PROP of V is given. A morphism of PROP's is a functor respecting the symmetric monoidal structures and also equal to the identity map on the objects.

An algebra over a PROP could have been called a *representation*, but since algebras over operads, which are similar objects, are nothing but familiar types of algebras, it is more common to use the term "algebra."

EXAMPLE 2.3. An example of an algebra over a PROP is a *Conformal Field Theory* (*CFT*), which may be defined (in the case of a vanishing central charge) as an algebra over the Segal PROP. The fact that the functor respects compositions of morphisms translates into the sewing axiom of CFT in the sense of G. Segal. Usually, one also asks for the functor to depend smoothly on the point in the moduli space $\mathcal{P}_{m,n}$. One needs to extend the Segal PROP by a line bundle to cover the case of an arbitrary charge, see [**Hua97**].

EXAMPLE 2.4 (Sullivan). Another example of an algebra over a PROP is a Lie bialgebra. Sullivan has shared with me a nice graph description of the corresponding PROP, see
http://www.math.umn.edu/~voronov/8390/lec4.pdf.

2.3. Operads. Now we are ready to deal with operads, which formalize the notion of a space of operations, as we mentioned in the introduction to Section 2. Informally, an operad is the part $\mathrm{Mor}(n, 1)$, $n \geq 0$, of a PROP. Of course, given only the collection of morphisms $\mathrm{Mor}(n, 1)$, it is not clear how to compose them. The idea is to take the union of m elements from $\mathrm{Mor}(n, 1)$ to be able to compose them with an element of Mor$(m,1)$. This leads to cumbersome notation and ugly axioms, compared to those of a PROP. However operads are in a sense more basic than the corresponding PROP's: the difference is similar to the difference between Lie algebras and the universal enveloping algebras.

DEFINITION 2.3. An *operad* $\mathcal{O}$ is a collection of sets (vector spaces, complexes, topological spaces, manifolds, ..., objects of a symmetric monoidal category) $\mathcal{O}(n)$, $n \geq 0$, with

(1) A composition law:
$$\gamma : \mathcal{O}(m) \otimes \mathcal{O}(n_1) \otimes \cdots \otimes \mathcal{O}(n_m) \to \mathcal{O}(n_1 + \cdots + n_m).$$
(2) A right action of the symmetric group S_n on $\mathcal{O}(n)$.
(3) A unit $e \in \mathcal{O}(1)$.

such that the following properties are satisfied:

(1) The composition is associative, *i.e.*, the following diagram is commutative:

$$\begin{array}{c}\left\{\begin{array}{c}\mathcal{O}(l) \otimes \mathcal{O}(m_1) \otimes \cdots \otimes \mathcal{O}(m_l) \\ \otimes \mathcal{O}(n_{11}) \otimes \cdots \otimes \mathcal{O}(n_{l,n_l})\end{array}\right\} \xrightarrow{\mathrm{id} \otimes \gamma^l} \mathcal{O}(l) \otimes \mathcal{O}(n_1) \otimes \cdots \otimes \mathcal{O}(n_l) \\ \gamma \otimes \mathrm{id} \downarrow \qquad\qquad\qquad\qquad\qquad\qquad\qquad \downarrow \gamma \\ \mathcal{O}(m) \otimes \mathcal{O}(n_{11}) \otimes \cdots \otimes \mathcal{O}(n_{m,n_m}) \xrightarrow{\gamma} \qquad \mathcal{O}(n)\end{array},$$

where $m = \sum_i m_i$, $n_i = \sum_j n_{ij}$, and $n = \sum_i n_i$.

(2) The composition is equivariant with respect to the symmetric group actions: the groups S_m, S_{n_1}, ..., S_{n_m} act on the left-hand side and map naturally to $S_{n_1+\cdots+n_m}$, acting on the right-hand side.
(3) The unit e satisfies natural properties with respect to the composition: $\gamma(e; f) = f$ and $\gamma(f; e, \ldots, e) = f$ for each $f \in \mathcal{O}(k)$.

The notion of a *morphism of operads* is introduced naturally.

REMARK 3. One can consider *non-Σ operads*, not assuming the action of the symmetric groups. Not requiring the existence of a unit e, we arrive at *nonunital operads*. Do not mix this up with operads with no $\mathcal{O}(0)$, algebras over which (see next section) have no unit. There are also good examples of operads having only $n \geq 2$ components $\mathcal{O}(n)$.

An equivalent definition of an operad may be given in terms of operations $f \circ_i g = \gamma(f; \mathrm{id}, \ldots, \mathrm{id}, g, \mathrm{id}, \ldots, \mathrm{id})$, $i = 1, \ldots, m$, for $f \in \mathcal{O}(m), g \in \mathcal{O}(n)$. Then the associativity condition translates as $f \circ_i (g \circ_j h) = (f \circ_i g) \circ_{i+j-1} h$ plus a natural symmetry condition for $(f \circ_i g) \circ_j h$, when g and h "fall into separate slots" in f, see *e.g.*, [**KSV96**].

EXAMPLE 2.5 (The Riemann surface and the endomorphism operads). $\mathcal{P}(n)$ is the space of Riemann spheres with $n+1$ boundary components, *i.e.*, n inputs and 1 output. Another example is the *endomorphism operad of a vector space V*: $\mathcal{E}nd_V(n) = \mathrm{Hom}(V^{\otimes n}, V)$, the space of n-linear mappings from V to V.

2.4. Algebras over an operad.

DEFINITION 2.4. An *algebra over an operad* $\mathcal{O}$ (in other terminology, a *representation of an operad*) is a morphism of operads $\mathcal{O} \to \mathcal{E}nd_V$, that is, a collection of maps
$$\mathcal{O}(n) \to \mathcal{E}nd_V(n) \qquad \text{for } n \geq 0$$

compatible with the symmetric group action, the unit elements, and the compositions. If the operad $\mathcal{O}$ is an operad of vector spaces, then we would usually require the morphism $\mathcal{O} \to \mathcal{E}nd_V$ to be a morphism of operads of vector spaces. Otherwise, we would think of this morphism as a morphism of operads of sets. Sometimes, we may also need a morphism to be continuous or respect differentials, or have other compatibility conditions.

2.4.1. *The commutative operad.* The *commutative operad* is the operad of k-vector spaces with the nth component $\mathcal{C}omm(n) = k$ for all $n \geq 0$. We assume that the symmetric group acts trivially on k and the compositions are just the multiplication of elements in the ground field k. The term "commutative operad" may seem confusing to some people, but it has been in use for a while. An algebra over the commutative operad is nothing but a commutative associative algebra with a unit, as we see from Exercise 2 below.

Another version of the commutative operad is $\mathcal{C}omm(n) = \{\text{point}\}$ for all $n \geq 0$. This is an operad of sets. It is equivalent to the previous version in the sense that an algebra over it is the same as a commutative associative unital algebra.

PUZZLE 1. Show that the operad $\mathcal{T}op(n) = \{$the set of diffeomorphism classes of Riemann spheres with n input holes and 1 output hole$\}$ is isomorphic to the commutative operad of sets.

EXERCISE 2. Prove that the structure of an algebra over the commutative operad $\mathcal{C}omm$ on a vector space is equivalent to the structure of a commutative associative algebra with a unit.

2.4.2. *The associative operad.* The *associative operad* $\mathcal{A}ssoc$ can be considered as a planar one-dimensional analogue of the commutative operad $\mathcal{T}op$. $\mathcal{A}ssoc(n)$ is the set of equivalence classes of connected planar binary (each vertex being of valence 3) trees that have a root edge and n leaves labeled by integers 1 through n:

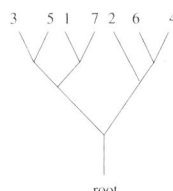

If $n = 1$, there is only one tree — it has no vertices and only one edge connecting a leaf and a root. If $n = 0$, the only tree is the one with no vertices and no leaves — it only has a root.

Two trees are equivalent if they are related by a sequence of moves of the kind

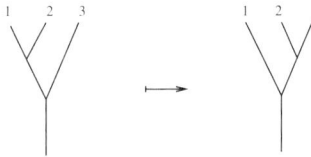

performed over pairs of two adjacent vertices of a tree. The symmetric group acts by relabeling the leaves, as usual. The composition is obtained by grafting the roots of m trees to the leaves of an m-tree, no new vertices being created at the grafting points. Note that this is similar to sewing Riemann surfaces and erasing the seam, just as we did to define operad composition in that case. By definition, grafting a 0-tree to a leaf just removes the leaf and, if this operation creates a vertex of valence 2, we should erase the vertex.

EXERCISE 3. Prove that the structure of an algebra over the associative operad $\mathcal{A}ssoc$ on a vector space is equivalent to the structure of an associative algebra with a unit.

2.4.3. *The Lie operad.* The *Lie operad* $\mathcal{L}ie$ is another variation on the theme of a tree operad. Consider the vector space spanned by the same planar binary trees as for the associative operad, except that we do not include a 0-tree, *i.e.*, the operad has only positive components $\mathcal{L}ie(n)$, $n \geq 1$, and there are now two kinds of equivalence relations:

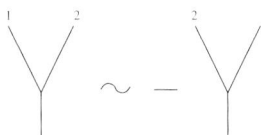

Skew Symmetry

and

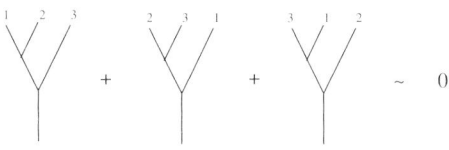

Jacobi Identity

Now that we have arithmetic operations in the equivalence relations, we consider the Lie operad as an operad of vector spaces. We also assume that the ground field is of a characteristic other than 2, because otherwise we will arrive at the wrong definition of a Lie algebra.

EXERCISE 4. Prove that the structure of an algebra over the Lie operad $\mathcal{L}ie$ on a vector space over a field of a characteristic other than 2 is equivalent to the structure of a Lie algebra.

EXERCISE 5. Describe algebraically an algebra over the operad $\mathcal{L}ie$, if we modify it by including a 0-tree, whose composition with any other tree is defined as (a) zero, (b) the one for the associative operad.

2.4.4. *The Poisson operad.* Recall that a *Poisson algebra* is a vector space V (over a field of characteristic zero) with a unit element e, a dot product ab, and a bracket $[a, b]$ defined, so that the dot product defines the structure of commutative associative unital algebra, the bracket defines the structure of a Lie algebra, and the bracket is a derivation of the dot product:

$$[a, bc] = [a, b]c + b[a, c] \qquad \text{for all } a, b, \text{ and } c \in V.$$

EXERCISE 6. Define the *Poisson operad*, using a tree model similar to the previous examples. Show that an algebra over it is nothing but a Poisson algebra. [*Hint*: Use two kinds of vertices, one for the dot product and the other one for the bracket.]

2.4.5. *The Riemann surface operad and vertex operator algebras.* Just for a change, let us return to the operad $\mathcal{P}$ of Riemann surfaces, more exactly, isomorphism classes of Riemann spheres with holomorphic holes. What is an algebra over it? Since there are infinitely many nonisomorphic pairs of pants, there are infinitely many (at least) binary operations. In fact, we have an infinite dimensional family of binary operations parameterized by classes of pairs of pants. However modulo the unary operations, those which correspond to cylinders, we have only one fundamental binary operation corresponding to a fixed pair of pants. An algebra over this operad $\mathcal{P}$ is part a CFT data. (For those who understand, this is the tree level, central charge $c = 0$ part). If we consider a holomorphic algebra over this operad, that is, require that the defining mappings $\mathcal{P}(n) \to \mathcal{E}nd_V(n)$, where V is a complex vector space, be holomorphic, then we get part of a chiral CFT, or an object which might have been called a *vertex operator algebra* (*VOA*) in an ideal world. This kind of object is not equivalent to what people use to call a VOA; according to Y.-Z. Huang's Theorem, a VOA is a holomorphic algebra over a "partial pseudo-operad of Riemann spheres with rescaling," which is a version of $\mathcal{P}$, where the disks are allowed to overlap. The fundamental operation $Y(a,z)b$ for $a,b \in V$, $z \in \mathbb{C}$ of a VOA is commonly chosen to be the one corresponding to a pair of pants which is the Riemann sphere with a standard holomorphic coordinate and three unit disks around the points 0, z, and ∞ (No doubt, these disks overlap badly, but we shrink them on the figure to look better):

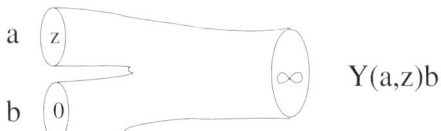

The famous associativity identity
$$Y(a, z-w)Y(b, -w)c = Y(Y(a,z)b, -w)c$$
for vertex operator algebras comes from the following natural isomorphism of the Riemann surfaces:

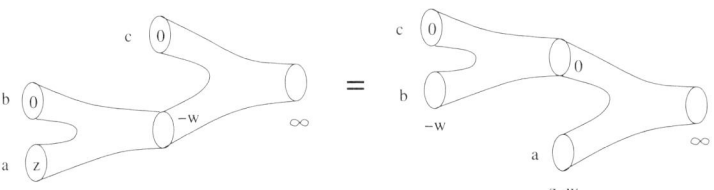

2.5. Operads via generators and relations. The tree operads that we looked at above, such as the associative and the Lie operads, are actually operads defined by generators and relations. Here is a way to define such operads in general. To fix notation, assume throughout this section that we work with operads $\mathcal{O}(n)$, $n \geq 1$, of vector spaces.

DEFINITION 2.5. An *ideal* in an operad $\mathcal{O}$ is a collection $\mathcal{I}$ of S_n-invariant subspaces $\mathcal{I}(n) \subset \mathcal{O}(n)$, for each $n \geq 1$, such that whenever $i \in \mathcal{I}$, $\gamma(\ldots, i, \ldots) \in \mathcal{I}$.

The intersection of an arbitrary number of ideals in an operad is also an ideal, and one can define the ideal generated by a subset in $\mathcal{O}$ as the minimal ideal containing the subset.

DEFINITION 2.6. For an operad ideal $\mathcal{I} \subset \mathcal{O}$, the *quotient operad* $\mathcal{O}/\mathcal{I}$ is the collection $\mathcal{O}(n)/\mathcal{I}(n)$, $n \geq 1$, with the structure of operad induced by that on $\mathcal{O}$.

The *free operad $F(S)$ generated by a collection* $S = \{S(n) \mid n \geq 1\}$ *of sets*, is defined as follows.
$$F(S)(n) = \bigoplus_{n\text{-trees } T} k \cdot S(T),$$
where the summation runs over all planar rooted trees T with n labeled leaves and
$$S(T) = \mathrm{Map}(v(T), S),$$
the set of maps from the set $v(T)$ of vertices of the tree T to the collection S assigning to a vertex v with $\mathrm{In}(v)$ incoming edges an element of $S(\mathrm{In}(v))$ (the edges are directed toward the root). In other words, an element of $F(S)(n)$ is a linear combination of planar n-trees whose vertices are decorated with elements of S. There is a special tree with no vertices:

The component $F(S)(1)$ contains, apart from $S(1)$, the one-dimensional subspace spanned by this tree.

The following data defines an operad structure on $F(S)$.

(1) The identity element is the special tree in $F(S)(1)$ with no vertices.
(2) The symmetric group S_n acts on $F(S)(n)$ by relabeling the inputs.
(3) The operad composition is given by grafting the roots of trees to the leaves of another tree. No new vertices are created.

DEFINITION 2.7. Now let R be a subset of $F(S)$, *i.e.*, a collection of subsets $R(n) \subset F(S)(n)$. Let (R) be the ideal in $F(S)$ generated by R. The quotient operad $F(S)/(R)$ is called the *operad with generators S and defining relations R*.

EXAMPLE 2.6. The associative operad $\mathcal{A}ssoc$ is the operad generated by a point $S = S(2) = \{\bullet\}$ with a defining relation given by the associativity condition, see Section 2.4.2, expressed in terms of trees. Note that equation $S = S(2)$ implies that $S(n) = \varnothing$ for $n \neq 2$.

EXAMPLE 2.7. The Lie operad $\mathcal{L}ie$ is the operad also generated by a point $S = S(2) = \{\bullet\}$ with defining relations given by the skew symmetry and the Jacobi identity, see Section 2.4.3.

EXAMPLE 2.8. The Poisson operad is the operad also generated by a two-point set $S = S(2) = \{\bullet, \circ\}$ with defining relations given by the commutativity and the associativity for simple trees decorated only with $\bullet$'s, the skew symmetry and the Jacobi identity for simple trees decorated with $\circ$'s, and the Leibniz identity for binary 3-trees with mixed decorations, see Section 2.4.4.

2.6. Homotopy algebra. The idea of a *homotopy "something" algebra* is to relax the axioms of the *"something" algebra*, so that the usual identities are satisfied up to homotopy. For example in a homotopy associative algebra, the associativity identity looks like

$$(ab)c - a(bc) \text{ is homotopic to zero.}$$

Or in a homotopy Poisson algebra, the Leibniz rule is

$$[a, bc] - [a, b]c - b[a, c] \text{ is homotopic to zero.}$$

Usually, a homotopy something algebra arises when one wants to lift the structure of a something algebra *a priori* defined on cohomology to the level of cochains.

This kind of relaxation seems to be too lax for many, practical and categorical, purposes, and one usually requires that the null-homotopies, regarded as new operations, satisfy their own identities, up to their own homotopy. These homotopies should also satisfy certain identities up to homotopy and so on. This resembles Hilbert's chains of syzygies in early homological algebra. Algebras with such chains of homotopies are called *strongly homotopy "something" algebras* or *"something"$_\infty$-algebras*.

Operads are especially helpful when one needs to work with something$_\infty$-algebras. We already know that defining the class of something algebras is equivalent to defining the something operad. Thus, if we have an operad $\mathcal{O}$, what is $\mathcal{O}_\infty$, *the* corresponding strongly homotopy operad? M. Markl's paper [**Mar96**] provides a satisfactory answer to this question: the *operad* $\mathcal{O}_\infty$ is a minimal model of the operad $\mathcal{O}$. A minimal model is unique up to isomorphism. The idea is borrowed from Sullivan's rational homotopy theory; a minimal model is, first of all, a free resolution of $\mathcal{O}$ in the category of operads of complexes, *i.e.*, an operad of complexes free as an operad of graded vector spaces, whose cohomology is $\mathcal{O}[0]$, the operad $\mathcal{O}$ sitting in degree zero, if it is an operad of vector spaces, or the operad $\mathcal{O}$ sitting in the original degrees, if it is already an operad of graded vector spaces. Second of all, a minimal model must satisfy a minimality condition: the image of the differential must be decomposable into a sum of compositions of elements from $\mathcal{O}(n)$, $n \geq 2$.

For certain specific classes of operads, one manages to describe a minimal model explicitly. For example, V. Ginzburg and M. Kapranov [**GK94**] do it (even earlier than the notion of a minimal model for an operad surfaced) for the so-called Koszul operads. Below we describe an example of such kind, giving rise to the notion of an A_∞-algebra and the A_∞ operad.

2.6.1. *A_∞-algebras.*

DEFINITION 2.8. An *A_∞-algebra*, or a *strongly homotopy associative algebra*, is a complex $V = \bigoplus_{i \in \mathbb{Z}} V^i$ with a differential d, $d^2 = 0$, of degree 1 and a collection of n-ary operations, called *products*:

$$M_n(v_1, \ldots, v_n) \in V, \qquad v_1, \ldots, v_n \in V, \ n \geq 2,$$

which are homogeneous of degree $2 - n$ and satisfy the relations

$$(4) \quad dM_n(v_1, \ldots, v_n) - (-1)^n \sum_{i=1}^n \epsilon(i) M_n(v_1, \ldots, dv_i, \ldots, v_n)$$

$$= \sum_{\substack{k+l=n+1 \\ k,l \geq 2}} \sum_{i=0}^{k-1} (-1)^{i+l(n-i-l)} \sigma(i) M_k(v_1, \ldots, v_i, M_l(v_{i+1}, \ldots, v_{i+l}), \ldots, v_n),$$

where $\epsilon(i) = (-1)^{|v_1|+\cdots+|v_{i-1}|}$ is the sign picked up by taking d through $v_1, \ldots, v_{i-1}$, $|v|$ denoting the degree of $v \in V$, and $\sigma(i)$ is the sign picked up by taking M_l through $v_1, \ldots, v_i$.

It is remarkable to look at these relations for $n = 2$ and 3:
$$dM_2(v_1, v_2) - M_2(dv_1, v_2) - (-1)^{|v_1|} M_2(v_1, dv_2) = 0,$$

$$dM_3(v_1, v_2, v_3) + M_3(dv_1, v_2, v_3) + (-1)^{|v_1|} M_3(v_1, dv_2, v_3)$$
$$+ (-1)^{|v_1|+|v_2|} M_3(v_1, v_2, dv_3)$$
$$= M_2(M_2(v_1, v_2), v_3) - M_2(v_1, M_2(v_2, v_3)),$$

which mean that the differential d is a derivation of the bilinear product M_2 and the trilinear product M_3 is a homotopy for the associativity of M_2, respectively.

A_∞-algebras can be described as algebras over a certain tree operad. This operad is the tree part of the graph complex, which will be the topic of the following sections.

2.6.2. *The A_∞ operad.* Let $A_\infty(n)$ be the linear span of the set of equivalence classes of connected planar trees that have a root edge and n leaves labeled by integers 1 through n, with vertices of a valence at least 3, $n \geq 2$. For $n = 1$ take one tree with a unique edge connecting a leaf and a root. Let us not include anything for $n = 0$, although one could do that similar to the associative operad case, so that the corresponding notion of an A_∞-algebra would have a unit.

We grade each vector space $A_\infty(n)$ by defining the degree $|T|$ of a tree $T \in A_\infty(n)$ via
$$|T| := v(T) + 1 - n = e(T) + 1 - 2n,$$
where $v(T)$ is the number of vertices and $e(T)$ the number of edges of T. Notice that $2 - n \leq |T| \leq 0$ for $n \geq 1$.

Let us define an operad structure on these spaces of trees. The symmetric group acts by relabeling the leaves, and the operad composition is obtained by *grafting*, as in the examples above, except one needs to take a sign into account. When we graft a tree T_2 to the ith leaf of a tree T_1, the result must be the grafted tree multiplied by a sign, which is (-1) to the power $(e(T_2) - 1)$(the number of edges to the right of the ith leaf in T_1), where the edges to the right of a leaf are the edges which are strictly on the right-hand side of a unique path from the leaf to the root. The reason for the sign above is that grafting must respect the differential, which is introduced below.

EXERCISE 7. Show that this operad is a free operad of graded vector spaces generated by the following trees for $n \geq 2$, which are sometimes called *corollas*.

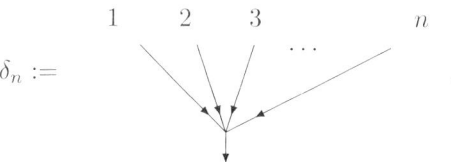

REMARK 4. There is no need to mark directions on the edges of a tree: from now on we will assume the edges are directed from top to bottom.

2.6.3. *The tree complex.* The above operad of trees is not yet the A_∞-operad, but only its underlying operad of graded vector spaces. The A_∞-operad is a DG operad, *i.e.*, an operad of complexes. The DG structure, or a differential, is defined as follows.

Before defining it, we will define the operation of internal-edge contraction on the set of trees.

DEFINITION 2.9. We use the notation T/e to denote the tree obtained from a tree T by contracting an internal edge e:

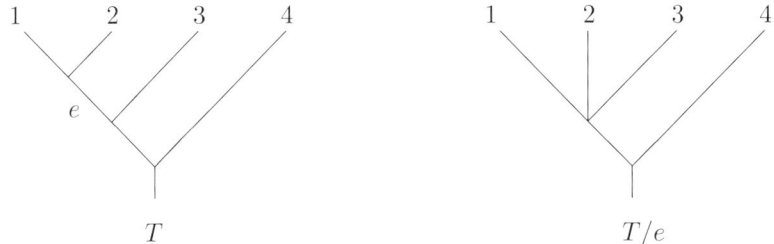

We can now define a *differential* $d : A_\infty(n) \to A_\infty(n)$ by the formula

$$dT := \sum_{T' : T = T'/e} \epsilon T',$$

where ϵ is the sign given by counting the number of edges below and to the left of the edge e in the tree T', not counting the root.

In particular,

(5)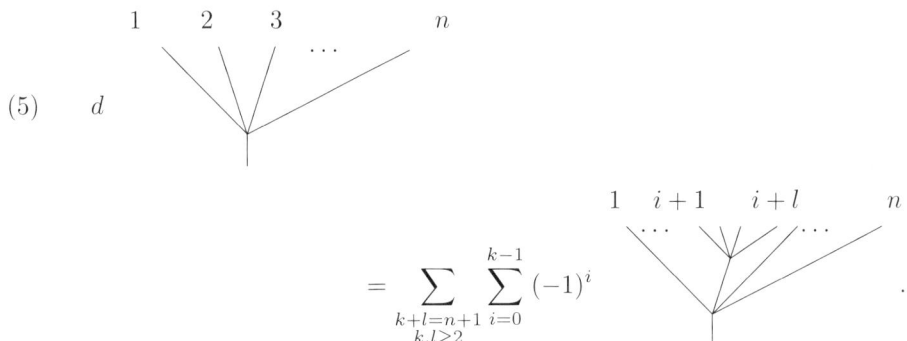

PROPOSITION 2.1. (1) *The operator d satisfies $d^2 = 0$ and $\deg d = 1$.*
(2) *The operad structure on $A_\infty = \{A_\infty(n) \mid n \geq 1\}$ is compatible with the differential d:*

$$d(T_1 \circ_i T_2) = dT_1 \circ_i T_2 + (-1)^{|T_1|} T_1 \circ_i dT_2,$$

i.e., A_∞ is a DG operad.

DEFINITION 2.10. We will call the DG operad A_∞ the A_∞ *operad*.

REMARK 5. The complex $A_\infty(n)$ is part of the (cochain) graph complex, see Section 3. A similar operad L_∞, based on abstract, *i.e.*, nonplanar trees, was introduced by V. Hinich and V. Schechtman [**HS93**]. The operad A_∞ is the DG dual operad in the sense of Ginzburg and Kapranov [**GK94**] of the associative operad $\mathcal{A}ssoc$. They also show that the cohomology of the operad A_∞ is the associative operad $\mathcal{A}ssoc$ of Section 2.4.2, implying that A_∞ is a free, and in fact, minimal, resolution of $\mathcal{A}ssoc$.

The following theorem shows that the A_∞ operad describes the class of A_∞-algebras.

THEOREM 2.2 ([**GK94**]). *An algebra over the A_∞ operad is an A_∞-algebra. Each A_∞-algebra admits a natural structure of an algebra over the A_∞ operad.*

PROOF. For a complex V of vector spaces with a differential d of degree 1, $d^2 = 0$, the structure of an algebra over the operad A_∞ on V is a morphism of DG operads:
$$\phi : A_\infty(n) \to \mathcal{E}nd_V(n), \qquad n \geq 1,$$
where $\mathcal{E}nd_V(n) := \operatorname{Hom}(V^{\otimes n}, V)$ is the *endomorphism operad*, which is also a DG operad (with the usual internal differential determined by d). Given such a morphism ϕ, we define the n-ary product on V:
$$M_n(v_1, \ldots, v_n) := \phi(\delta_n)(v_1 \otimes \cdots \otimes v_n).$$

Note that the degree of the product is equal to that of the corolla δ_n, which is $2 - n$. Since ϕ is a morphism of DG operads, $d\phi = \phi d$, and in view of (5), this is equivalent to the identity (4).

Conversely, given a collection of n-ary brackets on V, $n \geq 2$, we define a morphism ϕ on the generators δ_n by the above formula. The A_∞ operad is freely generated by the corollas δ_n, with a differential defined by (5), so the mappings ϕ define a morphism of DG operads, if the relations (5) are satisfied by the $\phi(\delta_n)$'s. Equations (4) show that this is the case. □

2.7. Cacti. One of the very recent applications of operads moves somewhat southwest and goes through cacti rather than trees. The word "homology" will mean rational homology throughout this section.

The cacti operad may be used to explain the structure of a BV-algebra on the homology of a free loop space on a compact oriented manifold discovered by M. Chas and D. Sullivan [**CS99**]. A *BV-algebra* is a graded vector space with the structure of a graded commutative algebra and a second-order derivation, called a *BV operator*, of degree one and square zero. The BV-algebra structure is also known as the algebraic structure induced on homology from the structure of an algebra over the famous *framed little disks operad* on a topological space, see [**Get94, SW03**]. In particular, the homology of a based double loop space is naturally a BV-algebra. Some of the missing details related to cacti are also captured by Ralph Cohen's Dennisfest lecture and his paper with John Jones [**CJ02**], where they develop a more civilized version of the cacti operad action in the category of spectra and show that Chas-Sullivan's BV-structure is the same as the BV-structure coming naturally after identifying the homology of the free loop space with the Hochschild homology of the cochain algebra of the target space.

The *cacti operad* is an operad $c = \{C(n) \mid n \geq 1\}$ of topological spaces. Its nth component $C(n)$ for $n \geq 1$ may be described as follows.

$C(n)$ is the set of ordered tree-like configurations of parameterized circles (lobes) of varying (positive) radii, along with a cyclic order of components at the intersection points and the choice of a point "0" on the whole configuration along with the choice of one of the circles on which this point 0 lies. The latter is, of course essential when 0 happens to be an intersection point. The topology on the set of configurations before choosing cyclic orders and marking a point 0 is induced from a natural embedding into $(S^1 \times S^1)^{\binom{n}{2}} \times \mathbb{R}_+^n$. The choice of cyclic orders defines a finite covering. After these choices are made, we can define a continuous flow on the cactus which goes along the parameters on the circles and jumps from one component to the next in the cyclic order at the intersection points. Two choices of a zero point are considered close to each other, if they are close in the sense of this flow. Thus the space of cacti with a marked point 0 is an S^1-bundle over a finite covering of the configuration space.

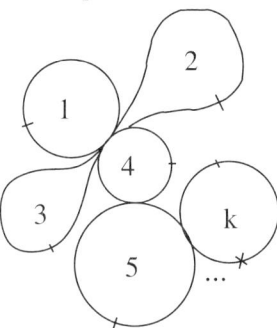

The operad structure on the cacti comes from the following observation. The choice of a point 0 and a component on a cactus gives a natural map from S^1 to the cactus. First rescale the radius of S^1 to match the sum of the radii of the lobes forming the cactus. Then wind this S^1 around the cactus and follow the flow along the lobes, starting with the chosen lobe at 0. Topologically, the constructed map will identify a few groups of points on S^1 and therefore will have a degree one on each lobe. Given two cacti and the ith lobe in the first one, the operad composition $\circ_i$ will be given by further collapsing the ith circle according to the map given by the second cactus.

The following theorem describes both the cacti operad C and how it produces the BV-structure on the homology of a free loop space LM in a compact oriented manifold M of dimension d.

THEOREM 2.3.   (1) *The cacti operad is homotopy equivalent to the framed little disks operad.*
(2) *The cacti operad C "acts" on LM in the following sense. The diagram*
$$C(n) \times (LM)^n \xleftarrow{i} L^{C(n)} M \xrightarrow{e} LM,$$
where $L^{C(n)}M$ is the space of continuous maps of n-component cacti to M, induces a composite map
$$H_\bullet(C(n)) \otimes H_\bullet(LM)^{\otimes n} \xrightarrow{i^!} (H_\bullet(L^{C(n)}M))[(1-n)d] \xrightarrow{e_*} H_\bullet(LM)[(1-n)d],$$
where $i^!$ denotes the pullback in homology, defined using the fact that i is a finite-codimensional embedding of manifolds. The collection of such maps for $n \geq 1$ is compatible with the operad structure on C.
(3) *The composite map $e_* i^!$ produces the structure of an algebra over the homology cacti operad $H_\bullet(C)$ on the space $H_\bullet(LM)[d]$.*

Combining Statements 1 and 3 of the theorem with a theorem of E. Getzler [**Get94**] which says that the homology framed little disks operad is the operad describing BV-algebras and checking what the basic operations (the dot product and the BV operator) really are, we obtain the following result.

COROLLARY 2.4. *The space $H_\bullet(LM)$ (after an appropriate degree shift) has the natural structure of a BV-algebra, coinciding with the one constructed by Chas and Sullivan.*

Cacti of higher dimensions are much subtler and are the topic of an upcoming paper with Sullivan. Here we will just mention one of the applications, which uses a generalization of Getzler's theorem to higher dimensions by P. Salvatore and N. Wahl [**SW03**].

THEOREM 2.5 (Sullivan-A.V. [**SV04**]). *Let M be a compact oriented manifold and $S^n M = \mathrm{Map}(S^n, M)$ the sphere space. Then the space $H_\bullet(S^n M)$ (after an appropriate degree shift) is naturally an algebra over the homology framed little $(n+1)$-disks operad. In particular, the homology of the sphere space has the following algebraic structure. For n odd, it is a BV_{n+1}-algebra, which is the same as a usual BV-algebra, except that the BV operator has degree n. For n even, this structure is the same as that of an $(n+1)$-algebra with a differential of degree $2n-1$.*

3. Graph homology

Even more interesting things start happening when you pass from trees to graphs. One of these things is that such a classical, analytic and algebraic geometric object as the moduli space of Riemann surfaces miraculously (to me) emerges in the horizon the very moment you say the word "graphs." This miracle is a deep discovery of K. Strebel and W. Thurston, which was later developed to the current level of understanding and usefulness by many other mathematicians, and is perhaps not a miracle at least to some of us. One of the several versions of the graph complex, the ribbon graph complex, is closely related to the moduli spaces of Riemann surfaces. The notion of a ribbon graph and the graph complex are due to R. Penner [**Pen86, Pen88**], who used the term "fatgraph." The ribbon graph complex is a generalization of the planar tree complex we considered in the previous section, except that we no longer allow any free legs. Other versions of the graph complex include those which do not require cyclic orders at vertices, see M. Culler and K. Vogtmann [**CV86**], or, on the contrary, have more complicated decorations at vertices, see Getzler-Kapranov's Feynman transform [**GK98**]. There is also a dual version of the graph complex, producing graph cohomology. Kontsevich [**Kon93**], who pointed out a common pattern in Penner's and Culler-Vogtmann's work, related different versions of the graph complex to different fundamental types of algebras (or operads that describe them) and proved a conjecture of Witten, had an enormous influence on the subject.

3.1. The graph complex. The right analogue of a planar tree is a *ribbon graph*, which is a nonempty connected graph Γ, given by an incidence relation on the sets of its vertices and edges, with the choice of a cyclic order on the set of half-edges around each vertex. Here the *half-edges* are the edges of the *edge refinement* of Γ, which is the graph obtained by splitting each edge of Γ into two separated by a new vertex in the middle of the edge. We also require that the valences of vertices must be greater than or equal to 3. A *boundary component* of a ribbon graph Γ is a cyclic sequence $\vec{e}_0, \vec{e}_1, \ldots, \vec{e}_q = \vec{e}_0$ of directed edges (*i.e.*, edges with directions chosen on each of them) of Γ, so that for each pair $(\vec{e}_i, \vec{e}_{i+1})$ of two subsequent edges, the tail of $\vec{e}_{i+1}$ is the half-edge that follows the head of $\vec{e}_i$ in the cyclic order at the common vertex of e_i and e_{i+1}. To have a more direct connection with moduli spaces of Riemann surfaces with labeled boundary components, we will assume that graphs under consideration will have their boundary components labeled. An *orientation* on a ribbon graph Γ is an ordering of the set of edges $e(\Gamma)$ of Γ, up to even permutation. We identify ribbon graphs which differ by an *isomorphism*, which is an isomorphism between the edge refinements of the graphs preserving the cyclic orders at vertices, orientation, and each of the boundary components (but not necessarily each edge forming a boundary component). The *group of automorphisms* of an oriented ribbon graph is denoted by $\operatorname{Aut}_\partial(\Gamma)$.

For each $m \geq 2$, let G_m be the free abelian group generated by the set of isomorphism classes of oriented ribbon graphs Γ as above with m edges, modulo the defining relations $\Gamma + (-\Gamma) = 0$, where $-\Gamma$ is the same graph as Γ, but with the opposite orientation.

Define a differential $d : G_m \to G_{m-1}$, so that $d^2 = 0$, as follows:

$$d\Gamma := \sum_{\substack{\text{edges } e \in e(\Gamma) \\ \text{which are not loops}}} \Gamma/e,$$

where Γ/e is the ribbon graph obtained from Γ by contracting edge e to a point. The cyclic order at the new vertex created by merging the two ends of e is obtained by a natural insertion. The orientation on Γ/e is defined so that the corresponding ordering of the edges of Γ/e with the edge e put after these edges gives the original ordering on the edges of Γ. The complex $G_\bullet$ with the differential d is called the *graph (chain) complex* and its homology is called *graph homology*.

Obviously, the differential preserves the number n of boundary components, as well as the *Euler characteristic* $\chi(\Gamma) := v(\Gamma) - e(\Gamma)$. It also preserves the *genus* g defined by the equation $\chi = 2 - 2g - n$. The solution $g = 1 - (\chi + n)/2$ is a nonnegative integer, because if we glue in n disks into the boundary components, we will get a compact orientable topological surface whose Euler characteristic equals $\chi + n = 2 - 2g$ for some nonnegative integer g. Thus, the graph complex splits into the direct sum $G_\bullet = G_\bullet^{g,n}$ of subcomplexes $G_\bullet^{g,n}$ with a fixed genus g and a number n of boundary components, $g \geq 0$, $n \geq 1$.

3.2. Metric ribbon graphs and moduli spaces. The graph complex above is in fact the chain complex of a certain cell complex, that of metric ribbon graphs. A *metric* on a ribbon graph is an assignment of a positive real number, a *length*, to each edge of the graph. Obviously, the space of isomorphism classes of metric ribbon graphs with an underlying graph Γ is the space $\mathbb{R}_+^{e(\Gamma)}/\operatorname{Aut}_\partial(\Gamma)$, which is, in fact, an orbifold. For different ribbon graphs Γ of a fixed genus g and a number n of boundary components, these orbifolds glue together by identifying metric graphs some of whose lengths degenerate to zero with the metric graphs obtained by contracting the zero length edges. Thus, these orbifolds become what is called the *rational cells* of an orbifold, the *space $G_{g,n}^{\mathrm{met}}$ of metric ribbon graphs*. If we forget the orbifold structure, these rational cells are just cells of a topological space, but not a cell complex, because only part of the boundary of each cell, the part obtained by letting the lengths of edges go to zero, is glued up to other cells. To get an honest cell complex, we may take the one-point compactification of the space of metric ribbon graphs, thus, gluing in a single point to all cells as lengths tend to positive infinity. This gives a nonorbifold-type singularity, and we get a compact smooth orbifold $\overline{G}_{g,n}^{\mathrm{met}}$ with one singular point. It may also be viewed as an ordinary cell complex with a base point. This space was designed with the following obvious proposition in mind.

PROPOSITION 3.1. *The cellular chain complex, computing the rational reduced homology of the orbifold $\overline{G}_{g,n}^{\mathrm{met}}$, is isomorphic to the ribbon graph complex $G_\bullet^{g,n} \otimes \mathbb{Q}$.*

REMARK 6. We had to take the rational coefficients to make sure that we were computing the orbifold homology. Over the integers, it would be the computation of the ordinary homology of the underlying space, which would be somewhat misleading.

The relevance of the graph complex and the metric ribbon graph space is explained by the following theorem, which is one of the deepest results in mathematics in the past twenty years.

THEOREM 3.2. *For $g, n > 0$ (or $g = 0$ and $n > 2$), there is an orbifold isomorphism*
$$\mathcal{M}_{g,n} \times \mathbb{R}_+^n \cong G_{g,n}^{\mathrm{met}},$$
where $\mathcal{M}_{g,n}$ is the moduli space of compact smooth Riemann surfaces of genus g with n labeled punctures.

REMARK 7. The idea of such combinatorial description of the moduli space belongs to Thurston. D. Mumford suggested a different approach, realized by J. Harer [**Har86**], using the theory of Strebel differentials on a Riemann surface, an analytic result producing a unique meromorphic quadratic differential q in such a way that a metrized ribbon graph emerges as the set of critical horizontal trajectories of q and the residues of $\sqrt{q}$ at punctures give elements in $\mathbb{R}_+^n$, see more detail in E. Looijenga [**Loo95**], R. Hain-Looijenga [**HL97**], and M. Mulase-M. Penkava [**MP98**]. R. Penner [**Pen87**] came up with quite a different way to obtain the above result. In Penner's work the points of $\mathbb{R}_+^n$ on the left-hand side were interpreted as the hyperbolic lengths of horocycles attached to the punctures in a hyperbolic model of the Riemann surface with the punctures removed. Kontsevich [**Kon92**] considered a natural compactification of the space of metric ribbon graphs and compared it with Deligne-Mumford's one for $\mathcal{M}_{g,n} \times \mathbb{R}_+^n$.

Combining this theorem with Proposition 3.1, one gets the following purely combinatorial way of computing the rational homology of the moduli spaces.

COROLLARY 3.3. *The homology of the ribbon graph complex $G_\bullet^{g,n} \otimes \mathbb{Q}$ is isomorphic to the rational reduced homology of the one-point compactification of $\mathcal{M}_{g,n} \times \mathbb{R}_+^n$.*

Applying Poincaré-Lefschetz duality, we see that
$$H_k(G_\bullet^{g,n} \otimes \mathbb{Q}) = H^{6g-6+3n-k}(\mathcal{M}_{g,n}; \mathbb{Q}).$$

The results of this section have been used to estimate the homological dimension of $\mathcal{M}_{g,n}$ and compute its virtual (*i.e.*, orbifold) Euler characteristic, see Hain-Looijenga [**HL97**]. Hain and Looijenga also mention an application to Harer-Ivanov's stability theorem. Unfortunately, the combinatorics of the graph complex is too complicated to produce more general, explicit computations of its homology as of yet. However, this combinatorial model was one of the key elements in Kontsevich's proof [**Kon92**] of the Witten conjecture [**Wit91**] on the intersection theory on moduli spaces and the KP hierarchy.

References

[Ada78]   J. F. Adams, *Infinite loop spaces*, Princeton University Press, Princeton, NJ, 1978.

[BFF+78]   F. Bayen, M. Flato, C. Frønsdal, A. Lichnerowicz, and D. Sternheimer, *Deformation theory and quantization. I. Deformations of symplectic structures*, Ann. Physics **111** (1978), no. 1, 61–110.

[CF00]   A. S. Cattaneo and G. Felder, *A path integral approach to the Kontsevich quantization formula*, Comm. Math. Phys. **212** (2000), no. 3, 591–611. MR1 779 159

[CFT02]   A. S. Cattaneo, G. Felder, and L. Tomassini, *From local to global deformation quantization of Poisson manifolds*, Duke Math. J. **115** (2002), no. 2, 329–352, math.QA/0012228.

[CJ02]   R. L. Cohen and J. D. S. Jones, *A homotopy theoretic realization of string topology*, Math. Ann. **324** (2002), no. 4, 773–798, math.GT/0107187. MR1 942 249

[CS99]   M. Chas and D. Sullivan, *String topology*, Preprint, CUNY, November 1999, math.GT/9911159.

[CV86]   M. Culler and K. Vogtmann, *Moduli of graphs and automorphisms of free groups*, Invent. Math. **84** (1986), no. 1, 91–119. MR87f:20048

[FGV95]   M. Flato, M. Gerstenhaber, and A. A. Voronov, *Cohomology and deformation of Leibniz pairs*, Letters in Mathematical Physics **34** (1995), no. 1, 77–90.

[Ger63]   M. Gerstenhaber, *The cohomology structure of an associative ring*, Ann. of Math. (2) **78** (1963), 267–288.

[Get94]   E. Getzler, *Batalin-Vilkovisky algebras and two-dimensional topological field theories*, Commun. Math. Phys. **159** (1994), 265–285, hep-th/9212043.

[GK94]   V. Ginzburg and M. M. Kapranov, *Koszul duality for operads*, Duke Math. J. **76** (1994), 203–272.

[GK98]   E. Getzler and M. M. Kapranov, *Modular operads*, Compositio Math. **110** (1998), no. 1, 65–126, dg-ga/9408003. MR99f:18009

[Har86]   J. L. Harer, *The virtual cohomological dimension of the mapping class group of an orientable surface*, Invent. Math. **84** (1986), no. 1, 157–176. MR87c:32030

[HL97]   R. Hain and E. Looijenga, *Mapping class groups and moduli spaces of curves*, Algebraic geometry—Santa Cruz 1995, Amer. Math. Soc., Providence, RI, 1997, pp. 97–142. MR99a:14032

[HS93]   V. Hinich and V. Schechtman, *Homotopy Lie algebras*, Adv. Studies Sov. Math. **16** (1993), 1–18.

[Hua97]   Y.-Z. Huang, *Two-dimensional conformal geometry and vertex operator algebras*, Birkhäuser Boston, Boston, MA, 1997. MR98i:17037

[Kon92]   M. Kontsevich, *Intersection theory on the moduli space of curves and the matrix Airy function*, Comm. Math. Phys. **147** (1992), no. 1, 1–23. MR93e:32027

[Kon93]   _____, *Formal (non)-commutative symplectic geometry*, The Gelfand Mathematics Seminars, 1990-1992 (L. Corwin, I. Gelfand, and J. Lepowsky, eds.), Birkhäuser Boston, 1993, pp. 173–187.

[Kon03]   _____, *Deformation quantization of Poisson manifolds*, Lett. Math. Phys. **66** (2003), no. 3, 157–216, math.QA/9709180. MR2 062 626

[KSV96]   T. Kimura, J. Stasheff, and A. A. Voronov, *Homology of moduli spaces of curves and commutative homotopy algebras*, The Gelfand Mathematics Seminars, 1993–1995 (I. Gelfand, J. Lepowsky, and M. M. Smirnov, eds.), Birkhäuser Boston, 1996, pp. 151–170.

[Loo95]   E. Looijenga, *Cellular decompositions of compactified moduli spaces of pointed curves*, The moduli space of curves (Texel Island, 1994), Birkhäuser Boston, Boston, MA, 1995, pp. 369–400. MR96m:14031

[Mar96]   M. Markl, *Models for operads*, Comm. Algebra **24** (1996), no. 4, 1471–1500.

[ML65]   S. Mac Lane, *Categorical algebra*, Bull. Amer. Math. Soc. **71** (1965), 40–106. MR30 #2053

[MP98]   M. Mulase and M. Penkava, *Ribbon graphs, quadratic differentials on Riemann surfaces, and algebraic curves defined over $\overline{\mathbf{Q}}$*, Asian J. Math. **2** (1998), no. 4, 875–919. MR2001g:30028

[Pen86]   R. C. Penner, *The moduli space of a punctured surface and perturbative series*, Bull. Amer. Math. Soc. (N.S.) **15** (1986), no. 1, 73–77. MR87i:32032

[Pen87] _____, *The decorated Teichmüller space of punctured surfaces*, Comm. Math. Phys. **113** (1987), no. 2, 299–339. MR89h:32044

[Pen88] _____, *Perturbative series and the moduli space of Riemann surfaces*, J. Differential Geom. **27** (1988), no. 1, 35–53. MR89h:32045

[Sta93] J. Stasheff, *The intrinsic bracket on the deformation complex of an associative algebra*, J. Pure Appl. Algebra **89** (1993), 231–235.

[SV04] D. Sullivan and A. A. Voronov, *Brane topology*, Preprint, University of Minnesota, September 2004.

[SW03] P. Salvatore and N. Wahl, *Framed discs operads and Batalin-Vilkovisky algebras*, Q. J. Math. **54** (2003), no. 2, 213–231, math.AT/0106242.

[Wit91] E. Witten, *Two-dimensional gravity and intersection theory on moduli space*, Surveys in differential geometry (Cambridge, MA, 1990), Lehigh Univ., Bethlehem, PA, 1991, pp. 243–310. MR93e:32028

SCHOOL OF MATHEMATICS, UNIVERSITY OF MINNESOTA, MINNEAPOLIS, MN 55455
E-mail address: `voronov@umn.edu`

The ring of differential operators on forms in noncommutative calculus

Dmitri Tamarkin and Boris Tsygan

Contents

1. Introduction
2. Statement of the main theorem
2.1. Gerstenhaber algebras
2.2. Enveloping algebra of a Gerstenhaber algebra
2.3. The Hochschild cochain complex
2.4. The Gerstenhaber algebra $\mathcal{V}^\bullet(A)$
2.5. Hochschild chains
2.6. The A_∞ algebra $C_\bullet(C^\bullet(A))$
2.7. Statement of the main theorem
3. Sketch of the proof of the main theorem
3.1. Operads G, G_∞, G_{alg}, G_{geom}
3.2. Calculi
3.3. Pairings between chains and cochains
3.4. The calculus $\text{Calc}(A)$
3.5. Two-colored operads
3.6. Two-colored operads Calc, Calc_∞, $\text{Calc}_{\text{geom}}$, Calc_{alg}
3.7. Enveloping algebra of an algebra over a two-colored operad
3.8. The A_∞ module structure on Hochschild chains
3.9. The algebra $Y_{\text{Calc}^0_{\text{alg}}}(C^\bullet(A))$ and the Hochschild complex
3.10. The algebra $Y_{\text{Calc}^0_\infty}(C^\bullet(A))$ and the Hochschild complex
3.11. Another proof of quasi-isomorphicity of the map (3.11)
3.12. (3.10) is a quasi-isomorphism
3.13. The algebra $Y_{\text{Calc}_\infty}(C^\bullet(A))$ and the cyclic complex
4. Appendix. Hochschild-Gerstenhaber homology
4.1. Introductory remarks
4.2. Hochschild-Gerstenhaber homology vs Hochschild homology
References

2000 *Mathematics Subject Classification.* 16E40.
Key words and phrases. Hochschild cohomology, Gerstenhaber algebras, operads.

©2005 American Mathematical Society

1. Introduction

Many standard geometric objects associated to a manifold M can be defined in terms of the algebra A of functions on M. Such definitions can be often made in a manner that makes sense for any associative algebra A, commutative or not. The study and applications of these generalized geometric constructions is the subject of noncommutative geometry [**C**], [**M**].

For example, a vector field on a smooth manifold M can be viewed as a derivation of the algebra $A = C^\infty(M)$. If we require such derivations to be local, i.e. to preserve supports, then the Lie algebra of all such derivations is precisely the Lie algebra $\text{Vect}(M)$ of vector fields. One can say that the noncommutative version of $\text{Vect}(M)$ is $\text{Der}(A)$, the Lie algebra of derivations of A. Depending on the nature of A, one can impose on derivations some conditions like locality, continuity, etc.

Now let us try to define in a similar way the algebra $\mathcal{V}^\bullet(M)$ of multivector fields on M. The space of multivector fields has a structure of a Gerstenhaber algebra. In other words, it is a graded commutative associative algebra, i.e. the multiplication satisfies

$$ba = (-1)^{|a||b|} ab;$$

$\mathcal{V}^\bullet[1]$ is a graded Lie algebra, i.e.

$$[b, a] = -(-1)^{(|a|-1)(|b|-1)}[a, b]$$

and

$$[a, [b, c]] = [[a, b], c] + (-1)^{(|a|-1)(|b|-1)}[b, [a, c]];$$

and the two operations satisfy the Leibnitz identity

$$[a, bc] = [a, b]c + (-1)^{(|a|-1)|b|} b[a, c]$$

(cf. [**G**]).

Throughout the paper, for a complex $\mathcal{V}^\bullet$ with differential d, $\mathcal{V}[1]^k = \mathcal{V}^{k+1}$ is the complex with the differential $-d$; the ground ring k will be of characteristic zero. For any associative algebra A, one can construct a Gerstenhaber algebra [**G**]; the underlying space of this algebra is the Hochschild cohomology of A. We denote it by $H^\bullet(A, A)$ or simply by $H^\bullet(A)$. It was essentially proven in [**HKR**] that, when $A = C^\infty(M)$, this Gerstenhaber algebra becomes $\mathcal{V}^\bullet(M)$ if one understands the Hochschild cohomology properly. More precisely, $H^\bullet(A, A)$ for $A = C^\infty(M)$ is the cohomology of the complex of Hochschild cochains given by multi-differential expressions.

The problem with the above construction is that the corresponding algebra shrinks considerably as soon as A becomes noncommutative. Indeed, $H^0(A, A)$ is the center of A, and $H^1(A, A) = \text{Der}^{\text{out}}(A)$. So, for example, $H^0 = \mathbb{C}$ and $H^1 = 0$ for such an important algebra as $A = D(\mathbb{R}^n)$, the ring of differential operators on $\mathbb{R}^n$. (In fact for this algebra $H^i = 0$ for all $i > 0$).

The problem of constructing a noncommutative analog of the algebra of multivector fields cannot be very easy because the algebra A of "zero-fields" is noncommutative. Consider, for example, the standard Hochschild cochain complex $C^\bullet(A, A)$ (see §2.3). We denote it also by $C^\bullet(A)$. It is well known that $C^\bullet(A)[1]$ carries a bracket (called the Gerstenhaber bracket) which makes it a dg (differential graded) Lie algebra; $C^\bullet(A)$ carries the cup product which makes it a dg associative algebra; at the level of cohomology these two operations induce the standard

Gerstenhaber algebra structure on $H^\bullet(A)$. At the cochain level, however, the associative algebra $C^\bullet(A)$ is not commutative (it contains $A = C^0$ as a subalgebra).

A solution to this problem was proposed in [**T**]. It was shown there that the Gerstenhaber bracket and the cup product on $C^\bullet(A)$ are part of a much richer algebraic structure, namely that of a G_∞ algebra whose underlying L_∞ structure is given by the Gerstenhaber bracket (cf. Theorem 2.4.1 of the present paper). There are two equivalent ways to say that a complex $\mathcal{C}^\bullet$ is a G_∞ algebra. One can define a G_∞ structure on $\mathcal{C}^\bullet$ explicitly in terms of some multilinear operations on $\mathcal{C}^\bullet$, subject to some quadratic relations. Or, equivalently, one can say that there is a differential graded Gerstenhaber algebra $\mathcal{V}^\bullet$, quasi-isomorphic to $\mathcal{C}^\bullet$ as a complex. Applying this to $C^\bullet(A)$, one gets a a dg Gerstenhaber algebra $\mathcal{V}^\bullet(A)$ together with a quasi-isomorphism $\mathcal{V}^\bullet(A)[1] \to C^\bullet(A)[1]$ of dg Lie algebras. The above quasi-isomorphism identifies the cohomology of the complex $\mathcal{V}^\bullet(A)$ with $H^\bullet(A)$, and this identification is a Gerstenhaber algebra isomorphism. If $A = C^\infty(M)$, then the dg Gerstenhaber algebra $\mathcal{V}^\bullet(A)$ is quasi-isomorphic to the dg Gerstenhaber algebra $\mathcal{V}^\bullet(M)$, the algebra of multivector fields on M with zero differential. (In particular one gets a chain of quasi-isomorphisms of dg Lie algebras

$$C^\bullet(A)[1] \leftarrow \mathcal{V}^\bullet(A)[1] \to \mathcal{V}^\bullet(M)[1]$$

which implies the formality theorem of Kontsevich [**K**]).

The Gerstenhaber algebra $\mathcal{V}^\bullet(A)$ is given by a standard tensor construction independent of anything but the vector space A. The difficult part is to construct the differential on $\mathcal{V}^\bullet(A)$. It is given by a universal formula involving the product on A and some universal coefficients. For these coefficients there seems to be no canonical choice; one can define them if one chooses a Drinfeld associator [**D**].

An alternative way to formulate the theorem from [**T**] is to say that $C^\bullet(A)$ is a G_∞ algebra. The notion of a G_∞ algebra was introduced in [**GJ**]. By definition, a complex $C^\bullet$ is a G_∞ algebra if it carries multi-linear operations

$$m_{k_1,\ldots,k_n}: (C^\bullet)^{\otimes(k_1+\ldots+k_n)} \to C^\bullet$$

for every $n > 0$, $k_1, \ldots, k_n > 0$; these operations are assumed to satisfy certain symmetry conditions under permutations and certain quadratic equations (which amount to the Maurer-Cartan equation in a certain dg Lie algebra).

For the Hochschild complex $C^\bullet(A)$, m_1 is the differential, $m_{1,1}$ is the Gerstenhaber bracket, and m_2 is the symmetrized cup product. The higher operations are defined by universal formulas involving the product in A and some coefficients. The choice of these coefficients depends on a choice of a Drinfeld associator.

One can see from this discussion how difficult, inexplicit, and non-canonical the construction of the dg Gerstenhaber algebra $\mathcal{V}^\bullet(A)$ is. In view of applications to index theory and other topics, it is natural to ask whether this algebra has some features that are explicit and canonical. For example, as a dg Lie algebra, $\mathcal{V}^\bullet(A)[1]$ is quasi-isomorphic to $C^\bullet(A)[1]$.

In this paper we propose an answer which has a clear geometric meaning. For any Gerstenhaber algebra $\mathcal{A}^\bullet$, one can define an enveloping algebra $Y(\mathcal{A}^\bullet)$, which is a graded associative algebra equipped with a differential d. If $\mathcal{A}^\bullet = \mathcal{V}^\bullet(M)$, the algebra of multivector fields on M, then $Y(\mathcal{A}^\bullet) = D(\Omega^\bullet(M))$, the algebra of differential operators on differential forms on M. The differential d acts by commuting an operator with the De Rham differential.

For a dg Gerstenhaber algebra $(\mathcal{A}^\bullet, \delta)$, $Y(\mathcal{A}^\bullet)$ inherits a differential which we still denote by δ. By a dg algebra $Y(\mathcal{A}^\bullet)$ we always mean $(Y(\mathcal{A}^\bullet), \delta)$ (the differential d being ignored).

The first new result of this paper is Theorem 2.6.1. We construct an explicit canonical A_∞ algebra whose underlying complex is the Hochschild **chain** complex $C_\bullet(C^\bullet(A))$ of the dg associative algebra $C^\bullet(A)$. (Here, the dg algebra structure on $C^\bullet(A)$ is given by the differential and the cup product). In other words, one can construct canonically a dg associative algebra $\mathrm{D}(A)$, together with a quasi-isomorphism of complexes $\mathrm{D}(A) \to C_\bullet(C^\bullet(A))$. We show (Theorem 2.7.1) that there is an A_∞ quasi-isomorphism $Y(\mathcal{V}^\bullet(A)) \to C_\bullet(C^\bullet(A))$. One can interpret that as a dg algebra quasi-isomorphism of dg algebras $\tilde{Y} \to \mathrm{D}(A)$ where $\tilde{Y}$ is a canonically constructed dg algebra quasi-isomorphic to $Y(\mathcal{V}^\bullet(A))$.

The above discussion does not take into account the differential d on $Y(\mathcal{V}^\bullet(A))$. To include d into the picture, note that the A_∞ structure on $C_\bullet(C^\bullet(A))$ can be extended to $C_\bullet(C^\bullet(A))[[u]]$, *the negative cyclic complex* of the dg algebra $C^\bullet(A)$. We show that there is an A_∞ quasi-isomorphism

$$(Y(\mathcal{V}^\bullet(A))[[u]], \delta + ud) \to C_\bullet(C^\bullet(A))[[u]]$$

Here δ is the differential on $Y(\mathcal{V}^\bullet(A))$ induced by the differential on $\mathcal{V}^\bullet(A)$.

In other words one can say that the cyclic differential B on the Hochschild complex extends to an A_∞ derivation of $C_\bullet(C^\bullet(A))$ (or, if one prefers, to a derivation of $\mathrm{D}(A)$). This derivation is intertwined with the differential d on $Y(\mathcal{V}^\bullet(A))$.

An extensive sketch of the proof of the main theorem 2.7.1 is given in section 3. A complete proof will be given in a more detailed exposition.

Let us give one example of computing the cohomology ring of the algebra $C_\bullet(C^\bullet(A))$ for a noncommutative algebra A. Let M be a symplectic manifold and $A = (C^\infty(M)[[h]], *)$ its deformation quantization ([**BFFLS**]). The following is contained in [**NT**].

THEOREM 1.0.1. *If M is simply connected, then the cohomology ring of the algebra $C_\bullet(C^\bullet(A))$ is isomorphic to $H^\bullet(M^{S^1})[[h]]$ where M^{S^1} is the free loop space of M.*

Let us say a few words about the A_∞ algebras $C_\bullet(C^\bullet(A))$ and $C_\bullet(C^\bullet(A))[[u]]$. The binary products in these algebras were first introduced in [**NT**]. They were applied to index theorems in [**BNT**] and in [**NT1**]. In fact their existence and properties were indications that the Gerstenhaber algebra $\mathcal{V}^\bullet(A)$ might exist (cf., for example, Theorem 4.3 from [**NT**]). Some work on the higher operations in $C_\bullet(C^\bullet(A))$ was done in [**Ma**].

In fact one constructs both a G_∞ structure on $C^\bullet$ and a canonical A_∞ structure on $C_\bullet(C^\bullet)[[u]]$ where $C^\bullet$ is a dg algebra of a special type, namely *a brace algebra*. An interpretation of the algebra $C_\bullet(C^\bullet)[[u]]$ which is more invariant than ours is given in [**Kh**]. When $C^\bullet$ is commutative, one gets the standard shuffle product on Hochschild chains, extending to the A_∞ product of Getzler-Jones on negative cyclic chains.

It would be interesting to recover the Gerstenhaber algebra $\mathcal{V}^\bullet(A)$ from a less subtle associative algebra $Y(\mathcal{V}^\bullet(A))$. Note that $Y(\mathcal{V}^\bullet(A))$ has an increasing filtration F_n, $n \geq 0$, such that:
- $F_0 = \mathcal{V}^\bullet(A)$; $F_m F_n \subset F_{m+n}$;
- $\mathrm{gr}_F Y$ is graded commutative;

- $dF_n \subset F_{n+1}$

Given any algebra with such filtration, one can recover the Gerstenhaber algebra structure on F_0: the product comes from the one on Y, and the bracket is the derived bracket $[a, db]$. It would be interesting to understand how to construct directly a family of filtrations on the canonical algebra Y, indexed by Drinfeld associators.

Let us finish by outlining a few possible areas of study in the future.

1. Connes-Moscovici type index theorems. Computations similar to those in $C_\bullet(C^\bullet(A))[[u]]$ are used in [**CM**], [**C1**]. It would be very interesting to find a unified framework for both approaches. In particular, the symmetry group acting on the space of possible choices of $\mathcal{V}^\bullet(A)$ is the Grothendieck-Teichmuller group which is closely related to $\mathrm{Gal}(\overline{\mathbb{Q}}/\mathbb{Q})$. In [**CM**] and [**C1**], the symmetry with respect to the renormalization group was used. Part of Connes' program of noncommutative geometry is to unite the renormalization group and $\mathrm{Gal}(\overline{\mathbb{Q}}/\mathbb{Q})$ (the second author thanks Alain Connes for helpful comments on this subject).

2. Quantum cohomology and the Fukaya category. Starting from a compact symplectic manifold M, one can construct the quantum cohomology ring $HQ^\bullet(M)$ (cf., for example, [**MDS**] or [**KM**]) and the A_∞ category $\mathcal{F}(M)$ (cf. [**F1**]). Conjecturally, the former is the Hochschild cohomology of the latter (cf. [**Sei**]).

One can generalize the construction of the algebra $Y(C^\bullet(A))$, or in fact the construction of both sides in Theorem 2.7.1, and replace an algebra A by an A_∞ category $\mathcal{F}$. Thus, one gets an associative dg algebra

$$Y(M) = Y(C^\bullet(\mathcal{F}(M)))$$

Its cohomology algebra, possibly noncommutative, we denote by $\mathcal{H}Q^\bullet(M)$. If the conjecture from [**Sei**] is true, then we get a morphism of algebras

$$HQ^\bullet(M) \to \mathcal{H}Q^\bullet(M)$$

The right hand side should be closely related to the cohomology of the free loop space of M, as suggested by Theorem 1.0.1.

3. Topological string theory of Chas-Sullivan. It is strongly believed that, for an oriented compact manifold X, the chain complex $C_\bullet(X^{S^1})$ is a G_∞ algebra, even a BV_∞ algebra, cf. [**CS**]. It would be interesting to study the enveloping algebra $Y(\mathcal{V}^\bullet(X))$ where $\mathcal{V}^\bullet(X)$ is the standard resolution of $C_\bullet(X^{S^1})$. Denote its cohomology algebra by $\mathcal{H}^\bullet_{\mathrm{loop}}(X)$. One gets a morphism of algebras

$$H_{n-\bullet}(X^{S^1}) \to \mathcal{H}^\bullet_{\mathrm{loop}}(X)$$

The algebra $\mathcal{H}^\bullet_{\mathrm{loop}}(X)$, possibly noncommutative, should be related to the double loop space of X.

A link between points **2** and **3** seems to be indicated in [**Sei**].

Acknowledgements. Both authors' work was partially supported by NSF grants. The authors would like to thank A. Beilinson, P. Bressler, A. Connes, V. Drinfeld, B. Feigin, K. Fukaya, E. Getzler, M. Khalkhali, R. Nest and A. Voronov for stimulating discussions. The second author thanks the Dennisfest organizers for their hospitality and for the wonderful scientific atmosphere.

2. Statement of the main theorem

2.1. Gerstenhaber algebras.
Let k be the ground ring of characteristic zero. A *Gerstenhaber algebra* is a graded space $\mathcal{V}^\bullet$ together with

- A graded commutative associative algebra structure on $\mathcal{V}^\bullet$;
- a graded Lie algebra structure on $\mathcal{V}^{\bullet+1}$ such that

$$[a, bc] = [a, b]c + (-1)^{deg(a)deg(b)} b[a, c]$$

EXAMPLE 2.1.1. Let $\mathfrak{g}$ be a Lie algebra. Then

$$C_\bullet(\mathfrak{g}) = \wedge^\bullet \mathfrak{g}$$

is a Gerstenhaber algebra.

The product is the exterior product, and the bracket is the unique bracket which turns $C_\bullet(\mathfrak{g})$ into a Gerstenhaber algebra and which is the Lie bracket on $\mathfrak{g} = \wedge^1(\mathfrak{g})$.

EXAMPLE 2.1.2. Let M be a smooth manifold. Then

$$\mathcal{V}_M^\bullet = \wedge^\bullet T_M$$

is a sheaf of Gerstenhaber algebras.

The product is the exterior product, and the bracket is the Schouten bracket. We denote by $\mathcal{V}^\bullet(M)$ the Gerstenhaber algebra of global sections of this sheaf. The previous example is the algebra of left-invariant multivector fields on the Lie group of $\mathfrak{g}$.

2.2. Enveloping algebra of a Gerstenhaber algebra.
The following construction is motivated by Example 2.1.2. For a Gerstenhaber algebra $\mathcal{V}^\bullet$, let $Y(\mathcal{V}^\bullet)$ be the associative algebra generated by two sets of generators i_a, L_a, $a \in \mathcal{V}^\bullet$, both i and L linear in a,

$$|i_a| = |a|; \quad |L_a| = |a| - 1$$

subject to relations

$$i_a i_b = i_{ab}; \quad [L_a, L_b] = L_{[a,b]};$$

$$[L_a, i_b] = i_{[a,b]}; \quad L_{ab} = L_a i_b + (-1)^{|a|} i_a L_b$$

The algebra $Y(\mathcal{V}^\bullet)$ is equipped with the differential d of degree one which is defined as a derivation sending i_a to L_a and L_a to zero.

For a smooth manifold M one has a homomorphism

$$Y(\mathcal{V}^\bullet(M)) \to \mathrm{D}(\Omega^\bullet(M))$$

The right hand side is the algebra of differential operators on differential forms on M. It is easy to see that this is in fact an isomorphism.

2.3. The Hochschild cochain complex.
Let A be a graded associative algebra with unit 1 over a commutative unital ring k of characteristic zero. A Hochschild d-cochain is a linear map $A^{\otimes d} \to A$. Put, for $d \geq 0$,

$$C^d(A) = C^d(A, A) = \mathrm{Hom}_k(\overline{A}^{\otimes d}, A)$$

where $\overline{A} = A/(k \cdot 1)$ is the quotient linear k-space. Elements of $C^d(A)$ are called *normalized cochains*. We prefer to work with normalized cochains because the formulas for pairings between chains and cochains are simpler.

Put
$$|D| = (\text{degree of the linear map } D) + d$$
Put for cochains D and E from $C^\bullet(A, A)$
$$(D \smile E)(a_1, \ldots, a_{d+e}) = (-1)^{|E| \sum_{i \leq d}(|a_i|+1)} D(a_1, \ldots, a_d) \times$$
$$\times E(a_{d+1}, \ldots, a_{d+e});$$
$$(D \circ E)(a_1, \ldots, a_{d+e-1}) = \sum_{j \geq 0} (-1)^{(|E|+1) \sum_{i=1}^{j}(|a_i|+1)}$$
$$D(a_1, \ldots, a_j, E(a_{j+1}, \ldots, a_{j+e}), \ldots);$$
$$[D, E] = D \circ E - (-1)^{(|D|+1)(|E|+1)} E \circ D$$

These operations define the graded associative algebra $(C^\bullet(A, A), \smile)$ and the graded Lie algebra $(C^{\bullet+1}(A, A), [\,,\,])$ (cf. [**CE**]; [**G**]). Let
$$m(a_1, a_2) = (-1)^{\deg a_1} a_1 a_2;$$
this is a 2-cochain of A (not in C^2). Put
$$\delta D = [m, D];$$
$$(\delta D)(a_1, \ldots, a_{d+1}) = (-1)^{|a_1||D|+|D|+1} \times$$
$$\times a_1 D(a_2, \ldots, a_{d+1})+$$
$$+ \sum_{j=1}^{d} (-1)^{|D|+1+\sum_{i=1}^{j}(|a_i|+1)} D(a_1, \ldots, a_j a_{j+1}, \ldots, a_{d+1})$$
$$+ (-1)^{|D| \sum_{i=1}^{d}(|a_i|+1)} D(a_1, \ldots, a_d) a_{d+1}$$

One has
$$\delta^2 = 0; \quad \delta(D \smile E) = \delta D \smile E + (-1)^{\deg D} D \smile \delta E$$
$$\delta[D, E] = [\delta D, E] + (-1)^{|D|+1} [D, \delta E]$$
($\delta^2 = 0$ follows from $[m, m] = 0$).

Thus $C^\bullet(A, A)$ becomes a complex; we will denote it also by $C^\bullet(A)$. The cohomology of this complex is $H^\bullet(A, A)$ or the *Hochschild cohomology*.

We denote it also by $H^\bullet(A)$. The $\smile$ product induces the *Yoneda product* on $H^\bullet(A, A) = Ext^\bullet_{A \otimes A^0}(A, A)$. The operation $[\,,\,]$ is the Gerstenhaber bracket [**G**].

If (A, ∂) is a differential graded algebra then one can define the differential ∂ acting on $C^\bullet(A)$ by:
$$\partial D = [\partial, D]$$

THEOREM 2.3.1. [**G**] *The cup product and the Gerstenhaber bracket induce a Gerstenhaber algebra structure on $H^\bullet(A)$.*

For cochains D and D_i define a new Hochschild cochain by the following formula of Gerstenhaber ([**G**]) and Getzler ([**G1**]):

$$D_0\{D_1, \ldots, D_m\}(a_1, \ldots, a_n) =$$
$$= \sum (-1)^{\sum_{k \leq i_p}(|a_k|+1)(|D_p|+1)} D_0(a_1, \ldots, a_{i_1}, D_1(a_{i_1+1}, \ldots), \ldots,$$
$$D_m(a_{i_m+1}, \ldots), \ldots)$$

PROPOSITION 2.3.2. *One has*
$$(D\{E_1,\ldots,E_k\})\{F_1,\ldots,F_l\} = \sum (-1)^{\sum_{q\leq i_p}(|E_p|+1)(|F_q|+1)} \times$$
$$\times D\{F_1,\ldots,E_1\{F_{i_1+1},\ldots,\},\ldots,E_k\{F_{i_k+1},\ldots,\},\ldots,\}$$

The above proposition can be restated as follows. For a cochain D let $D^{(k)}$ be the following k-cochain of $C^\bullet(A)$:
$$D^{(k)}(D_1,\ldots,D_k) = D\{D_1,\ldots,D_k\}$$

PROPOSITION 2.3.3. *The map*
$$D \mapsto \sum_{k\geq 0} D^{(k)}$$
is a morphism of differential graded algebras
$$C^\bullet(A) \to C^\bullet(C^\bullet(A))$$

2.4. The Gerstenhaber algebra $\mathcal{V}^\bullet(A)$. Below is the theorem from [**T**]. We sketch its proof in 3.1.

THEOREM 2.4.1. *For every associative algebra A there exists a dg Gerstenhaber algebra $\mathcal{V}^\bullet(A)$ such that:*

- *There is a quasi-isomorphism of dg Lie algebras*
$$\mathcal{V}^\bullet(A)[1] \to C^\bullet(A)[1]$$

- *The above quasi-isomorphism induces an isomorphism of Gerstenhaber algebras*
$$H^\bullet(\mathcal{V}^\bullet(A)) \to H^\bullet(A)$$
where the Gerstenhaber structure on the right hand side is the standard one from 2.3.

- *For $A = C^\infty(M)$ there is a quasi-isomorphism of dg Gerstenhaber algebras*
$$\mathcal{V}^\bullet(A) \to \mathcal{V}^\bullet(M)$$

2.5. Hochschild chains. Let A be an associative dg algebra with unit 1 over a ground ring k. The differential on A is denoted by δ. Recall that by definition
$$\overline{A} = A/(k \cdot 1)$$
Set
$$C_p(A,A) = C_p(A) = A \otimes \overline{A}^{\otimes p}$$
Define the differentials $\delta: C_\bullet(A) \to C_\bullet(A)$, $b: C_\bullet(A) \to C_{\bullet-1}(A)$, $B: C_\bullet(A) \to C_{\bullet+1}(A)$ as follows.

$$\delta(a_0 \otimes \cdots \otimes a_p) = \sum_{i=1}^{p} (-1)^{\sum_{k<i}(|a_k|+1)+1}(a_0 \otimes \cdots \otimes \delta a_i \otimes \cdots \otimes a_p)$$

(2.1) $$b(a_0 \otimes \ldots \otimes a_p) = \sum_{k=0}^{p-1} (-1)^{\sum_{i=0}^{k}(|a_i|+1)+1} a_0 \ldots \otimes a_k a_{k+1} \otimes \ldots a_p$$
$$+ (-1)^{|a_p|+(|a_p|+1)\sum_{i=0}^{p-1}(|a_i|+1)} a_p a_0 \otimes \ldots \otimes a_{p-1}$$

$$(2.2) \quad B(a_0 \otimes \ldots \otimes a_p) = \sum_{k=0}^{p} (-1)^{\sum_{i \leq k}(|a_i|+1) \sum_{i \geq k}(|a_i|+1)} 1 \otimes a_{k+1} \otimes \ldots a_p \otimes$$
$$\otimes a_0 \otimes \ldots \otimes a_k$$

The complex $C_\bullet(A)$ is the total complex of the double complex with the differential $b + \delta$.

Let u be a formal variable of degree -2. The complex $(C^\bullet(A)[[u]], b + \delta + uB)$ is called *the negative cyclic complex* of A.

One can define a product

$$(2.3) \quad \mathrm{sh}: C^\bullet(A) \otimes C^\bullet(A) \to C^\bullet(A)$$

and its extension

$$(2.4) \quad \mathrm{sh} + u\,\mathrm{sh}' : C^\bullet(A)[[u]] \otimes C^\bullet(A)[[u]] \to C^\bullet(A)[[u]]$$

[**L**] by the following explicit formulas:

$$(2.5) \quad (a_0 \otimes \ldots \otimes a_p) \otimes (c_0 \otimes \ldots \otimes c_q) \stackrel{\mathrm{sh}}{\mapsto} a_0 c_0 \otimes \mathrm{sh}_{pq}(a_1, \ldots, a_p, c_1, \ldots, c_q)$$

where

$$(2.6) \quad \mathrm{sh}_{pq}(x_1, \ldots, x_{p+q}) = \sum_{\sigma \in \mathrm{Sh}(p,q)} \mathrm{sgn}(\sigma) x_{\sigma^{-1}1} \otimes \ldots \otimes x_{\sigma^{-1}(p+q)}$$

and

$$\mathrm{Sh}(p,q) = \{\sigma \in \Sigma_{p+q} \,|\, \sigma 1 < \ldots < \sigma p;\ \sigma(p+1) < \ldots < \sigma(p+q)\}$$

In the graded case, $\mathrm{sgn}(\sigma)$ gets replaced by the sign computed by the following rule: in all transpositions, the parity of a_i is equal to $|a_i| + 1$ if $i \neq 0$ and $|a_0|$ if $i = 0$, and similarly for c_i. A transposition contributes a product of parities.

$$(2.7) \quad (a_0 \otimes \ldots \otimes a_p) \otimes (c_0 \otimes \ldots \otimes c_q) \stackrel{\mathrm{sh}'}{\mapsto} 1 \otimes \mathrm{sh}'_{p+1,q+1}(a_0, \ldots, a_p, c_0, \ldots, c_q)$$

where

$$(2.8) \quad \mathrm{sh}'_{p+1,q+1}(x_0, \ldots, x_{p+q+1}) = \sum_{\sigma \in \mathrm{Sh}'(p+1,q+1)} \mathrm{sgn}(\sigma) x_{\sigma^{-1}0} \otimes \ldots \otimes x_{\sigma^{-1}(p+q+1)}$$

and $\mathrm{Sh}'(p+1, q+1)$ is the set of all permutations $\sigma \in \Sigma_{p+q+2}$ such that $\sigma 0 < \ldots < \sigma p$, $\sigma(p+1) < \ldots < \sigma(p+q+1)$, and $\sigma 0 < \sigma(p+1)$

2.6. The A_∞ algebra $C_\bullet(C^\bullet(A))$. Recall [**LS**], [**St1**] that an A_∞ algebra is a graded vector space $\mathcal{C}$ together with a Hochschild cochain m of total degree 1,

$$m = \sum_{n=1}^{\infty} m_n$$

where $m_n \in C^n(\mathcal{C})$ and

$$[m, m] = 0$$

Consider the Hochschild cochain complex of a graded algebra A as a differential graded associative algebra $(C^\bullet(A), \smile, \delta)$. Consider the Hochschild *chain* complex

of this differential graded algebra. The total differential in this complex is $b + \delta$; the degree of a chain is given by
$$|D_0 \otimes \ldots \otimes D_n| = |D_0| + \sum_{i=1}^{n}(|D_i| + 1)$$
where D_i are Hochschild cochains.

The complex $C_\bullet(C^\bullet(A))$ contains the Hochschild cochain complex $C^\bullet(A)$ as a subcomplex (of zero-chains) and has the Hochschild chain complex $C_\bullet(A)$ as a quotient complex:
$$C^\bullet(A) \xrightarrow{i} C_\bullet(C^\bullet(A)) \xrightarrow{\pi} C_\bullet(A)$$
(this sequence is not by any means exact). The projection on the right splits if A is commutative. If not, $C_\bullet(A)$ is naturally a graded subspace but not a subcomplex.

THEOREM 2.6.1. *There is an A_∞ structure on $C_\bullet(C^\bullet(A))[[u]]$ such that:*
- *All m_n are $k[[u]]$-linear, (u)-adically continuous*
- $m_1 = b + \delta + uB$
  For $x, y \in C_\bullet(A)$:
- $(-1)^{|x|} m_2(x,y) = (\mathrm{sh} + u\,\mathrm{sh}')(x,y)$
  For $D, E \in C^\bullet(A)$:
- $(-1)^{|D|} m_2(D, E) = D \smile E$
- $m_2(1 \otimes D, 1 \otimes E) + (-1)^{|D||E|} m_2(1 \otimes E, 1 \otimes D) = (-1)^{|D|} 1 \otimes [D, E]$
- $m_2(D, 1 \otimes E) + (-1)^{(|D|+1)|E|} m_2(1 \otimes E, D) = (-1)^{|D|+1}[D, E]$

Here is an explicit description of the above A_∞ structure. We define for $n \geq 2$
$$m_n = m_n^{(1)} + m_n^{(2)} + u m_n^{(3)}$$
where, for
$$a^{(k)} = D_0^{(k)} \otimes \ldots \otimes D_{N_k}^{(k)},$$
$$m_n^{(1)} = 0$$
for $n \geq 3$;
$$m_2^{(1)}(a^{(1)}, a^{(2)}) = (-1)^{|a^{(1)}|} \sum \pm D_0^{(1)} \smile D_0^{(2)}\{\ \} \otimes \underline{\ \ldots\ } \otimes$$
$$\otimes D_1^{(2)}\{\ \} \otimes \underline{\ \ldots\ } \ldots \otimes D_{N_2}^{(2)}\{\ \} \otimes \underline{\ \ldots\ }$$

The space designated by _ is filled with $D_1^{(1)}, \ldots, D_{N_1}^{(1)}$ whose order is preserved. The sign rule is as follows: the parity of $D_j^{(i)}$ is $|D_j^{(i)}|$ for $j = 0$ and $|D_j^{(i)}| + 1$ otherwise.

$$m_n^{(2)}(a^{(1)}, \ldots, a^{(n)}) =$$
$$= (-1)^{\sum_{i=1}^{n-1}|a_1|+n} \sum \pm D_{N_n}^{(n)}\{\ldots, D_0^{(1)}, \ldots, D_0^{(n-1)}\{\}, \ldots\} \smile$$
$$\smile D_0^{(n)}\{\_\} \otimes \underline{\ \ldots\ } \otimes D_1^{(n)}\{\ \} \otimes \underline{\ \ldots\ } \ldots \otimes D_{N_n-1}^{(n)}\{\ \} \otimes \underline{\ \ldots\ }$$

The space designated by _ is filled with $D_i^{(j)}$ for $j < n$ in such a way that:
- the cyclic order of each group $D_0^{(k)}, \ldots, D_{N_k}^{(k)}$ is preserved
- $D_0^{(1)}, \ldots, D_0^{(n-1)}$ are all inside the braces in $D_{N_n}^{(n)}\{\ \}$
- $D_0^{(i)}$ is to the left of $D_0^{(j)}$ for $i < j$

- any cochain $D_j^{(i)}$ may contain some of its neighbors on the right inside the braces, provided that all of these neighbors are of the form $D_q^{(p)}$ with $p < i$

The parity of $D_j^{(i)}$ is $|D_j^{(i)}|$ if $i = n$ and $j = 0$; it is $|D_j^{(i)}| + 1$ otherwise. Note that the formula for $m_n^{(2)}$ gives the Hochschild chain differential b for $n = 1$.

Finally, define
$$m_n^{(3)}(a^{(1)}, \ldots, a^{(n)}) = (-1)^{n+1} \sum \pm 1 \otimes \underline{\ldots \otimes D_0^{(0)} \otimes \ldots \otimes D_0^{(n)}\{\} \otimes \ldots}$$

The underlined space is filled with $D_i^{(j)}$ in such a way that:
- the cyclic order of each group $D_0^{(k)}, \ldots, D_{N_k}^{(k)}$ is preserved
- $D_0^{(i)}$ is to the left of $D_0^{(j)}$ for $i < j$
- any cochain $D_j^{(i)}$ may contain some of its neighbors on the right inside the braces, provided that all of these neighbors are of the form $D_q^{(p)}$ with $p < i$. The parity of $D_j^{(i)}$ is always $|D_j^{(i)}| + 1$.

One checks by a direct computation that the above formulas provide an A_∞ structure on $C_\bullet(C^\bullet(A))[[u]]$.

REMARK 2.6.2. Let A be a commutative algebra. Then $C_\bullet(A)[[u]]$ is not only a subcomplex but an A_∞ subalgebra of $C_\bullet(C^\bullet(A))[[u]]$. This A_∞ structure on $C_\bullet(A)[[u]]$ was introduced in [**GJ1**].

2.7. Statement of the main theorem.

THEOREM 2.7.1. *There is a A_∞ quasi-isomorphism*
$$Y(\mathcal{V}^\bullet(A)) \to C_\bullet(C^\bullet(A))$$
which extends to a $k[[u]]$-linear, (u)-adically continuous A_∞ quasi-isomorphism
$$(Y(\mathcal{V}^\bullet(A))[[u]], \delta + ud) \to C_\bullet(C^\bullet(A))[[u]]$$

3. Sketch of the proof of the main theorem

We will start by introducing several operads used in the proof of Theorem 2.4.1. We will then extend the notion of a Gerstenhaber algebra to that of a *calculus* (3.2). Next we will introduce some basic pairings between Hochschild chains and cochains, extending the definitions from 2.5. Then we will state Theorem 3.4.1 which asserts that the Gerstenhaber algebra $\mathcal{V}^\bullet(A)$ can be extended to a calculus $\mathrm{Calc}(A)$.

Next, we will introduce a notion of a two-colored operad which is suitable for working with objects like calculi (3.5). After that we extend the contents of 3.1 by introducing corresponding two-colored operads and stating relations among them (3.6).

Next, we define a notion of the enveloping algebra of an algebra $\mathcal{A}^\bullet$ over an operad $\mathcal{O}$, provided that $\mathcal{O}$ is part of a two-colored operad $\mathcal{P}$. This is an associative dg algebra which we denote by $Y_\mathcal{P}(\mathcal{A}^\bullet)$. Because of Theorem 3.4.1, one has morphisms of dg algebras

$$(3.1) \quad Y_{\mathrm{Calc}^0_{\mathrm{alg}}}(\mathcal{V}^\bullet(A)) \xleftarrow{\phi} Y_{\mathrm{Calc}^0_\infty}(C^\bullet(A)) \xleftarrow{\tau} Y_{\mathrm{Calc}^0_\infty}(\mathcal{V}^\bullet(A)) \xrightarrow{\psi} Y_{\mathrm{Calc}^0}(\mathcal{V}^\bullet(A))$$

where two-colored operads $\mathrm{Calc}^0_{\mathrm{alg}}$, Calc^0_∞, Calc^0 are defined in 3.6.

Next, we note that $C_\bullet(A)$ is an A_∞ module over the A_∞ algebra $C_\bullet(C^\bullet(A))[[u]]$. We interpret this result as the existence of an A_∞ morphism

$$C_\bullet(C^\bullet(A)) \to Y_{\text{Calc}^0_{\text{alg}}}(\mathcal{V}^\bullet(A))$$

Then we observe that this is a quasi-isomorphism. The maps ψ and τ in (3.1) are also quasi-isomorphisms; therefore we get an A_∞ morphism from $Y(\mathcal{V}^\bullet(A))$ to $C_\bullet(C^\bullet(A))$. To prove that this is a quasi-isomorphism, we are reduced to proving that ϕ is a quasi-isomorphism.

To that end, we study in some more detail the homology of the algebra $Y_{\text{Calc}^0_\infty}(\mathcal{A}^\bullet)$ for any Gerstenhaber algebra $\mathcal{A}^\bullet$ (Appendix). We show that this is a twisted version of the Hochschild homology, i.e. there is a spectral sequence starting with the latter and converging to the former. We observe that the map ϕ is filtered with respect to some filtration, and the fact that it is a quasi-isomorphism follows from considering the corresponding spectral sequence.

In the last subsection before Appendix, we modify the above arguments to prove the cyclic case of the main theorem.

3.1. Operads G, G_∞, G_{alg}, G_{geom}. Here we recall the scheme of the proof of Theorem 2.4.1 which was used in [**T**], [**T1**].

Gerstenhaber algebras are algebras over an operad which we will denote by G.. In other words, $G(n)$ is the graded k-module of all n-ary operations composed of the product and the bracket in a Gerstenhaber algebra, subject to all relations following from Gerstenhaber algebra axioms. The operad G is often denoted also by e_2.

By G_∞ we denote the standard free resolution of G. This is an operad in the category of complexes. One description of it is as follows. Consider a graded space $\mathcal{A}^\bullet$. Let us pretend for a moment that $\mathcal{A}^\bullet$ is finite-dimensional. Consider the free graded Lie algebra $\text{Lie}(\mathcal{A}^\bullet[1]^*)$ generated by the dual space to $\mathcal{A}^\bullet[1]$. Then the space $\mathcal{F}^\bullet(\mathcal{A}^\bullet) = \wedge^\bullet \text{Lie}(\mathcal{A}^\bullet[1]^*)$ carries the structure of a Gerstenhaber algebra (Example 2.1.1). In fact $\mathcal{F}^\bullet(\mathcal{A}^\bullet)$ is the free Gerstenhaber algebra generated by $\mathcal{A}^\bullet[1]^*$. A G_∞ structure on $\mathcal{A}^\bullet$ is by definition a derivation δ of the Gerstenhaber algebra $\mathcal{F}^\bullet(\mathcal{A}^\bullet)$ such that $|\delta| = 1$ and $\delta^2 = 0$.

REMARK 3.1.1. As stated, this definition has a problem if $\mathcal{A}^\bullet$ is infinite-dimensional: it involves various linear maps from $\mathcal{A}^\bullet[1]^*$ to tensor powers of $\mathcal{A}^\bullet[1]^*$, satisfying certain relations. What we actually mean are dual maps from tensor powers of $\mathcal{A}^\bullet[1]$ to $\mathcal{A}^\bullet[1]$, satisfying dual relations. A rigorous definition can be given in the dual language of coalgebras. The same remark applies to the definitions and constructions of 3.6, 3.10, 4 below.

Another, equivalent way of defining the operad G_∞ is to put

$$G_\infty = \text{Cobar}(G^{\text{dual}})$$

where G^{dual} is the Koszul dual operad as in [**GK**]. Note that the operad G is Koszul [**GJ**].

The operad G_{geom} is the little discs operad [**May**]. To pass from it to an operad in the category of complexes, one defines the operad $C_\bullet(G_{\text{geom}})$. It is known that the homology operad $H_\bullet(G_{\text{geom}})$ is isomorphic to G, see [**A**], [**Co**], [**GJ**].

The operad in the category of complexes G_{alg} is the operad of all universal operations on the Hochschild cochain complex of an associative algebra. It has various versions each of which suits our purposes ([**KS**], [**MS**], [**T1**]). Let us start with the version from [**MS**]. Consider all multi-linear operations on Hochschild cochains which are linear combinations of iterated compositions of the following elementary operations:

$$op(D)(a_1, \ldots, a_d) = a_1 D(a_2, \ldots, a_d)$$

$$op(D)(a_1, \ldots, a_d) = D(a_1, \ldots, a_i a_{i+1}, \ldots, a_d)$$

$$op(D)(a_1, \ldots, a_d) = D(a_1, \ldots, a_{d-1})a_d$$

$$op(D, E)(a_1, \ldots, a_{d+e-1}) = D(a_1, \ldots, a_i, E(a_{i+1}, \ldots, a_{i+e}), \ldots)$$

$$op(D, E)(a_1, \ldots, a_{d+e}) = D(a_1, \ldots, a_d) E(a_{d+1}, \ldots, a_{d+e})$$

(a minor technical point: this construction makes sense if one works with non-normalized Hochschild cochains $C^\bullet(A) = \operatorname{Hom}(A^{\otimes n}, A)$). One can arrange these operations into an operad in the category of complexes, cf. [**MS**]. In [**KS**], a different version of this operad is proposed, namely the minimal operad $\mathcal{M}$. It consists of all universal operations on Hochschild cochains which are linear combinations of compositions of the brace operations from 2.5, the cup product, and the higher cup products which are defined if A is an A_∞ algebra. Such n-ary operations are naturally indexed by rooted trees whose vertices are labeled by symbols $1, \ldots, n, m_i$, $i \geq 2$, in such a way that each label from 1 to n enters exactly once, and m_i may only label vertices with i outgoing edges. Finally, by $\mathcal{G}_{\text{alg}}$ we denote the standard free resolution of G_{alg}, the operad $\operatorname{Bar} \operatorname{Cobar}(G_{\text{alg}})$. For any operad $\mathcal{O}$ in the category of complexes, the operad $\operatorname{Bar} \operatorname{Cobar}(\mathcal{O})$ admits an explicit description as in [**KS**] (cf. also 3.6).

The relation between the above operads is as follows.

(3.2) $$G_\infty \xrightarrow{f_1} G_{\text{alg}} \xleftarrow{g_1} \mathcal{G}_{\text{alg}} \xrightarrow{g_2} C_\bullet(G_{\text{geom}}) \xrightarrow{F} G$$

The quasi-isomorphism g_2 can be deduced from [**KS**], from [**MS**], or [**T1**]; g_1 is the standard quasi-isomorphism between a resolution of an operad and the operad itself. F is the formality quasi-isomorphism from [**K1**], [**T1**]; it depends on a choice of a Drinfeld associator. The existence of f_1 follow from the fact that , thanks to the existence of F, G_∞ and $\mathcal{G}_{\text{alg}}$ are two free resolutions of G.

Therefore, since $\mathcal{G}_{\text{alg}}$ acts on $C^\bullet(A)$, we see that $C^\bullet(A)$ is a G_∞ algebra. This summarizes one of the versions of the proof of Theorem 2.4.1.

3.2. Calculi.

DEFINITION 3.2.1. *A precalculus is a pair of a Gerstenhaber algebra $\mathcal{V}^\bullet$ and a graded space $\Omega^\bullet$ together with*

- *a structure of a graded module over the graded commutative algebra $\mathcal{V}^\bullet$ on $\Omega^{-\bullet}$ (corresponding action is denoted by i_a, $a \in \mathcal{V}^\bullet$);*
- *a structure of a graded module over the graded Lie algebra $\mathcal{V}^{\bullet+1}$ on $\Omega^{-\bullet}$ (corresponding action is denoted by L_a, $a \in \mathcal{V}^\bullet$) such that*

$$[L_a, i_b] = i_{[a,b]}$$

and

$$L_{ab} = L_a i_b + (-1)^{|a|} i_a L_b$$

DEFINITION 3.2.2. *A calculus is a precalculus together with an operator d of degree 1 on $\Omega^\bullet$ such that $d^2 = 0$ and*
$$[d, i_a] = L_a.$$

EXAMPLE 3.2.3. For any manifold one defines a calculus $\mathrm{Calc}(M)$ with $\mathcal{V}^\bullet$ being the algebra of multivector fields, $\Omega^\bullet$ the space of differential forms, and d the De Rham differential.

EXAMPLE 3.2.4. For any associative algebra A one defines a calculus $\mathrm{Calc}_0(A)$ by putting $\mathcal{V}^\bullet = H^\bullet(A, A)$ and $\Omega^\bullet = H_\bullet(A, A)$. The five operations from Definition 3.2.2 are the cup product, the Gerstenhaber bracket, the pairings i_D and L_D, and the differential B, as in 3.3 below.

A differential graded (dg) calculus is a calculus with extra differentials δ of degree 1 on $\mathcal{V}^\bullet$ and b of degree -1 on $\Omega^\bullet$ which are derivations with respect to all the structures.

3.3. Pairings between chains and cochains.
For a graded algebra A, for $D \in C^d(A, A)$, define

$$(3.3) \quad i_D(a_0 \otimes \ldots \otimes a_n) = (-1)^{|D|\sum_{i \leq d}(|a_i|+1)} a_0 D(a_1, \ldots, a_d) \otimes a_{d+1} \otimes \ldots \otimes a_n$$

PROPOSITION 3.3.1.
$$[b, i_D] = i_{\delta D}$$
$$i_D i_E = (-1)^{|D||E|} i_{E \smile D}$$

Now, put

$$(3.4) \quad L_D(a_0 \otimes \ldots \otimes a_n) = \sum_{k=1}^{n-d} \epsilon_k a_0 \otimes \ldots \otimes D(a_{k+1}, \ldots, a_{k+d}) \otimes \ldots \otimes a_n +$$

$$\sum_{k=n+1-d}^{n} \eta_k D(a_{k+1}, \ldots, a_n, a_0, \ldots) \otimes \ldots \otimes a_k$$

(The second sum in the above formula is taken over all cyclic permutations such that a_0 is inside D). The signs are given by

$$\epsilon_k = (|D| + 1)\sum_{i=0}^{k}(|a_i| + 1)$$

and

$$\eta_k = |D| + 1 + \sum_{i \leq k}(|a_i| + 1) \sum_{i \geq k}(|a_i| + 1)$$

PROPOSITION 3.3.2.
$$[L_D, L_E] = L_{[D,E]}$$
$$[b, L_D] + L_{\delta D} = 0$$
$$[L_D, B] = 0$$

Now let us extend the above operations to the cyclic complex. Define
$$(3.5) \quad S_D(a_0 \otimes \ldots \otimes a_n) = \sum_{j \geq 0;\; k \geq j+d} \epsilon_{jk} 1 \otimes a_{k+1} \otimes \ldots a_0 \otimes \ldots \otimes$$
$$\otimes D(a_{j+1}, \ldots, a_{j+d}) \otimes \ldots \otimes a_k$$

(The sum is taken over all cyclic permutations for which a_0 appears to the left of D). The signs are as follows:
$$\epsilon_{jk} = |D|(|a_0| + \sum_{i=1}^{n}(|a_i|+1)) + (|D|+1)\sum_{j+1}^{k}(|a_i|+1) + \sum_{i \leq k}(|a_i|+1)\sum_{i \geq k}(|a_i|+1)$$

As we will see later, all the above operations are partial cases of a unified algebraic structure for chains and cochains, cf. 3.8; the sign rule for this unified construction was explained in 2.6.

PROPOSITION 3.3.3. *([R])*
$$[b + uB, i_D + uS_D] - i_{\delta D} - uS_{\delta D} = L_D$$

PROPOSITION 3.3.4. *([DGT])* *There exists a linear transformation $T(D, E)$ of the Hochschild chain complex, bilinear in $D, E \in C^\bullet(A,A)$, such that*
$$[b + uB, T(D,E)] - T(\delta D, E) - (-1)^{|D|}T(D, \delta E) =$$

$$= [L_D, i_E + uS_E] - (-1)^{|D|+1}(i_{[D,E]} + uS_{[D,E]})$$

3.4. The calculus Calc(A).

THEOREM 3.4.1. **[TT]** *For any associative algebra A one can define a dg calculus Calc(A) such that:*

1). As dg Lie algebras, $\mathcal{V}^{\bullet+1}(A)$ is quasi-isomorphic to the Hochschild cochain complex $C^{\bullet+1}(A)$. As dg modules over the dg Lie algebra $\mathcal{V}^{\bullet+1}(A)$, $\Omega^\bullet(A)[[u]]$ with the differential $\delta + uB$ is quasi-isomorphic to the negative cyclic complex $C_\bullet^-(A)[[u]]$ with the differential $b + uB$.

2). If $A = C^\infty(M)$ then there is a quasi-isomorphism of calculi
$$\mathrm{Calc}(A) \to \mathrm{Calc}(M)$$

3). For any A, the calculus $H^\bullet(\mathrm{Calc}(A))$ is isomorphic to the calculus $\mathrm{Calc}_0(A)$ from Example 3.2.4.

3.5. Two-colored operads.
The notion of a two-colored operad formalizes the situation when one has an algebra A over an operad $\mathcal{O}$, an object B, and a set $\mathcal{M}(n)$ of operations $A^{\otimes n} \otimes B \to B$. The union of $\mathcal{M}(n)$ is supposed to be closed under some natural operations.

More precisely, a two-colored operad $(\mathcal{O}, \mathcal{M})$ consists of an operad $\mathcal{O}$ and a collection of objects $\mathcal{M}(n)$, $n \geq 0$, together with an action of the symmetric group S_n on $\mathcal{M}(n)$ for all n, and with the operations

$$(3.6) \quad \mathcal{M}(k) \otimes \mathcal{O}(n_1) \otimes \ldots \otimes \mathcal{O}(n_k) \to \mathcal{M}(n_1 + \ldots + n_k)$$

$$(3.7) \quad \mathcal{M}(k) \otimes \mathcal{M}(l) \to \mathcal{M}(k+l)$$

subject to natural conditions of associativity and symmetry with respect to permutations.

An algebra over a two-colored operad $(\mathcal{O}, \mathcal{M})$ is a pair of objects (A, B) such that A is an $\mathcal{O}$-algebra, together with operations

(3.8) $$\mathcal{M}(n) \otimes A^{\otimes n} \otimes B \to B$$

subject to natural relations.

One can easily adapt the basic notions of the theory of operads to this context. For example, the free two-colored operad generated by a collection of $k[S_n]$-modules $\mathcal{Q}(n)$, $\mathcal{R}(n)$ is the pair $(\mathcal{O}, \mathcal{M})$ where $\mathcal{O}$ is the free operad generated by $\{\mathcal{Q}(n)\}$ and $\mathcal{M}(n)$ is described as follows. Consider all rooted trees with a chosen path $\mathbf{p}_0$ from the root to an external vertex $\mathbf{v}_0$. Let the external vertices other than $\mathbf{v}_0$ be numbered by $1, \ldots, n$. We will call such objects *two-colored trees*. For such a tree, put

$$\mathcal{M}(T) = \bigotimes_{\text{internal vertices } v} \mathcal{M}(v),$$

where
$$\mathcal{M}(v) = \mathcal{Q}(\#(\text{edges outgoing from } v))$$

if v is not on $\mathbf{p}_0$;

$$\mathcal{M}(v) = \mathcal{R}(\#(\text{edges outgoing from } v))$$

if v is on $\mathbf{p}_0$. $\mathcal{M}(n)$ is the direct sum of $\mathcal{M}(T)$ over all isomorphism classes of such trees.

If $(\mathcal{O}, \mathcal{M})$ is a two-colored dg operad, then the cofree two-colored cooperad cogenerated by $(\mathcal{O}, \mathcal{M})$ acquires a differential. This differential is the sum of the one induced from $(\mathcal{O}, \mathcal{M})$ and the new one, which sends an element of the direct summand corresponding to a (two-colored) tree T to the sum of elements in direct summands corresponding to trees T' obtained from T by contracting an internal edge. These elements are obtained from the original element by applying an appropriate composition in the operad $(\mathcal{O}, \mathcal{M})$. Thus one gets a two-colored cooperad in the category of complexes $\text{Cobar}(\mathcal{O}, \mathcal{M})$.

Dually, if $(\mathcal{O}, \mathcal{M})$ is a two-colored dg cooperad, one constructs a two-colored dg operad $\text{Bar}(\mathcal{O}, \mathcal{M})$. Its underlying space is a direct sum over (two-colored) trees, and the new component of the differential consists of inserting an internal edge in all possible positions, combined with an appropriate cooperadic cocomposition.

Composing these two constructions, one produces for a dg two-colored operad $(\mathcal{O}, \mathcal{M})$ a new dg two-colored operad $\text{Bar}\,\text{Cobar}(\mathcal{O}, \mathcal{M})$. This is the standard free resolution of $(\mathcal{O}, \mathcal{M})$, which means that it is free as an operad in the category of graded vector spaces and that there is a canonical quasi-isomorphism of operads

$$\text{Bar}\,\text{Cobar}(\mathcal{O}, \mathcal{M}) \to (\mathcal{O}, \mathcal{M}).$$

Explicitly (compare [**KS**]), this resolution is the direct sum of components numbered by (two-colored) trees whose edges are labeled by one of the two labels, *finite* or *infinite*. All external edges are infinite. The terms in the new differential are of two types: a) contracting a finite edge, combined with an operadic composition; b) making a finite edge infinite.

One can also define, following [**GK**], a Koszul dual cooperad of a two-colored operad, and extend the notion of a Koszul operad into the two-colored setting. For a Koszul two-colored operad $(\mathcal{O}, \mathcal{M})$, the dg operad $\text{Bar}((\mathcal{O}, \mathcal{M})^{\text{dual}})$ is a free resolution of $(\mathcal{O}, \mathcal{M})$.

3.6. Two-colored operads Calc, Calc_∞, $\text{Calc}_{\text{geom}}$, Calc_{alg}. In this section we will extend the method that was outlined in 3.1. This will enable us to prove both Theorem 3.4.1, 1)-3), and Theorem 2.7.1. To prove statement 4) of Theorem 3.4.1, some additional work is needed; we will give a proof in a subsequent paper.

By Calc, resp. Calc^0, we denote the two-colored operad in the category of graded spaces such that algebras over them are calculi (resp. precalculi). In other words, $\mathcal{O} = G$ and $\mathcal{M}(n)$ consists of all n-ary operations composed of i_a, L_a, and d (resp. i_a and L_a).

By Calc_∞ we denote the standard resolution of Calc. One can write it as

$$\text{Calc}_\infty = \text{Bar}\,\text{Calc}^{\text{dual}}$$

Similarly,

$$\text{Calc}^0_\infty = \text{Bar}\,\text{Calc}^{0\,\text{dual}}$$

(one can show that Calc and Calc^0 are Koszul).

Alternatively, one can give the following explicit definition.

A Calc_∞ algebra is a pair of graded vector spaces $(\mathcal{A}^\bullet, \mathcal{B}^\bullet)$ where $\mathcal{A}^\bullet$ is a G_∞ algebra, together with the following extra data. As in 3.1, let us pretend that $\mathcal{A}^\bullet, \mathcal{B}^\bullet$ are finite-dimensional (cf. Remark 3.1.1). Recall from 3.1 that one can define the dg Gerstenhaber algebra $\mathcal{F}^\bullet(\mathcal{A}^\bullet)$. Let $\Omega^{-\bullet}_{\mathcal{A}^\bullet, \mathcal{B}^\bullet}$ be the free graded $Y(\mathcal{F}^\bullet(\mathcal{A}^\bullet))$-module generated by $(\mathcal{B}^{-\bullet}[1])^*$. In other words, let $(\mathcal{F}^\bullet(\mathcal{A}^\bullet), \Omega^\bullet_{\mathcal{A}^\bullet, \mathcal{B}^\bullet})$ be the free precalculus generated by $(\mathcal{A}^{-\bullet}[1])^*$ and $(\mathcal{B}^{-\bullet}[1])^*$.

A Calc^0_∞ algebra structure on $(\mathcal{A}^\bullet, \mathcal{B}^\bullet)$ is a linear operator d on $\Omega^\bullet_{\mathcal{A}^\bullet, \mathcal{B}^\bullet}$ of square zero and degree one, such that $(\Omega^{-\bullet}_{\mathcal{A}^\bullet, \mathcal{B}^\bullet}, d)$ is a dg module over the dg algebra $(Y(\mathcal{F}^\bullet(\mathcal{A}^\bullet)), \delta)$.

Let u be a formal parameter of degree two. A Calc_∞ algebra structure on $(\mathcal{A}^\bullet, \mathcal{B}^\bullet)$ is a $k[[u]]$-linear, (u)-adically continuous operator d on $\Omega^\bullet_{\mathcal{A}^\bullet, \mathcal{B}^\bullet}[[u]]$ of square zero and degree one, such that $(\Omega^{-\bullet}_{\mathcal{A}^\bullet, \mathcal{B}^\bullet}[[u]], d)$ is a dg module over the dg algebra $(Y(\mathcal{F}^\bullet(\mathcal{A}^\bullet))[[u]], \delta + ud)$.

We define $\text{Calc}^0_{\text{geom}} = (\mathcal{O}, \mathcal{M})$ where $\mathcal{O}$ is the little discs operad and $\mathcal{M}$ is the configuration space, up to dilations, of cylinders $[0, r] \times S^1$ with n disjoint discs in the interior. The operations of type (3.6) consist of inserting little discs into the discs on the cylinder. The operations of type (3.7) consist of putting one cylinder on top of the other.

As for $\text{Calc}_{\text{geom}}$, $\mathcal{O}$ is the little discs operad and $\mathcal{M}$ is the configuration space, up to dilations and horizontal rotations, of cylinders $[0, r] \times S^1$ with a marked point on each component of the boundary and with n disjoint discs in the interior. The operations of type (3.6) consist of inserting little discs into the discs on the cylinder. The operations of type (3.7) consist of aligning marked points and then putting one cylinder on top of the other.

Now let us indicate how one defines $\text{Calc}^0_{\text{alg}}$ and Calc_{alg}. For them, $\mathcal{O} = G_{\text{alg}}$. To describe $\mathcal{M}$, consider all universal operations $C^\bullet(A)^{\otimes n} \otimes C_\bullet(A) \to C_\bullet(A)$ which are linear combinations of iterated compositions of the operations from G_{alg} with the following elementary operations:

$$a_0 \otimes \ldots \otimes a_p \mapsto 1 \otimes a_0 \otimes \ldots \otimes a_p$$

$$a_0 \otimes \ldots \otimes a_p \mapsto a_p \otimes a_0 \otimes \ldots \otimes a_{p-1}$$

$$a_0 \otimes \ldots \otimes a_p \mapsto a_0 a_1 \otimes a_2 \otimes \ldots \otimes a_p$$

$$(D, a_0 \otimes \ldots \otimes a_p) \mapsto a_0 \otimes D(a_1, \ldots, a_d) \otimes a_{d+1} \otimes \ldots \otimes a_p$$

(As in 3.1, to make it correct, one has to work with non-normalized chains and cochains).

For the two-colored operad $\text{Calc}_{\text{alg}}^0$, the $\mathcal{M}$ part consists only of the operations for which the term containing a_0 remains on the position number zero.

For example, the operation

$$(D_1, D_2, D_3, a_0 \otimes \ldots \otimes a_7) \mapsto D_1(a_7, a_0) D_3(a_1, D_2(a_2, a_3)) a_4 \otimes a_5 a_6$$

is in $\mathcal{M}$ for both Calc_{alg} and $\text{Calc}_{\text{alg}}^0$; the operation

$$(D_1, D_2, a_0 \otimes \ldots \otimes a_5) \mapsto a_1 a_2 \otimes a_3 \otimes D_2(a_4, D_1(a_5), a_0))$$

is in $\mathcal{M}$ for Calc_{alg} but not for $\text{Calc}_{\text{alg}}^0$ (the term containing a_0 is on the position number two); the operation

$$(D_1, D_2, D_3, a_0 \otimes \ldots \otimes a_5) \mapsto D_1(a_5 a_0) a_1 \otimes a_2 D_3(a_4, D_2(a_3))$$

is not in $\mathcal{M}$ for either (because the cyclic order of a_i's is broken).

As in 3.1, a minimal version in the manner of [**KS**] can be defined (and is necessary for the current version of the proof of the main theorem). We will discuss it in full in a more detailed exposition.

Finally, put

$$\mathcal{C}_{\text{alg}} = \text{Bar} \, \text{Cobar}(\text{Calc}_{\text{alg}})$$

Adapting the arguments from [**MS**], [**T1**] or from [**KS**] to our purposes, one proves that the following chain of quasi-isomorphisms extends that of 3.1:

$$(3.9) \qquad \text{Calc}_\infty \xrightarrow{f_1} \text{Calc}_{\text{alg}} \xleftarrow{g_1} \mathcal{C}_{\text{alg}} \xrightarrow{g_2} C_\bullet(\text{Calc}_{\text{geom}}) \xrightarrow{F} \text{Calc}$$

and similarly for Calc^0. As a corollary, one gets Theorem 3.4.1.

3.7. Enveloping algebra of an algebra over a two-colored operad. Let $\mathcal{P} = (\mathcal{O}, \mathcal{M})$ be a two-colored dg operad. For an $\mathcal{O}$-algebra $\mathcal{A}^\bullet$ put

$$Y_\mathcal{P}(\mathcal{A}^\bullet) = \oplus_{n>0} \mathcal{A}^{\bullet \otimes n} \otimes_{k[S_n]} \mathcal{M}(n) / \sim$$

where the equivalence relation $\sim$ is generated by the following:

for $a_1, \ldots, a_n \in \mathcal{A}^\bullet$, $m \in \mathcal{M}(k)$, $o \in \mathcal{O}(l)$, $k + l - 1 = n$, and for all i

$$(a_1 \otimes \ldots \otimes a_n) \otimes (o \circ_i m) \sim (a_1 \otimes \ldots \otimes a_{i-1} \otimes o(a_i, \ldots, a_{i+l-1}) \otimes \ldots \otimes a_n) \otimes m$$

taken with the appropriate sign, where $\circ_i : \mathcal{O}(l) \otimes \mathcal{M}(k) \to \mathcal{M}(n)$ is the ith elementary composition.

The operations $\mathcal{M}(k) \otimes \mathcal{M}(l) \to \mathcal{M}(k+l)$ turn $Y_\mathcal{P}(\mathcal{A}^\bullet)$ into an associative algebra.

EXAMPLE 3.7.1. . Let $\mathcal{P} = (\text{As}, \mathcal{M})$ be the two-colored operad algebras over which are pairs (A, B), where A is an associative algebra and B is a left A-module (resp. a right module, resp. a bimodule). Then $Y_\mathcal{P}(A) = A$ (resp. A^{op}, resp. $A \otimes A^{\text{op}}$).

EXAMPLE 3.7.2. . Let $\mathcal{P} = (\text{Lie}, \mathcal{M})$ be the two-colored operad algebras over which are pairs $(\mathfrak{g}, B)$ where $\mathfrak{g}$ is a Lie algebra and B is a $\mathfrak{g}$-module. Then $Y_\mathcal{P}(\mathfrak{g}) = U(\mathfrak{g})$.

EXAMPLE 3.7.3. Let $\mathcal{P} = \text{Calc}^0$. Then, for a Gerstenhaber algebra $\mathcal{A}^\bullet$, $Y_\mathcal{P}(\mathcal{A}^\bullet) = Y(\mathcal{A}^\bullet)$.

EXAMPLE 3.7.4. Let $\mathcal{P} = \text{Calc}$. Denote by $\mathfrak{a}$ a one-dimensional Abelian graded Lie algebra concentrated in degree one. This algebra acts on $Y(\mathcal{A}^\bullet)$ by derivations, the generator acting by d. One can form a cross product $U(\mathfrak{a}) \ltimes Y(\mathcal{A}^\bullet)$. For a Gerstenhaber algebra $\mathcal{A}^\bullet$,

$$Y_\mathcal{P}(\mathcal{A}^\bullet) \simeq U(\mathfrak{a}) \ltimes Y(\mathcal{A}^\bullet).$$

3.8. The A_∞ module structure on Hochschild chains. Recall the definition of A_∞ modules over A_∞ algebras. First, note that for a graded space $\mathcal{M}$, the Gerstenhaber bracket $[\,,\,]$ can be extended to the space

$$\text{Hom}(\overline{\mathcal{C}}^{\otimes \bullet}, \mathcal{C}) \oplus \text{Hom}(\mathcal{M} \otimes \overline{\mathcal{C}}^{\otimes \bullet}, \mathcal{M})$$

For a graded k-module $\mathcal{M}$, a structure of an A_∞ module over an A_∞ algebra $\mathcal{C}$ on $\mathcal{M}$ is a cochain

$$\mu = \sum_{n=1}^\infty \mu_n$$

$$\mu_n \in Hom(\mathcal{M} \otimes \overline{\mathcal{C}}^{\otimes n-1}, \mathcal{M})$$

such that

$$[m + \mu, m + \mu] = 0$$

THEOREM 3.8.1. *On $C_\bullet(A)[[u]]$, there exists a structure of an A_∞ module over the A_∞ algebra $C_\bullet(C^\bullet(A))[[u]]$ such that:*

- *All μ_n are $k[[u]]$-linear, (u)-adically continuous*
- *$\mu_1 = b + uB$ on $C_\bullet(A)[[u]]$*
  For $a \in C_\bullet(A)[[u]]$:
- *$\mu_2(a, D) = (-1)^{|a||D|+|a|}(i_D + uS_D)a$*
- *$\mu_2(a, 1 \otimes D) = (-1)^{|a||D|} L_D a$*
  For $a, x \in C_\bullet(A)[[u]]$:

$$(-1)^{|a|} \mu_2(a,x) = (\text{sh} + u\,\text{sh}')(a,x)$$

To obtain formulas for the structure of an A_∞ module from Theorem 3.8.1, one has to assume that, in the formulas for the A_∞ structure from Theorem 2.6.1, all $D_j^{(1)}$ are elements of A; then one has to replace braces $\{\ \}$ by the usual parentheses $(\ )$ symbolizing evaluation of a multi-linear map at elements of A.

3.9. The algebra $Y_{\text{Calc}^0_{\text{alg}}}(C^\bullet(A))$ and the Hochschild complex. Note that the construction from 3.8 can be interpreted as existence of an A_∞ morphism

(3.10) $$C_\bullet(C^\bullet(A)) \to Y_{\text{Calc}^0_{\text{alg}}}(C^\bullet(A))$$

From (3.9) one gets an algebra homomorphism

(3.11) $$Y_{\text{Calc}^0_\infty}(C^\bullet(A)) \to Y_{\text{Calc}^0_{\text{alg}}}(C^\bullet(A))$$

It is easy to show that both maps

(3.12) $$Y_{\text{Calc}^0_\infty}(C^\bullet(A)) \leftarrow Y_{\text{Calc}^0_\infty}(\mathcal{V}^\bullet(A)) \to Y_{\text{Calc}^0}(\mathcal{V}^\bullet(A))$$

are quasi-isomorphisms. To prove the first part of Theorem 2.7.1, it remains to show that (3.11) and (3.10) are quasi-isomorphisms under our assumptions.

3.10. The algebra $Y_{\mathrm{Calc}_\infty^0}(C^\bullet(A))$ and the Hochschild complex.

To show that (3.11) is a quasi-isomorphism, we introduce the following filtrations:

on $C_\bullet(C^\bullet(A))$, let

$$\text{(3.13)} \quad \mathrm{filt}_n = C_{\leq n}(C^\bullet(A)) = C^\bullet(A) \otimes \overline{C^\bullet(A)}^{\otimes \leq n}$$

On $Y_{\mathrm{Calc}_{\mathrm{alg}}^0}(C^\bullet(A))$ and $Y_{\mathrm{Calc}_\infty^0}(C^\bullet(A))$, let

$$\text{(3.14)} \quad \mathrm{filt}_n = \sum_{m \leq n} \mathcal{Q}(m) \otimes_{k[S_m]} C^\bullet(A)^{\otimes m} / \sim$$

It is easy to see that (3.10) preserves the filtration and therefore induces an isomorphism of associated graded quotients. The morphism (3.11) preserves the filtration by definition. At the level of associated graded quotients, (3.11) induces a morphism of complexes

$$\text{(3.15)} \quad Y_{\mathrm{Calc}_\infty^0}(H^\bullet(A)) \to C_\bullet(H^\bullet(A))$$

where, in the left hand side, $H^\bullet(A)$ is the Hochschild cohomology, viewed as a Gerstenhaber algebra on which all the operations are zero. The fact that (3.15) is an isomorphism follows from Proposition 4.2.4 in the Appendix.

3.11. Another proof of quasi-isomorphicity of the map (3.11).

We are going to prove a slightly more general statement.

PROPOSITION 3.11.1. *Let U be an arbitrary G_{alg}-algebra. Then the map $Y_{\mathrm{Calc}_\infty^0}(U) \to Y_{\mathrm{Calc}_{\mathrm{alg}}^0}(U)$ constructed in the same way as the map (3.11) is a quasi-isomorphism.*

This proof is based on two facts. The first one is that the map of colored operads f_1 from (3.9) is a quasi-isomorphism. The second fact says that M_{alg}^0 is free over G_{alg}. This means that one can choose S_n-equivariant subspaces $E(n) \subset M_{\mathrm{alg}}^0(n)$ (for any n) in such a way that for all N the insertion map

$$\bigoplus_{n, M_1 + M_2 + \cdots + M_n = N} M_{\mathrm{alg}}^0(n) \otimes_{S_n} G_{\mathrm{alg}}(M_1) \otimes \cdots \otimes G_{\mathrm{alg}}(M_n)$$

$$\otimes_{S_{M_1} \times \cdots \times S_{M_n}} k[S_N] \to M_{\mathrm{alg}}^0(N)$$

is an isomorphism. This fact follows from the explicit construction of $\mathrm{Calc}_{\mathrm{alg}}$.

Having these two facts, we prove the Proposition as follows. First, let us translate the definition of a universal enveloping algebra into the language of PROPs. Let $(\mathcal{O}, M)$ be a two-colored operad and U be an $\mathcal{O}$-algebra. Let $P_\mathcal{O}$ be the PROP generated by $\mathcal{O}$. Then the structure maps

$$\mathcal{O}(n_1) \otimes \cdots \otimes \mathcal{O}(n_m) \otimes M(m) \to M(n_1 + \cdots + n_m)$$

endow M with a structure of a functor $M' : P_\mathcal{O}^{\mathrm{op}} \to \mathrm{Complexes}$, where for $[n] \in \mathrm{Ob} P_\mathcal{O}$, $M'([n]) = M(n)$. Further on, we will denote M' by M. Put $F_U([n]) = U^{\otimes n}$. Then, since U is a $P_\mathcal{O}$-algebra, F_U is a functor $P_\mathcal{O} \to \mathrm{Complexes}$.

We see that

$$Y_{\mathcal{O}, M}(U) \cong F_U \otimes_{P_\mathcal{O}} M,$$

where on the right hand side we use the MacLane tensor product.

We have a quasi-isomorpism $f_1 : \mathrm{Calc}_\infty^0 \to \mathrm{Calc}_{\mathrm{alg}}^0$. It produces a symmetric monoidal functor $F : P_{G_\infty} \to P_{G_{\mathrm{alg}}}$ which induces a quasi-isomorphism on the spaces of homomorphisms; also, f_1 gives rise to a natural transformation $G : M_\infty^0 \to M_{\mathrm{alg}}^0 \circ F$ which is a quasi-isomorphism.

Let us come back to our G_{alg}-algebra U. Let U' be U considered as a G_∞-algebra, where the corresponding structure is induced by f_1. We have:

(3.16) $$F_{U'} \cong F_U \circ F.$$

In this light, the map $Y_{\mathrm{Calc}_\infty^0}(U') \to Y_{\mathrm{Calc}_{\mathrm{alg}}^0}(U)$, which coincides with the map (3.11), is described as the composition:

(3.17) $$M_\infty^0 \otimes_{P_{G_\infty}} F_{U'} \to M_{\mathrm{alg}}^0 \circ F \otimes_{P_{G_\infty}} F_U \circ F \to M_{\mathrm{alg}}^0 \otimes_{P_{G_{\mathrm{alg}}}} F_U,$$

where the first map is induced by G and (3.16).

We need to show that this composition produces a quasi-isomorphism. To this end, we introduce a functor $F_!$ from the category of functors $P_{G_\infty} \to$ Complexes to the category of functors $P_{G_{\mathrm{alg}}} \to$ Complexes. Denote $h_{[m]}([n]) = \hom_{P_{G_{\mathrm{alg}}}}([n], [m])$. Each $h_{[m]}$ is a functor $P_{G_{\mathrm{alg}}}^{\mathrm{op}} \to$ Complexes. For a functor $N : P_{G_\infty} \to$ Complexes set $F_! N([m]) := N \otimes_{P_{G_\infty}} h_{[m]}$. We have canonical isomorphisms

$$\hom_{P_{G_{\mathrm{alg}}}}(F_! N, M) \cong \hom_{P_{G_\infty}}(N, M \circ F)$$

for any $M : P_{G_{\mathrm{alg}}} \to$ Complexes; and

$$F_! M \otimes_{P_{G_{\mathrm{alg}}}} L \cong M \otimes_{P_{G_\infty}} (L \circ F).$$

In particular, the map G induces a map $G' : F_! M_\infty^0 \to M_{\mathrm{alg}}^0$. Since F is a quasi-isomorphism and M_∞^0 is semi-free and has a set of generators centered in non-positive degrees, it follows that G' is a quasi-isomorphism. One sees that the map in (3.17) can be rewritten as the following composition:

$$M_\infty^0 \otimes_{P_{G_\infty}} F_{U'} \cong M_\infty^0 \otimes_{P_{G_\infty}} F_U \circ F \cong F_! M_\infty^0 \otimes_{P_{G_{\mathrm{alg}}}} F_U \to M_{\mathrm{alg}}^0 \otimes_{P_{G_{\mathrm{alg}}}} F_U.$$

where the second map is induced by G'. Since G' is a quasi-isomorphism and M_{alg}^0 is semi-free and has a set of generators centered in non-positive degrees,the second map is a quasi-isomorphism, therefore, the whole composition is a quasi-isomorphism.

3.12. (3.10) is a quasi-isomorphism. Again, we we will replace $C^\bullet(A, A)$ with an arbitrary G_{alg}-algebra U. U has an associative cup-product, and we can form the Hochschild chain complex $C_\bullet(U)$ with respect to this product. The map (3.10) is then generalized to a map $\phi := \phi_U : C_\bullet(U) \to Y_{\mathrm{Calc}^0}(U)$, and we need to show that this map is a quasi-isomorphism. We are going to make a couple of reductions. Firstly, we can use the filtration as in (3.13) by the number of tensor factors of U. Then it suffices to check that ϕ induces a quasi-isomorphism of the associated graded spaces. This implies that it suffices to consider the case in which all operations on U vanish. To prove this statement we need one more reduction. Note that any commutative algebra B can be considered as a G_{alg}-algebra in which the cup product is commutative and all braces vanish. Denote thus obtained G_{alg}-algebra by B'. We have $U = B'$, where B has zero product. We now want to reduce this case to the case in which $U = (SV)'$, where SV is a free commutative algebra. To this end we notice that for B having zero product, there exists a semi-free resolution $p : SW \to B$ having the property that SW can be endowed with an increasing exhausting filtration F such that the associated graded algebra $\mathrm{Gr}\, SW$ is free. Since the functors $C_\bullet$ and Y_{Calc^0} preserve quasi-isomorphisms, it suffices to show that ϕ_{SW} is a quasi-isomorphism; using the filtration induced by F, we reduce the statement to quasi-isomorphicity of $\phi_{\text{Gr } SW}$, where Gr SW is free.

Thus, from now on we set $U = (SV)'$. In this case we have the Kostant-Hochschild-Rosenberg quasi-isomorphism $\omega : SV \otimes S(V[1]) \to C_\bullet(SV)$. Compute the composition $\omega\phi : SV \otimes S(V[1]) \to Y_{\text{Calc}^0}(SV)$. To this end, we are going to use the elements $i, L \in M^0_{\text{alg}}(1)$ defined in (3.2.1). Then it is easy to check that $\omega\phi$ is homotopy equivalent to the map ψ, which on an element $u \otimes (a_1 \wedge \cdots \wedge a_n)$, where $u \in SV$ and $a_i \in V$, takes the value

$$i_u \text{Alt}(L_{a_{i_1}} L_{a_{i_2}} ... L_{a_{i_n}}),$$

where Alt means alternation.

Let us now use the quasi-isomorphisms $f_1 : \text{Calc}^0_\infty \to \text{Calc}^0_{\text{alg}}$ from (3.10) and $f : Y_{\text{Calc}^0_\infty}(SV) \to Y_{\text{Calc}^0_{\text{alg}}}(SV)$ as in (3.11). Let $I' \in M^0_\infty(1)$, $L' \in M^0_\infty(1)$ be such that $f_1(I')$ (resp. $f_1(L')$) is homologous in $M^0_{\text{alg}}(1)$ to i (resp. L).

Define a map $\chi : SV \otimes S(V[1]) \to Y_{\text{Calc}^0_\infty}(SV)$ by setting

$$\chi(u \otimes (a_1 \wedge \cdots a_n)) = I'_u \text{Alt}(L'_{a_{i_1}} L'_{a_{i_2}} ... L'_{a_{i_n}}).$$

We see that ψ is homotopy equivalent to $f\chi$. Hence, our task is to show that χ is a quasi-isomorphism. To this end we will use the natural map $r : Y_{\text{Calc}^0_\infty}(SV) \to Y_{\text{Calc}^0}(SV)$ induced by the map of calculi $\text{Calc}^0_\infty \to \text{Calc}^0$. It is easy to see that this map is a quasi-isomorphism and that $r\chi$ is an isomorphism, therefore χ is a quasi-isomorphism, hence the statement.

3.13. The algebra $Y_{\text{Calc}_\infty}(C^\bullet(A))$ and the cyclic complex. To adapt the above arguments to the cyclic case, notice first that one can interpret the cyclic part of Theorem 2.6.1 as follows: there is an L_∞ action of the graded Lie algebra $\mathfrak{a}$ (cf. 3.7.4) on the A_∞ algebra $C_\bullet(C^\bullet(A))$, and one can form a corresponding A_∞ algebra cross product $U(\mathfrak{a}) \ltimes C_\bullet(C^\bullet(A))$. One of the ways to explain this is the following. The A_∞ algebra $C_\bullet(C^\bullet(A))$ is quasi-isomorphic to a dg associative algebra $\mathcal{R}$; the free dg Lie algebra resolution $\mathcal{L}$ of $\mathfrak{a}$ acts on $\mathcal{R}$ by derivations; form a cross product $U(\mathcal{L}) \ltimes \mathcal{R}$. As a complex, it is quasi-isomorphic to $U(\mathfrak{a}) \otimes C_\bullet(C^\bullet(A))$; from this, one can recover the A_∞ structure on the latter. For example, the binary product of the generator of $\mathfrak{a}$ with an element $c \in C_\bullet(C^\bullet(A))$ is Bc where B is the cyclic differential.

By Example 3.7.4,

$$Y_{\text{Calc}}(\mathcal{V}^\bullet(A)) \simeq U(\mathfrak{a}) \ltimes Y(\mathcal{V}^\bullet(A))$$

One extends (3.10) to a quasi-isomorphism

(3.18) $$U(\mathfrak{a}) \ltimes C_\bullet(C^\bullet(A)) \to Y_{\text{Calc}_{\text{alg}}}(C^\bullet(A))$$

One can also construct an A_∞ quasi-isomorphism

$$U(\mathfrak{a}) \ltimes Y_{\text{Calc}^0_{\text{alg}}}(C^\bullet(A)) \to Y_{\text{Calc}_{\text{alg}}}(C^\bullet(A))$$

Both maps

(3.19) $$Y_{\text{Calc}_\infty}(C^\bullet(A)) \leftarrow Y_{\text{Calc}_\infty}(\mathcal{V}^\bullet(A)) \to Y_{\text{Calc}}(\mathcal{V}^\bullet(A))$$

are quasi-isomorphisms. Thus, we have an A_∞ quasi-isomorphism from $U(\mathfrak{a}) \ltimes Y(\mathcal{V}^\bullet(A))$ to $U(\mathfrak{a}) \ltimes C_\bullet(C^\bullet(A))$. From this it is easy to deduce the statement of Theorem 2.7.1 regarding the cyclic complexes.

4. Appendix. Hochschild-Gerstenhaber homology

4.1. Introductory remarks. Let $\mathcal{A}^\bullet$ be a Gerstenhaber algebra (or, more generally, a G_∞ algebra). In this section we will construct $HG_\bullet(\mathcal{A}^\bullet)$, a new homology functor of $\mathcal{A}^\bullet$. It is defined by means of an explicit complex and is a limit of a spectral sequence whose E_1 term is the Hochschild homology of the graded associative algebra $\mathcal{A}^\bullet$. This spectral sequence degenerates at E_1 in important partial cases. The homology $HG_\bullet(\mathcal{A}^\bullet)$ is an associative algebra. When the spectral sequence does degenerate, one can view the associative algebra $HG_\bullet(\mathcal{A}^\bullet)$ as a deformation of the graded commutative algebra $H_\bullet(\mathcal{A}^\bullet)$ with the shuffle product sh, cf. (2.3).

One can extend the definition of $HG_\bullet$ and define the negative cyclic Gerstenhaber homology $HGC_\bullet^-(\mathcal{A}^\bullet)$. It relates to $HG_\bullet$ exactly as the Hochschild homology to the negative cyclic homology.

4.2. Hochschild-Gerstenhaber homology vs Hochschild homology. To define $HG_\bullet(\mathcal{A}^\bullet)$, recall the two-colored operad Calc^0 and its canonical free resolution Calc^0_∞ from 3.6.

DEFINITION 4.2.1. *For a G_∞ algebra $\mathcal{A}^\bullet$, let $HG_\bullet(\mathcal{A}^\bullet)$ be the homology of the complex $Y_{\mathrm{Calc}^0_\infty}(\mathcal{A}^\bullet)$.*

Let us start by realizing $Y_{\mathrm{Calc}^0_\infty}(\mathcal{A}^\bullet)$ as an explicit complex. Recall from 3.1 that a G_∞ algebra structure on $\mathcal{A}^\bullet$ determines a derivation δ of the Gerstenhaber algebra $\mathcal{F}^\bullet(\mathcal{A}^\bullet) = \wedge^\bullet \mathrm{Lie}(\mathcal{A}^\bullet[1]^*)$. (Here and in all the computations below, as usual, one has to understand the duals properly; cf. Remark 3.1.1). For an associative augmented dg algebra Y, we denote by $\mathrm{Bar}(Y)$ its standard bar (bi)complex which computes $\mathrm{Ext}^\bullet(k,k)$.

PROPOSITION 4.2.2. *There is a natural isomorphism of complexes*
$$Y_{\mathrm{Calc}^0_\infty}(\mathcal{A}^\bullet) \to \mathrm{Bar}(Y(\wedge^\bullet \mathrm{Lie}(\mathcal{A}^\bullet[1]^*)))$$

This can be seen directly from the definitions. The right hand side in (4.2.2) stands for the enveloping algebra of the dg Gerstenhaber algebra. It is an associative dg algebra, with the differential induced by δ.

In what follows, A_∞ is a Gerstenhaber algebra.

THEOREM 4.2.3. *There is a natural spectral sequence converging to $HG_\bullet(\mathcal{A}^\bullet)$ for which*
$$E_1 = E_2 = H_\bullet(\mathcal{A}^\bullet),$$
the Hochschild homology of the graded algebra A_∞.

An important partial case is the following:

PROPOSITION 4.2.4. *Let A_∞ is a Gerstenhaber algebra on which both operations are zero. Then*
$$HG_\bullet(\mathcal{A}^\bullet) \xrightarrow{\sim} H_\bullet(\mathcal{A}^\bullet) = C_\bullet(\mathcal{A}^\bullet)$$

Proof of Theorem 4.2.2. Introduce a filtration on the standard complex $Y_{\mathrm{Calc}^0_\infty}(\mathcal{A}^\bullet)$: note that $Y(\wedge^\bullet \mathrm{Lie}(\mathcal{A}^\bullet[1]^*))$ is graded by the number of generators i_a occurring in a monomial, and let F_n consist of all those cochains that annihilate all elements of degree greater than n.

PROPOSITION 4.2.5. *The spectral sequence associated to the filtration F has the properties as in Theorem 4.2.3.*

Denote
$$\mathcal{L}(\mathcal{A}^\bullet) = \text{Lie}(\mathcal{A}^\bullet[1]^*)$$
To prove the above Proposition, let us start by reinterpreting $Y(\wedge^\bullet \mathcal{L}(\mathcal{A}^\bullet))$.

Let ϵ be a formal parameter of degree -1 and square zero. Consider the dg Lie algebra $\mathcal{L}(\mathcal{A}^\bullet)[\epsilon]$. The universal enveloping algebra $U(\mathcal{L}(\mathcal{A}^\bullet)[\epsilon])$ admits a derivation which is characterized by the following. On $U(\mathcal{L}(\mathcal{A}^\bullet)\epsilon) = \wedge^\bullet(\mathcal{L}(\mathcal{A}^\bullet))$, this is just the derivation δ determined by the G_∞ structure; and it is the only such derivation which commutes with the derivation induced by $\frac{\partial}{\partial \epsilon}$. Thus, $U(\mathcal{L}(\mathcal{A}^\bullet)[\epsilon])$ becomes a dg associative algebra.

LEMMA 4.2.6. *There is a natural isomorphism of dg algebras*
$$Y(\wedge^\bullet(\mathcal{L}(\mathcal{A}^\bullet))) \to U(\mathcal{L}(\mathcal{A}^\bullet)[\epsilon])$$

The proof is straightforward.

Now we have an identification
$$Y_{\text{Calc}_\infty^0}(\mathcal{A}^\bullet) \simeq \text{Bar}(U(\mathcal{L}(\mathcal{A}^\bullet)[\epsilon]))$$
Under this identification, the filtration F becomes as follows. Note that $U(\mathcal{L}(\mathcal{A}^\bullet)[\epsilon])$ is graded by the number of factors ϵ in a monomial, and F_n consists of cochains annihilating all elements of degree greater than n.

In the situation of Proposition 4.2.4, the differential on $U(\mathcal{L}(\mathcal{A}^\bullet)[\epsilon])$ is zero. So, to finish the proof of Proposition 4.2.5, and therefore of Theorem 2.7.1, on can go directly to Lemma 4.2.10.

Note that the graded Lie algebra $\mathcal{L}(\mathcal{A}^\bullet)$ possesses a Lie algebra derivation determined by the commutative product on $\mathcal{A}^\bullet$ (or, in the general G_∞ case, from the C_∞ structure on $\mathcal{A}^\bullet$). Thus $\mathcal{L}(\mathcal{A}^\bullet)$, as well as $\mathcal{L}(\mathcal{A}^\bullet)[\epsilon]$, becomes a dg Lie algebra. One sees easily that the following is true.

LEMMA 4.2.7. *The first term of the spectral sequence associated to the filtration F is equal to $H^\bullet(\mathcal{L}(\mathcal{A}^\bullet)[\epsilon])$, the cohomology of the differential graded Lie algebra with trivial coefficients.*

LEMMA 4.2.8. *In the above spectral sequence $E_2 = E_1$.*

To prove this, we need some notation. Let $\mathfrak{g}$ be a dg Lie algebra. By PBW, there is an $\text{ad}(\mathfrak{g})$- invariant isomorphism $U(\mathfrak{g}) \to S(\mathfrak{g})$. For any i, let $U^{>i}(\mathfrak{g})$ be the pre-image of $S^{>i}(\mathfrak{g})$ under this isomorphism. In particular, $U^{>1}(\mathfrak{g})$ is a $\text{ad}(\mathfrak{g})$-complement of $\mathfrak{g}$ in the augmentation ideal.

The differential on $\mathfrak{g}$ induces a differential on $U(\mathfrak{g})$. Denote this differential by δ. Let δ_1 be another derivation of degree one of $U(\mathfrak{g})$. Assume that, for any $X \in \mathfrak{g}$,

(4.1) $$\delta X - \delta_1 X \in U^{>1}(\mathfrak{g})$$

Since the cohomology $H^\bullet(\mathfrak{g})$ is computed by the bar complex $\text{Bar}(U(\mathfrak{g}))$, the new differential δ_1 acts on $H^\bullet(\mathfrak{g})$.

LEMMA 4.2.9. *Under the assumption (4.1), the action of δ_1 on $H^\bullet(\mathfrak{g})$ is trivial.*

Proof. Recall how the isomorphism between $H^\bullet(\mathfrak{g})$ and the cohomology of the bar complex is constructed. One has two standard resolutions of the trivial $U(\mathfrak{g})$-module k: The bar resolution $\text{Bar}_n = U(\mathfrak{g}) \otimes \overline{U(\mathfrak{g})}^{\otimes n}$ and the Koszul resolution

$K_n = U(\mathfrak{g}) \otimes \wedge^n(\mathfrak{g})$. One has the standard embedding of complexes $i : K_\bullet \to \mathrm{Bar}_\bullet$ defined by
$$f \otimes (x_1 \wedge \ldots \wedge x_n) \mapsto f \otimes \mathrm{Alt}(x_1 \otimes \ldots \otimes x_n)$$
for $f \in U(\mathfrak{g})$ and $x_i \in \mathfrak{g}$. By the standard techniques of homological algebra, one can split this embedding and construct a projection $j : \mathrm{Bar}_\bullet \to K_\bullet$, so that $ji = 1$ and ij is homotopic to the identity as a morphism of complexes of $U(\mathfrak{g})$-modules. It is not difficult to see that one can pick j that annihilates modulo $U^{>0}(\mathfrak{g}) \cdot K_\bullet$ all elements $f \otimes \mathrm{Alt}(x_1 \otimes \ldots \otimes x_n)$ where $f \in U(\mathfrak{g})$, $x_1 \in U^{>1}(\mathfrak{g})$, and $x_i \in \mathfrak{g}$ for $i > 1$. But the differential induced by δ_1 sends the image of i precisely to such elements. This shows that δ_1 acts by zero on the cohomology of the complex $\mathrm{Hom}_{U(\mathfrak{g})}(\mathrm{Bar}_\bullet, k)$ which computes $H^\bullet(\mathfrak{g})$.

It remains to show that

LEMMA 4.2.10. *For a unital Gerstenhaber algebra* $\mathcal{A}^\bullet$
$$H^\bullet(\mathcal{L}(\mathcal{A}^\bullet)[\epsilon]) \simeq H_\bullet(\mathcal{A}^\bullet)$$

Proof. First, $C^\bullet(\mathcal{L}(\mathcal{A}^\bullet)[\epsilon])$ is isomorphic to $C^\bullet(\mathcal{L}(\mathcal{A}^\bullet), S^\bullet(\mathcal{L}(\mathcal{A}^\bullet))^*)$, the standard cochain complex with coefficients in $S^\bullet(\mathcal{L}(\mathcal{A}^\bullet))^*$. Since $\mathcal{L}(\mathcal{A}^\bullet)$ is free as a graded Lie algebra, the embedding into $C^\bullet(\mathcal{L}(\mathcal{A}^\bullet), S^\bullet(\mathcal{L}(\mathcal{A}^\bullet))^*)$ of the subcomplex

(4.2) $$C^o \to \mathrm{Ker}(C^1 \to C^2)$$

is a quasi-isomorphism where
$$C^i = \mathrm{Hom}(\wedge^i(\mathcal{L}(\mathcal{A}^\bullet)), S^\bullet(\mathcal{L}(\mathcal{A}^\bullet))^*)$$

Finally, the complex (4.2) can be written explicitly: let, as before, $C_\bullet(\mathcal{A}^\bullet)$ be the usual Hochschild complex, and let $C'_\bullet(\mathcal{A}^\bullet)$ be the same complex equipped with the bar differential b', cf. [**L**]. Then one can identify C_0 with $C'_\bullet(\mathcal{A}^\bullet)$, $\mathrm{Ker}(C^1 \to C^2)$ with $C_\bullet(\mathcal{A}^\bullet)$, and the differential between the two with the map $1-t$ from [**L**]. The identification is done as follows:
$$S(\mathcal{L}(\mathcal{A}^\bullet)) \xrightarrow{PBW} U(\mathcal{L}(\mathcal{A}^\bullet)) \simeq T(\mathcal{A}^\bullet[1]^*);$$
therefore
$$C^0 \simeq T(\mathcal{A}^\bullet[1]) = C'_\bullet(\mathcal{A}^\bullet)$$
and
$$\mathrm{Ker}(C^1 \to C^2) \simeq \mathcal{A}^\bullet[1] \otimes T(\mathcal{A}^\bullet[1]) \simeq C_\bullet(\mathcal{A}^\bullet)[1]$$
Since $\mathcal{A}^\bullet$ is unital, $C'_\bullet(\mathcal{A}^\bullet)$ is contractible, and the theorem is proven.

References

[A] V. Arnold, *The cohomology ring of the colored braid group*, Mat. Zametki **5** (1969), 227-231.

[BB] A. Beilinson and J. Bernstein, *A proof of Jantzen conjectures*, Advances in Soviet Mathematics, **16**, Part 1 (1993), 1-50.

[BV] J. Boardman, R. Vogt, *Homotopy invariant algebraic structures on topological spaces*, LNM **347** (1973).

[BFFLS] F. Bayen, M. Flato, C. Fronsdal, A. Lichnerowicz, D. Sternheimer, *Deformation Theory and Quantization*, Ann. Phys. **111** (1977). p. 61-151

[BNT] P. Bressler, R. Nest, B. Tsygan, *Riemann-Roch theorems via deformation quantization I and II*, Advances in Mathematics **167** (2002), 1, 1-73.

[CE] H. Cartan, S. Eilenberg, *Homological Algebra*, Princeton, 1956

[CS] M. Chas, D. Sullivan, *String topology*, preprint GT/9911159.

[Co] F. Cohen, *The cohomology of C_{n+1} spaces*, $n \geq 0$, Lecture Notes in Math. **533** (1976), 207-351.

[C] A. Connes, *Noncommutative Geometry*, New York-London, Academic Press, 1994.

[C1] A. Connes, *Noncommutative geometry - year 2000*, GAFA 2000, special volume, Part II, 481-559.

[CM] A. Connes, H. Moscovici, *Cyclic homology and Hopf algebras*, Lett. Math. Phys **48** (1999), 97-108.

[CM1] A. Connes, H. Moscovici, *Hopf algebras, cyclic homology and the transverse index theorem*, Com. Math. Phys. **198** (1998), 199-246.

[CM2] A. Connes, H. Moscovici, *The local index formula in noncommutative geometry*, GAFA **5** (1995), 174-243.

[D] V. Drinfeld, *On quasi-triangular Hopf algebras and on a group that is closely connected with* $\mathrm{Gal}(\bar{\mathbb{Q}}/\mathbb{Q})$, Leningrad J. Math. **2** (1991), 4, 829-860.

[DT] Yu. Daletski, B. Tsygan, *Operations on cyclic and Hochschild complexes*, Methods Funct. Anal. Topology **5** (1999), 4, 62-86.

[DGT] Yu. Daletski, I. Gelfand and B. Tsygan, *On a variant of noncommutative geometry*, Soviet Math. Dokl. **40** (1990), 2, 422-426.

[FM] W. Fulton and R. MacPherson, *A compactification of the configuration spaces*, Ann. Math. **139** (1994), 183-225.

[FT] B. Feigin and B. Tsygan, Additive K-theory, LMN 1289 (1987), 66-220.

[FT3] B. Feigin and B. Tsygan, *Additive K-theory and crystalline cohomology*, Funct. Anal. and Appl. **19** (1985).

[F] K. Fukaya, *Floer homology of Lagrangian foliation and noncommutative mirror symmetry*, Kyoto University preprint 98-08.

[F1] K. Fukaya, *Deformation theory, homological algebra, and mirror symmetry*, Kyoto University preprint, December, 2001.

[G] M. Gerstenhaber, *The cohomology structure of an associative ring*, Ann. Math. **78** (1963), 267-288.

[GS] M. Gerstenhaber, S. Schack, *The shuffle bialgebra and the cohomology of commutative algebras*, Journal of Pure and Applied Algebra **70** (1991), 263-272.

[GV] M. Gerstenhaber, A. Voronov, *Homotopy G-algebras and moduli space operad*, IMRN (1995), 141-153

[G1] E. Getzler, *Cartan homotopy formulas and the Gauss-Manin connection in cyclic homology*, Israel Math. Conf. Proc., **7**, 65-78.

[GJ] E. Getzler, J. Jones, *Operads, homotopy algebra and iterated integrals for double loop spaces*, preprint hep-th9403055.

[GJ1] E. Getzler and J. Jones, *A_∞ algebras and the cyclic bar complex*, Illinois J. of Math. **34** (1990), 256-283.

[GK] V. Ginzburg and M. Kapranov, *Koszul duality for operads*, Duke Mathematical Journal **76** (1994), 1, 203-272.

[HJ] C.E. Hood and J.D.S. Jones, *Some algebraic properties of cyclic homology groups*, K - Theory, **1** (1987), 361-384.

[HKR] G. Hochschild, B. Kostant, and A. Rosenberg, *Differential forms on regular affine algebras*, Transactions AMS **102** (1962), 383-408.

[Kh] M. Khalkhali, *An approach to operations in cyclic homology*, Journal of Pure Appl. Alg. **107** (1996), 1, 47-59.

[K1] M. Kontsevich, *Operads and motives in deformation quantization*, Lett. Math. Phys. **48** (1999), no. 1, 35-72.

[K] M. Kontsevich, *Deformation quantization of Poisson manifolds I*, preprint QA/9709040.

[KM] M. Kontsevich and Yu. Manin, *Gromov-Witten classes, quantum cohomology and enumerative geometry*, Comm. Math. Phys. **164** (1994), 525-562.

[KS] M. Kontsevich and Y. Soibelman, *Deformations of algebras over operads and Deligne conjecture*, EuroConference Moshé Flato 1999, Vol.I (Dijon), 255-307.

[KS1] M. Kontsevich and Y. Soibelman, *Deformations of algebras over operads and Deligne conjecture*, EuroConference Moshé Flato 2000, Part III (Dijon), Lett. Math. Phys. 56 (2001), 3, 271-294.

[LS]   T. Lada, J.D. Stasheff, *Introduction to sh algebras for physicists*, International Journal of Theor. Physics **32** (1993), 1087-1103

[L]   J.-L. Loday, *Cyclic Homology*, Springer Verlag, 1993.

[LSV] J.-L. Loday, J. Stasheff, and A. Voronov (ed.), Operads: Proceedings of Renaissance Conferences, Contemporary Mathematics **202**, AMS, Providence RI, 1997.

[M]   Yu. Manin, *Topics in noncommutative Geometry*, M. B. Porter Lectures, Princeton University Press, Princeton, NJ, 1991.

[Ma]   M. Mata, *PhD thesis*, Penn State, 1998.

[May] P. May, Infinite loop space theory, Bull. AMS **83** (1977), 4, 456-494.

[MDS] D. McDuff and J. Salamon, *J-holomorphic curves and quantum cohomology*, AMS, Providence, RI, University Lecture Series, **6**, 1994.

[MS]   J. McClure and J. Smith, *A solution of Deligne's Hochschild cohomology conjecture*, Recent Progress in Homotopy Theory (Baltimore, MD, 2000), Contemp. Math. **293**, 2002.

[NT]   R. Nest, B. Tsygan, *On the cohomology ring of an algebra*, Advances in Geometry, in: Progress in Mathematics, Birkhäuser, **172** (1997).

[NT1] R. Nest, B. Tsygan, *Algebraic Index theorem for families*, Adv.Math. **113**, 2, pp. 151-205.

[NT2] R. Nest, B. Tsygan, *Fukaya type categories for associative algebras*, Deformation Theory and Symplectic Geometry, Mathematical Physics Studies, vol. 20, Klüwer Acad. Publ. (1997)

[P]   S. Priddy, *Koszul resolutions*, Trans. AMS **152** (1970), 239-60.

[R]   G. Rinehart, *Differential forms on general commutative algebras*, Trans. AMS **108** (1963), 139-174.

[Sei] P. Seidel, *Fukaya categories and deformations*, preprint math.SG/0206155

[St]   J. Stasheff, *Closed string theory, strong homotopy Lie algebras and the operad actions of moduli space*, Perspectives in Mathematical Physics, Conf. Proc. Lect. Notes Math. Phys. III, International Press, Cambridge, MA, 1994, 265-288.

[St1] J. Stasheff, *Homotopy associativity of H-spaces, I and II*, Trans. AMS **108** (1963), 275-312.

[S]   D. Sullivan, *Infinitesimal computations in topology*, Publ. Math. IHES **47** (1977), 269-331.

[T]   D. Tamarkin, *Another proof of M.Kontsevich formality theorem*, preprint QA/9803025

[T1]   D. Tamarkin, *Formality of chain operads of small squares*, preprint QA/9809164

[TT]   D.Tamarkin, B.Tsygan, *Cyclic formality and index theorems*, EuroConference Moshé Flato 2000 Part II (Dijon), Letters in Mathematical Physics, **56** (2001), 2, 85-97.

[TT1] D.Tamarkin, B.Tsygan, *Noncommutative differential calculus, homotopy BV algebras and formality conjectures*, Methods Funct. Anal. and Topology **6** (2000), 2, 85-100.

[Ts]   B. Tsygan *Homology of Lie algebras of matrices over rings and Hochschild homology*, Uspekhi Mat. Nauk **38**, 2 (1983), 217-218.

[Ts1] B. Tsygan *Formality conjectures for chains*, Amer. Math. Soc. Transl., **194**, 2 (1999), 261-274.

Twisted chiral de Rham algebras on $\mathbb{P}^1$

Vassily Gorbounov, Fyodor Malikov, and Vadim Schechtman

CONTENTS

1. Preface: A few words on the chiral de Rham complex
2. Introduction
3. The sheaf $\Omega^{ch}_{X,(\alpha)}$
4. Global sections
5. The first cohomology and a chiral Serre duality
6. Free field realization

References

1. Preface: A few words on the chiral de Rham complex

The editors of these Proceedings told us that Dennis Sullivan asked us to write some introductory remarks (for non-experts) to this paper. So we present motivation for an amusing object called the "chiral de Rham complex" of a complex variety, and some of its relatives.

1.1. Physical motivation. The story starts from attempts to understand what physicists call "the sigma model with values in a variety M". This model is a quantum theory describing fluctuations of "strings" in the "space-time" M. One believes that each quantum theory is described by a linear space ("the space of states"), on which a lot of linear operators act. So, to each variety M should correspond a space $\Sigma(M)$ - the space of states of the corresponding sigma-model.

We should explain what is meant by a "variety". According to General relativity, space-time is a Riemannian manifold. However, in our situation, one postulates more structure - namely, we suppose that M is a complex manifold equipped with a Kählerian metric; moreover, the physicists assume that M has a holomorphic volume element (the Calabi-Yau condition).

In the paper [**LVW**] some desired good features of the corresponding space $\Sigma(M)$ are described. Roughly speaking, it should be an infinite dimensional version of the Dolbeault complex of M. This space should carry a rich algebraic structure

---

2000 *Mathematics Subject Classification.* 17B69, 17B67, 17B37.

Key words and phrases. vertex algebras, Kac-Moody algebras, quantum groups.

coming from the fact that it is the space of states of a model of two-dimensional (super) Conformal field theory, [**BPZ**]. In mathematical language, $\Sigma(M)$ should be a "vertex (super)algebra", [**B**]. It admits an enormous Lie algebra of symmetries: "$N=2$ superextension of the Virasoro algebra" — the algebra of vector fields on the "supercircle".

The question arises, *how to define $\Sigma(M)$?*

1.2. Chiral de Rham complex. Now let us pass to a more algebraic situation: instead of Kählerian varieties, consider (nonsingular) complex algebraic varieties and replace the Dolbeaut complex by the algebraic de Rham complex $\Omega_M^{\cdot}$. Then a miracle happens: the corresponding "string" object indeed exists. It is the "chiral de Rham complex" of M, cf. [**MSV**]. More precisely, one defines a sheaf of vertex superalgebras Ω_M^{ch} containing $\Omega_M^{\cdot}$ and then one suggests that the resolution complex $R\Gamma(M, \Omega_M^{ch})$ might be a good approximation for $\Sigma(M)$. If M is Calabi-Yau then Ω_M^{ch} admits an $N=2$ super Virasoro symmetry, as it should be.

Of course, $R\Gamma(M, \Omega_M^{ch})$ is a well defined object of the derived category. We owe to L. Borisov the important remark that, when M is compact, one can replace this complex by its cohomology $H^{\cdot}(M, \Omega_M^{ch})$, cf. [**Bor**].

The next important point is that Ω_M^{ch} is a "stringy" version not of the de Rham complex itself but of the superalgebra $\text{Diff}(\Omega_M^{\cdot})$ of differential operators acting on it. Here we see a very interesting analogy with the work of D. Tamarkin and B. Tsygan in these proceedings, [**TT**], where the authors construct noncommutative analogs of $\text{Diff}(\Omega_M^{\cdot})$.

1.3. Vertex (or chiral) algebras of differential operators. What about a stringy version $\mathcal{D}_M^{ch}$ of a simpler algebra $\mathcal{D}_M = \text{Diff}(\mathcal{O}_M)$? It turns out that there is an obstruction of cohomological nature to the existence of such vertex algebras (a physicist would say "a quantum anomaly"); it is essentially the second Chern character of the tangent bundle $ch_2(\mathcal{T}_M)$, where $ch_2 = c_1^2/2 - c_2$, cf. [**GMS1**]. More generally, given a vector bundle E on M, the obstruction to the existence of vertex superalgebras $Diff^{ch}(\Lambda^{\cdot}E)$ of differential operators acting on the exterior algebra $\Lambda^{\cdot}E$, equals $ch_2(\mathcal{T}_M) - ch_2(E)$, cf. [**GMS2**]. Here E has odd parity, so we add "fermions" to the model.

In particular, this obstruction disappears when $E = \Omega_M^1$ or $E = \mathcal{T}_M$ ("cancellation of anomalies"). In the first case one arrives at the chiral de Rham complex. In the second case one arrives at the "mirror dual" object: the vertex superalgebra $\Lambda^{ch}\mathcal{T}_M$ - "the chiral algebra of polyvector fields". Let M be a Calabi-Yau hypersurface in a projective space. In [**GM**] the vertex superalgebra $H^{\cdot}(M, \Lambda^{ch}\mathcal{T}_M)$ is computed as an "orbifold" of a "Landau - Ginzburg" (aka "chiral Koszul complex") vertex algebras. These computations provide serious evidence that the space $\Sigma(M)$ for a Calabi-Yau M coincides with $H^{\cdot}(M, \Lambda^{ch}\mathcal{T}_M)$. So we have closed the circle.

1.4. What is a vertex algebroid? Classically each algebra of differential operators may be defined as the enveloping algebra of a "Lie algebroid"; for example, $Diff_M$ is the envelope of $\mathcal{T}_M$. Similarly, each vertex algebra of differential operators is the envelope of a "vertex algebroid", cf. [**GMS1**]. It is a finite dimensional, though rather tricky object, whose definition follows from Borcherds' axioms of a vertex algebra. It turns out that one can give (cf. [**S**]) a completely different definition of a vertex algebroid, which does not mention vertex algebras at all, but uses instead a certain mixture of very classical objects: complexes of de Rham, Hochschild and Koszul.

1.5. Interesting references. At the moment we know several definitions of the chiral de Rham complex, see [**MSV, GMS2, BD, KV, Br**]. In the paper [**BL**] it is proved that the character of the vertex algebra $H^{\cdot}(M, \Omega_M^{ch})$ is equal to the elliptic genus of M. The paper [**Bor**], although difficult to read, is quite important. It gives (for the case of Calabi-Yau complete intersections) a "free field" realization of the last algebra (i.e., as the cohomology of a complex which is, with the differential forgotten, the simplest possible vertex algebra). These results form the starting point for the computations of [**GM**].

The first examples of vertex algebras of differential operators appeared in the work of B. Feigin and E. Frenkel, cf. [**FF1**]; the ideas of these authors played decisive role in the further development of the subject. Their algebras lived on the open Schubert cell of a flag space G/B (G being the semisimple Lie group, B its Borel subgroup), and had been invented for the needs of representation theory of Kac-Moody Lie algebras. For a continuation of this theme, see [**AG, GMS3**].

The general definition of a chiral algebra of differential operators, and a deep study of these objects, is due to A. Beilinson and V. Drinfeld, [**BD**]. The last book may be judged as a fundamental source on Conformal Field Theory for those who like Algebraic Geometry.

2. Introduction

2.1. Let G be a semisimple complex Lie group, $B \subset G$ be a Borel subgroup, and $X = G/B$ be the flag space. Let $\mathfrak{g} = Lie(G)$ and $\widehat{\mathfrak{g}}$ be the corresponding Kac-Moody Lie algebra. B. Feigin and E. Frenkel [**FF1, FF2, FF3**] have constructed (generalizing the previous construction by M. Wakimoto [**W**]) a family of representations of $\widehat{\mathfrak{g}}$, called *Wakimoto modules*, which are "loop space" analogs of contragradient Verma modules over $\mathfrak{g}$, realized in the spaces of differential operators over an open Schubert cell in X.

Let us recall this construction. From now on we restrict ourselves to the case $G = SL(2)$, so $\mathfrak{g} = sl(2)$, $X = \mathbb{P}^1$. Below we will use the language of vertex algebras. For the background on them we refer the reader to the books [**K1, FBZ**].

Let L be a Lie superalgebra with a base a_n, b_n, C (even vectors) and ψ_n, ϕ_n (odd vectors), $n \in \mathbb{Z}$, and supercommutators

$$[a_m, b_n] = \delta_{m,-n} C; \quad [\psi_m, \phi_n] = \delta_{m,-n} C \qquad (2.1.1)$$

all other commutators being trivial. Let M denote the "vacuum" representation of L, generated by a single vector $\mathbf{1}$, subject to relations

$$C\mathbf{1} = \mathbf{1}; \quad a_n \mathbf{1} = \psi_n \mathbf{1} = 0 \ (n \geq 0); \quad b_n \mathbf{1} = \phi_n \mathbf{1} = 0 \ (n > 0) \qquad (2.1.2)$$

As a vector space, M is a polynomial algebra in infinite set of variables,

$$M = \mathbb{C}[a_{-i-1}\mathbf{1}, b_{-i}\mathbf{1}, \phi_{-i}\mathbf{1}, \psi_{-i-1}\mathbf{1}]_{i \geq 0}$$

It is $\mathbb{Z}$-graded by "fermion number": $M = \oplus_{i \in \mathbb{Z}} M^i$ where $M^i \subset M$ is the subspace generated by all monomials such that the number of ϕ's minus the number of ψ's is equal to i.

This space admits a structure of a vertex superalgebra, cf. [**K1**], 3.6. This means that to each vector $v \in M$ corresponds a "field" $v(z)$ which is a formal power series (infinite in both directions) $\sum_{n \in \mathbb{Z}} v_n z^{-n}$ where $v_n \in End(M)$; the correspondence $v \mapsto v(z)$ satisfies a set of "Borcherds axioms", cf. [**K1**], 4.8 (b). For example,

$$a_{-1}(z) = \sum_n a_n z^{-n-1}; \quad \psi_{-1}(z) = \sum_n \psi_n z^{-n-1};$$

$$b_0(z) = \sum_n b_n z^{-n}; \quad \phi_0(z) = \sum_n \phi_n z^{-n}$$

To simplify the notations, we denote these fields by $a(z), \psi(z), b(z)$ and $\phi(z)$.

The commutator relations (2.1.1) are rewritten in the language of fields as "operator product expansions" (OPE)

$$a(z)b(w) \sim \frac{1}{z-w}; \quad \psi(z)\phi(w) \sim \frac{1}{z-w} \qquad (2.1.3)$$

cf. [**K1**], (2.3.7b), (3.6.2).

Let e, f, h be the standard generators of $\mathfrak{g}$. Let $V_{0,k}$ denote the vacuum module over $\widehat{\mathfrak{g}}$ of level k. This is a vertex algebra generated by the fields $e(z), h(z), f(z)$ subject to standard relations, cf. [**K1**], Example 4.9(b).

Let α be a complex number. A correspondence

$$e(z) \mapsto -a(z), \quad h(z) \mapsto -2 : a(z)b(z) : -2\alpha : \phi(z)\psi(z) :,$$

$$f(z) \mapsto : b(z)^2 a(z) : +2\alpha : b(z)\phi(z)\psi(z) : -2(1-\alpha^2)b(z)' \qquad (2.1.4)$$

defines a morphism of vertex algebras

$$V_{0,2\alpha^2-2} \longrightarrow M \qquad (2.1.5)$$

and makes M (in fact, all M^i) a $\widehat{\mathfrak{g}}$-module of central charge $2\alpha^2 - 2$. In particular, the fermion number zero component M^0 is a Wakimoto module to be denoted $W_{0,2\alpha^2-2}$.

The above construction coincides with the one from [**FF1**], except for one modification: we have replaced the "Cartan" boson from *op. cit.* by the product of two fermions $\phi\psi$ using the boson-fermion correspondence (cf., for example, [**K1**], 5.2).

2.2. As was mentioned above, the classical analogs of Wakimoto modules are contragradient Verma modules over $\mathfrak{g}$. According to Beilinson-Bernstein, [**BB**], they may be realized as spaces of sections of certain "standard" $\mathcal{D}$-modules over X corresponding to Schubert cells; in our case they correspond to open cells.

The first aim of the present note is to "localize" the definition from 2.1, more precisely to construct sheaves of vertex algebras over $X = \mathbb{P}^1$ whose spaces of sections over an open Schubert cell would contain the Wakimoto modules $W_{0,2\alpha^2-2}$.

In the case $\alpha = 1$ this problem is solved by the chiral de Rham complex Ω_X^{ch} introduced in [**MSV**]. Let us explain this. Choose a coordinate b on $X = \mathbb{P}^1$ and consider the covering $X = U_0 \cup U_1$ by two affine lines, where $U_0 = \{x \in X | b(x) \neq \infty\}$, $U_1 = \{x \in X | b(x) \neq 0\}$, so $b : U_0 \xrightarrow{\sim} \mathbb{A}^1$ and $b^{-1} : U_1 \xrightarrow{\sim} \mathbb{A}^1$.

We identify $b = b_0 \mathbf{1}$ and consider the space M from 2.1 as a $\mathbb{C}[b] = \mathcal{O}(U_0)$-module in the obvious way. Let $\mathcal{M}_0 = M^\sim$ be the corresponding quasicoherent sheaf over U_0. According to the Zariski localization procedure described in [**MSV**] (cf. also [**GMS1**], 1.10), $\mathcal{M}_0$ admits a structure of a sheaf of vertex (super)algebras over U_0. Similarly using the coordinate b^{-1} we get a sheaf of vertex algebras $\mathcal{M}_1$ over U_1. To get a sheaf over X, we want to glue $\mathcal{M}_0$ and $\mathcal{M}_1$ over the intersection

$U_{01} = U_0 \cap U_1 \xrightarrow{\sim} \mathbb{G}_m$, i.e. to define an isomorphism $\mathcal{M}_0(U_{01}) \xrightarrow{\sim} \mathcal{M}_1(U_{01})$. This is done by the formulas

$$b(z) \mapsto b(z)^{-1}, \ a(z) \mapsto -:b(z)^2 a(z): -2:b(z)\phi(z)\psi(z):$$
$$\phi(z) \mapsto\ :b(z)^{-2}\phi(z):, \ \psi(z) \mapsto\ :b(z)^2\psi(z): \qquad (2.2.1)$$

The meaning of these formulas is very simple. We consider the superalgebra $\mathbb{C}[a,b,\phi,\psi]$ as a superalgebra of derivations of the de Rham algebra $\Omega^{\cdot}_{\mathbb{C}[b]}$ where $a = \partial_b, \phi = db$ and $\psi = \partial_\phi$. Then we make a change of variables $b \mapsto b^{-1}$ and see how the variables a, ϕ, ψ transform. The formulas (2.2.1) are nothing but the same transformation law translated into the universe of fields. The fact that they define an automorphism of the *vertex algebra* $M[b^{-1}] = \mathcal{M}_0(U_{01})$ (the "absence of quantum anomaly") is not banal - it is one of the main observations in [**MSV**].

By definition, the chiral de Rham algebra Ω^{ch}_X is a sheaf of vertex superalgebras over X obtained by the above gluing. By a general fact (cf. [**MS1**], Part III, Corollary 1.3) the Kac-Moody algebra $\hat{\mathfrak{g}}$ acts on Ω^{ch}_X by derivations, with central charge zero. The space of sections of the fermionic zero component $\Omega^{ch,0}_X$ is the Wakimoto module $W_{0,0}$. A description of the spaces $\Gamma(X, \Omega^{ch,i}_X)$ as $\hat{\mathfrak{g}}$-modules is given in *op. cit.*, Part. III, 2.2 (cf. Proposition 4.3 below).

In order to obtain the other Wakimoto modules $W_{0,2\alpha^2-2}$ we want to "twist" the definition of Ω^{ch}_X. More precisely, in Ω^{ch}_X the fields ϕ (resp. ψ) are sections of the sheaf of one-forms $\Omega^1_X = \mathcal{O}_{\mathbb{P}^1}(-2)$ (resp. of the tangent sheaf $\mathcal{T}_X = \mathcal{O}_{\mathbb{P}^1}(2)$). Assume that $2\alpha \in \mathbb{Z}$. In a "twisted" de Rham algebra $\Omega^{ch}_{X,(\alpha)}$ they will be sections of $\mathcal{O}_{\mathbb{P}^1}(-2\alpha)$ and $\mathcal{O}_{\mathbb{P}^1}(2\alpha)$ respectively. In fact, $\Omega^{ch}_{X,(\alpha)}$ is an example of a twisted sheaf of chiral differential operators appeared in [**GMS2**], cf. 3.5 below.

The Lie algebra $\hat{\mathfrak{g}}$ acts on $\Omega^{ch}_{X,(\alpha)}$ with central charge $k = 2\alpha^2 - 2$. Similarly to the case $\alpha = 1$ these sheaves are $\mathbb{Z}$-graded by fermionic number. We describe the spaces $\Gamma(X, \Omega^{ch,i}_{X,(\alpha)})$ as extensions of Weyl modules.

We also introduce, following the pattern of [**MS2**], a sort of *chiral Serre duality* on the category of $\Omega^{ch}_{X,(\alpha)}$-modules and use it to get a similar description of $H^1(X, \Omega^{ch,i}_{X,(\alpha)})$. This allows to compute their characters.

2.3. The next topic of this note will be a "free field resolution" of the vertex superalgebra $H^*(X, \Omega^{ch}_{X,(\alpha)})$, i.e. a description of the last algebra (which is a complicated object) as a cohomology of a simple "free field" algebra with respect to certain square zero odd derivation (which is a Fourier component of a field, and called a "screening charge"). At this point we use the fact that $\mathbb{P}^1$ is a toric variety. In [**Bor**], L. Borisov constructs for an arbitrary smooth toric variety Y a remarkable free field resolution of $H^*(Y, \Omega^{ch}_Y)$ (cf. also [**FMS**]). We show how to twist the Borisov construction to obtain a free field resolution of $H^*(X, \Omega^{ch}_{X,(\alpha)})$. It turns out that this "twisted" Borisov resolution essentially coincides with a resolution discussed by B. Feigin in an unpublished manuscript [**F**]. Introduced in [**F**] are several free field resolutions of the format from [**FF4**]. According to the philosophy of *op. cit.*, these should be considered as "quantizations" of standard Bernstein-Gelfand-Gelfand complexes of a maximal nilpotent Lie subalgebra $\mathfrak{n}'$ in a semisimple Lie algebra $\mathfrak{g}'$ (with coefficients in a Heisenberg module). "Quantization" means that instead of $\mathfrak{g}'$ one deals with corresponding quantum group. B. Feigin has discovered that in the case at hand the quantum group that governs the resolution corresponds to the Lie superalgebra $sl(2,1)$, $\mathfrak{n}'$ being generated by two odd elements. In our text this manifests in that the operators $D, \bar{D}$ appearing in (6.9.3) generate an algebra isomorphic to $\mathfrak{n}'$, and the complex (6.9.3) is exactly the Bernstein-Gelfand-Gelfand complex associated with V_L^0.

To be more precise, Feigin discusses the analog of this resolution in the case of a *generic* α (and here the quantum group arises). In our picture, the full sheaf $\Omega^{ch}_{X,(\alpha)}$ makes no sense for generic α; however, its component of *fermionic charge* 0 does (and has a vacuum $\widehat{sl}_2$-module as the space of global sections). So we can argue that the present construction gives a natural extension of Feigin's resolution in the case of non-generic α.

The "chiral Serre duality" makes it clear why Feigin's resolution is indeed a resolution and exactly what higher cohomology appear for non-generic α. This is probably the only real advantage of our approach over Feigin's.

The "linear algebra data" underlying our resolution is a 3-dimensional lattice L equipped with a one-parameter family of "hyperbolic reflections" ϕ_α, cf. *infra* (6.7.1). These data are in some sense associated with the root system of $sl(2)$. Can one define similar data for other root systems?

Remark. One can define similar twisted de Rham algebras which localize the Wakimoto modules for an arbitrary semisimple Lie group G. However, we do not know the analogs of Feigin-Borisov resolutions for an arbitrary G — this is a very interesting open question.

2.4. Acknowledgments. During the work on this note the authors enjoyed the hospitality of Max-Planck-Institut für Mathematik in Bonn and Korteweg-de Vries Institute in Amsterdam. We are grateful to Don Zagier who taught us the computations 4.5-4.6. We thank the referee who drew our attention to the article [**FMS**].

3. The sheaf $\Omega^{ch}_{X,(\alpha)}$

We use all the notations of the Introduction.

3.1. Consider the vertex algebra $M[b^{-1}] = \mathcal{M}_0(U_{01})$. It is generated by the fields $b(z), b(z)^{-1}, a(z), \phi(z), \psi(z)$. By composing the Wakimoto map (2.1.4) with the canonical localisation map $M \longrightarrow M[b^{-1}]$ we get a morphism of vertex algebras

$$V_{0,2\alpha^2-2} \longrightarrow M[b^{-1}] \qquad (3.1.1)$$

Each vacuum module $V_{0,k}$ affords an involution defined by

$$e(z) \mapsto f(z), \ f(z) \mapsto e(z) \qquad (3.1.2)$$

(note that $V_{0,k}$ is generated by two fields $e(z), f(z)$).

On the classical level, if we realize $\mathfrak{g}$ by means of vector fields acting on X and identify e with $-\partial_b$ and f with $b^2 \partial_b$ then the interchange of e with f corresponds to the coordinate change $b \mapsto b^{-1}$. The next lemma chiralizes this automorphism.

3.2. Lemma. *Assume that $\alpha \in \frac{1}{2}\mathbb{Z}$. The assignment*

$$b(z) \mapsto b(z)^{-1}, \ a(z) \mapsto -:b(z)^2 a(z): -2\alpha :b(z)\phi(z)\psi(z): -2(1-\alpha^2)b(z)',$$
$$\phi(z) \mapsto :b(z)^{-2\alpha}\phi(z):, \ \psi(z) \mapsto :b(z)^{2\alpha}\psi(z): \qquad (3.2.1)$$

defines an automorphism of $M[b^{-1}]$ which is compatible with the involution (3.1.2) via the map (3.1.1).

If α is not necessarily a half-integer, the involution (3.2.1) is still well defined on the fermionic number zero component $M[b^{-1}]^0$. △

3.3. Remark. Talking about fields is equivalent to talking about the states. In the language of states (3.2.1) is equivalent to

$$b_0\mathbf{1} \mapsto b_0^{-1}\mathbf{1}, \ a_{-1}\mathbf{1} \mapsto -b_0^2 a_{-1}\mathbf{1} - 2\alpha b_0 \phi_0 \psi_{-1}\mathbf{1} - 2(1-\alpha^2)b_{-1}\mathbf{1}$$

$$\phi_0\mathbf{1} \mapsto b_0^{-2\alpha}\phi_0\mathbf{1}, \ \psi_{-1}\mathbf{1} \mapsto b_0^{2\alpha}\psi_{-1}\mathbf{1} \quad (3.3.1)$$

3.4. On the neighborhood of infinity U_1 with coordinate $\bar{b} = b^{-1}$ we have the sheaf $\mathcal{M}_1$ generated by the fields $\bar{a}(z), \bar{b}(z), \bar{\phi}(z), \bar{\psi}(z)$. By sending these barred fields to the corresponding right hand side expressions from (3.2.1) we get an isomorphism

$$g: \mathcal{M}_1|_{U_{01}} \xrightarrow{\sim} \mathcal{M}_0|_{U_{01}} \quad (3.4.1)$$

By the way, to get g^{-1} one simply puts the bars over the fields in the right-hand side of (3.2.1).

Gluing $\mathcal{M}_0$ with $\mathcal{M}_1$ over U_{01} by means of g we get a sheaf of vertex superalgebras on X, to be denoted by $\Omega^{ch}_{X,(\alpha)}$. This sheaf is defined if $\alpha \in \frac{1}{2}\mathbb{Z}$. For $\alpha = 1$ we get the old chiral de Rham algebra $\Omega^{ch}_{X,(1)} = \Omega^{ch}_X$.

For an arbitrary α the whole sheaf is not defined but the fermion number zero component $\Omega^{ch,0}_{X,(\alpha)}$ is.

3.5. One can easily explain the meaning of the above sheaves in the framework of [**GMS2**]. Namely, consider a supervariety $X_{\mathcal{O}(-2\alpha)}$ with the same underlying space as X and with the structure sheaf $\Omega_{X,(\alpha)} = \mathcal{O}_X \oplus \mathcal{O}_X(-2\alpha)$ being the symmetric (in the super sense) algebra of the line bundle $\mathcal{O}_X(-2\alpha)$ where this line bundle is equipped the odd parity. So this superalgebra is a twisted analog of the de Rham algebra.

The sheaf $\Omega^{ch}_{X,(\alpha)}$ is nothing but *the* sheaf of chiral differential operators on $X_{\mathcal{O}(-2\alpha)}$. Note that since X is a curve, such a sheaf is unique up to a unique isomorphism, by the general cohomological formalism of *op. cit.*

3.6. By definition, $\Omega^{ch}_{X,(\alpha)}$ is a sheaf of $\widehat{\mathfrak{g}}$-modules of level $k = 2\alpha^2 - 2$. Therefore it carries a conformal structure where the Virasoro field is provided by the Sugawara construction. Locally on U_0 it is given by

$$L(z) = \frac{1}{k+2}\{: e(z)f(z): + :f(z)e(z): + \frac{1}{2}:h(z)^2:\} =$$

$$=: b(z)'a(z): + \frac{1-\alpha}{2\alpha}:\phi(z)\psi(z)': + \frac{1+\alpha}{2\alpha}:\phi(z)'\psi(z): \quad (3.6.1)$$

The grading given by the operator L_0 of this field differs from the obvious one: e.g.,

$$L_0(\phi_0\mathbf{1}) = \frac{\alpha-1}{2\alpha}\phi_0\mathbf{1}, \ L_0(\psi_{-1}\mathbf{1}) = -\frac{\alpha+1}{2\alpha}\psi_{-1}\mathbf{1} \quad (3.6.2)$$

However, the old grading (but not the old conformal structure) which assigns to $x_i\mathbf{1}$ the degree $-i$, still makes sense for the whole sheaf $\Omega^{ch}_{X,(\alpha)}$. Below when we speak about "conformal weight" we mean always this *old* grading. Thus, the conformal weight zero component $\Omega^{ch}_{X,(\alpha)0} = \mathcal{O}_X \oplus \mathcal{O}_X(-2\alpha)$, $\phi_0\mathbf{1}$ being a local section of $\mathcal{O}_X(-2\alpha)$. Similarly, $\Omega^{ch}_{X,(\alpha)1}$ admits $\mathcal{O}_X(2\alpha)$ as a canonical quotient, $\psi_{-1}\mathbf{1}$ being its local generator, etc.

The fermionic charge grading on $\Omega^{ch}_{X,(\alpha)}$ is defined as in the untwisted case. Our algebra admits a canonical filtration whose graded quotient is given by (cf. [**BL**], [**GMS2**], 7.6)

$$[\mathrm{Gr}\,\Omega^{ch}_{X,(\alpha)}] = \prod_{n\geq 0}[S_{q^n}\Omega^1_X][S_{q^{n+1}}\mathcal{T}_X][\Lambda_{yq^n}\mathcal{O}_X(-2\alpha)][\Lambda_{y^{-1}q^{n+1}}\mathcal{O}_X(2\alpha)] \qquad (3.6.3)$$

Here the square brackets denote the class in the K-theory; the power at q (resp. y) counts the conformal weight (resp. fermionic charge). In the right hand side we use the notation

$$[S_x(E)] = \sum_{i=0}^{\infty}[S^i_{\mathcal{O}_X}(E)]x^i \in K(X)[[x]], \ [\Lambda_x(E)] = \sum_{i=0}^{\infty}[\Lambda^i(E)]x^i \in K(X)[x] \qquad (3.6.4)$$

for a vector bundle E over X.

4. Global sections

4.1. In this section we give a description of $\Gamma(X, \Omega^{ch}_{X,(\alpha)})$ as a $\hat{\mathfrak{g}}$-module and compute its character.

Let m be a nonnegative integer and k an arbitrary complex number. Let L_m denote the $(m+1)$-dimensional irreducible $\mathfrak{g}$-module. The Kac-Moody algebra $\hat{\mathfrak{g}} = \mathfrak{g}[t, t^{-1}] \oplus \mathbb{C}c$ contains the Lie subalgebra $\hat{\mathfrak{g}}_+ = \mathfrak{g}[t] \oplus \mathbb{C}c$ equipped with an obvious "evaluation at 0" projection $p: \hat{\mathfrak{g}}_+ \longrightarrow \mathfrak{g} \oplus \mathbb{C}c$. Let $L_{m,k}$ denote a $\mathfrak{g} \oplus \mathbb{C}c$-module equal to L_m as a $\mathfrak{g}$-module, where c acts as multiplication by k; using p we can regard $L_{m,k}$ as a $\hat{\mathfrak{g}}_+$-module. By definition, the *Weyl module* $V_{m,k}$ is the induced $\hat{\mathfrak{g}}$-module

$$V_{m,k} = \mathrm{Ind}^{\hat{\mathfrak{g}}}_{\hat{\mathfrak{g}}_+} L_{m,k} \qquad (4.1.1)$$

4.2. Let $\Omega^{ch,i}_{X,(\alpha)} \subset \Omega^{ch}_{X,(\alpha)}$ denote the component of fermionic charge $i \in \mathbb{Z}$. By construction,

$$\Gamma(U_0, \Omega^{ch,0}_{X,(\alpha)}) = W_{0, 2\alpha^2 - 2} \qquad (4.2.1)$$

Similarly,

$$\Gamma(U_0, \Omega^{ch,i}_{X,(\alpha)}) = W_{-2\alpha i, 2\alpha^2 - 2} \qquad (4.2.2)$$

where in the right hand side we have the Wakimoto module having the highest weight vector on which h acts as $2\alpha i$.

It is clear that just as in the untwisted ($\alpha = 1$) case discussed in [**MS1**], Part III.2 we have

$$\Gamma(X, \Omega^{ch}_{X,(\alpha)}) = \Gamma(U_0, \Omega^{ch}_{X,(\alpha)})^{\mathfrak{g}-\mathrm{int}} \qquad (4.2.3)$$

where the superscript in the right hand side denotes the maximal $\mathfrak{g}$-integrable submodule.

The diagram on p. 301 of [**FF3**] implies:

4.3. Proposition. (i) *If $\alpha \in \mathbb{Z}_{\geq 0}, i \leq 0$, then $\Gamma(X, \Omega^{ch,i}_{X,(\alpha)})$ admits a filtration with quotients equal to Weyl modules $V_{-2i\alpha + 4j\alpha^2, 2\alpha^2 - 2}$, $j = 0, 1, 2, \ldots$;*

(ii) *if $\alpha \in \mathbb{Z}_{\geq 0}, i > 0$, then $\Gamma(X, \Omega^{ch,i}_{X,(\alpha)})$ admits a filtration with quotients equal to dual Weyl modules $V^*_{2i\alpha - 2 + 4(j+1)\alpha^2, 2\alpha^2 - 2}$, $j = 0, 1, 2, \ldots$;*

(iii) *if α is irrational then $\Omega^{ch}_{X,(\alpha)}$ does not exist but $\Omega^{ch,0}_{X,(\alpha)}$ does, and we have*

$$\Gamma(X, \Omega^{ch,0}_{X,(\alpha)}) = W^{\mathfrak{g}-int}_{0, 2\alpha^2 - 2} = V_{0, 2\alpha^2 - 2} \qquad (4.3.1)$$

△

4.4. Let us write down the character of our module. First consider the case $\alpha = 1$. We define

$$ch\ \Gamma(X, \Omega_X^{ch}) := \sum_{i \geq 0} \dim \Gamma(X, \Omega_X^{ch})_i \cdot q^i \qquad (4.4.1)$$

where the subscript i denotes the component of conformal weight i.

Due to Proposition 4.3, parts (i) and (ii), we have:

$$ch\ \Gamma(X, \Omega_X^{ch}) =$$

$$\prod_{n=1}^{\infty} (1-q^n)^{-3} \Big\{ \sum_{m=0}^{\infty} (4m+1)(2m+1)q^{(2m+1)m} + \sum_{m=0}^{\infty} (4m+3)(2m+1)q^{(2m+1)(m+1)} \Big\}$$

$$= \prod_{n=1}^{\infty} (1-q^n)^{-3} \sum_{m \in \mathbb{Z}} (4m+1)(2m+1)q^{(2m+1)m}. \qquad (4.4.2)$$

We want to transform this to a more familiar expression.

4.5. Introduce the standard notation

$$\eta(q) = q^{1/24} \prod_{n=1}^{\infty} (1-q^n) \qquad (4.5.1)$$

$$G_2(q) = -\frac{1}{24} + \sum_{n=1}^{\infty} \sigma_1(n)q^n, \ \sigma_1(n) := \sum_{d|n} d \qquad (4.5.2)$$

and (with D. Zagier, [**Z**]) a less standard one

$$\theta(q) = \sum_{m=0}^{\infty} q^{(2m+1)^2/8} \qquad (4.5.3)$$

for the generating function counting the triangular numbers.

We have the Gauß formula (cf. [**K2**], Exercise 12.4, p. 176)

$$\phi(q^2)^2/\phi(q) = \sum_{m \in \mathbb{Z}} q^{m(2m+1)} \qquad (4.5.4)$$

where

$$\phi(q) = \prod_{n=1}^{\infty} (1-q^n) \qquad (4.5.5)$$

from which one deduces easily, using the identity

$$m(2m+1) = \frac{(4m+1)^2 - 1}{8} \qquad (4.5.6)$$

that

$$\theta(q) = \eta(q^2)^2/\eta(q) \qquad (4.5.7)$$

Similarly, using the Jacobi formula (cf. *loc. cit*)

$$\phi(q)^3 = \sum_{m \in \mathbb{Z}} (4m+1)q^{m(2m+1)} \qquad (4.5.8)$$

one deduces that

$$\eta(q)^3 = \sum_{m \in \mathbb{Z}, m \equiv 1 \pmod 4} m q^{m^2/8} \qquad (4.5.9)$$

where we have used that
$$\sum_{m \text{ odd}, m>0} m^i q^{m^2/8} = \sum_{m=1(\text{mod}4),\ m\in\mathbb{Z}} m^i q^{m^2/8} \qquad (4.5.10)$$

We also have
$$q\partial_q \log \eta(q) = -G_2(q) \qquad (4.5.11)$$
(see [**HBJ**], Appendix I, Lemma 4.9).

4.6. It follows from (4.5.7) and (4.5.11) that
$$q\partial_q \log \theta(q) = G_2(q) - 2G_2(q^2) \qquad (4.6.1)$$
On the other hand, using again (4.5.10),
$$q\partial_q \theta(q) = \frac{1}{8} \sum_{m\in\mathbb{Z}, m=1(\text{mod}4)} m^2 q^{m^2/8} \qquad (4.6.2)$$

Returning to our character (4.4.2) we deduce (using (4.5.6))
$$ch\ \Gamma(X, \Omega_X^{ch}) = q^{-1/8} \sum_{n=1(\text{mod}4),\ n\in\mathbb{Z}} \frac{n(n+1)}{2} q^{n^2/8} =$$
$$= \frac{1}{2}\{8q\partial_q \theta(q) + \eta^3(q)\}\eta(q)^{-3} = \frac{1}{2} + \frac{4\theta(q)}{\eta(q)^3}\{G_2(q) - 4G_2(q^2)\} \qquad (4.6.3)$$
Note that
$$(\theta(q)/\eta(q)^3)^4 = \eta(q^2)^8/\eta(q)^{16} = (16\epsilon(q))^{-1} \qquad (4.6.4)$$
where $\epsilon = (e_1 - e_2)(e_1 - e_3)$ is a standard modular form of weight 4 with respect to $\Gamma_0(2)$, cf. [**HBJ**], Appendix I, Cor. 4.11.

4.7. Now let us consider the case of an arbitrary nonnegative half-integer α. From 4.3 we have
$$ch\ H^0(X, \Omega_{X,(\alpha)}^{ch}) = \sum_{(m,j)\in Q_+\cup -Q_-} (4j\alpha^2 + 2m\alpha + 1) q^{(m\alpha+2j\alpha^2)(1+m\alpha+2j\alpha^2)/2\alpha^2} \phi(q)^{-3}$$
where $Q_+ = \{(m,j)\in\mathbb{Z}^2 | m\geq 0,\ j\geq 0\}$; $Q_- = \{(m,j)\in\mathbb{Z}^2 | m<0,\ j<0\}$; now make change of variables $m + 2\alpha j = d + 2\alpha k$, $0 \leq d < 2\alpha$,
$$= q^{-1/8\alpha^2} \sum_{d=0}^{2\alpha-1} \sum_{k\in\mathbb{Z}} (k+1)(2\alpha(2\alpha k + d) + 1) q^{2\alpha^2(k+\frac{2\alpha d+1}{4\alpha^2})^2} \cdot \phi(q)^{-3} =$$
$$= q^{-1/8\alpha^2} \sum_{d=0}^{2\alpha-1} \sum_{l\in\mathbb{Z}+\frac{2\alpha d+1}{4\alpha^2}} \{4\alpha^2 l + 4\alpha^2 - 2\alpha d - 1\} l q^{2\alpha^2 l^2} \cdot \phi(q)^{-3} \qquad (4.7.1)$$
Introduce the Jacobi theta functions
$$\theta_{n,m}(\tau, z) = \sum_{l\in\mathbb{Z}+n/2m} e^{2\pi i m(l^2\tau - lz)},\ m\in\frac{1}{2}\mathbb{Z},\ m\geq 0,\ n\in\mathbb{Z}/2m\mathbb{Z} \qquad (4.7.2)$$
cf. [**K2**], Example 13.3; as usual, $q = e^{2\pi i \tau}$.

We see that the numerator in (4.7.1) is a finite linear combination of the values $\partial_z \theta_{2\alpha d+1, 2\alpha^2}(\tau, 0)$ and $\partial_\tau \theta_{2\alpha d+1, 2\alpha^2}(\tau, 0)$. This implies that it is a quasimodular form with respect to a finite index subgroup $\Gamma \subset SL(2;\mathbb{Z})$, cf. [**KZ**].

5. The first cohomology and a chiral Serre duality

In this section we give a similar description of $H^1(X, \Omega^{ch}_{X,(\alpha)})$ as a $\widehat{\mathfrak{g}}$-module, the tool of trade being the observation that the chiral Serre duality [**MS2**] carries over to the twisted case word for word. Let us explain the last point very briefly.

5.1. Recall that we have defined in 3.5 the sheaf $\Omega_{X,(\alpha)} = \mathcal{O}_X \oplus \mathcal{O}_X(-2\alpha)$ of supercommutative $\mathcal{O}_X$-algebras. Let $D_{\Omega_{X,(\alpha)}}$ be the algebra of superdifferential operators acting on $\Omega_{X,(\alpha)}$.

Consider the category of left $D_{\Omega_{X,(\alpha)}}$-modules, to be denoted $D_{\Omega_{X,(\alpha)}} - \mathrm{Mod}^l$, and of *restricted* left $\Omega^{ch}_{X,(\alpha)}$-modules, to be denoted $\Omega^{ch}_{X,(\alpha)} - \mathrm{Mod}^l$; restricted means that the subsheaves of negative conformal weights are equal to 0.

To compare $D_{\Omega_{X,(\alpha)}} - \mathrm{Mod}^l$ with $\Omega^{ch}_{X,(\alpha)} - \mathrm{Mod}^l$ introduce, by analogy with [**MSV**], 6.11, or [**MS1**], I.4.5., the *Weyl* functor

$$\mathbb{W}^l : D_{\Omega_{X,(\alpha)}} - \mathrm{Mod}^l \to \Omega^{ch}_{X,(\alpha)} - \mathrm{Mod}^l \tag{5.1.1}$$

to be the left adjoint to the functor of taking the conformal weight zero component.

5.2. Example. $\mathbb{W}^l(\Omega_{X,(\alpha)}) = \Omega^{ch}_{X,(\alpha)}$.

5.3. Proposition. $\mathbb{W}^l$ *is an equivalence of categories.* △

5.4. Now we want to play a similar game with *right* modules. Consider the category of right $D_{\Omega_{X,(\alpha)}}$-modules, to be denoted $D_{\Omega_{X,(\alpha)}} - \mathrm{Mod}^r$, and of restricted right $\Omega^{ch}_{X,(\alpha)}$-modules, to be denoted $\Omega^{ch}_{X,(\alpha)} - \mathrm{Mod}^r$; here restricted means that the subsheaves of positive conformal weights are equal to 0. (We leave the definition of a right module over a vertex algebra to the reader.)

Define the right Weyl functor

$$\mathbb{W}^r : D_{\Omega_{X,(\alpha)}} - \mathrm{Mod}^r \to \Omega^{ch}_{X,(\alpha)} - \mathrm{Mod}^r \tag{5.4.1}$$

to be the left adjoint to the functor of taking the conformal weight zero component.

5.5. Proposition. $\mathbb{W}^r$ *is an equivalence of categories.* △

5.6. Let us call a $D_{\Omega_{X,(\alpha)}}$-module *smooth* if it is a locally free $\mathcal{O}_X$-module of finite rank. Given a smooth $D_{\Omega_{X,(\alpha)}}$-module N, set

$$N^\vee = \mathcal{H}om_{\mathcal{O}_X}(N, \Omega^1_X) \tag{5.6.1}$$

By virtue of the Serre duality

$$H^1(X, N) \xrightarrow{\sim} H^0(X, N^\vee)^* \tag{5.6.2}$$

The $\mathcal{O}_X$-module $N^\vee$ admits a canonical structure of a right $D_{\Omega_{X,(\alpha)}}$-module.

Property (4.6.2) chiralizes as follows (cf. [**MS2**] formula (11.6)).

5.7. Proposition. *If N is a smooth $D_{\Omega_{X,(\alpha)}}$-module, then*

$$H^1(X, \mathbb{W}(N)) \xrightarrow{\sim} H^0(X, \mathbb{W}^r(N^\vee))^* \tag{5.7.1}$$

△

5.8. Denote $\Omega^{ch\vee}_{X,(\alpha)} = \mathbb{W}^r(\Omega^\vee_{X,(\alpha)})$. Example 5.2 and Proposition 5.7 combined give
$$H^1(X, \Omega^{ch}_{X,(\alpha)}) = H^0(X, \Omega^{ch\vee}_{X,(\alpha)})^* \tag{5.8.1}$$
It is clear that the dual sheaf $\Omega^{ch\vee}_{X,(\alpha)}$ is very similar to $\Omega^{ch}_{X,(\alpha)}$ because both are defined as left adjoint functors. $\Omega^{ch\vee}_{X,(\alpha)}$ inherits the fermion number grading from $\Omega^{ch}_{X,(\alpha)}$ if we place Ω^1_X in, say, the fermion number 1 component:
$$\Omega^{ch\vee}_{X,(\alpha)} = \oplus_{i \in \mathbb{Z}} \Omega^{ch\vee,i}_{X,(\alpha)} \tag{5.8.2}$$
Further, $\Omega^{ch\vee}_{X,(\alpha)}$ carries a right action of $\widehat{sl}_2$, and this action composed with a canonical anti-involution of $\widehat{sl}_2$ makes $\Omega^{ch\vee}_{X,(\alpha)}$ a sheaf of $\widehat{sl}_2$-modules of level $2\alpha^2 - 2$.

The analogs of (4.2.1)-(4.2.3) are as follows:
$$\Gamma(U_0, \Omega^{ch\vee,1}_{X,(\alpha)}) = W_{-2,2\alpha^2-2}, \tag{5.8.3}$$
$$\Gamma(U_0, \Omega^{ch\vee,i}_{X,(\alpha)}) = W_{-2(i-1)\alpha i-2, 2\alpha^2-2}, \tag{5.8.4}$$
$$\Gamma(X, \Omega^{ch\vee}_{X,(\alpha)}) = \Gamma(U_0, \Omega^{ch\vee}_{X,(\alpha)})^{\mathfrak{g}-int} \tag{5.8.5}$$
Comparing duality (5.8.1) with (5.8.3)-(5.8.5), we arrive at the following analog of Proposition 4.3.

5.9. Proposition. (i) *If $\alpha \in \mathbb{Z}_{>0}, i \geq 1$, then $H^1(X, \Omega^{ch,i}_{X,(\alpha)})$ admits a filtration with quotients equal to the dual Weyl modules $V^*_{2i\alpha-2+4j\alpha^2, 2\alpha^2-2}$, $j = 0, 1, 2, \ldots$;*

(ii) *if $\alpha \in \mathbb{Z}_{>0}, i < 0$, then $H^1(X, \Omega^{ch,i}_{X,(\alpha)})$ admits a filtration with quotients equal to the Weyl modules $V_{-2i\alpha+4(j+1)\alpha^2, 2\alpha^2-2}$, $j = 0, 1, 2, \ldots$;*

(iii) *if α is irrational then $\Omega^{ch}_{X,(\alpha)}$ does not exist but $\Omega^{ch,0}_{X,(\alpha)}$ does, and we have*
$$H^1(X, \Omega^{ch,0}_{X,(\alpha)}) = W^{\mathfrak{g}-int}_{-2,2\alpha^2-2} = 0. \tag{5.9.1}$$

$\triangle$

6. Free field realization

In this section we want to write down a free field resolution of the vertex superalgebra $H^*(X, \Omega^{ch}_{X,(\alpha)})$. From now on unless mentioned otherwise *assume that $\alpha \in \mathbb{Z}$.*

6.1. Consider a 3-dimensional integer lattice L with generators μ, β, ρ and an inner product $(\mu, \beta) = (\beta, \mu) = (\rho, \rho) = 1$, the other products of basic elements being zero. One associates with it a lattice vertex superalgebra V_L. Recall (cf. [**K1**], 5.5) that as a vector space
$$V_L = Vac_{\mathfrak{h}} \otimes \mathbb{C}[L] \tag{6.1.1}$$
where $Vac_{\mathfrak{h}}$ is the vacuum module over the Heisenberg Lie algebra $\widehat{\mathfrak{h}} = \mathfrak{h}[t, t^{-1}] \oplus \mathbb{C}c$, $\mathfrak{h} := \mathbb{C} \otimes_{\mathbb{Z}} L$, $\mathbb{C}[L]$ being the group algebra of L whose basic elements are denoted by e^γ, $(\gamma \in L)$. The parity of e^γ is equal to
$$p(e^\gamma) = (\gamma, \gamma) \mod 2 \tag{6.1.2}$$
the factor $Vac_{\mathfrak{h}}$ is purely even.

According to *loc. cit.*, the definition of V_L depends also on a choice of certain 2-cocycle $\epsilon: L \times L \longrightarrow \mathbb{C}^*$. However, the algebras corresponding to different choices of ϵ are canonically isomorphic. We fix one ϵ and banish it from the notation.

As a vertex algebra V_L is generated by fields $\gamma(z) := (\gamma \otimes 1)(z)$ and $e^\gamma(z) := (1 \otimes e^\gamma)(z)$ ($\gamma \in L$).

Thus, V_L is $\mathbb{Z}^3$-graded

$$V_L = \oplus_{(i,j,k)\in\mathbb{Z}^3} V_L^{ijk}, \quad V_L^{ijk} := Vac_{\mathfrak{h}} \otimes e^{i\beta+j\mu+k\rho} \tag{6.1.3}$$

We will use the notation V_L^γ for V_L^{ijk} where $\gamma = i\beta + j\mu + k\rho$, so

$$V_L = \oplus_{\gamma \in L} V_L^\gamma \tag{6.1.4}$$

We also use the $\mathbb{Z}$-grading

$$V_L = \oplus_{j\in\mathbb{Z}} V_L^j, \quad V_L^j := \oplus_{(i,k)\in\mathbb{Z}^2} V_L^{ijk} \tag{6.1.5}$$

The subspace $Vac_{\mathfrak{h}} \otimes 1 = V_L^{000}$ is purely even vertex subalgebra of V_L, to be denoted simply by $Vac_{\mathfrak{h}}$. Each V_L^γ is a vertex module over $Vac_{\mathfrak{h}}$; it is an irreducible Heisenberg $\hat{\mathfrak{g}}$-module on whose highest weight $\mathfrak{h}$ acts by means of the character $(\gamma,)$.

For $\gamma' = i'\beta + j'\mu + k'\rho \in L$, the field $e^\gamma(z)$ has degree (i', j', k'), that is, all its Fourier components act from $V_L^\gamma = V_L^{ijk}$ to $V_L^{\gamma+\gamma'} = V_L^{i+i',j+j',k+k'}$.

6.2. Remark. The Heisenberg modules V_L^γ make perfect sense for arbitrary $\gamma \in \mathfrak{h} = L_{\mathbb{C}}$. Also one can write down the usual formula (cf. [**K1**] (5.5.9)) for vertex operators $e^{\gamma'}(z)$ for an arbitrary $\gamma' \in \mathfrak{h}$.

These "fields" act also from V_L^γ to $V_L^{\gamma+\gamma'}$. However, they are multivalued: they have "monodromy" (whose logarithm is equal to $2\pi i(\gamma, \gamma')$). (Note that for $\gamma = 0$, or more generally for such γ that $(\gamma, \gamma') \in \mathbb{Z}$ this monodromy disappears). Their composition acquires monodromy as well. These monodromies is a source of appearance of quantum groups in the picture.

6.3. According to [**FMS**] (cf. also [**FF5**, **Bor**]), the assignment

$$b(z) \mapsto e^\beta(z),\ b(z)^{-1} \mapsto e^{-\beta}(z),$$

$$a(z) \mapsto\ :(\mu(z) - \rho(z))e^{-\beta}(z):,\ \phi(z) \mapsto e^{\beta+\rho}(z),\ \psi(z) \mapsto e^{-\beta-\rho}(z) \tag{6.3.1}$$

determines an embedding of vertex superalgebras

$$M[b^{-1}] \hookrightarrow V_L \tag{6.3.2}$$

which induces an isomorphism

$$M[b^{-1}] \xrightarrow{\sim} V_L^0 \tag{6.3.3}$$

6.4. Consider an odd (since $(\mu - \rho, \mu - \rho) = 1$) field $e^{\mu-\rho}(z)$ and denote its zeroth component

$$d_M = \int e^{\mu-\rho}(z) \tag{6.4.1}$$

Here and below we will use the notation $\int c(z)$ for the component $c_{(0)}$ of a field $c(z) = \sum c_{(i)} z^{-i-1}$.

The operator d_M has degree 1 with respect to the grading (6.1.5), and $d_M^2 = 0$.

6.5. Lemma. *The sequence*

$$0 \longrightarrow M \longrightarrow V_L^0 \xrightarrow{d_M} V_L^1 \xrightarrow{d_M} \ldots \qquad (6.5.1)$$

is exact.

This follows from the results of [**FMS**] together with the observation that the operator d_M is the residue of a fermionic field; cf. [**FF5**]. (We are thankful to the referee for this remark).

Here the first arrow is the composition of the obvious embedding $M \hookrightarrow M[b^{-1}]$ with the isomorphism (6.3.3).

6.6. We can compute the cohomology superalgebra $H^*(X, \Omega^{ch}_{X,(\alpha)})$ by means of the Cech complex of the covering $X = U_0 \cup U_1$:

$$0 \longrightarrow M \oplus \bar{M} \longrightarrow M[b^{-1}] \longrightarrow 0 \qquad (6.6.1)$$

Here $\bar{M} = \Gamma(U_1, \Omega^{ch}_{X,(\alpha)})$ is simply another copy of M (whose fields we will denote by barred letters). The differential is the obvious embedding on the first component and (taken with the minus sign) composition of the obvious embedding $\bar{M} \hookrightarrow M[b^{-1}]$ and the automorphism of $M[b^{-1}]$ described in Lemma 3.2.

Now we want to replace $M, \bar{M}$ by their Borisov resolutions (6.5.1) (denote them V_L and $V_{\bar{L}}$) and $M[b^{-1}]$ by the isomorphic algebra V_L^0, cf. (6.3.3).

The only problem is to lift the differential (6.5.1) to a map of resolutions

$$V_L \oplus V_{\bar{L}} \longrightarrow V_L^0 \qquad (6.6.2)$$

where we consider V_L^0 as a complex sitting in degree zero.

6.7. Consider an automorphism $\phi_\alpha : L \xrightarrow{\sim} L$ given by

$$\phi_\alpha(\beta) = -\beta, \ \phi_\alpha(\mu) = -\mu - 2(\alpha-1)\rho + 2(\alpha-1)^2\beta, \ \phi_\alpha(\rho) = \rho - 2(\alpha-1)\beta \quad (6.7.1)$$

One checks that ϕ_α respects the inner product defined in 6.1 and that $\phi_\alpha^2 = Id$.

Hence it defines an involution of V_L which we will denote by the same letter

$$\phi_\alpha : V_L \xrightarrow{\sim} V_L \qquad (6.7.2)$$

6.8. Lemma. *If we identify $M[b^{-1}]$ with V_L^0 by means of the isomorphism (6.3.3), then the involution ϕ_α (6.7.2) induces the automorphism (3.2.1) of $M[b^{-1}]$.* △

6.9. This lemma provides the necessary tool for defining (6.6.2). We have to define a morphism

$$\widehat{d} : V_L^0 \oplus V_{\bar{L}}^0 \longrightarrow V_L^0 \qquad (6.9.1)$$

It will be the identity on the first component and taken with the minus sign composition of the tautological identification $V_{\bar{L}} = V_L$ and the involution (6.7.2).

Since bicomplex (6.6.2) is a resolution of the Cech complex (6.6.1), the cohomology algebra $H^*(X, \Omega^{ch}_{X,(\alpha)})$ is canonically identified with the cohomology of the complex

$$0 \longrightarrow \operatorname{Ker} \widehat{d} \xrightarrow{d_M \oplus d_{\bar{M}}} V_L^1 \oplus V_{\bar{L}}^1 \xrightarrow{d_M \oplus d_{\bar{M}}} V_L^2 \oplus V_{\bar{L}}^2 \longrightarrow \ldots \qquad (6.9.2)$$

Now let us identify $V_{\bar{L}}$ with V_L using the composition of the tautological identification and of (6.6.2). Then (6.9.2) becomes

$$0 \longrightarrow V_L^0 \xrightarrow{(D, \bar{D})} V_L^1 \oplus V_L^{-1} \xrightarrow{D \oplus \bar{D}} V_L^2 \oplus V_L^{-2} \longrightarrow \ldots \qquad (6.9.3)$$

where
$$D = d_M = \int e^{\mu-\rho}(z) : V_L^i \longrightarrow V_L^{i+1} \qquad (6.9.4)$$
cf. (6.3.1), and
$$\bar{D} = \int e^{\bar{\mu}-\bar{\rho}}(z) = \int e^{-\mu+(1-2\alpha)\rho+2\alpha(\alpha-1)\beta}(z) : V_L^i \longrightarrow V_L^{i-1} \qquad (6.9.5)$$

6.10. It follows from (6.9.3) that
$$H^0(\mathbb{P}^1, \Omega_{X,(\alpha)}^{ch}) = \mathrm{Ker}(D : V_L^0 \longrightarrow V_L^1) \cap \mathrm{Ker}(\bar{D} : V_L^0 \longrightarrow V_L^{-1}) \qquad (6.10.1)$$

Thus our space of global sections is the intersection of kernels of two screening operators connected with lattice vectors $\mu - \rho$ and $\bar{\mu} - \bar{\rho}$ respectively. Let us compute the angle between these two vectors:
$$(\mu-\rho, \bar{\mu}-\bar{\rho}) = (\mu-\rho, -\mu+(1-2\alpha)\rho+2\alpha(\alpha-1)\beta) = 2\alpha^2 - 1 = k+1 \qquad (6.10.2)$$
(Recall that $k = 2\alpha^2 - 2$ is the central charge of $\hat{\mathfrak{g}}$ acting on our modules.)

6.11. Set $V_L'^0 = \oplus_{i\in\mathbb{Z}} V_L^{i00} \subset V_L^0$. The operators $D, \bar{D}$ make sense on $V_L'^0$ for all $\alpha \in \mathbb{C}$ not necessarily integer, cf. 5.2. Intersection of their kernels is equal to $H^0(X, \Omega_{X,(\alpha)}^{ch,0})$, which due to Proposition 4.3 (iii) is $V_{0,2\alpha^2-2}$ (when α is generic); as far as we can judge, this is Feigin's result [**F**]. The restriction of (6.9.2) to $V_L'^0$ is the resolution appearing, perhaps implicitly, in [**F**]. It is indeed a resolution due to Proposition 5.9, (iii). If α is not generic, then at most the 1st cohomology arise and its structure is provided by Proposition 5.9, parts (i)-(ii). The size of the 0th cohomology group also jumps at special values of α as Proposition 4.3, parts (i)-(ii) show.

References

[AG] S. Arkhipov, D. Gaitsgory, *Differential operators on the loop group via chiral algebras*, Int. Math. Res. Notes, **2002**, no. 4, 165-210.

[BB] A. Beilinson, J. Bernstein, *Localisation de $\mathfrak{g}$-modules*, C. R. Acad. Sci. Paris, Sér. I Math., **292** (1981), no. 1, 15-18.

[BD] A. Beilinson, V. Drinfeld, *Chiral algebras*, Amer. Math. Society, Colloquium Publications, **51**, 2004.

[BPZ] A. Belavin, A. Polyakov, A. Zamolodchikov, *Infinite conformal symmetry in two-dimensional quantum field theory*, Nucl. Phys. **B 241** (1984), 333-380.

[B] R. Borcherds, *Vertex algebras, Kac-Moody algebras and the Monster*, Proc. Natl. Acad. Sci. USA, **83** (1986), 3068-3071.

[Bor] L. Borisov, *Vertex algebras and mirror symmetry*, Comm. Math. Phys., **215** (2001), 517-557.

[BL] L. Borisov, A. Libgober, *Elliptic genera of toric varieties and applications to mirror symmetry*, Inv. Math., **140** (2000), 453-485.

[Br] P. Bressler, *Vertex algebroids I, II*. math.AG/0202185, 0304115.

[F] B. Feigin, *Super quantum groups and the algebra of screening for $\widehat{sl}_2$-algebra*, Preprint.

[FF1] B. Feigin, E. Frenkel, *A family of representations of affine Lie algebras*, Uspechi Math. Nauk, **43** (1988), no. 5 (263), 227-228 (in Russian).

[FF2] B. Feigin, E. Frenkel, *Affine Kac-Moody algebras and semi-infinite flag manifolds*, Comm. Math. Phys., **128** (1990), 161-189.

[FF3] B. Feigin, E. Frenkel, *Representations of affine Kac-Moody algebras and bosonisation*, V.G. Knizhnik Memorial volume, L. Brink, D. Friedan, A.M. Polyakov (eds.), World Scientific, Singapore (1990), 271-316.

[FF4] B. Feigin, E. Frenkel, *Integrals of motion and quantum groups*, Integrable systems and quantum groups, R. Donagi, B. Dubrovin, E. Frenkel, E. Previato, Lect. Notes in Math., **1620** (1995), 349-418.

[FF5] B. Feigin, E. Frenkel, Semi-infinite Weil complex and the Virasoro algebra, *Comm. Math. Phys.*, **137** (1991), 617-639.

[FBZ] E. Frenkel, D. Ben-Zvi, Vertex algebras and algebraic curves, *Mathematical Surveys and Monographs*, **88**, Amer. Math. Soc., Providence, RI, 2001.

[FMS] D. Friedan, E. Martinec, S. Schenker, Conformal invariance, supersymmetry and string theory, *Nucl. Phys.*, **B271** (1986), 93-165.

[GM] V. Gorbounov, F. Malikov, Vertex algebras and the Landau - Ginzburg/ Calabi-Yau correspondence, math.AG/0308114

[GMS1] V. Gorbounov, F. Malikov, V. Schechtman, Gerbes of chiral differential operators. II, math.AG/0003170.

[GMS2] V. Gorbounov, F. Malikov, V. Schechtman, Gerbes of chiral differential operators. III, *The orbit method in Geometry and Physics*, In honor of A.A.Kirillov, Progress in Mathematics, **213** (2003), 73-100.

[GMS3] V. Gorbounov, F. Malikov, V. Schechtman, On chiral differential operators over homogeneous spaces, *Int. J. Math. Sci.* **26** (2001), 83-106.

[HBJ] F. Hirzebruch, Th. Berger, R. Jung, Manifolds and modular forms, 2nd ed., *Aspects of Mathematics*, Vieweg, 1994.

[K1] V. Kac, Vertex algebras for beginners, Second Edition, University Lecture Series, **10**, American Mathematical Society, Providence, Rhode Island, 1998.

[K2] V. Kac, Infinite dimensional Lie algebras, 2nd ed., Cambridge University Press, Cambridge et al., 1985.

[KZ] M. Kaneko, D. Zagier, A generalized Jacobi theta function and quasimodular forms, The moduli space of curves (Texel Island, 1994), *Progr. Math.*, **129** (1995), Birkhäuser Boston, MA, 165-172.

[KV] M. Kapranov, E. Vasserot, Vertex algebras and the formal loop space, math.AG/0107143.

[LVW] W. Lerche, C. Vafa, P. Warner, Chiral rings in $N = 2$ superconformal field theory, *Nucl. Phys.* **B 324** (1989), 427-474.

[MS1] F. Malikov, V. Schechtman, Chiral de Rham complex. II, *Amer. Math. Soc. Transl.*, **194** (1999), 149-188.

[MS2] F. Malikov, V. Schechtman, Chiral Poincaré duality, *Math. Res.Lett* **6** (1999), 533-546

[MSV] F. Malikov, V. Schechtman, A. Vaintrob, Chiral de Rham complex, *Comm. Math. Phys.*, **204** (1999), 439-473.

[S] V. Schechtman, Definitio nova algebroidis verticiani, Prépublication du lab. Emile Picard, **271**, Toulouse (2003).

[TT] D. Tamarkin, B. Tsygan, The ring of differential operators on forms in non-commutative calculus, these Proceedings.

[W] M. Wakimoto, Fock representations of the affine Lie algebra A_1^1, *Comm. Math. Phys.*, **104** (1986), 605-609.

[Z] D. Zagier, A proof of the Kac-Wakimoto affine denominator formula for the strange series, *Math. Res. Let.*, **7** (2000), 597-604.

DEPARTMENT OF MATHEMATICS, UNIVERSITY OF KENTUCKY, LEXINGTON, KY 40506, USA
E-mail address: vgorb@ms.uky.edu

DEPARTMENT OF MATHEMATICS, UNIVERSITY OF SOUTHERN CALIFORNIA, LOS ANGELES, CA 90089, USA
E-mail address: fmalikov@mathj.usc.edu

LABORATOIRE EMILE PICARD, UFR MIG, UNIVERSITÉ PAUL SABATIER, 118 ROUTE DE NARBONNE, 31062 TOULOUSE, FRANCE
E-mail address: schechtman@picard.ups-tlse.fr

MANIFOLDS: INVARIANTS AND MIRROR SYMMETRY

Invariants of tangles with flat connections in their complements

R. Kashaev and N. Reshetikhin

ABSTRACT. Let G be a simple complex algebraic group. By using a notion of a G-category we define invariants of tangles with flat G-connections in their complements. We also show that quantized universal enveloping algebras at roots of unity provide examples of G-categories.

CONTENTS

Introduction
1. Tangles with flat G-connections in their complements
2. The category of G-colored diagrams
3. Braided G-categories
4. Invariants of framed G-tangles
5. Quantized universal enveloping algebra of gl_2
6. The algebra $\mathcal{U}$
7. The algebra $\mathcal{U}_\varepsilon$
8. Conclusion
References

Introduction

A breakthrough in the theory of invariants of knots came with the discovery of the Jones polynomial [**J**] and its generalizations (HOMFLY and Kauffman). Then it was shown in [**RT**] that such invariants as well as invariants of tangled graphs can be obtained from quantized universal enveloping algebras of simple Lie algebras.

In this paper we extend the construction of invariants of links from [**RT**] (for details see [**T1**]). We construct invariants of tangles in $\mathbb{R}^2 \times [0,1]$ with a flat connection in a principal G-bundle (where G is a simple complex algebraic group) over the complement of a tangle. The invariant is a functor from the category of G-tangles to a given G-category (see sections 1–3 for definitions). The invariant of a tangle with a gauge class of flat connections in the complement is defined as the image of the corresponding G-tangle.

In section 1 of this paper first we describe the moduli space of flat connections in the complement to a tangle in $\mathbb{R}^2 \times [0,1]$. Then we define the category of tangles with gauge classes of flat connections in their complements. The construction is essentially the same as π-tangles introduced and studied by Turaev in [**T2**]. In section 2 we define the category of G-colored tangle diagrams and show that it is naturally equivalent to the category of tangles with flat G-connections in their complements. In section 3 we introduce the notion of a G-category, which is very similar but different from that of [**T2**]. In section 4 we construct invariants of G-tangles. In sections 5–7 we show that representations of $\mathcal{U}_\varepsilon$ provide examples of GL_2^*-categories and therefore provide invariants of tangles with GL_2-connections in their complements.

We are grateful to R. Kirby, D. Thurston, V. Turaev, and M. Yakimov for interesting discussions and useful comments. Both authors were partly supported by the NSF grant DMS-0070931 and by the CRDF grant RM1-2244 and by the Swiss National Science Foundation.

1. Tangles with flat G-connections in their complements

1.1. Tangles and diagrams. Let $I \equiv [0,1]$ be the closed unit interval. A *geometric tangle* $t \subset \mathbb{R}^2 \times I$ is the image of a smooth embedding $\underbrace{I \sqcup \ldots \sqcup I}_{k \text{ times}} \sqcup \underbrace{S^1 \sqcup \ldots \sqcup S^1}_{l \text{ times}} \to \mathbb{R}^2 \times I$ with oriented components such that $\partial t \subset \mathbb{R}^2 \times \partial I$, and t intersects the boundary $\mathbb{R}^2 \times \partial I$ perpendicularly. We have $\partial t = \partial_+ t \sqcup \partial_- t$ where $\partial_+ t = (\mathbb{R}^2 \times \{1\}) \cap \partial t$ and $\partial_- t = (\mathbb{R}^2 \times \{0\}) \cap \partial t$. The geometric tangle t is called *standard* if $\partial_+ t = \{(0,1,1), (0,2,1) \ldots (0,n,1)\}$ and $\partial_- t = \{(0,1,0), (0,2,0) \ldots (0,m,0)\}$ for some $n, m \in \mathbb{Z}_{>0}$. We will say that such tangle has type (m, n).

Denote by D_t the image of t under the projection

(1) $$p \colon \mathbb{R}^2 \times I \to \mathbb{R} \times I, \ (x, y, z) \mapsto (y, z)$$

together with the additional information at each self-crossing point about underpassing and overpassing segments. The projection D_t is called *regular* if its only singularities are double points at which the images of the corresponding components of t intersect transversally. The ambient isotopy class of a regular projection D_t is called a *diagram* of the tangle t. When it is not confusing we will use the same notation D_t for diagrams and regular projections.

The diagram of a tangle can be regarded as the isotopy class of a smooth embedding of a graph with four-valent vertices of two types and one-valent vertices. These are the double points and the boundary points, respectively. The edges of the graph are the segments running between the vertices. A vertex of D_t is called *positive* if it is a double point and the angle between upper and lower components is positive. If the angle is negative, the vertex is called *negative*, see Fig. 1. Each such embedding is always a diagram of some tangle. Two diagrams are called *Reidemeister equivalent* if one can be obtained from another by a finite sequence of Reidemeister moves (see Figs 2–4).

A *tangle* is the isotopy class of a geometric tangle. Tangles are in bijection with Reidemeister classes of diagrams.

A *geometric framing* of a geometric tangle t is a continuous nonsingular section of the normal bundle to t. A framing of a tangle is the isotopy class of a geometric

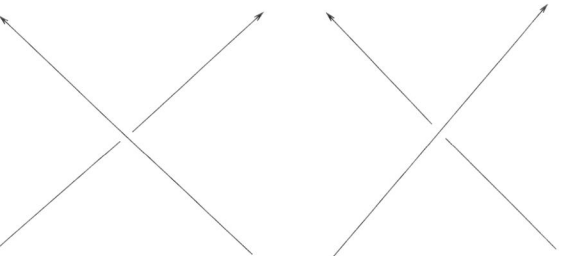

(a) Negative intersection (b) Positive intersection

FIGURE 1

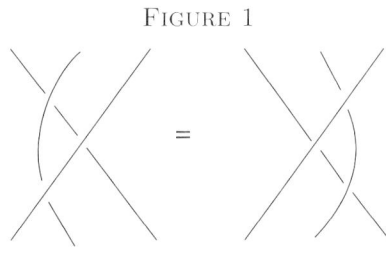

FIGURE 2

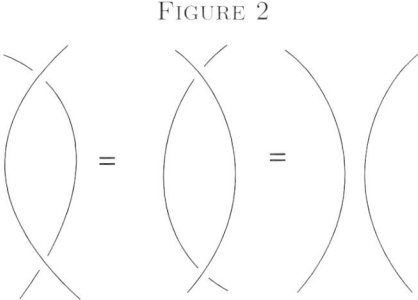

FIGURE 3

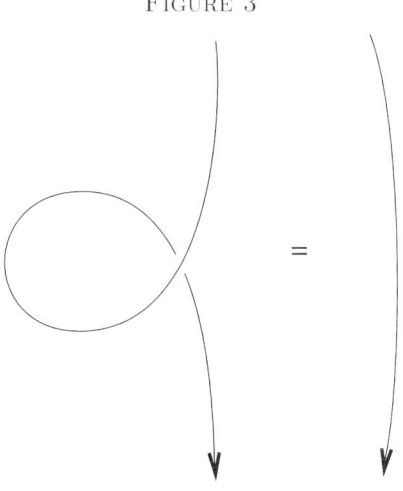

FIGURE 4

framing. A framing is *standard* if it is parallel to the x-axis at $\partial_\pm t$ with positive projection to this axis. It is clear that every framed geometric tangle with such framing at the boundary is isotopically equivalent to a framed geometric tangle with the framing parallel to x-axis. Framed tangle is the isotopy class of a framed geometric tangle with the standard framing.

Two diagrams of tangles are called *framed Reidemeister equivalent* if they are connected by a sequence of moves in Figs 2, 3, 5.

Framed Reidemeister classes of diagrams are in bijection with the framed tangles.

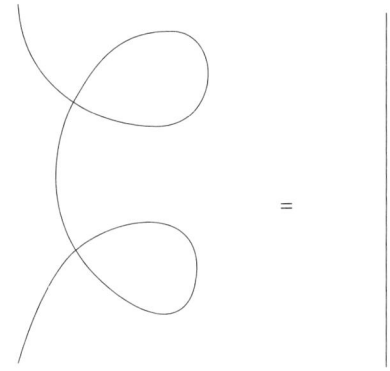

FIGURE 5

1.2. Flat G-connections in the complement of a tangle and G-tangles. Let E be a trivial principal G-bundle over $\mathbb{R}^2 \times I$ and $A_t \in \Omega^1(\mathbb{R}^2 \times I \backslash t, \mathfrak{g})$ be the 1-form representing a flat connection in E over the complement to a standard geometric framed tangle t of type (m, n). The corresponding parallel transport operators along paths give an equivalent description of the flat connection as a representation of the fundamental groupoid of $(\mathbb{R}^2 \times I)\backslash t$ in G.

Let E_m be a trivial principal G-bundle over $\mathbb{R}^2$, and α be the $\mathfrak{g}$-valued 1-form of a flat connection in E' over $\mathbb{R}^2\backslash\{(0,1),\ldots,(0,m)\}$. Assume that $\alpha(x,y)$ decays sufficiently fast when y goes to $-\infty$ so that the parallel transport operator of α along a path connecting (x_0, y_0) with (x, y) in $\mathbb{R}^2$ has a limit as $y_0 \to -\infty$. Let $[\alpha]$ be the gauge class of α with respect to gauge transformations trivial at ∞. Let γ_i be a path which starts at $(0, -\infty)$ encircles points $(0, 1), \ldots, (0, i)$ and then returns to $(0, -\infty)$ (see Fig. 6). We can identify the class $[\alpha]$ with an element of $G^m = G \times \cdots \times G$ so that the group element g_i in the collection $(g_1, \ldots, g_m) \in G^m$ is the holonomy along the path γ_i.

Let α and β be flat connections in E_m and E_n, respectively. Denote by $\mathcal{A}_t(\alpha, \beta)$ the space of flat G-connections over the complement of t which, when restricted to

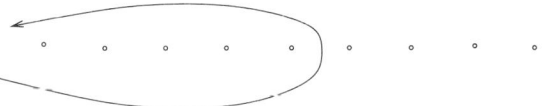

FIGURE 6

$\mathbb{R}^2 \times \{0\} \setminus \{(0,1,0),\ldots,(0,m,0)\}$ and $\mathbb{R}^2 \times \{1\} \setminus \{(0,1,1),\ldots,(0,n,1)\}$, coincide with α and β, respectively, and which decay sufficiently fast when $y \to -\infty$ (so that the parallel transport operator along a path connecting (x_0, y_0, z_0) and (x, y, z) has a finite limit when $y_0 \to -\infty$).

If G_t is the group of gauge transformations trivial at infinity, the quotient space $\mathcal{M}_t([\alpha], [\beta]) = \mathcal{A}_t(\alpha, \beta)/G_t$ is by definition the moduli space of those flat G-connections in the complement of t which, when restricted to two boundary components $\mathbb{R}^2 \times \{0\} \setminus \partial t$ and $\mathbb{R}^2 \times \{1\} \setminus \partial t$, give representatives of the gauge classes $[\alpha]$ and $[\beta]$ respectively.

Following [**T2**], denote $C_t = (\mathbb{R}^2 \times I) \setminus t$ and consider the fundamental group of C_t with the base point in $\{0\} \times [-\infty, L] \times I$ with $t \subset \mathbb{R} \times [L+1, \infty] \times I$. The set of such base points is contractible and we will write $\pi_1(C_t)$ for the fundamental group, suppressing the base point. We will call such base points left base points. It is clear that an element of $\mathcal{M}_t([\alpha], [\beta])$ defines a representation of $\pi_1(C_t)$. Gauge classes of boundary values $[\alpha]$ and $[\beta]$ become elements of G^m and G^n respectively, defined as holonomies along γ_i (see above). Thus, the space $\mathcal{M}_t([\alpha], [\beta])$ can be identified with a subspace of G-tangles [**T2**], where a G-tangle is defined as a pair consisting of a framed oriented tangle t and a group homomorphism $\rho \colon \pi_1(C_t) \to G$.

Let t and t' be two standard isotopically equivalent geometric tangles. An isotopy bringing t to t' lifts to an isomorphism between G-connections over the complements of t and t' and therefore to an isomorphism between corresponding moduli spaces of flat G-connections. The isomorphism class of the moduli spaces generated by these isomorphisms can be identified with the equivalence class of pairs (t, A_t) with respect to isotopies of tangles, their pull-backs acting on connections and gauge transformations from G_t. We will denote the obtained set of classes by $\mathcal{M}_{[t]}([\alpha], [\beta])$.

From now on we will work with G-tangles. The geometric picture with flat connections will be used only once below when we shall assign a geometrical meaning to G-colorings of diagrams.

Let $v = (x, y, z) \in t \subset \mathbb{R}^2 \times I$ and $\gamma_v \subset C_t$ be a (homotopy) path which starts at a left base point then goes to the point $(x, y + \delta, z)$ "over" the tangle, then returns to the base point "under" the tangle. Here δ is sufficiently small and we assume that t is transversal to (x,y)-plane at (x, y, z). We will call such path (the homotopy class thereof) *standard* for $v = (x, y, z)$.

1.3. The category of G-tangles. *Objects* of the category $\mathcal{T}(G)$ of G-tangles are finite sequences $\{(\varepsilon_1, g_1), \ldots, (\varepsilon_n, g_n)\}$ with $\varepsilon_i = \pm 1$, $g_i \in G$. *Morphisms* from $\{(\varepsilon_1, g_1), \ldots, (\varepsilon_m, g_m)\}$ to $\{(\sigma_1, h_1), \ldots, (\sigma_n, h_n)\}$ are G-tangles.

Here m is the number of connected components of $(\mathbb{R}^2 \times \{0\}) \cap t$ and n is the number of connected components of $(\mathbb{R}^2 \times \{1\}) \cap t$. The signs ε_i and σ_i show the orientation of the boundary components. If $"+"$, the component is oriented upward and if $"-"$, it is oriented downward. The representation $\rho : \pi_1(C_t) \to G$ should agree with $\{g_i\}$ and $\{h_i\}$ in the following way. If γ_i^+ is a path from a left base point encircling boundary points $(0, 1, 0), \ldots (0, i, 0)$ in the vicinity of $\mathbb{R}^2 \times \{0\}$ then $\rho(\gamma_i^+) = g_i$. Similarly for a path γ_i^- encircling points $(0, 1, 1), \ldots, (0, i, 1)$ in the vicinity of $\mathbb{R}^2 \times \{1\}$ we have $\rho(\gamma_i^-) = h_i$.

The composition of morphisms is defined by gluing tangles. The identity morphism of

$\{(\varepsilon_1, g_1), \ldots, (\varepsilon_n, g_n)\}$ to itself is the trivial braid with the representation of the fundamental group defined by $g_1, \ldots, g_n$.

REMARK 1. *The category $\mathcal{T}(G)$ can also be defined in more geometrical terms of gauge classes of flat connections. Objects are sequences $(\varepsilon_1, \ldots, \varepsilon_n; [\alpha])$ where $[\alpha]$ is a gauge class of a flat connection over $\mathbb{R}^2 \setminus \{(0,1) \ldots (0,n)\}$ in the trivial principal G-bundle over $\mathbb{R}^2$ and $\varepsilon_i = \pm 1$. Morphisms between $(\varepsilon_1, \ldots, \varepsilon_m; [\alpha])$ and $(\sigma_1, \ldots, \sigma_n, [\beta])$ are elements of $\mathcal{M}_{[t]}([\alpha], [\beta])$ i.e. equivalence classes of pairs (t, A_t) described in the previous subsection.*

2. The category of G-colored diagrams

2.1. Factorizable groups and Lie groups. We say that group G is *factorizable* into two subgroups $G_\pm \subset G$ if any element $g \in G$ can be represented in a unique way as

$$g = g_+ g_-^{-1} \tag{2}$$

where $g_\pm \in G_\pm$.

If G is a complex algebraic Lie group (later we will focus on this case) we will say that it is factorizable if there exists a Zariski open neighborhood $G' \subset G$ of 1 such that every element of G' has a unique factorization (2). Notice that in this case any $l \in \mathfrak{g} = Lie(G)$ has a unique decomposition

$$l = l_+ - l_-$$

where $l_\pm \in \mathfrak{g}_\pm = Lie(G_\pm)$.

REMARK 2. *We can choose $G_+ = \{e\}$ and $G_- = G$. We call it trivial factorizability.*

Let G be a factorizable group. Define a binary operation

$$g \star h = g_+ h_+ (g_- h_-)^{-1}$$

which obviously defines a group structure on G with the same identity element as for the original group structure. The inverse of g in this group is $i(g) = g_+^{-1} g_-$. This operation corresponds to the multiplication of the group $G_+ \times G_-$ under the mapping $G_+ \times G_- \to G$, $(g_+, g_-) \mapsto g_+ (g_-)^{-1}$. In what follows, we shall denote this group G^*.

2.2. G-colorings of diagrams. Let t be a standard geometric tangle and D_t, its diagram with the set of edges $E(D_t)$. Assume that G is a factorizable group.

DEFINITION 1. *The map $E(D_t) \to G$ which associates to edge e element $x_e \in G$ is called a G-coloring of diagram D_t if at each double point it satisfies the relations*

$$x_{b_v} = (x_{a_v})_\pm^{-1} x_{c_v} (x_{a_v})_\pm, \quad x_{a_v} = (x_{c_v})_\mp x_{d_v} (x_{c_v})_\mp^{-1},$$

depending on whether the intersection is positive or negative. Here the enumeration of edges is the same as on Fig. 7.

Let $x_L, x_R \colon G \times G \to G$ be mappings acting as

$$x_L(x, y) = x_- y x_-^{-1}, \quad x_R(x, y) = x_L(x, y)_+^{-1} x x_L(x, y)_+ \tag{3}$$

In terms of these maps the definition above means that at positive double points we have $(x_a, x_b) = (x_L(x_c, x_d), x_R(x_c, x_d))$ and at negative double points $(x_c, x_d) = (x_L(x_a, x_b), x_R(x_a, x_b))$.

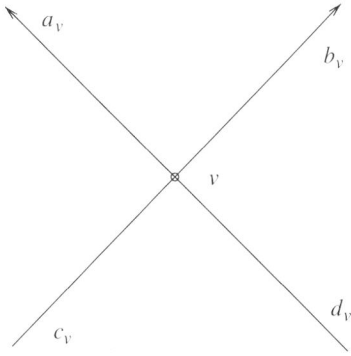

FIGURE 7

The following proposition is due to Weinstein and Xu [**WX**] in the context of factorizable Poisson–Lie groups.

PROPOSITION 1. *The map* $\mathcal{R}\colon G \times G \to G \times G$ *acting as* $(x,y) \mapsto (x_L(y,x), x_R(y,x))$ *satisfies the set-theoretical Yang–Baxter equation:*

$$\mathcal{R}_{12}\mathcal{R}_{13}\mathcal{R}_{23} = \mathcal{R}_{23}\mathcal{R}_{13}\mathcal{R}_{12}$$

Here all mappings act from $G^{\times 3}$ to itself. The mapping $\mathcal{R}_{12}$ acts as $\mathcal{R}$ in the first two factors and trivially in the last one. The other mappings act in a similar way. The subindices indicate in which factors the mapping acts non-trivially.

It follows from this proposition that if the G-coloring of lower edges of both diagrams on Fig. 2 are x, y, z, then the colorings of upper edges of the diagrams (which are determined by the diagrams and the coloring of lower edges) are the same for both diagrams. In other words, the G-coloring is compatible with the third Reidemeister move. It is easy to see that the G-coloring is also compatible with other framed Reidemeister moves.

Let (D, c) and (D', c') be two G-colored diagrams which are Reidemeister equivalent. Then, since in each Reidemeister move the coloring of a new diagram is uniquely defined by the coloring of the initial diagram, c' is uniquely determined by c. Thus, we have Reidemeister classes of G-colored diagrams.

2.3. The category of G-colored diagrams. Let G be a factorizable group. Define category $\mathcal{D}(G)$ of G-colored diagrams as follows.

Objects of the category are sequences $\{(\varepsilon_1, x_1), \ldots, (\varepsilon_n, x_n)\}$ where $\varepsilon_i = \pm$ and $x_i \in G$ and the empty set.

Morphisms between $\{(\varepsilon_1, x_1), \ldots, (\varepsilon_n, x_n)\}$ and $\{(\sigma_1, y_1), \ldots, (\sigma_m, y_m)\}$ are Reidemeister classes of G-colored diagrams with the orientation of the boundary edges (adjacent to 1-valent vertices) defined by ε_i and σ_j as it is shown on Fig. 8 and with the G-colorings of the boundary edges given by $\varepsilon_i(x_i)$ and $\sigma_j(y_j)$ where $\varepsilon(x)$ is defined as

(4) $$\varepsilon(x) = \begin{cases} x = x_+ x_-^{-1} & \text{for } \varepsilon = +1 \\ i(x) = x_+^{-1} x_- & \text{for } \varepsilon = -1 \end{cases}$$

The operation $x \to i(x)$ is taking the inverse in G^*. The identity morphism is shown on Fig. 9.

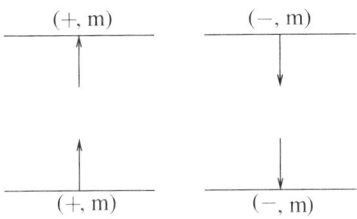

FIGURE 8

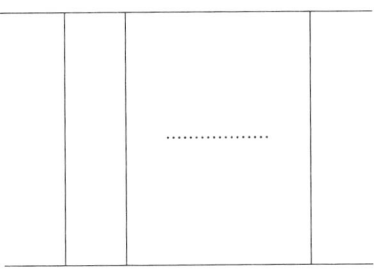

FIGURE 9

Composition of Reidemeister classes of G-colored diagrams (D_1, c_1) and (D_2, c_2) is the Reidemeister class of the G-colored diagram $(D_1 \circ D_2, c)$ where $D_1 \circ D_2$ is the diagram obtained by gluing D_1 and D_2 and the coloring c is induced by colorings c_1 and c_2,

2.4. The equivalence of categories. Here we will prove the equivalence of categories $\mathcal{T}(G) \simeq \mathcal{D}(G)$. Consider the map $F : \mathcal{T}(G) \to \mathcal{D}(G)$ acting on objects as

$$F((\varepsilon_1, g_1), \ldots, (\varepsilon_n, g_n)) = \{(\varepsilon_1, x_1) \ldots (\varepsilon_n, x_n)\}$$

Here $x_i \in G$ are related to $g_i \in G$ via

(5) $$g_i = (x_1)_+^{\varepsilon_1} \ldots (x_i)_+^{\varepsilon_i} (x_i)_-^{-\varepsilon_i} (x_1)_-^{-\varepsilon_1}$$

For a G-tangle (t, ρ) define

$$F((t, \rho)) = [(D_t, c)]$$

where D_t is a diagram of the tangle t, c is the coloring of D_t which we define below, and $[(D_t, c)]$ the colored framed Reidemeister class of (D, c).

Consider a standard path γ_v associated with point $v \in t$. Let $e_1, \ldots, e_i$ be the edges of D_t intersected by the projection of γ_v, and $\varepsilon_1, \ldots, \varepsilon_i$ the signs of the projections of their orientations to the vertical axis. Then the holonomy $g_v = g_i$ along γ_v associated to ρ is given by formula (5), where $x_1, \ldots, x_i$ now are the colors of edges $e_1, \ldots, e_i$. It is easy to see that this correspondence between ρ and the edge colors is one to one, and thus the mapping F is invertible.

It is easy to see that the map F is a functor. To prove this it remains to show that $F(fg) = F(f)F(g)$ for morphisms f and g, which is obvious. It is also clear that this functor is an equivalence of categories.

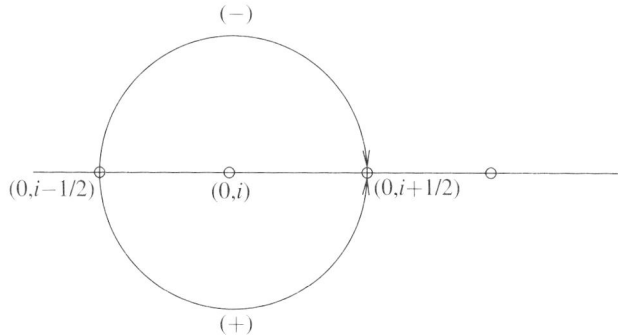

FIGURE 10

2.5. A geometric version of the functor F. Now consider a geometric description of the category $\mathcal{T}(G)$ and describe the functor F in this terms.

To define the action of F on objects we consider a special representative $\tilde{\alpha}$ of the gauge class $[\alpha]$. Namely, $\tilde{\alpha}$ is continuous and vanishes outside the strips $\{(x,y)|\ x \in \mathbb{R},\ i-\epsilon < y < i+\epsilon\}$ for some $0 < \epsilon < 1/2$. Let $\gamma_i^{\pm}$ be paths connecting points $(0, i-1/2)$ and $(0, i+1/2)$ which go around $(0, i)$ in the clock-wise direction for γ_i^- and in the counter clock-wise direction for γ_i^+. These paths are shown as $(-)$ and $(+)$ paths respectively on Fig. 10.

Define the action of F on objects as:

$$F: (\varepsilon_1, \ldots, \varepsilon_n; [\alpha]) \mapsto \{(\varepsilon_1, x_1), \ldots, (\varepsilon_n, x_n)\}$$

where $x_i = hol_{\gamma_i}(\tilde{\alpha})$ and $\gamma_i = \gamma_i^+ \circ (\gamma_i^-)^{-1}$.

To define the action of F on morphisms, first, let us look at the geometry of the projection p. The preimages of edges of D_t form a system of "walls" $p^{-1}(e) \subset \mathbb{R}^2 \times I$, $e \in E(D_t)$, intersecting at lines which are preimages of the vertices of D_t. Let A_t be a flat connection over the complement of t. If the tangle is not a link all chambers bounded by walls including the "outer" chambers are simply connected so the flat connection A_t can be trivialized inside these chambers, except a thin neighborhood of walls. If the tangle is a link, the outer chamber is not simply connected, but we still can trivialize the flat connection inside this chamber since the link can be placed into a ball and the outside of a ball is a simply connected region in $\mathbb{R}^2 \times I$. Thus, we can trivialize the flat connection in each chamber outside of a thin neighborhood of walls. Assume that this thin neighborhood is such that at the boundary it is inside of strips $x \in \mathbb{R},\ i - \epsilon < y < i + \epsilon$. The tangle t separates each wall into two semi-infinite parts. Let e be an edge of D_t, denote by $p^{-1}(e)_-$ the part of the wall $p^{-1}(e)$ which is semi-infinite in the negative x-direction and by $p^{-1}(e)_+$ the other part of this wall. Since the connection is flat and since now it is trivial inside the chambers (except of a thin neighborhood of walls), the holonomy through the wall $p^{-1}(e)$ along any path that is based on a pair of points separated by this wall and which are outside of a thin neighborhood of the wall, depends only on the homotopy class of the path. Let us call the path positive if its orientation together with the orientation of the edge e and with the direction of the projection p form positively oriented triple of vectors in $\mathbb{R}^3$ with the standard orientation. This produces two elements $x_{\pm}(e)$ of G which we can assign to the edge e corresponding to holonomies along positive paths through $p^{-1}(e)_{\pm}$. The product

$x(e) = x_+(e).x_-(e)^{-1}$ is the holonomy along a closed path that crosses first the wall $p^{-1}(e)_+$ in the positive direction and then $p^{-1}(e)_-$ in the negative direction. If G is factorizable there exists unique pair $x(e)_\pm$ which factorizes $x(e)$ as above with $x(e)_\pm \in G_\pm$, and one can choose the connection A_t so that $x_\pm(e) = x(e)_\pm$. This gives us an assignment $e \mapsto x(e)$ of group elements to edges of the diagram. Notice that this assignment does not depend on base points inside chambers and depends only on the gauge class of the flat connection.

The holonomy along a path connecting two points based inside chambers is the product of holonomies through the walls intersected by this path. Given a tangle t and its diagram D_t these holonomies can be computed by the *wall crossing rule* (see Fig. 11). Horizontal edges on Fig. 11 represent parts of the path. Vertical edges represent parts of the tangle (walls). Under-crossings and over-crossings show whether the path went through $p^{-1}(e)_-$ or $p^{-1}(e)_+$ respectively where e is the corresponding edge of the diagram. The holonomy gained at the crossing is given in terms of the G coloring of e.

To show that the map assigning group elements to edges constructed above gives G-colorings, one should consider pairs of paths from Figs 12–15. Isotopy equivalence of these paths implies the equality of the corresponding holonomies. Computation of them according to the "wall-crossing rules" described above gives the identities in the definition of the G-coloring.

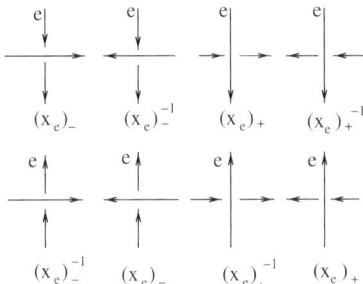

FIGURE 11

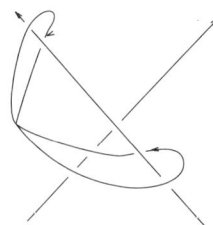

FIGURE 12

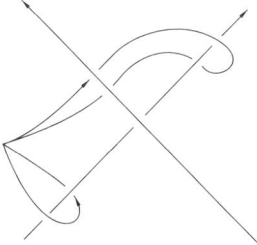

FIGURE 13

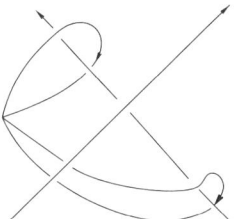

FIGURE 14

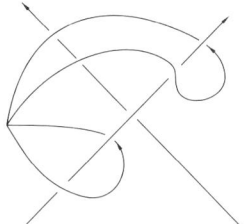

FIGURE 15

3. Braided G-categories

3.1. Braided G-categories. A *braided group* is a pair $(G, \mathcal{R} : G \times G \to G \times G)$ where G is a group and the map $\mathcal{R}$ satisfies the following requirements
  (1) $m \circ \mathcal{R} = m'$
  (2) $\mathcal{R} \circ (m \times \mathrm{id}) = (m \times \mathrm{id}) \circ \mathcal{R}_{23} \circ \mathcal{R}_{13}$
  (3) $\mathcal{R} \circ (\mathrm{id} \times m) = (\mathrm{id} \times m) \circ \mathcal{R}_{12} \circ \mathcal{R}_{13}$

where m is the multiplication and m' is the opposite multiplication in G. In particular, $\mathcal{R}$ satisfies the set-theoretical Yang–Baxter equation

$$\mathcal{R}_{12} \circ \mathcal{R}_{13} \circ \mathcal{R}_{23} = \mathcal{R}_{23} \circ \mathcal{R}_{13} \circ \mathcal{R}_{12}.$$

An example of a braided group is the pair $(G^*, \mathcal{R})$ where G^* is a factorized group G with multiplication $m(x,y) = x \star y$ and $\mathcal{R}$ is given by (3).

For a set A we will say that category $\mathcal{C}$ *is fibered over* A if
  - There is a projection $\pi : \mathrm{Ob}(\mathcal{C}) \to A$
  - $\mathrm{Hom}_{\mathcal{C}}(X, Y) = \emptyset$, if $\pi(X) \neq \pi(Y)$

Recall that a monoidal category is a category with a functor $\otimes : \mathcal{C} \times \mathcal{C} \to \mathcal{C}$. This functor is given together with the natural transformations $a : \otimes \circ \otimes \times id \simeq \otimes \circ id \times \otimes$ which is called an associativity constraint. This natural transformation should satisfy the pentagon identity [**ML**]. In addition to this, a monoidal category has an identity object $\mathbb{1}$ given together with a system of functorial isomorphisms $l_X : \mathbb{1} \otimes X \simeq X$ and $r_X : \mathbb{1} \otimes X \simeq X$ which satisfy some natural conditions [**ML**].

A monoidal category fibered over a group G is called a *G-category* if

- $\pi(X \otimes Y) = \pi(X)\pi(Y)$
- $\pi(\mathbb{1}) = e \in G$

Associativity constraint and functorial morphisms r_X and l_X should act fiber-wise. From now on we will work only with strict monoidal categories: we assume that the associativity constrain is trivial (see [**ML**] for details on what this exactly means).

Recall that a monoidal category is called rigid if any object X has a left dual object X^*, the injection and evaluation mappings $i_X : \mathbb{1} \to X \otimes X^*$ and $e_X : X^* \otimes X \to \mathbb{1}$, and if the triple X^*, i_X, e_X is unique up to an isomorphism. In a rigid monoidal category double dual is not necessary isomorphic to the object itself, and, in particular, each object has left and right duals.

A G-category $\mathcal{C}$ is called *rigid* if it is a rigid monoidal category and in addition to the properties listed above one has

$$\pi(X^*) = \pi(^*X) = \pi(X)^{-1}$$

where $^*X$ and X^* are left and right duals to X respectively. The evaluation and injection morphisms act fiber-wise.

Now assume that the group G is a braided group. A category $\mathcal{C}$ is called braided rigid G-category if it is a rigid G-category and in addition to this it has the following properties:

(1) There exists a functor $B : \mathcal{C} \times \mathcal{C} \to \mathcal{C} \times \mathcal{C}$ such that the following diagram is commutative

$$\begin{array}{ccc} \mathcal{C} \times \mathcal{C} & \xrightarrow{B} & \mathcal{C} \times \mathcal{C} \\ \pi \times \pi \downarrow & & \downarrow \pi \times \pi \\ G \times G & \xrightarrow{\check{\mathcal{R}}} & G \times G \end{array}$$

where $\check{\mathcal{R}} = \mathcal{R} \circ P$, and $P(x,y) = (y,x)$. We will write $B : (X,Y) \to (X_L(X,Y), X_R(X,Y))$ for the action of B. This property is the lifting of property (1) of braided group G to the category $\mathcal{C}$.

(2) The functor B satisfies the following identities (for functors $\mathcal{C} \times \mathcal{C} \times \mathcal{C} \to \mathcal{C} \times \mathcal{C} \times \mathcal{C}$)

$$B \circ (\otimes \times id) = (id \times \otimes) \circ (B \times id) \circ (id \times B)$$

and the same identity for B^{-1}. This properties of B are liftings of properties (2) and (3) of the braided group G to the category $\mathcal{C}$.

(3) There exists an isomorphism of functors c which makes the following diagram commutative

$$\begin{array}{ccc} \mathcal{C} \times \mathcal{C} & \xrightarrow{B} & \mathcal{C} \times \mathcal{C} \\ \otimes \searrow & \stackrel{c}{\Rightarrow} & \swarrow \otimes \\ & \mathcal{C} & \end{array}$$

In other words, there exists a system of functorial isomorphisms
$$c^{X,Y} : X \otimes Y \to X_L(X,Y) \otimes X_R(X,Y).$$

(4) The commutativity constraint should satisfy the hexagon axioms
$$c^{X \otimes Y, Z} = (c^{X, X_L(Y,Z)} \otimes \mathrm{id})(\mathrm{id} \otimes c^{Y,Z})$$
$$c^{X, Y \otimes Z} = (c^{X,Y} \otimes \mathrm{id})(c^{X_R(X,Y), Z} \otimes \mathrm{id})$$

A braided G-category is called a *ribbon category* if it in addition to being a G-category has a system of functorial morphisms $\{\mu_X : X \to X^{**}\}_{X \in \mathrm{Ob}(\mathcal{C})}$ such that

- $\mu_{X \otimes Y} = \mu_X \otimes \mu_Y$
- $\mu_{X^*} = (\mu_X^*)^{-1}$
- $\mu_{\mathbb{1}} = id$

When it will not be misleading we will shorten the name " rigid braided ribbon G-category" to "ribbon G-category". The theorem 1 provides an example of a G-category.

When $G_+ = e$ are $G_- = G$ the notion of the G-category introduced above is equivalent to the one introduced in [**T2**].

3.2. The category of G-colored diagrams is a ribbon G-category.

THEOREM 1. *The category of G-colored framed diagrams is a ribbon G^*-category where $G^* = G_+ \times G_-$.*

PROOF. First, let us check that $\mathcal{R}$ satisfies the required properties. The first identity for $\mathcal{R}$ is equivalent to
$$x_L(x,y) \cdot x_R(x,y) = x \cdot y$$
The second identity for $\mathcal{R}$ is equivalent to
$$x_L(x, y \cdot z) = x_L(x,y) \cdot x_L(x_R(x,y), z)$$
and
$$x_R(x, y \cdot z) = x_R(x_R(x,y), z)$$
and similar identities assure the last property of $\mathcal{R}$. Here the multiplication is taken in G^*. All these identities are easy to check.

Now let us describe the structure of a G^*-category explicitly. Define the G^* structure on $\mathcal{D}(G)$ as
$$\pi((\varepsilon_1, x_1), \ldots, (\varepsilon_n, x_n)) = \varepsilon_1(x_1) \cdots \varepsilon_n(x_n)$$
Here the product is taken in G^*, $x \cdot y = x_+ y_+ y_-^{-1} x_-^{-1}$, and $\varepsilon(x)$ is defined in (4).

The monoidal structure is the same as for the category of diagrams. The tensor product of objects is
$$\{(\varepsilon_1, x_1), \ldots, (\varepsilon_n, x_n)\} \otimes \{(\sigma_1, y_1), \ldots, (\sigma_m, y_m)\}$$
$$= \{(\varepsilon_1, x_1), \ldots, (\varepsilon_n, x_n), (\sigma_1, y_1), \ldots, (\sigma_m, y_m)\}$$

The tensor product of morphisms is shown on Fig. 16. The identity object is the empty set.

The object dual to $\{(\varepsilon_1, x_1), \ldots, (\varepsilon_n, x_n)\}$ is $\{(-\varepsilon_n, i(x_n)), \ldots, (-\varepsilon_1, i(x_1))\}$ with the evaluation and the injection morphisms given by diagrams from Fig. 17 and Fig. 18 with the G-colorings induced by objects.

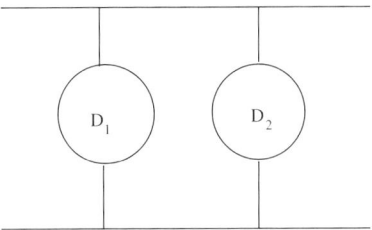

FIGURE 16

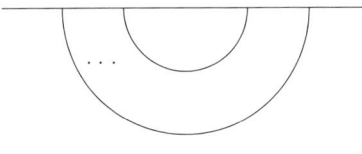

FIGURE 17

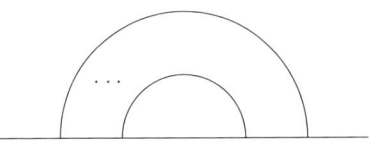

FIGURE 18

To describe the braiding, let us first define the functor $B : \mathcal{D}(G) \times \mathcal{D}(G) \to \mathcal{D}(G) \times \mathcal{D}(G)$ as follows. On objects it acts as

$$B(\{(\varepsilon_1, x_1), \ldots, (\varepsilon_n, x_n)\}, \{(\sigma_1, y_1), \ldots, (\sigma_m, y_m)\})$$
$$= (\{(\sigma_1, y_1^L), \ldots, (\sigma_m, y_m^L)\}, \{(\varepsilon_1, x_1^R), \ldots, (\varepsilon_n, x_n^R)\})$$

where $(x_1^L, \ldots, x_m^L, y_1^R, \ldots, y_n^R)$ is the image of $(x_1, \ldots, x_n, y_1, \ldots, y_m)$ with respect to the map

$$(s_n \cdots s_{n+m-1})(s_{n-1} \cdots s_{n+m-2}) \cdots (s_2 \cdots s_{n+1})(s_1 \cdots s_n) : G^{\times(n+m)} \to G^{\times(n+m)}$$

where $s_i = \check{\mathcal{R}}_{ii+1}$.

If $[(D_i, c_i)]$ is a morphism $(\varepsilon^{(i)}, x^{(i)}) \to (\sigma^{(i)}, y^{(i)})$,

$$B((D_1, c_1), (D_2, c_2)) = ((D_2, c_2'), (D_1, c_1')).$$

Here colorings c_1' and c_2' are determined by c_1 and c_2 and by the corresponding objects.

The commutativity morphism is represented by the diagram on Fig. 19 and it is a mapping

$$\{(\varepsilon_1, x_1), \ldots, (\varepsilon_n, x_n)\} \otimes \{(\sigma_1, y_1), \ldots, (\sigma_m, y_m)\}$$
$$\to \{(\sigma_1, y_1^L), \ldots, (\sigma_m, y_m^L)\} \otimes \{(\varepsilon_1, x_1^R), \ldots, (\varepsilon_n, x_n^R)\}$$

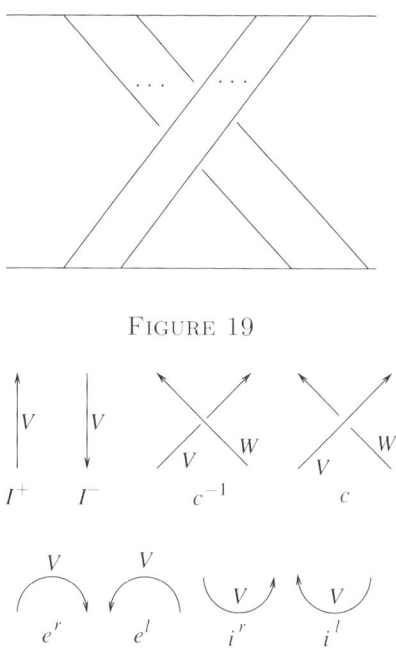

FIGURE 19

FIGURE 20

The coloring of the diagram on Fig. 19 is determined by the objects. □

3.3. Elementary diagrams. The following fact is a key for construction of invariants of tangles via braided monoidal categories.

PROPOSITION 2. *All morphisms in the category $\mathcal{D}(G)$ are compositions of tensor products of elementary diagrams. Elementary diagrams are given on Fig. 20.*

The proof of this proposition and the definition of elementary diagrams are the same as for the category of framed tangles (see [**T1**]).

3.4. The category of $\mathcal{C}$-diagrams. Let $\mathcal{C}$ be a braided G-category.

DEFINITION 2. *A pair (D, a) where D is a diagram and $a\colon E(D) \to \mathrm{Ob}(\mathcal{C})$ is called a $\mathcal{C}$-diagram if at each double point the values of the map a on adjacent edges satisfy the following conditions (the edges are enumerated as in Fig. 7):*
- *If a double point is positive, $X_a = X_L(X_c, X_d)$, $X_b = X_R(X_c, X_d)$.*
- *If the double point is negative, $X_c = X_L(X_a, X_b)$, $X_d = X_R(X_a, X_b)$.*

If G is factorizable and $\mathcal{R}$ is given by (3), then it is clear that each $\mathcal{C}$-diagram defines a G-colored diagram defined by the composition map $\pi \circ a \colon E(D) \to G$.

Now let us define the category $\mathcal{D}(\mathcal{C})$ of $\mathcal{C}$-diagrams.

Objects of $\mathcal{D}(\mathcal{C})$ are finite sequences $\{(\varepsilon_1, X_1), \ldots, (\varepsilon_n, X_n)\}$, where $\varepsilon_i = \pm$ and $X_i \in \mathrm{Ob}(\mathcal{C})$.

Morphisms from $\{(\varepsilon_1, X_1), \ldots, (\varepsilon_n, X_n)\}$ to $\{(\sigma_1, Y_1), \ldots, (\sigma_m, Y_m)\}$ are framed $\mathcal{C}$-diagrams (D, a) with $a(e_i^+) = Y_i^{\varepsilon_i}$, $a(e_i^-) = X_i^{\sigma_i}$. Here e_i^+ are edges of D adjacent to the upper boundary, enumerated from left to right $i = 1, \ldots, m$, and e_i^- are edges adjacent to the lower boundary, with $i = 1, \ldots, n$.

The identity morphism of $(\varepsilon_1, X_1), \ldots, (\varepsilon_n, X_n)$ is the trivial braid with the orientation of components defined by ε_i and with $X_i \in \mathrm{Ob}(\mathcal{C})$ assigned to i-th strand.

The following theorem is a generalization of Theorem 1 about G-colored diagrams.

THEOREM 2. *The category $\mathcal{D}(\mathcal{C})$ is a ribbon G-category.*

The proof is parallel to the one of Theorem 1.

4. Invariants of framed G-tangles

4.1. The functor $\Phi \colon \mathcal{D}(\mathcal{C}) \to \mathcal{C}$. As in the case of framed tangles and tangled framed graphs the first step in the construction of invariants of G-tangles will be the construction of a rigid G-braided monoidal functor from the category of $\mathcal{C}$-diagrams to the category $\mathcal{C}$.

Let $\mathcal{C}$ be a ribbon G-category.

THEOREM 3. *There exists a unique covariant functor $\Phi \colon \mathcal{D}(\mathcal{C}) \to \mathcal{C}$ such that*
- $\Phi(\{(\varepsilon_1, X_1), \ldots (\varepsilon_n, X_n)\}) = X_1^{\varepsilon_1} \otimes \cdots \otimes X_n^{\varepsilon_n}$, *where* $X^+ = X$, $X^- = X^*$.
- Φ *is a monoidal functor, i.e.*

$$\Phi((D_1, a_1) \otimes (D_2, a_2)) = \Phi((D_1, a_1)) \otimes \Phi((D_2, a_2))$$

  where (D_i, a_i) are $\mathcal{C}$-diagrams.
- *Values of Φ on elementary diagrams are:*
  (1) *If $I^\varepsilon \colon (\varepsilon, X) \to (\varepsilon, X)$ is the identity morphism then*

  $$\Phi(I^\varepsilon) = id_{X^\varepsilon}.$$

  (2) *For morphism $e_X^{r,l} \colon (\pm, X) \otimes (\mp, X) \to 1$ we have:*

  $$\Phi(e_X^r) = e_{X^*} \circ (\mu_X \otimes id) \colon X \otimes X^* \to X^{**} \otimes X^* \to \mathbb{1},$$

  $$\Phi(e_X^l) = e_X \colon X^* \otimes X \to \mathbb{1}$$

  (3) *For $i^{r,l} \colon \mathbb{1} \to (\mp, X) \otimes (\pm, X)$ we have*

  $$\Phi(i_X^r) = (\mu_X^{-1} \otimes id) \circ i_{X^*} \colon \mathbb{1} \to X \otimes X^*$$

  $$\Phi(i_X^l) = i_X \colon \mathbb{1} \to X^* \otimes X,$$

  (4) *If $c \colon (+, X) \otimes (+, Y) \to (+, X_L(X, Y)) \otimes (+, X_R(X, Y))$ is the braiding morphism,*

  $$\Phi(c) = c^{X,Y} \colon X \otimes Y \to X_L \otimes X_R.$$

  (5) *For the inverse morphism $c^{-1} \colon (+, X_L) \otimes (+, X_R) \to (+, X) \otimes (+, Y)$ we have:*

  $$\Phi(c^{-1}) = (c^{X,Y})^{-1} \colon X_L \otimes X_R \to X \otimes Y.$$

- *This functor Φ is rigid monoidal and G-braided.*

The proof of this theorem is completely parallel to the corresponding theorem describing invariants of framed tangles.

4.2. Invariants of framed G-tangles. Let $\mathcal{C}$ be a ribbon G-category and $A = \{A_x\}_{x \in G}$ be a family of objects such that $\pi(A_x) = x$. In other words A is a section of $\pi : \mathcal{C} \to G$. Assume that objects from this family have the following property:
$$B : (A_x, A_y) \to (A_{x_L(x,y)}, A_{x_R(x,y)})$$
This implies that the braiding morphisms act as:
$$c^{A_x, A_y} : A_x \otimes A_y \to A_{x_L(x,y)} \otimes A_{x_R(x,y)}$$
It also implies that
$$c^{A_x^*, A_y} : A_x^* \otimes A_y \to A_{x_R(y,x_1)} \otimes A_{x_1}^*$$
where $x_1 \in G$ is such that $x = x_L(y, x_1)$. This can be derived from the previous formula and from the axioms for the evaluation and injection morphisms. The action of the braiding on other duals can be computed similarly.

Now, let us fix such family of objects in $\mathcal{C}$ to define a $\mathcal{C}$-coloring of D for a given G-coloring c of the diagram D. We construct invariants of G-tangles as follows:
- given a G-tangle (t, ρ) define the G-colored diagram $[(D, c)]$ as in section 2 using the equivalence of categories of G-tangles and G-diagrams.
- given a G-colored diagram $[(D, c)]$ define a $\mathcal{C}$-diagram $[(D, a)]$ as above.
- apply the functor Φ to the $\mathcal{C}$-diagram $[(D, a)]$.

It is clear, from the definition of every step here, that the composition map is an invariant of (t, ρ) with values in morphisms of the category $\mathcal{C}$.

In the following section we will describe a $GL_2(\mathbb{C})$-category associated with the quantized universal enveloping algebra of $gl_2(\mathbb{C})$.

5. Quantized universal enveloping algebra of gl_2

5.1. The algebra $U_h(gl_2)$. The algebra $U_h(gl_2)$ over the ring $\mathbb{C}[[h]]$ is generated by elements H, G, X, and Y with defining relations
$$[H, G] = 0, [H, X] = 2X, [H, Y] = -2Y,$$
$$[G, X] = 2X, [G, Y] = -2Y,$$
$$[X, Y] = \frac{e^{\frac{hH}{2}} - e^{-\frac{hG}{2}}}{e^{\frac{h}{2}} - e^{-\frac{h}{2}}}$$

The Hopf algebra structure on $U_h(gl_2)$ is defined by the action of the comultiplication on generators:

(6) $\quad \Delta H = H \otimes 1 + 1 \otimes H, \ \Delta G = G \otimes 1 + 1 \otimes G,$

(7) $\quad \Delta X = X \otimes e^{\frac{hH}{2}} + 1 \otimes X, \ \Delta Y = Y \otimes 1 + e^{-\frac{hG}{2}} \otimes Y$

Elements H, G, X and Y "correspond to" elements $2e_{11}, 2e_{22}, e_{12}$ and e_{21} respectively.

The algebra $U_h(gl_2)$ is the Drinfeld double of the quantized universal enveloping algebra $U_h(b) \subset U_h(sl_2)$ where b is a Borel subalgebra in sl_2. As the double of a Hopf algebra it is quasitriangular with the universal R-matrix
$$R = \exp\left(\frac{h}{4} H \otimes G\right) \prod_{n \geq 0} (1 + e^{\frac{h}{2}}(e^{\frac{h}{2}} - e^{-\frac{h}{2}})^2 X \otimes Y e^{-nh})$$

This is an element of $U_h(gl_2)^{\otimes 2}$ which one should consider as a formal power series in h.

Since R is the universal R-matrix it satisfies the following identities:

(8) $$R\Delta(a)R^{-1} = \sigma \cdot \Delta(a)$$
$$(\Delta \otimes id)(R) = R_{13}R_{23}$$
$$(id \otimes \Delta)(R) = R_{13}R_{12}$$

where σ is the permutation operator $\sigma(a \otimes b) = b \otimes a$. In particular, R satisfies the Yang-Baxter equation

$$R_{12}R_{13}R_{23} = R_{23}R_{13}R_{12}$$

5.2. The inner automorphism $\mathcal{R}$. Define the inner automorphism $\mathcal{R} : U_h(gl_2)^{\otimes 2}[[h]] \to U_h(gl_2)^{\otimes 2}[[h]]$ as

(9) $$\mathcal{R}(x \otimes y) = R(x \otimes y)R^{-1}$$

It is easy to compute the action of $\mathcal{R}$ on generators.

THEOREM 4. *The following identities hold.*

$$\mathcal{R}(1 \otimes e^{\frac{hH}{2}}) = (1 \otimes e^{\frac{hH}{2}})(1 + e^{\frac{h}{2}}(e^{\frac{h}{2}} - e^{-\frac{h}{2}})^2 e^{-\frac{hH}{2}} X \otimes Y e^{\frac{hG}{2}})^{-1}$$

$$\mathcal{R}(1 \otimes e^{\frac{hG}{2}}) = (1 \otimes e^{\frac{hG}{2}})(1 + e^{\frac{h}{2}}(e^{\frac{h}{2}} - e^{-\frac{h}{2}})^2 e^{-\frac{hH}{2}} X \otimes Y e^{\frac{hG}{2}})^{-1}$$

$$\mathcal{R}(X \otimes 1) = X \otimes e^{\frac{hG}{2}}$$

$$\mathcal{R}(1 \otimes Y) = e^{-\frac{hH}{2}} \otimes Y$$

The theorem follows immediately from the commutation relations between generators and from the equation

$$f(zq^{-1};q) = (1 + zq^{-1})f(z;q)$$

for the function

$$f(z;q) = \prod_{n=0}^{\infty} (1 + zq^n).$$

The action of $\mathcal{R}$ on elements $1 \otimes X$, $Y \otimes 1$, $e^{\frac{hH}{2}} \otimes 1$, and $e^{\frac{hG}{2}} \otimes 1$ can be derived from the formulae above and from the identity (8).

The Yang-Baxter equation for R implies the Yang–Baxter equation for $\mathcal{R}$:

$$\mathcal{R}_{12} \cdot \mathcal{R}_{13} \cdot \mathcal{R}_{23} = \mathcal{R}_{23} \cdot \mathcal{R}_{13} \cdot \mathcal{R}_{12}$$

6. The algebra $\mathcal{U}$

The algebra $\mathcal{U}$ is generated over $\mathbb{C}[t, t^{-1}]$ by elements K, L, E and F with the following defining relations

$$KL = LK, \ KE = t^2 EK, \ KF = t^{-2}FK,$$
$$LE = t^2 EL, \ LF = t^{-2}FL,$$
$$EF - FE = (t - t^{-1})(K - L^{-1})$$

The center of $\mathcal{U}$ is generated freely by Laurent polynomials in KL^{-1} and

(10) $$c = EF + Kt^{-1} + L^{-1}t$$

This is a Hopf algebra with

$$\Delta(K) = K \otimes K, \ \Delta(L) = L \otimes L,$$
$$\Delta(E) = E \otimes K + 1 \otimes E, \ \Delta(F) = F \otimes 1 + L^{-1} \otimes F \ .$$

The map $\phi : \mathcal{U} \to U_h(gl_2)$ acting on generators as

$$\phi(K) = \exp(\frac{hH}{2}),\ \phi(L) = \exp(\frac{hG}{2}), \phi(t) = e^{\frac{h}{2}}$$

$$\phi(E) = (e^{\frac{h}{2}} - e^{-\frac{h}{2}})X,\ \phi(F) = (e^{\frac{h}{2}} - e^{-\frac{h}{2}})Y$$

extends to a homomorphism of Hopf algebras.

The algebra $\mathcal{U}$ is not quasitriangular. Instead, there is an outer automorphism of the division ring $\mathcal{U}^{\bar{\otimes}2}$ of $\mathcal{U}^{\otimes 2}$ which we denote by the same letter $\mathcal{R}$ as the automorphism (9) which acts on generators as

$$\mathcal{R}(1 \otimes K) = (1 \otimes K)(1 + tK^{-1}E \otimes FL)^{-1}$$

$$\mathcal{R}(1 \otimes L) = (1 \otimes L)(1 + tK^{-1}E \otimes FL)^{-1}$$

$$\mathcal{R}(E \otimes 1) = E \otimes L$$

$$\mathcal{R}(1 \otimes F) = K^{-1} \otimes F$$

Define its action on generators $K \otimes 1$, $L \otimes 1$, $1 \otimes E$, and $F \otimes 1$ such that

$$\mathcal{R}(\Delta(a)) = \sigma \circ \Delta(a)$$

where a is one of the generators of $\mathcal{U}$.

It is clear that the homomorphism ϕ brings the outer automorphism $\mathcal{R}$ to (9).

7. The algebra $\mathcal{U}_\varepsilon$

Let ε be a primitive root of 1 of an odd degree ℓ. Denote by $\mathcal{U}_\varepsilon$ the specialization of $\mathcal{U}$ to $t = \varepsilon$. The following theorem is a version of the corresponding facts for simple Lie algebras proved in [**DC-K**].

THEOREM 5.
- Elements E^ℓ, F^ℓ, $K^{\pm \ell}$, and $L^{\pm \ell}$ generate a central subalgebra $Z_0 \subset \mathcal{U}_\varepsilon$.
- Z_0 is a Hopf subalgebra with

$$\Delta(K^\ell) = K^\ell \otimes K^\ell,\ \Delta(L^\ell) = L^\ell \otimes L^\ell,$$

$$\Delta(E^\ell) = E^\ell \otimes K^\ell + 1 \otimes E^\ell,\ \Delta(F^\ell) = F^\ell \otimes 1 + L^{-\ell} \otimes F^\ell.$$

- The algebra $\mathcal{U}_\varepsilon$ is a free Z_0-module of dimension ℓ^4.
- The center $Z(\mathcal{U}_\varepsilon)$ is generated by Z_0 and by the element (10) modulo the relation

$$\prod_{j=0}^{\ell-1}(c - K\varepsilon^{j+1} - L^{-1}\varepsilon^{-j-1}) = E^\ell F^\ell$$

- Let α, β, a and b be coordinates on the group $B_+ \times B_-$ such that for $b_\pm \in B_\pm$ we have:

$$b_+ = \begin{pmatrix} 1 & \beta \\ 0 & \alpha \end{pmatrix},\ b_- = \begin{pmatrix} a & 0 \\ b & 1 \end{pmatrix}$$

Then the map $F^\ell \to ba^{-1}$, $E^\ell \to \beta$, $K^\ell \to \alpha$, $L^\ell \to a$ is a homomorphism of Hopf algebras $Z_0 \to C(B_+ \times B_-)$
- $\mathcal{U}_\varepsilon$ is semisimple over a Zariski open subvariety of $Spec(Z_0) \simeq B_+ \times B_-$.

Let $x \in \tilde{GL}_2^*$ be an irreducible Z_0-character and $I_x \subset \mathcal{U}_\varepsilon$ be the corresponding ideal. The quotient algebra
$$A_x = \mathcal{U}_\varepsilon/I_x$$
is finite-dimensional of dimension ℓ^4. There are three natural structures of a left module on A_x. For $a \in \mathcal{U}_\varepsilon$ denote by $[a]$ the class of a in A_x. Then these three actions are:
- $\pi(a)[b] = [ab]$,
- $\phi(a)[b] = [bS(a)]$,
- $\psi(a)[b] = [bS^{-1}(a)]$.

Assume that $x \in GL_2^*$ is generic, i.e. that A_x is semisimple. Fix an isomorphism of algebras $\phi_x : A_x \simeq \oplus_{i=1}^n Mat(k_i)$. For the algebra $\mathcal{U}_\varepsilon$ it is known [**DC-K**] that $n = \ell^2$ and $k_i = \ell$ for all $i = 1, \ldots, n$. Define
$$t : A_x \to \mathbb{C}, \ t(a) = \sum_{i=1}^n t_i Tr(\phi_x^i(a))$$
where Tr is the matrix trace in $Mat(k_i)$ and $\phi_x^i : A_x \to Mat(k_i)$ is the i-th component of ϕ_x. It is clear that $t(a)$ does not depend on a particular choice of ϕ_x. Indeed, any other such isomorphism differs from ϕ_x by an inner automorphism of $\oplus_{i=1}^n Mat(k_i)$. Since trace is cyclically invariant, the value of $t(a)$ for such isomorphism will be the same as for ϕ_x. Thus, for generic x, we have an invariant bilinear form on A_x:
$$(a,b) = t(ab).$$
It is a scalar product if $t_i \neq 0$ for each $i = 1, \ldots, n$.

Fix a scalar product on A_x as above. This gives an isomorphism of vector spaces $A_x^* \simeq A_x$. It is easy to verify that the pairing between $\mathcal{U}_\varepsilon$-modules (A_x, ϕ) and (A_x, π) given by the map

(11) $$e_x : (A_x, \phi) \otimes (A_x, \pi) \to \mathbb{C}$$

acting as $a \otimes b \mapsto t(ab)$ is $\mathcal{U}_\varepsilon$-invariant with respect to the diagonal action. Indeed, using Sweedler's notation $\Delta(c) = \sum_c c^{(1)} \otimes c^{(2)}$ for the comultiplication of element c, we have:
$$e_x(\sum_c aS(c^{(1)}) \otimes c^{(2)}b) = \sum_c t(aS(c^{(1)})c^{(2)}b) = \varepsilon(c)t(ab)$$
Similarly
$$e_x(\sum_c c^{(1)}a \otimes bS^{-1}(c^{(2)})) = \sum_c t(c^{(1)}abS^{-1}(c^{(2)})) = \varepsilon(c)t(ab)$$

Therefore the map $e_x : (A_x, \pi) \otimes (A_x, \psi) \to \mathbb{C}$ defined as in (11) is also $\mathcal{U}_\varepsilon$-invariant.

Let linear mapping $i_x : \mathbb{C} \to A_x \otimes A_x$ be defined by the formula
$$i_x(1) \mapsto \sum_i e_i \otimes e^i,$$

$\{e_i\}$ is a linear basis of A_x and $\{e^i\}$ the corresponding dual basis. It is easy to see that it is a morphism of $\mathcal{U}_\varepsilon$-modules $\mathbb{C} \to (A_x, \pi) \otimes (A_x, \phi)$ and $\mathbb{C} \to (A_x, \psi) \otimes$

(A_x, π). Indeed, let $a_i^j = t(ae_ie^j)$ for any $a \in A_x$, then

$$\sum_a \sum_i a^{(1)} e_i \otimes e^i S(a^{(2)}) = \sum_a \sum_{i,j} (a^{(1)})_i^j e_j \otimes e^i (S(a^{(2)}))$$

$$= \sum_a \sum_j e_j \otimes e^j a^{(1)} S(a^{(2)}) = \varepsilon(a) \sum_i e_i \otimes e^i$$

which implies the first statement and the second statement can be proved similarly.

Thus, for the object (A_x, π) we have the left dual (A_x, ϕ) (and the right dual (A_x, ψ)).

THEOREM 6. *The subspace $Z_0 \otimes Z_0 \subset \mathcal{U}_\varepsilon \otimes \mathcal{U}_\varepsilon$ is invariant with respect to the action of the automorphism $\mathcal{R}$.*

PROOF. From the action of $\mathcal{R}$ on generators of $\mathcal{U}_\varepsilon$ and from the relations between generators we have:

$$\mathcal{R}(1 \otimes K^\ell) = (1 \otimes K^\ell)(1 + K^{-\ell} E^\ell \otimes F^\ell L^\ell)^{-1}$$
$$\mathcal{R}(1 \otimes L^\ell) = (1 \otimes L^\ell)(1 + K^{-\ell} E^\ell \otimes F^\ell L^\ell)^{-1}$$
$$\mathcal{R}(E^\ell \otimes 1) = E^\ell \otimes L^\ell$$
$$\mathcal{R}(1 \otimes F^\ell) = K^{-\ell} \otimes F^\ell .$$

The comultiplication acts on ℓ-th powers of generators as:

$$\Delta(K^\ell) = K^\ell \otimes K^\ell, \quad \Delta(L^\ell) = L^\ell \otimes L^\ell$$
$$\Delta(E^\ell) = E^\ell \otimes K^\ell + 1 \otimes E^\ell$$
$$\Delta(F^\ell) = F^\ell \otimes 1 + L^{-\ell} \otimes F^\ell .$$

These formulae and the defining property $\mathcal{R}(\Delta(a)) = \sigma \circ \Delta(a)$ describe the action of $\mathcal{R}$ on generators of $Z_0 \otimes Z_0$. In particular, it is clear that the image is in $Z_0 \otimes Z_0$. □

Comparing the action of $\mathcal{R}$ on generators of $Z_0 \otimes Z_0$ with the identification of Z_0 and $C(GL_2^*)$ we have the following statement.

THEOREM 7. *The automorphism $\mathcal{R}$ is the pull-back of the mapping $b: GL_2^* \times GL_2^* \to GL_2^* \times GL_2^*$ acting as follows. First, identify GL_2^* with the Zariski open subvariety in GL_2 via the factorization mapping. Then the mapping b acts as*

$$(x, y) \mapsto (x_L(x, y), x_R(x, y))$$

where $x_L(x, y) = x_- y x_-^{-1}$ and $x_R(x, y) = (x_L(x, y))_+^{-1} x (x_L(x, y))_+$.

Let I_x be the ideal in $\mathcal{U}_\varepsilon$ with the Z_0 character $x \in GL_2^*$. From now on we will consider only generic points x and therefore we can identify GL_2 with GL_2^*. The two theorems have an important corollary. The mapping $\mathcal{R}$ acts as:

$$\mathcal{R}(I_x \otimes I_y) \subset I_{x_R(x,y)} \otimes I_{x_L(x,y)}$$

This implies that the mapping $\mathcal{R}$ induces the isomorphism of algebras:

$$\mathcal{R}(x, y) : A_x \otimes A_y \to A_{x_R(x,y)} \otimes A_{x_L(x,y)}$$

This mapping is also an isomorphism of the tensor product of left $\mathcal{U}_\varepsilon$-modules (A_x, π).

Finally, consider the category of $\mathcal{U}_\varepsilon$-modules generated by tensor products of (A_x, π) and their duals. It is clear that this category is a braided rigid monoidal G-category with $G = GL_2^*$ and with the braiding given by the composition $\sigma \circ \mathcal{R}$.

8. Conclusion

In this paper we have constructed invariants of tangles with flat connections in their complements. An example of such construction is described for $GL_2(\mathbb{C})$. This is rather simple example related to quantum invariant constructed in [**Ka**]. More interesting examples related to irreducible representations of quantized universal enveloping algebras of simple Lie algebras will be analyzed in a separate paper.

The complement of a tangle is a rather special 3-manifold. The construction of invariants of 3-manifolds with G-flat connections in them for any simple Lie group G is the next step. The case of $G = PSL_2(\mathbb{C})$ was studied in [**BB**].

An interesting, but somewhat speculative question: how to relate the constructed invariants to a topological quantum field theory defined "phenomenologically" in terms of functional integrals. We expect that this theory will be Chern–Simons theory with complex simple G. The corresponding boundary conformal field theory should be complex Wess–Zumino–Witten theory on a surface with boundary with boundary operators "parametrized" by elements of G.

References

[BB]   S. Baseilhac and R. Benedetti. *Quantum Hyperbolic Invariants Of 3-Manifolds With PSL(2,C)-Characters*, math.GT/0306280

[DC-K] C. De Concini and V. Kac. *Representations of quantum groups at roots of 1*, in Progress in Math., **92**, Birkäuser, 1990, pp. 471-506.

[DP]   C. DeConcini,C.Procesi. Quantum Groups. In: *D-Modules, representation theory, and quantum groups* (Venice, 1992), 31–140, Lecture Notes in Math., 1565, Springer, Berlin, 1993.

[J]   V.F.R. Jones. *A polynomial invariant of knots via von Neumann algebras*, Bull. Amer. Math. Soc., **12** (1985), 103-111.

[Ka]   R. Kashaev. *A link invariant from quantum dilogarithm*, Modern Physics Letters, **A 10** (1995), no. 19, 1409–1418.

[ML]   S. MacLane. *Categories for the Working Mathematician*, Graduate Text in Mathematics, (1971), Springer, Berlin.

[RT]   N. Reshetikhin and V. Turaev. Ribbon graphs and their invariants derived from quantum groups. *Comm. Math. Phys.* **127** (1990), no. 1, 1–26

[Ro]   M. Rosso. Quantum groups at a root of 1 and tangle invariants. *Topological and geometrical methods in field theory (Turku, 1991)*, 347–358, World Sci. Publishing, River Edge, NJ, 1992

[T1]   V. Turaev. *Quantum invariants of knots and 3-manifolds*. de Gruyter Studies in Mathematics, 18. Walter de Gruyter & Co., Berlin, 1994.

[T2]   V. Turaev. *Homotopy field theory in dimension 3 and crossed group-categories*, math.GT/0005291.

[WX]   A. Weinstein and P. Xu. Classical solutions of the quantum Yang-Baxter equation. *Comm. Math. Phys.* **148** (1992), no. 2, 309–343.

SECTION DE MATHÉMATIQUES, UNIVERSITÉ DE GENÈVE CP 240, 2-4 RUE DU LIÈVRE, CH 1211 GENÈVE 24, SUISSE
E-mail address: `Rinat.Kashaev@math.unige.ch`

DEPARTMENT OF MATHEMATICS, UNIVERSITY OF CALIFORNIA, BERKELEY, CA 94720, USA
E-mail address: `reshetik@math.berkeley.edu`

Tree-level invariants of three-manifolds, Massey products and the Johnson homomorphism

Stavros Garoufalidis and Jerome Levine

ABSTRACT. We show that the tree-level part of a theory of finite type invariants of 3-manifolds (based on surgery on objects called claspers, Y-graphs or clovers) is essentially given by classical algebraic topology in terms of the Johnson homomorphism and Massey products, for arbitrary 3-manifolds. A key role of our proof is played by the notion of a homology cylinder, viewed as an enlargement of the mapping class group, and an apparently new Lie algebra of graphs colored by $H_1(\Sigma)$ of a closed surface Σ, closely related to deformation quantization on a surface [**AMR1, AMR2, Ko3**] as well as to a Lie algebra that encodes the symmetries of Massey products and the Johnson homomorphism. In addition, we give a realization theorem for Massey products and the Johnson homomorphism by homology cylinders.

CONTENTS

1. Introduction
2. Statement of the results
3. Finite type invariants of 3-manifolds
4. Massey products and the Johnson homomorphism
5. Questions

References

1. Introduction

1.1. A brief summary. In this paper we investigate relations between three different phenomena in low-dimensional topology:

(a) Massey products on the first cohomology $H^1(M)$ with integer coefficients of 3-manifolds M.
(b) the Johnson homomorphism on the mapping class group of an orientable surface
(c) the Goussarov-Habiro theory of finite-type invariants of 3-manifolds.

1991 *Mathematics Subject Classification.* 55S30, 57R65, 57M27.
Key words and phrases. Massey products, Johnson homomorphism, homology cylinders, finite-type invariants.

A key point of the connection between (a) and (b) is the notion of a *homology cylinder*, i.e., a homology cobordism between an orientable surface and itself. This notion generalizes the mapping class group of a surface (in that case the cobordism is a product). We will construct an extension of the Johnson homomorphism to homology cylinders and use it to completely determine, in an explicit fashion, the possible Massey products at the first non-trivial level in a closed 3-manifold (assuming that the first homology H_1 is torsion-free)—see Theorem 1, Corollary 2.2 and Theorem 4.

This generalizes the known relationship between the Johnson homomorphism and Massey products in the mapping torus of a diffeomorphism of a surface to the more general situation of homology cylinders—see Theorems 2, 3 and Remark 4.12.

For historical reasons, we should mention early work of Sullivan [**Su**] on a relation between (a) and (b), and, for an alternative point of view, work of Turaev [**Tu**].

With regards to the connection between (b) and (c), the main idea is to consider Massey products as finite-type invariants of 3-manifolds, and to interpret them by a graphical calculus on *trees*—see Theorem 7—in much the same way that Vassiliev invariants of links have a graphical representation and that Milnor's μ-invariants are known to be exactly the Vassiliev invariants of (concordance classes of) string links which are represented by trees, see [**HM**]. A by-product of this investigation is a curious Lie algebra structure on a vector space of the graphs which describe finite type invariants of homology cylinders—see Proposition 2.8 and Theorem 5—that corresponds to the stacking of one homology cylinder on top of another, and is closely related to deformation quantization on a surface [**AMR1, AMR2, Ko3**].

1.2. History.

Years ago, Johnson introduced a homomorphism (the so-called Johnson homomorphism) which he used to study the mapping class group, [**Jo1, Mo4**]. Morita [**Mo1**] discovered a close relation between the Johnson homomorphism and the simplest finite type invariant of 3-manifolds, namely the Casson invariant; this relation was subsequently generalized by the authors [**GL1, GL2**] to all finite type invariants of integral homology 3-spheres (i.e., 3-manifolds M with $H_1(M, \mathbb{Z}) = 0$). This generalization posed the question of understanding the Johnson homomorphism (crucial to the structure of the mapping class group) from the point of view of finite type invariants. Unfortunately, this question is rather hard to answer if we confine ourselves to invariants of integral homology 3-spheres. This difficulty is overcome by using a theory of finite type invariants based on the notion of surgery on Y-links, see [**Gu1, Gu2, Hb, Oh, GGP**]. Using this theory we will show that the Johnson homomorphism is contained in its tree-level part, and we conjecture that an extension of the Johnson homomorphism to homology cylinders (i.e., 3-manifolds with boundary that homologically look like the product of a surface with $[0,1]$), which we define below, gives the full tree-level part; thus answering questions raised by Hain and Morita [**Ha, Mo4**].

En route to answering the above question, we were led to study this theory of invariants for homology cylinders (studied also from a slightly different perspective by Goussarov [**Gu1, Gu2**] and Habiro [**Hb**]) and discovered an apparently new Lie algebra of graphs colored by $H_1(\Sigma)$ of a closed surface Σ, closely related to deformation quantization on a surface [**AMR1, AMR2**], and to the curious graded group $\mathsf{D}(A) \stackrel{\text{def}}{=} \text{Ker}(A \otimes \mathsf{L}(A) \to \mathsf{L}(A))$, (where $\mathsf{L}(A)$ denotes the free Lie ring of a torsion-free abelian group A) studied independently by several authors with a variety of motivations [**Jo2, Mo2, Ih, Dr, Ko1, Ko2, O1, O2, HM**].

It turns out that Massey products of 3-manifolds naturally take values in $\mathsf{D}(A)$, and so does the Johnson homomorphism, which is also closely related to Massey products— a fact well-known to Johnson [**Jo2**], and later proved by Kitano [**Ki**]. However it is now known that the Johnson homomorphism cannot realize all elements of $\mathsf{D}(A)$, but we will see that one can achieve this realizability by replacing surface diffeomorphisms by homology cylinders.

The generalized Johnson homomorphism actually provides universally-defined invariants of homology cylinders, lifting the only partially-defined Massey products. These are our explicit candidates for the full tree-level part of the Goussarov-Habiro theory (for homology cylinders). This phenomenon was already observed when one replaces 3-manifolds by string-links up to homotopy, see Bar-Natan [**B-N**] or by string-links up to concordance, see Habegger-Masbaum [**HM**]. On the other hand, Massey products apply to more general manifolds and they should provide partially-defined finite-type invariants.

2. Statement of the results

2.1. Conventions. F will always stand for a free group and H for a torsion-free abelian group. The *lower central series* of a group G is inductively defined by $G_1 = G$ and $G_{n+1} = [G, G_n]$. A group homomorphism $p : K \to G$ is called an *n-equivalence* if it induces an isomorphism $K/K_n \cong G/G_n$. All manifolds will be oriented, and all maps between them will preserve orientation, unless otherwise mentioned. The boundary of an oriented manifold is oriented with the "outward normal first" convention.

2.2. Massey products. Recall the notion of Massey product, as formulated in [**Dw**] (also see [**FS**]). Suppose $\alpha_1, \ldots, \alpha_n$ are cohomology classes in $H^*(X)$. A *defining set* for $\langle \alpha_1, \ldots, \alpha_n \rangle$ is a collection of cochains a_{ij} for $1 \leq i \leq j \leq n$, except for $i = 1$, $j = n$, satisfying:

- a_{ii} is a cocycle representing α_i,
- $\delta a_{ij} = \sum_{k=i}^{j-1} a_{ik} \smile a_{k+1,j}$.

It is useful to picture the a_{ij} as entries of an $n \times n$ upper triangular matrix. Then $\sum_{k=1}^{n-1} a_{1k} \smile a_{k+1,n}$ is a cocycle whose cohomology class is called the *value* of this defining set. The Massey product $\langle \alpha_1, \ldots, \alpha_n \rangle$ is defined if there exists at least one defining set, and is the set of all values of defining sets. If α_i has dimension d_i then the dimension of any value is $d_1 + \cdots + d_n + 2 - n$. One can see from the definition that $\langle \alpha_1, \ldots, \alpha_n \rangle$ is defined if and only if each $\langle \alpha_1, \ldots, \hat{\alpha}_i, \ldots \alpha_n \rangle$ is defined and contains 0.

In this work we will only be interested in *Massey products* of length $n \geq 2$ in $H^2(\pi)$, i.e. $\langle \alpha_1, \ldots, \alpha_n \rangle \in H^2(\pi)$ for $\alpha_i \in H^1(\pi)$, which are defined assuming that the ones of length $n-1$ are defined and vanish. We have the following theorem on universal Massey products:

THEOREM 1. *(i) Given a connected topological space X and 2-equivalence $p : F \to \pi \stackrel{def}{=} \pi_1(X)$, then X has vanishing Massey products of length less than n if and only if p is an n-equivalence.*

(ii) In that case, we have a short exact sequence[1]
$$H_2(X,\mathbb{Z}) \to \mathsf{L}_n(H_1(X,\mathbb{Z})) \to \pi_n/\pi_{n+1} \to 0,$$
where the first map determines and is determined by all length n Massey products (for a precise expression, see Corollary 4.3) and the second is induced by the Lie bracket.

(iii) In addition, we have that
$$\alpha_1 \smile \langle \alpha_2, \ldots, \alpha_{n+1} \rangle = \langle \alpha_1, \ldots, a_n \rangle \smile \alpha_{n+1} \in H^2(\pi),$$
for any $\alpha_1, \ldots, \alpha_{n+1} \in H^1(\pi)$.

REMARK 2.1. For the dependence of the short exact sequence in the above theorem on the map p, see Remark 4.4. If X satisfies the hypothesis of Theorem 1 we have dually, over $\mathbb{Q}$:
$$0 \to (\pi_n/\pi_{n+1})^*_\mathbb{Q} \to \mathsf{L}_n(H^1(X,\mathbb{Q})) \to H^2(X,\mathbb{Q}).$$
Note that the first part of Theorem 1 appears in [**O2**, Lemma 16], and that the exact sequence above was first suggested by Sullivan in [**Su**] for $n = 2$, and subsequently proven by Lambe in [**La**] for $n = 2$ using different techniques involving minimal models.

COROLLARY 2.2. *Given an n-equivalence $p : F \to \pi \stackrel{def}{=} \pi_1(M)$, where M is a closed 3-manifold, we have the exact sequence*
$$H^* \to \mathsf{L}_n(H) \to \pi_n/\pi_{n+1} \to 0$$
where $H = H_1(M,\mathbb{Q})$. If $\mu_n(M,p) \in H \otimes \mathsf{L}_n(H)$ denotes the first map (abbreviated by $\mu_n(M)$ if p is clear), then we have that
$$\mu_n(M,p) \in \mathsf{D}_n(H).$$
In particular, $\mu_n(M,p) = 0$ if and only if $p : F \to \pi$ is an $(n+1)$-equivalence.

Given an integer n and a torsion-free abelian group H, it is natural to ask which elements of $\mathsf{D}_n(H)$ are realized by 3-manifolds as above. For this see Theorem 4 below.

2.3. The Johnson homomorphism. We now discuss the relation between Massey products and the Johnson homomorphism.

Let $\Gamma_{g,1}$ denote the mapping class group of a surface $\Sigma_{g,1}$ of genus g with one boundary component (i.e., the group of surface diffeomorphisms that pointwise preserve the boundary), and let $\Gamma_{g,1}[n]$ denote its subgroup that consists of surface diffeomorphisms that induce the identity on π/π_{n+1}, where $\pi = \pi_1(\Sigma_{g,1})$. In [**Jo1**], Johnson defined a homomorphism
$$\tau_n : \Gamma_{g,1}[n] \to \mathsf{D}_{n+1}(H),$$
where $H = H_1(\Sigma_{g,1}, \mathbb{Z})$, which he further extended to the case of a closed surface. We recall the definition of τ_n (see [**Jo1**], [**Mo2**] for more detail). Let $x_1, \ldots, x_g, y_1, \ldots y_g$ be the canonical basis of π. If $\phi \in \Gamma_{g,1}[n]$, then we can write
$$\phi(x_i) = x_i \alpha_i \quad \phi(y_i) = y_i \beta_i$$

---
[1] Note that π_n denotes the nth commutator subgroup of $\pi_1(X)$ and not the nth homotopy group of X.

where $\alpha_i, \beta_i \in \pi_{n+1}$. Let a_i, b_i be the elements of $\mathsf{L}_{n+1}(H)$ corresponding to α_i, β_i under the identification $\mathsf{L}_{n+1}(H) \cong \pi_{n+1}'/\pi_{n+2}$. Then $\tau_n(\phi)$ is defined to be
$$\sum_i (\bar{x}_i \otimes b_i - \bar{y}_i \otimes a_i)$$
where $\bar{x}_i, \bar{y}_i$ are the homology classes of x_i, y_i. It is proved in [**Mo2**] that $\tau_n(\phi) \in \mathsf{D}_{n+1}(H)$.

Johnson was well-aware of the relation between his homomorphism and Massey products on mapping torii, i.e., on twisted surface bundles over a circle; see [**Jo2**, p. 171], further elucidated by Kitano [**Ki**]. In the present note, we extend this relation to Massey products that come from an arbitrary pair (Σ, M) of an imbedding $\iota : \Sigma \hookrightarrow M$ of a closed surface (not necessarily separating) in a 3-manifold. Fix a closed 3-manifold M and an $(n+1)$-equivalence $F \to \pi \stackrel{\mathrm{def}}{=} \pi_1(M)$. Given a pair (Σ, M), and $\phi \in \Gamma[n]$, let M_ϕ denote the result of cutting M along Σ, twisting by the element ϕ of its mapping class group and gluing back. In this case, there exists a canonical cobordism N_ϕ between M and M_ϕ such that the maps $\pi_1(M) \to \pi_1(N_\phi) \leftarrow \pi_1(M_\phi)$ (induced by the inclusions $M, M_\phi \hookrightarrow N_\phi$) are $(n+1)$-equivalences; thus by Theorem 1, M_ϕ has vanishing Massey products of length less than $n+1$. The ones of length $n+1$ on M_ϕ are determined in terms of those of M and the Johnson homomorphism as follows:

THEOREM 2. *With the above assumptions, we have*
$$\mu_{n+1}(M_\phi) = \mu_{n+1}(M) + \iota_* \tau_n(\phi).$$

See also Remark 4.12.

2.4. Homology cylinders and realization. It is well known [**Mo3**, **Ha**] that the Johnson homomorphism τ_n is not onto, in other words not every element of $\mathsf{D}_{n+1}(H)$ can be realized by surface diffeomorphisms. Generalizing surface diffeomorphisms to a more general notion of homology cylinders (defined below) allows us to define an *ungraded* version of the Johnson homomorphism, which then induces, on the associated graded level, generalizations of the Johnson homomorphisms. We will show that all of these are onto, see Theorem 3. As an application of this result, we will show that we can realize every element in $\mathsf{D}_n(H)$ by 3-manifolds as in Corollary 2.2 and, in addition give a proof, free of spectral sequences, of the isomorphism (1), as mentioned above.

Let $\Sigma_{g,1}$ denote the compact orientable surface of genus g with one boundary component. A *homology cylinder* over $\Sigma_{g,1}$ is a compact orientable 3-manifold M equipped with two imbeddings $i^-, i^+ : \Sigma_{g,1} \to \partial M$ so that i^+ is orientation-preserving and i^- is orientation-reversing and if we denote $\Sigma^\pm = \mathrm{Im}\, i^\pm(\Sigma_{g,1})$, then $\partial M = \Sigma^+ \cup \Sigma^-$ and $\Sigma^+ \cap \Sigma^- = \partial \Sigma^+ = \partial \Sigma^-$. We also require that $i^\pm$ be homology isomorphisms. We can multiply two homology cylinders by identifying Σ^- in the first with Σ^+ in the second via the appropriate $i^\pm$. Thus $\mathcal{H}_{g,1}$, the set of orientation-preserving diffeomorphism classes of homology cylinders over $\Sigma_{g,1}$ is a semi-group with an obvious identity.

There is a canonical homomorphism $\Gamma_{g,1} \to \mathcal{H}_{g,1}$ that sends ϕ to $(I \times \Sigma_{g,1}, 0 \times \mathrm{id}, 1 \times \phi)$. Nielsen showed that the natural map $\Gamma_{g,1} \to \mathrm{A}_0(F)$ is an isomorphism, where F is the free group on $2g$ generators $\{x_i, y_i\}$, identified with the fundamental group of $\Sigma_{g,1}$ (with base-point on $\partial \Sigma_{g,1}$), and $\mathrm{A}_0(F)$ is the group of automorphisms of F which fix the element $\omega_g = [x_1, y_1] \cdots [x_g, y_g]$, representing the boundary of

$\Sigma_{g,1}$. It is natural to ask whether there exists an analogous isomorphism for the semigroup $\mathcal{H}_{g,1}$. Below, we construct for every n a homomorphism $\sigma_n : \mathcal{H}_{g,1} \to A_0(F/F_n)$, where $A_0(F/F_n)$ is the group of automorphisms ϕ of F/F_n such that a lift of ϕ to an endomorphism $\bar\phi$ of F fixes ω_g mod F_{n+1}. It is easy to see that this condition is independent of the lift. For example $A_0(F/F_2) = \mathrm{Sp}(g, \mathbb{Z})$.

Given $(M, i^+, i^-) \in \mathcal{H}_{g,1}$ consider the homomorphisms $i^\pm_* : F \to \pi_1(M)$, where the base-point is taken in $\partial\Sigma^+ = \partial\Sigma^-$. In general, $i^\pm_*$ are not isomorphisms— however, since $i^\pm$ are homology isomorphisms, it follows from Stallings [**St**] that they induce isomorphisms $i^\pm_n : F/F_n \to \pi_1(M)/\pi_1(M)_n$. We then define $\sigma_n(M, i^\pm) = (i^-_n)^{-1} \circ i^+_n$. It is easy to see that $\sigma_n(M, i^\pm) \in A_0(F/F_n)$.

THEOREM 3. *The map $\sigma_n : \mathcal{H}_{g,1} \to A_0(F/F_n)$ is surjective.*

REMARK 2.3. We can convert $\mathcal{H}_{g,1}$ into a group $\mathcal{H}^c_{g,1}$ by considering *homology cobordism classes* of homology cylinders. The inverse of an element is just the reflection in the I coordinate. It is easy to see that the invariants σ_n just depend on the homology bordism class and so define homomorphisms $\mathcal{H}^c_{g,1} \to A_0(F/F_n)$. The natural homomorphism $\Gamma_{g,1} \to \mathcal{H}^c_{g,1}$ is seen to be injective by the existence of the σ_n and the fact that the homomorphism $\Gamma_{g,1} \to A_0(F)$ is an isomorphism.

In addition, we can combine the maps σ_n, for all n, to a single map $\sigma^{\mathrm{nil}} : \mathcal{H}_{g,1} \to A_0(F^{\mathrm{nil}})$, where F^{nil} is the nilpotent completion of F. Unlike σ, σ^{nil} is not one-to-one, i.e., $\cap_n \mathrm{Ker}\,\sigma_n \neq \{1\}$. For example, if P is any homology sphere, then the connected sum $(I \times \Sigma_{g,1}) \sharp P$ defines an element in the kernel. Also σ^{nil} is not onto, even though each σ_n is. To identify the image of σ^{nil} we have to consider the *algebraic closure* $\bar F \subseteq F^{\mathrm{nil}}$, see [**Le2**]. Using the arguments of [**Le2**], we can show that any element of $\mathrm{Im}(\sigma^{\mathrm{nil}})$ restricts to an automorphism of $\bar F$ and, by arguments similar to the proof of Theorem 3, it can be proved that $\mathrm{Im}(\sigma^{\mathrm{nil}})$ consists precisely of those $\phi \in A_0(F^{\mathrm{nil}})$ which restrict to an automorphism of $\bar F$ and such that the element of $H_2(\bar F)$ associated to ϕ (see the proof of Theorem 3) is zero. But since we do not know whether $H_2(\bar F) = 0$, this result does not seem very useful at this time.

REMARK 2.4. The $\{\sigma_n\}$ can be described by numerical invariants if we consider the coefficients of the Magnus expansion of $\sigma_n(M)(x_i), \sigma_n(M)(y_i)$. This is analogous to the definition of the μ-invariants of a string link. We can refer to these as *μ-invariants of homology cylinders*.

It will be useful for us to consider the filtration defined by the maps σ_n, namely we define a decreasing *weight filtration* on $\mathcal{H}_{g,1}$ and on $\mathcal{H}^c_{g,1}$ by setting $\mathcal{H}_{g,1}[n] = \mathrm{Ker}(\sigma_n)$.

PROPOSITION 2.5. *We have an exact sequence*

$$1 \to \mathsf{D}_n(H) \to A_0(F/F_{n+1}) \to A_0(F/F_n) \to 1$$

and a commutative diagram

$$\begin{array}{ccc} \Gamma_{g,1}[n] & \longrightarrow & \mathcal{H}_{g,1}[n] \\ & \searrow{\tau_n} \quad \swarrow{\varsigma_n} & \\ & \mathsf{D}_n(H) & \end{array}$$

where the map ς_n, induced by σ_n, is onto. It follows that
$$0 \to \mathcal{H}^c_{g,1}[n+1] \to \mathcal{H}^c_{g,1}[n] \xrightarrow{\varsigma_n} \mathsf{D}_n(H) \to 0$$
is exact.

REMARK 2.6. A major problem in the study of the mapping class group is to determine the image of the Johnson homomorphism τ_n, which largely determines the algebraic structure of the mapping class group since $\cap_n \Gamma_{g,1}[n] = 1$. In contrast, Theorem 3 largely determines the structure of $\mathcal{H}_{g,1}/\mathcal{H}_{g,1}[\infty]$, but in this case $\mathcal{H}_{g,1}[\infty] = \cap_n \mathcal{H}_{g,1}[n]$ is not trivial—see Question 7 at the end of the paper.

REMARK 2.7. It is instructive to consider the analogy between, on the one hand, the mapping class group, homology cylinders and the invariant σ_n and the Johnson homomorphism, and, on the other hand, the pure braid group, string links and the Milnor μ-invariants. There is an injection of the pure braid group on g strands into the mapping class group $\Gamma_{g,1}$, first defined by Oda and studied in [**Le2**], which preserves the weight filtrations and induces a monomorphism of the associated graded Lie algebras. This can, in fact, be generalized to an injection of the semi-group $\mathcal{S}_g$ of string links on g strands into the semi-group $\mathcal{H}_{g,1}$ (and of the string-link concordance group $\mathcal{S}^c_g$ into $\mathcal{H}^c_{g,1}$), under which σ_n and the μ-invariants correspond. We will explain this in a future paper.

THEOREM 4. *Every element in $\mathsf{D}_n(H)$ is realized by an n-equivalence $F \to \pi_1(M^3)$, for some closed 3-manifold M, as in Corollary 2.2. In addition, a 2-equivalence $F \to H$ gives rise to a map $H_3(F/F_n) \to H \otimes \mathsf{L}_n(H)$ inducing the isomorphism of Equation (1).*

We will give two different proofs of this theorem. One approach is to apply results of Orr [**O1**] and Igusa-Orr [**IO**] on $H_3(F/F_n)$ and, in particular, the isomorphism

(1) $$\operatorname{cok}(H_3(F/F_{n+1}) \to H_3(F/F_n)) \cong \mathsf{D}_n(H).$$

A very similar argument appears in [**CGO**].

Alternatively we will see that this realizability is a consequence of Theorem 3. This approach has the advantage of being "spectral-sequence-free" and also gives another proof of (1).

2.5. Homology cylinders and finite type invariants of 3-manifolds. Goussarov and Habiro [**Gu1, Gu2, Hb**] have studied two rather dual notions: an n-equivalence relation among 3-manifolds, and a theory of invariants of 3-manifolds with values in an abelian group. Since their work is recent and not yet fully written, we will, for the benefit of the reader, give a short introduction using terminology and notation from [**GGP**] (to which we refer the reader for detailed proofs). Both notions are intimately related to that of surgery M_Γ along a Y-link Γ in a 3-manifold M, i.e., surgery along an imbedded link associated to an imbedding of an appropriately oriented, framed graph with trivalent and univalent vertices so that the univalent ones end in "leaves" (explained below). Two manifolds are n-*equivalent* if one can pass from one to the other by surgery on a Y-link associated to a *connected* Y-graph of degree (i.e., number of trivalent vertices) at least n. For example a theorem of Matveev [**M**] says that two closed manifolds are 1-equivalent if and only if there is an isomorphism between their first homology groups which preserves the linking form on the torsion subgroups. Similarly, a *finite type invariant* λ ought to be the analog of a polynomial on the set of 3-manifolds, in other words for some integer n it satisfies a difference equation

$$\sum_{\Gamma' \subseteq \Gamma} (-1)^{|\Gamma'|} \lambda(M_\Gamma) = 0$$

where Γ is a Y-link in M of more than n components and the sum is over all Y-sublinks Γ' of Γ. In view of the above definition, it is natural to consider the free abelian group $\mathcal{M}$ generated by homeomorphism classes of closed oriented 3-manifolds, and to define a decreasing filtration $\mathcal{F}^Y \mathcal{M}$ on $\mathcal{M}$ in such a way that λ is an invariant of type n if and only if it vanishes on $\mathcal{F}^Y_{n+1} \mathcal{M}$. Thus the question of how many invariants of degree n there are translates into a question about the size of the graded quotients $\mathcal{G}^Y_n \mathcal{M} \stackrel{\text{def}}{=} \mathcal{F}^Y_n \mathcal{M} / \mathcal{F}^Y_{n+1} \mathcal{M}$. One traditionally approaches this problem by giving independently an upper bound and a lower bound, which hopefully match. In this theory, an upper bound has been obtained in terms of an abelian group of decorated graphs as follows. One observes first that surgery along Y-links preserves the homology and linking form of 3-manifolds, as well as the boundary. Define an equivalence relation on compact 3-manifolds: $M \sim N$ if there exists an isomorphism $\rho : H_1(M) \to H_1(N)$ inducing an isometry of the linking forms, and a homeomorphism $\partial M \to \partial N$ consistent with ρ. Thus if we let $\mathcal{M}(M)$ denote the subgroup of $\mathcal{M}$ generated by equivalent 3-manifolds, we have a direct sum decomposition $\mathcal{M} = \oplus_\sim \mathcal{M}(M)$ (and also, $\mathcal{F}^Y \mathcal{M} = \oplus_\sim \mathcal{F}^Y \mathcal{M}(M)$), where the sum is over a choice of one manifold M from each equivalence class. In fact for closed 3-manifolds Matveev's theorem tells us that $\mathcal{G}^Y_0 \mathcal{M}(M) = \mathbb{Z}$. After we fix a 3-manifold M, and an oriented link $\mathfrak{b}$ in M that represents a basis of $H_1(M, \mathbb{Z})/\text{torsion}$, together with a framing of $\mathfrak{b}$ (i.e., a choice of a trivialization of the normal bundle of each component of $\mathfrak{b}$), it turns out that there is a map[2]

(2) $$W^{\mathfrak{b}}_n : \mathcal{A}_n(M) \to \mathcal{G}^Y_n \mathcal{M}(M),$$

which is onto over $\mathbb{Q}$ (actually, onto over $\mathbb{Z}[1/(2|\text{torsion}|)]$), where $\mathcal{A}(M)$ is the group generated by graphs with univalent and trivalent vertices, with a cyclic order along each trivalent vertex, decorated by an element of $H_1(M, \mathbb{Z})$ on each univalent vertex, modulo some relations, see [**GGP**]. Here $\mathcal{A}_n(M)$ is the subgroup generated by graphs of degree n, i.e., with n trivalent vertices; thus we have $\mathcal{A}(M) = \oplus_n \mathcal{A}_n(M) = \mathcal{A}^t(M) \oplus \mathcal{A}^l(M)$, where $\mathcal{A}^t(M)$ (resp. $\mathcal{A}^l(M)$) is the subgroup of $\mathcal{A}(M)$ generated by trees (resp. graphs with nontrivial first homology). For a detailed discussion of the map $W^{\mathfrak{b}}$, see also Section 3.

We should point out that for $M = S^3$ (i.e., for integral homology 3-spheres) one can construct sufficiently many invariants of integral homology 3-spheres to show that $W^{\mathfrak{b}}$ is an isomorphism, over $\mathbb{Q}$, see [**LMO**]. The same is true for finite type (i.e., Vassiliev) invariants of links in S^3, over $\mathbb{Q}$, see [**Ko2**]. However, it is at present unknown whether the map (2) is one-to-one (and thus, an isomorphism), over $\mathbb{Q}$, for all 3-manifolds.

We now discuss a well-known isomorphism [**Ih, O2, Dr, HM**], over $\mathbb{Q}$, for a torsion-free abelian group A:

(3) $$\Psi_n : \mathcal{A}^t_n(A) \cong_\mathbb{Q} \mathsf{D}_{n+1}(A),$$

---

[2] a more precise notation, which we will not use, would be $W^{M, \mathfrak{b}}_n$.

which will help us relate the Johnson homomorphism to the tree-level part of finite type invariants of 3-manifolds. This map is defined as follows: Fix an oriented uni-trivalent tree T of degree n (thus with $n+2$ legs, i.e., univalent vertices) and let $c : \mathrm{Leg}(T) \to A$ be a coloring of its legs. Given a leg l of T, (T,l) is a rooted colored tree to which we can associate an element (T,l) of $\mathsf{L}_{n+1}(A)$. Due to the IHX relation (see Figure 1), the function

$$T \to \sum_{l \in \mathrm{Leg}(T)} c(l) \otimes (T,l)$$

descends to one $\mathcal{A}_n^t(A) \to A \otimes \mathsf{L}_{n+1}(A)$ so that its composition with $A \otimes \mathsf{L}_{n+1}(A) \to \mathsf{L}_{n+2}(A)$ vanishes, thus defining the map Ψ_n. There is a map $A \otimes \mathsf{L}_{n+1}(A) \to \mathcal{A}_n^t(A)$ (defined by sending $a \otimes b \in A \otimes \mathsf{L}_{n+1}(A)$ to the rooted tree with one root colored by a and n additional legs colored by c), which shows that Ψ_n is one-to-one; and by counting ranks it follows that it is in fact a vector space isomorphism. It is unknown to the authors whether Ψ_n is an isomorphism over $\mathbb{Z}[1/6]$.

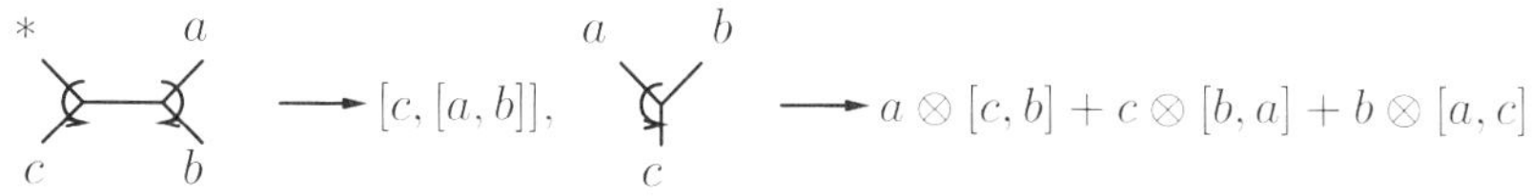

Figure 1. On the left, the map from rooted vertex-oriented trees to the free Lie algebra; on the right the map Ψ_1.

It turns out that a skew-symmetric form $\mathfrak{c} : A \otimes A \to \mathbb{Q}$ equips $\mathcal{A}^t(A)$ with the structure of a graded Lie algebra, by defining the Lie bracket

(4) $$[\Gamma, \Gamma']^{\mathfrak{c}} = \sum_{a,b} \mathfrak{c}(a,b)(\Gamma_a \mathrm{glue} \Gamma'_b),$$

where the sum is over each leg a of Γ and b of Γ' and $\Gamma_a \mathrm{glue} \Gamma'_b$ is the graph obtained by gluing the legs a and b of Γ and Γ' respectively, with the understanding that the sum over an empty set is zero. In other words, $[\Gamma, \Gamma']^{\mathfrak{c}}$ is the sum of all *contractions* of a leg of Γ with a leg of Γ'. This Lie bracket is not new, it has been observed and used by Morita [**Mo2**] and Kontsevich [**Ko1**] on a close relative of $\mathcal{A}^t(A)$, namely $\mathsf{D}'(A) \stackrel{\mathrm{def}}{=} A^* \otimes \mathsf{L}(A)$ (which carries a bracket of degree -1).

We now explain the Lie bracket on $\mathcal{A}^t(A)$ from the point of view of finite type invariants of 3-manifolds. Fixing a compact surface $\Sigma_{g,1}$ of genus g with one boundary component, it follows by definition that $\mathcal{M}(\Sigma_{g,1} \times I)$ is generated by homology cylinders over $\Sigma_{g,1}$ and is a ring with multiplication $M_1 * M_2$ defined by stacking M_1 below M_2. Fix a framed oriented link $\mathfrak{b}$ in $\Sigma_{g,1} \times I$ that represents a basis of $H_1(\Sigma_{g,1} \times I; \mathbb{Z})$ and consider the associated onto map $W^{\mathfrak{b}} : \mathcal{A}(\Sigma_{g,1} \times I) \to \mathcal{G}^Y \mathcal{M}(\Sigma_{g,1} \times I)$ from (2), which is expected to be an isomorphism. Thus, $\mathcal{A}(\Sigma_{g,1} \times I)$ should be equipped with a ring structure. This is the content of the following

PROPOSITION 2.8. *(i)* $\mathfrak{b}$ *induces a homomorphism*

$$\langle \cdot, \cdot \rangle^{\mathfrak{b}} : H_1(\Sigma_{g,1}, \mathbb{Z}) \otimes H_1(\Sigma_{g,1}, \mathbb{Z}) \to \mathbb{Z}$$

satisfying[3]

$$\langle a, b \rangle^{\mathfrak{b}} - \langle b, a \rangle^{\mathfrak{b}} = a \cdot b,$$

---

[3]$\langle \cdot, \cdot \rangle^{\mathfrak{b}}$ will often be denoted by $\langle \cdot, \cdot \rangle$ if $\mathfrak{b}$ is clear from the context.

where $\cdot$ is the natural symplectic form on $H_1(\Sigma_{g,1}, \mathbb{Z})$.

(ii) $\mathcal{A}(\Sigma_{g,1} \times I)$ is a ring with $*$-multiplication (depending on $\flat$) defined as follows: for $\Gamma, \Gamma' \in \mathcal{A}(\Sigma_{g,1} \times I)$,
$$\Gamma * \Gamma' = \sum_{l=0}^{\infty} \langle \Gamma, \Gamma' \rangle_l,$$
where
$$\langle \Gamma, \Gamma' \rangle_l = (-1)^l \sum_{a,b} \prod_{i=1}^{l} \langle a_i, b_i \rangle^{\flat} (\Gamma_a \mathrm{glue} \Gamma'_b)$$
is the sum over all ordered subsets $a = (a_1, \ldots, a_l)$ and $b = (b_1, \ldots, b_l)$ of the set of legs of Γ and Γ' respectively, $\Gamma_a\mathrm{glue}\Gamma'_b$ is the graph obtained by gluing the a_i-leg of Γ to the b_i-leg of Γ', for every i, with the understanding that a sum over the empty set is zero (thus the multiplication $*$ is a finite sum).

(iii) $\mathcal{A}^c(\Sigma_{g,1} \times I)$ is a Lie subring of $\mathcal{A}(\Sigma_{g,1} \times I)$ with bracket defined by
$$[\Gamma, \Gamma'] = \Gamma * \Gamma' - \Gamma' * \Gamma = \sum_{l=1}^{\infty} \langle \Gamma, \Gamma' \rangle_l - \langle \Gamma', \Gamma \rangle_l.$$

(iv) Over $\mathbb{Q}$, there is an algebra isomorphism $\mathsf{U}(\mathcal{A}^c(\Sigma_{g,1} \times I)) \cong_{\mathbb{Q}} \mathcal{A}(\Sigma_{g,1} \times I)$, where U is the universal enveloping algebra functor.

REMARK 2.9. The leading term $\langle \cdot, \cdot \rangle_1$ of the $*$-multiplication that involves contracting a single leg is independent of $\flat$ (see also part (iii) of Theorem 5), whereas the subleading terms $\langle \cdot, \cdot \rangle_l$ for $l \geq 2$ depend on $\flat$. This is a common phenomenon in mathematical physics, analogous to the fact that differential operators such as the Laplacian or the Dirac depend on a Riemannian metric, but have symbols independent of it.

REMARK 2.10. It is interesting to compare $\langle a, b \rangle^{\flat}$ to the Seifert matrix of a knot. Both notions depend on "linking numbers" of "stacked" curves, i.e. curves pushed in a positive or negative direction and the relation of $\langle \cdot, \cdot \rangle^{\flat}$ and the Seifert matrix to the symplectic structure on $H_1(\Sigma)$ is the same. The noncommutativity of the stacking is reflected by the fact that the form $\langle \cdot, \cdot \rangle^{\flat}$ is not symmetric. For a related appearance of this noncommutativity, see also [**AMR1, AMR2**]. Over $\mathbb{Q}$, the Lie algebra $\mathcal{A}^c(\Sigma_{g,1} \times I)$ is closely related to a Lie algebra of chord diagrams on Σ considered by Andersen-Mattes-Reshetikhin in relation to deformation quantization, loc. cit. We will postpone an explanation of this relation to a subsequent publication.

The ring structure on $\mathcal{A}(\Sigma_{g,1} \times I)$ would be of little interest were it not compatible with the one of $\mathcal{G}^Y \mathcal{M}(\Sigma_{g,1} \times I)$ and $\mathcal{A}^t(\Sigma_{g,1} \times I)$; this is the content of the following

THEOREM 5. (i) The map $W^{\flat} : \mathcal{A}(\Sigma_{g,1} \times I) \to \mathcal{G}^Y \mathcal{M}(\Sigma_{g,1} \times I)$ preserves the ring structure.

(ii) $\mathcal{A}^l(\Sigma_{g,1} \times I)$ is a Lie ideal of $\mathcal{A}^c(\Sigma_{g,1} \times I)$.

(iii) The Lie bracket of the quotient $\mathcal{A}^c(\Sigma \times I)/\mathcal{A}^l(\Sigma_{g,1} \times I) \cong \mathcal{A}^t(\Sigma_{g,1} \times I)$ is equal to $(-1)^{\deg - 1}$ times the Lie bracket of Equation (4) using the symplectic form on $H_1(\Sigma_{g,1} \times I) \cong H_1(\Sigma_{g,1})$. In particular, it is independent of the basis $\flat$.

From now on, we will work over $\mathbb{Q}$. We now show that the Johnson homomorphism $\tau : \mathcal{GT}_{g,1} \to \mathcal{A}^t(\Sigma_{g,1} \times I)$, or rather its signed version $\overline{\tau} \stackrel{\mathrm{def}}{=} (-1)^{\deg - 1} \tau$, can be recovered from the Lie algebra structure on $\mathcal{A}^c(\Sigma_{g,1} \times I)$, where $\mathcal{T}_{g,1}(n) \subseteq \Gamma_{g,1}[n]$ is the subgroup of the Torelli group generated by n-fold commutators, and $\mathcal{G}_n \mathcal{T}_{g,1}$ denotes the quotient $\mathcal{T}_{g,1}(n)/\mathcal{T}_{g,1}(n+1)$. Recall the map $\mathcal{T}_{g,1} \to \mathcal{M}(\Sigma_{g,1} \times I)$ defined by changing the parametrization of the top part of homology cylinders, its linear extension $(I\mathcal{T}_{g,1})^n \to \mathcal{F}_n^Y \mathcal{M}(\Sigma_{g,1} \times I)$, where $I\mathcal{T}_{g,1}$ is the augmentation ideal of the group ring $\mathbb{Q}\Gamma_{g,1}$, and the induced algebra map $\mathcal{G}\mathcal{T}_{g,1} \to \mathcal{G}^Y\mathcal{M}(\Sigma_{g,1} \times I)$. The theorem below explains the statement that the Johnson homomorphism is contained in the tree-level part of a theory of invariants in $\Sigma_{g,1} \times I$.

THEOREM 6. *Given a surface $\Sigma_{g,1}$ as above of genus at least 6, there exists a map $\Phi : \mathcal{G}\mathcal{T}_{g,1} \to \mathcal{A}^c(\Sigma_{g,1} \times I)$ and commutative diagrams of graded Lie algebras:*

$$\begin{array}{ccc} \mathcal{G}\mathcal{T}_{g,1} & & \mathcal{G}\mathcal{T}_{g,1} \\ \Phi \downarrow \searrow^{\overline{\tau}} & \text{and} & \Phi \downarrow \searrow \\ \mathcal{A}^c(\Sigma_{g,1} \times I) \to \mathcal{A}^t(\Sigma_{g,1} \times I) & & \mathcal{A}^c(\Sigma_{g,1} \times I) \xrightarrow{W^\flat} \mathcal{G}^Y\mathcal{M}(\Sigma_{g,1} \times I) \end{array}$$

where the left horizontal map is the natural projection on the tree part. In other words, for $\phi \in \mathcal{T}_{g,1}(n)$ we have

$$\Phi_n(\phi) = (-1)^{n-1} \overline{\tau}_n(\phi) + \text{loops}$$

in $\mathcal{A}^c(\Sigma_{g,1} \times I)$.

REMARK 2.11. For a closed surface Σ_g of genus g, there is an identical version of Proposition 2.8 and Theorem 5 above. As for Theorem 6, given a closed surface Σ_g of genus at least 6, there exists commutative diagrams

$$\begin{array}{ccc} \mathcal{G}\mathcal{T}_g & & \mathcal{G}\mathcal{T}_g \\ \Phi \downarrow \searrow^{\overline{\tau}} & \text{and} & \Phi \downarrow \searrow \\ \mathcal{A}^c(\Sigma_g \times I)/\Gamma_\omega \to \mathcal{A}^t(\Sigma_g \times I)/\Gamma_\omega & & \mathcal{A}^c(\Sigma_g \times I)/\Gamma_\omega \xrightarrow{W^\flat} \mathcal{G}^Y\mathcal{M}(\Sigma_{g,1} \times I) \end{array}$$

where Γ_ω is the ideal of $\mathcal{A}(\Sigma_g \times I)$ which is generated by all elements of the form $\sum_i \Gamma_{x_i,y_i,a}$ where $\{x_i,y_i\}$ is a standard symplectic basis of $H_1(\Sigma_g)$, $a \in H_1(\Sigma_g)$ and $\Gamma_{b,c,d}$ denote the degree 1 graph

$${}^b\mathsf{Y}^c_d$$

with counterclockwise orientation.

It is natural to ask for a statement of the above theorem involving general (closed) 3-manifolds M. How does one construct elements in $\mathcal{F}^Y\mathcal{M}(M)$? Given an imbedding $\iota : \Sigma \hookrightarrow M$ of a closed surface Σ and $\phi \in \mathcal{T}(n)$, it was shown in [**GGP**] that $M - M_\phi \in \mathcal{F}_n^Y\mathcal{M}(M)$. The following theorem relates the Johnson homomorphism and the tree-level part of finite type invariants of 3-manifolds.

THEOREM 7. *With the above assumptions, we can find $c_{\phi,n}^\flat \in \mathcal{A}_n^l(M)$ so that we have in $\mathcal{G}_n^Y\mathcal{M}(M)$:*

$$W_n^\flat(\Psi_n^{-1} \iota_* \overline{\tau}_n(\phi) + c_{\phi,n}^\flat) = M - M_\phi.$$

The above theorem should be compared with [**HM**, Theorem 6.1], where they show that if a string link L has vanishing μ-invariants of length less than n, then the degree n tree-level part of the Kontsevich integral of L is given by the degree n μ-invariants of L. Note that these μ-invariants are Massey products on the closed 3-manifold obtained by 0-surgery along the closure of the string link.

2.6. Plan of the proof. The paper consists of two, largely independent sections; the reader could easily skip one of them without any loss of understanding of the results of the other. Two notions that jointly appear in Sections 3 and 4 are the Johnson homomorphism and the notion of homology cylinders.

In Section 3, we use combinatorial techniques that are usually grouped under the name of finite type invariants (of knotted objects such as braids, links, string links or 3-manifolds) or graph cohomology. A key aspect is the introduction of a Lie algebra $\mathcal{A}(\Sigma_{g,1} \times I)$ of graphs and its relation to the Johnson homomorphism, via Proposition 2.8 and Theorems 5, 6 and 7.

In Section 4, we use standard techniques from algebraic and geometric topology to prove Theorems 1 concerning Massey products in general spaces and closed 3-manifolds,in particular, and Theorem 2 which relates the Johnson homomorphism to Massey products. In addition, we use standard surgery techniques adapted to homology cylinders to prove the two realization Theorems 3 and 4.

Finally, in Section 5 we pose a set of questions that naturally arise in our present study.

3. Finite type invariants of 3-manifolds

This section concentrates on the proof of Theorems 2.8, 5, 6 and 7. The techniques that we use are a combination of geometric and combinatorial arguments.

PROOF. (of Proposition 2.8) We only explain the first part. Statements (ii) and (iii) are obvious and (iv) follows by a theorem of Milnor-Moore [**MM**] regarding the structure of cocommutative graded connected Hopf algebras.

First we arrange that the components of $\mathfrak{b}$ project to immersions in Σ with transverse self-intersections and so that the framing has its first componenet vector field pointing in the I direction. We call such a link *generic*. Given a two-component sublink $\{b_1, b_2\}$ of the framed oriented link $\mathfrak{b}$ in $\Sigma_{g,1} \times I$, let $\{p(b_1), p(b_2)\}$ denote its projection on $\Sigma \times 0$. Then $p(b_1)$ and $p(b_2)$ intersect transversely at double points. Define $\langle b_1, b_2 \rangle^{\mathfrak{b}}$ to be the sum with signs over all points in $p(b_1) \cap p(b_2)$ that $p(b_1)$ overcrosses $p(b_2)$, according to the convention

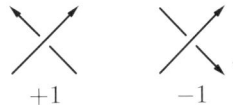

If $b_1 = b_2$, then we define $\langle b_1, b_1 \rangle^{\mathfrak{b}}$ by counting the self-intersections of b_1 in the above manner. Since $\mathfrak{b}$ is a basis of $H_1(\Sigma_{g,1} \times I, \mathbb{Z}) \cong H_1(\Sigma, \mathbb{Z})$, this defines, by linearity, the desired map $\langle \cdot, \cdot \rangle^{\mathfrak{b}}$.

Since $p(b_1) \cdot p(b_2)$ is the sum with signs over all points $p(b_1) \cap p(b_2)$, it follows that $\langle a, b \rangle - \langle b, a \rangle = a \cdot b$ for $a, b \in H_1(\Sigma, \mathbb{Z})$. □

Before we proceed with the proof of Theorems 5 and 7, we need to recall the definition of the map $W^{\mathfrak{b}} : \mathcal{A}(M) \to \mathcal{G}^Y \mathcal{M}(M)$: Given a colored graph Γ, we will construct an imbedding of it in M in two steps.

First, we imbed the leaves, as follows. Given the decoration $x \in H_1(M, \mathbb{Z})$, consider its projection $x^{tf} \in H_1(M, \mathbb{Z})/\text{torsion}$ and write $x^{tf} = \sum_{b \in \mathfrak{b}} n_b[b]$, for integers n_b. Consider the oriented link L_x obtained by the union (over b) of n_b parallel copies of b, where parallel copies of a component of $\mathfrak{b}$ are obtained by pushing off using the framing of $\mathfrak{b}$. Choose a basing for $\mathfrak{b}$, i.e., a set or meridians on each component of $\mathfrak{b}$ together with a path to a base point. Join the components of L_x using this basing to construct a knot K_x in M. Apply this construction to every leaf of Γ. Of course, the resulting link depends on the above choices of basing and joining.

Second, imbed the edges of Γ arbitrarily in M.

This defines an imbedding of Γ in M (which we denote by the same name) that also depends on the above choices; however the associated element $[M, \Gamma]$ in $\mathcal{G}^Y \mathcal{M}(M)$, where

$$[M, \Gamma] = \sum_{\Gamma' \subseteq \Gamma} (-1)^{|\Gamma'|} M_{\Gamma'}$$

is the alternating sum over all Y-sublinks Γ' of Γ, is well-defined, depending only on the framed link $\mathfrak{b}$. This follows from the following equalities in $\mathcal{G}^Y \mathcal{M}(M)$ (for detailed proofs see [**Gu2**, **Hb**] and also [**GGP**]):

(5)

(6)

(7)

where in the above equalities $[M, \Gamma]$ is abbreviated by $[\Gamma]$. Using further identities in $\mathcal{G}^Y \mathcal{M}(M)$, one can show loc.cit. that the map $\Gamma \to [M, \Gamma]$ factors through further relations to define a map $W^{\mathfrak{b}} : \mathcal{A}(M) \to \mathcal{G}^Y \mathcal{M}(M)$.

PROOF. (of Theorem 5) For the first part of the theorem, we begin by choosing $\mathfrak{b}$ in $\Sigma_{g,1} \times I$ to be a generic link. Note that $\mathfrak{b}$ can be recovered from its projection $p(\mathfrak{b})$ together with a knowledge of the signs at each overcrossing.

Given $\Gamma \in \mathcal{A}(\Sigma_{g,1} \times I)$, let $L^{\mathfrak{b}}(\Gamma)$ be an associated Y-link in $\Sigma_{g,1} \times I$ such that $W^{\mathfrak{b}}(\Gamma) = [\Sigma_{g,1} \times I, L^{\mathfrak{b}}(\Gamma)]$. Without loss of generality, we can assume that the leaves $l^{\mathfrak{b}}(\Gamma)$ of $L^{\mathfrak{b}}(\Gamma)$ form a generic link, and by abuse of notation, we can write that $W^{\mathfrak{b}}(\Gamma) = [\Sigma_{g,1} \times I, p(L^{\mathfrak{b}}(\Gamma))]$, with the understanding that we have fixed the signs on the overcrossings of $p(L^{\mathfrak{b}}(\Gamma))$.

Now, given $\Gamma, \Gamma' \in \mathcal{A}(\Sigma_{g,1} \times I)$, let $L^{\mathfrak{b}}(\Gamma)$ and $L^{\mathfrak{b}}(\Gamma')$ be the associated Y-links in $\Sigma \times [0, 1/2]$ and $\Sigma \times [1/2, 1]$ respectively, and let $p_i : \Sigma_{g,1} \times I \to \Sigma \times \{i\}$ denote the canonical projection. Then we have that

$$\begin{aligned}
W^\flat(\Gamma) \cdot W^\flat(\Gamma') &= [\Sigma_{g,1} \times I, p_0(l^\flat(\Gamma)) \cup p_{1/2}(l^\flat(\Gamma'))] \\
&= [(\Sigma_{g,1} \times I)_C, p_0(l^\flat(\Gamma)) \cup p_0(l^\flat(\Gamma'))] \\
&= [(\Sigma_{g,1} \times I)_C, L^\flat(\Gamma) \cup L^\flat(\Gamma')]
\end{aligned}$$

where C is a unit-framed trivial link in $\Sigma_{g,1} \times I$, whose components encircle some crossings of $p_0(l^\flat(\Gamma))$ and $p_0(l^\flat(\Gamma'))$, so that surgery on C brings $p_0(l^\flat(\Gamma))$ below $p_0(l^\flat(\Gamma'))$. The following lemma implies that the result of changing an overcrossing of $L^\flat(\Gamma)$ over $L^\flat(\Gamma')$ to an undercrossing can be achieved as a difference of the disjoint union of two graphs minus the disjoint union of two graphs with a leg glued. Together with Equation (6) and the definition of the multiplication $\Gamma * \Gamma'$, it implies that

$$[(\Sigma_{g,1} \times I)_C, L^\flat(\Gamma) \cup L^\flat(\Gamma')] = W^\flat(\Gamma * \Gamma')$$

which finishes the proof of the first part of Theorem 5.

LEMMA 3.1. *The following identity holds in $\mathcal{G}^Y\mathcal{M}(\Sigma_{g,1} \times I)$:*

$$\left[\succ\!\!-\!\!\bigcirc\!\!\bigcirc\!\!-\!\!\prec \right] = \left[\succ\!\!-\!\!\bigcirc\!\!-\!\!\prec \right] - \left[\succ\!\!-\!\!-\!\!\prec \right]$$

where the framing of the unknot on the left hand side of the equation is $+1$ and where we alternate with respect to the Y-links of the figure. The vertical arcs are arbitrary tubes.

PROOF. This follows from the second equality above in $\mathcal{G}^Y\mathcal{M}(\Sigma_{g,1} \times I)$ and from

$$\left[\succ\!\!-\!\!\bigcirc\!\!\bigcirc\!\!-\!\!\prec \right] = -\left[\succ\!\!-\!\!-\!\!\prec \right]$$

$$2\left[\succ\!\!-\!\!\bigcirc \right] = 0,$$

see [**GGP**]. □

The second part of Theorem 5 is obvious from the definition of the Lie bracket.

For the third part, notice that the Lie algebra structure on the quotient $\mathcal{A}^t(\Sigma_{g,1} \times I) \cong \mathcal{A}^c(\Sigma_{g,1} \times I)/\mathcal{A}^l(\Sigma_{g,1} \times I)$ is given by

$$[\Gamma, \Gamma'] = \langle \Gamma, \Gamma' \rangle_1 - \langle \Gamma, \Gamma' \rangle_1 = -[\Gamma, \Gamma']^\omega,$$

where the last equality follows from the first part of Proposition 2.8 and where ω is the symplectic form. □

PROOF. (of Theorem 6 and Remark 2.11) For a surface $\Sigma_{g,1}$ of genus $g \geq 3$ with one boundary component, Johnson [**Jo1**] introduced a homomorphism $\tau_1 : \mathcal{G}_1\mathcal{T}_{g,1} \to \Lambda^3(H)$, where, following the conventions of [**Jo1**, Chapter 4], $H = H_1(\Sigma_{g,1}, \mathbb{Z})$ and $\Lambda^m(H)$ is identified with a submodule of the m-th tensor power $\mathsf{T}^k(H)$ by defining

$$a_1 \wedge \cdots \wedge a_m = \sum_{\pi \in \mathrm{Sym}_m} a_{\pi(1)} \otimes \cdots \otimes a_{\pi(m)}.$$

In subsequent work, Johnson showed that modulo 2-torsion his homomorphism coincides with the abelianization of $\mathcal{T}_{g,1}$, thus one gets, over $\mathbb{Q}$, an onto map of Lie algebras $\mathsf{L}(\Lambda^3(H)) \to \mathcal{GT}_{g,1}$.

For the rest of the proof we will work over $\mathbb{Q}$. In [**Ha**, Section 11] Hain proved that for genus $g \geq 6$, the above map of Lie algebras has kernel generated by quadratic relations $R_{g,1}$ (Hain's notation for $\mathcal{G}_n \mathcal{T}_{g,1}$ is $\mathsf{t}_{g,1}(n)$). Combining the proof of [**Ha**] Proposition 10.3 with Theorem 11.1 and Proposition 11.4], it follows that the relation set $R_{g,1}$ is the symplectic submodule of $\mathsf{L}_2(\Lambda^3(H)) = \Lambda^2(\Lambda^3(H))$ generated by

$$[x_1 \wedge x_2 \wedge y_2, x_3 \wedge x_4 \wedge y_4] = 0$$

in terms of a standard symplectic basis $\{x_i, y_i\}$ of H.

Using the isomorphism $\Lambda^3(H) \cong \mathcal{A}_1^t(\Sigma_{g,1} \times I)$ given by mapping $a \wedge b \wedge c \in \Lambda^3(H)$ to the degree 1 graph $\Gamma_{a,b,c}$ as in remark 2.11, we obtain a map of Lie algebras $\mathsf{L}(\Lambda^3(H)) \to \mathcal{A}^c(\Sigma_{g,1} \times I)$.

Since for every choice of $a \in \{x_1, x_2, y_2\}$ and $b \in \{x_3, x_4, y_4\}$ we have $a \cdot b = 0$, the first part of Proposition 2.8 implies for every basis $\mathfrak{b}$ we have $\langle a, b \rangle^{\mathfrak{b}} = \langle b, a \rangle^{\mathfrak{b}}$. This implies, by definition, that $[\Gamma_{x_1,x_2,y_2}, \Gamma_{x_3,x_4,y_4}] = 0 \in \mathcal{A}_2^c(\Sigma_{g,1} \times I)$, thus obtaining the desired map $\Phi : \mathcal{GT}_{g,1} \cong \mathsf{L}(\Lambda^3(H))/(R_{g,1}) \to \mathcal{A}^c(\Sigma_{g,1} \times I)$.

Since $\mathcal{GT}_{g,1}$ is generated by its elements of degree 1, the commutativity of the two diagrams follows by their commutativity in degree 1; the later follows by definition for the first diagram, and by the fact that surgery on a Y-link of degree 1 with counterclockwise orientation and leaves decorated by a, b, c is equivalent to cutting, twisting and gluing by an element of the Torelli group (of a surface of genus 3 with one boundary component, imbedded in $\Sigma_{g,1}$) whose image under the Johnson homomorphism is equal to $a \wedge b \wedge c$, see [**GGP**]. This concludes the proof of Theorem 6.

We now prove the statements in Remark 2.11. For a closed surface Σ_g of genus $g \geq 3$, Johnson [**Jo1**] gave a version of his homomorphism $\tau_1 : \mathcal{GT}_g \to \Lambda_0^3(H)$ where $\Lambda_0^3(H)$ is defined to be the cokernel of the homomorphism $H \to \Lambda^3(H)$ that sends x to $\omega \wedge x$, where $\omega = \sum_i x_i \wedge y_i$ is the symplectic form of Σ_g, for a choice of symplectic basis. Working, from now on, over $\mathbb{Q}$, Johnson [**Jo1**, Chapter 4] gave an identification of $\Lambda_0^3(H)$ with $\mathrm{Ker}(C)$, where $C : \Lambda^3(H) \to H$ is given by

$$C(x \wedge y \wedge z) = 2((x \cdot y)z + (y \cdot z)x + (z \cdot x)y)$$

and $\cdot$ denotes the symplectic form. Explicitly, we will think of $\Lambda_0^3(H)$ as the submodule of $\Lambda^3(H)$ which is generated by elements of the form $2(g-1)a - \omega \wedge C(a)$ for $a \in \Lambda^3(H)$. In [**Ha**, Theorems 1.1 and 10.1] Hain proved that for genus $g \geq 6$, there is an isomorphism of graded Lie algebras $\mathsf{L}(\Lambda_0^3(H))/(R_g) \to \mathcal{GT}_g$ which, in degree 1, is the inverse of the Johnson homomorphism, where R_g is the symplectic submodule of $\mathsf{L}_2(\Lambda_0^3(H)) = \Lambda^2(\Lambda_0^3(H))$ generated by the relations

$$[2(g-1)x_1 \wedge x_2 \wedge y_2 - x_1 \wedge \omega, 2(g-1)x_3 \wedge x_4 \wedge y_4 - x_3 \wedge \omega] = 0.$$

The slight difference of 2 in the relations that Hain gave and the ones mentioned above are due to the difference in the normalization of the $\wedge$-product between Hain and Johnson. The restriction of the map $\mathsf{L}(\Lambda^3(H)) \to \mathcal{A}^c(\Sigma_{g,1} \times I) \cong \mathcal{A}^c(\Sigma_g \times I) \to \mathcal{A}^c(\Sigma_g \times I)/(\Gamma_\omega)$ to $\mathsf{L}(\Lambda_0^3(H))$ gives a map $\mathsf{L}(\Lambda_0^3(H)) \to \mathcal{A}^c(\Sigma_g \times I)/(\Gamma_\omega)$ which sends the relations R_g to zero (this really follows from the calculation of the surface with one boundary component together with the fact that $a \wedge \omega \in \Lambda^3(H)$ is sent into the ideal Γ_ω of $A_1^t(\Sigma_g \times I)$), thus inducing the desired map $\Phi : \mathcal{GT}_g \to \mathcal{A}^c(\Sigma_g \times I)/(\Gamma_\omega)$.

We claim that $W^{\flat}(M) : \mathcal{A}(\Sigma_g \times I) \to \mathcal{G}^Y \mathcal{M}(\Sigma_g \times I)$ maps Γ_ω to zero. This follows from the identity
$$(\tau_\partial)^{2g-2} = \prod_i [\tau_{x_i}, \tau_{y_i}]$$
of Dehn twists on the mapping class group of $\Sigma_{g,1}$ [**Mo1**, Theorem 5.3], where x_i, y_i refer to the standard meridian, longitude pairs associated with a symplectic basis of $H_1(\Sigma_{g,1})$ and ∂ is the boundary curve of $\Sigma_{g,1}$. Thus we have the relation
$$1 = \prod_i \tau_a [\tau_{x_i}, \tau_{y_i}] \tau_{a'}^{-1}$$
on the mapping class group of Σ_g (where a, a' are simple closed curves in $\Sigma_{g,1}$ with isotopic images in Σ_g), together with the fact surgery along the Y-link $\Gamma_{x_i, y_i, a}$ corresponds to the Dehn twist $\tau_a [\tau_{x_i}, \tau_{y_i}] \tau_{a'}^{-1}$ in $\mathcal{T}_g$, [**GGP**].

Since $\mathcal{GT}_g$ is generated by its elements of degree 1, the commutativity of the two diagrams follows by their commutativity in degree 1; this is shown in the same way as for a surface with one boundary component. This concludes the proof of Remark 2.11. □

PROOF. (of Theorem 7) For a closed surface Σ of genus at least 6, Theorem 7 follows from Remark 2.11 and the following Lemma 3.3, perhaps of independent interest. For a closed surface Σ of genus less than 6, fix a disk and consider an imbedding of its complement to a surface Σ' in M of genus at least 6. Choose a lifting of ϕ to a diffeomorphism of the punctured surface that preserves the boundary and extend it trivially to a diffeomorphism of ϕ' of Σ'. Since the Johnson homomorphism is stable with respect to increase in genus, and since $M_\phi = M_{\phi'}$, the result follows from the previous case. □

LEMMA 3.2. *If $\Gamma \subseteq M$ is an imbedded graph in M with a distinguished leaf that bounds a surface disjoint from the other leaves of Γ, then $[M, \Gamma] = 0 \in \mathcal{G}^Y \mathcal{M}(M)$.*

PROOF. First of all, recall that $[M, \Gamma] = 0$ if any leaf bounds a disk disjoint from the other leaves of Γ. As explained in [**GGP**], an alternative way of writing Equation (6) is as follows:

(8) $\quad \left[\begin{array}{c} \end{array} \right] = \left[\begin{array}{c} \end{array} \right] + \left[\begin{array}{c} \end{array} \right]$

for arbitrary disjoint imbeddings of two based oriented knots in M. Given a based knot α in M, let $\overline{\alpha}$ denote the based knot obtained by a push-off of α in its normal direction (any will do) followed by reversing the orientation.

The above identity implies that $[M, \Gamma_{\overline{\alpha}}] = -[M, \Gamma_\alpha]$ in $\mathcal{G}^Y \mathcal{M}(M)$, where Γ_κ is any imbedded graph in M with a distinguished leaf the based oriented knot κ in M.

Given Γ as in the statement of the lemma, it follows that its distinguished leaf is the connected sum of disjoint based knots of the form $\alpha_i \sharp \beta_i \sharp \overline{\alpha_i} \sharp \overline{\beta_i}$; thus it follows from the above discussion that $[M, \Gamma] = 0$ in $\mathcal{G}^Y \mathcal{M}(M)$. □

LEMMA 3.3. *Given an imbedding $\iota : N \to M$ of (not-necessarily closed) 3-manifolds, and links $\mathfrak{b}(N)$ (resp. $\mathfrak{b}(M)$) in N (resp. M) representing a basis of*

$H_1(N)$ (resp. $H_1(M)$), there is an induced map $\iota_* : \mathcal{A}(N) \to \mathcal{A}(M)$ induced by $\iota_* : H_1(N) \to H_1(M)$ on the colorings of the legs of the graphs and a diagram

$$\begin{array}{ccc} \mathcal{A}(N) & \xrightarrow{\iota_*} & \mathcal{A}(M) \\ W^{\mathfrak{b}(N)} \downarrow & & \downarrow W^{\mathfrak{b}(M)} \\ \mathcal{G}^Y \mathcal{M}(N) & \xrightarrow{\iota_*} & \mathcal{G}^Y \mathcal{M}(M) \end{array}$$

that commutes up to $W^{\mathfrak{b}(M)}(\mathcal{A}^l(M))$. In particular, for $\Gamma \in \mathcal{A}^t(N)$ we have

$$\iota_* W^{\mathfrak{b}(N)}(\Gamma) = W^{\mathfrak{b}(M)}(\iota_*(\Gamma)) + \text{loops}.$$

PROOF. Fix a graph $\Gamma \in \mathcal{A}^t(N)$. Let Γ_L (resp. $\Gamma_{L'}$) denote the Y-links in M (with leaves L (resp. L')) such that

$$\iota_* W^{\mathfrak{b}(N)}(\Gamma) = [M, \Gamma_L] \text{ and } W^{\mathfrak{b}(M)}(\iota_* \Gamma) = [M, \Gamma_{L'}].$$

It follows by definition that L and L' are homologous links in M. After choosing a common base point for each pair (L_i, L'_i) of components of L and L', it follows that the connected sum $L_i \sharp \overline{L'_i}$ is nullhomologous in M and thus bounds a surface Σ_i in M. The surface Σ_i might intersect the other components of L or L' at finitely many points; however by deleting disks around the points of intersection of Σ_i with $L \cup L'$, we can find a nullhomotopic based link L''_i and a surface Σ'_i disjoint from $L \cup L'$ with based boundary such that $L_i \sharp \overline{L'_i} = L''_i \sharp \partial \Sigma'_i$. Equation (8) and Lemma 3.2 imply that $[M, \Gamma_L] - [M, \Gamma_{L'}]$ is a sum of terms over Y-links Γ_κ in M which are trees, with at least one component κ being nullhomotopic. By choosing a sequence of crossing changes (represented by a unit-framed trivial link $C(\kappa)$) that trivialize κ and using Lemma 3.1, it follows that $[M, \Gamma_\kappa] = [M, \Gamma_{\text{trivial}}] = 0$ modulo terms that involve joining some legs of Γ (thus modulo terms that involve graphs with loops), which concludes the proof. $\square$

4. Massey products and the Johnson homomorphism

4.1. Universal Massey products.
In this section, homology will be with integer coefficients, unless otherwise stated. A useful tool in the proof of Theorem 1 is a *five-term* exact sequence of Stallings [**St**]: given a short exact sequence of groups $1 \to H \to G \to K \to 1$, there is an associated five-term exact sequence

$$H_2(G) \to H_2(K) \to H/[G,H] \to H_1(G) \to H_1(K) \to 1.$$

Applying the five-term sequence to the exact sequence $1 \to R \to F \to G \to 1$, (where F is a free group) we get Hopf's theorem [**Ho**]

$$H_2(G) \cong (R \cap [F,F])/[R,F], \text{ and in particular, } H_2(F/F_n) \cong F_n/F_{n+1}.$$

In the rest of this section, we will give a proof of Theorem 1 and its corollaries. We will follow a rather traditional notation involving local coordinates, [**Ma, FS, Dw**]. Let F denote the free group with basis $(x_1, \ldots, x_m)$; we will denote by the same name the corresponding basis of $H = H_1(F)$. Let $(u_1, \ldots, u_m)$ be the dual basis of $H^1(F) \cong H^1(F/F_n)$. The graded vector space $\bigoplus_n F_n/F_{n+1}$ has the structure of a Lie algebra induced by commutator, and is naturally identified with the free Lie algebra $\mathsf{L}(H) = \bigoplus_n \mathsf{L}_n(H)$.

We consider the *Magnus* expansion, [**MKS**]. Let $\mathbb{Z}[\![t_1,\ldots,t_m]\!]$ denote the power series ring in non-commuting variable $\{t_1,\ldots,t_m\}$. Define $\delta : F \to \mathbb{Z}[\![t_1,\ldots,t_m]\!]$ to be the multiplicative map defined by $\delta(x_i) = 1 + t_i$. This is an imbedding and induces imbeddings $\delta_n : \mathsf{L}_n(H) = F_n/F_{n+1} \to \mathbb{Z}[\![t_1,\ldots,t_m]\!]_n$, where $\mathbb{Z}[\![t_1,\ldots,t_m]\!]_n$ is the subspace of homogeneous polynomials of degree n.

We also recall the isomorphism $H_2(F/F_n) \cong F_n/F_{n+1} = \mathsf{L}_n(H)$ from above. We now describe the Massey product structure on $H^1(F/F_n)$. For a sequence $I = (i_1,\ldots,i_r)$ of numbers $i_j \in \{1,\ldots,m\}$ (of length $|I| \stackrel{\text{def}}{=} r$), we will let u_I denote the sequence $(u_{i_1},\ldots,u_{i_r})$ and, if each $u_i \in H^1(F/F_n)$ we let $\langle u_I \rangle$ denote the length r Massey product; we also let t_I denote the element $\prod_{j=1}^r t_{i_j}$.

PROPOSITION 4.1. *Any Massey product $\langle u_I \rangle$ of F/F_n vanishes if $|I| < n$. The action of any Massey product $\langle u_I \rangle$ on $H_2(F/F_n) \cong F_n/F_{n+1} = \mathsf{L}_n(H) \subseteq \mathbb{Z}[\![t_1,\ldots,t_m]\!]$ is determined by the formula*

$$\langle u_I \rangle \cdot t_J = \begin{cases} 1 & \text{if } I = J \text{ and } |I| = n \\ 0 & \text{otherwise} \end{cases} \tag{9}$$

In other words, the set $\{\langle u_I \rangle | |I| = n\}$ defines the basis of $\mathbb{Z}[\![t_1,\ldots,t_m]\!]_n^*$ dual to $\{t_I | |I| = n\}$.

See [**O2**] for a slightly less explicit version of this theorem.

PROOF. This follows easily from [**FS**]. Suppose $w \in F_n$ is some n-fold commutator. Then consider the one-relator group $G = F/\langle w \rangle$ and the projection $p : F/F_n \to G$. Consider I of length $r \leq n$. By induction we can assume that $\langle u_I \rangle$ is uniquely defined in F/F_n and by [**FS**], it is well-defined in G. Moreover, by naturality under p^*, they take the same value on w. If $r < n$ this is zero by [**FS**]. Since this holds for all w it follows that $\langle u_I \rangle = 0$. If $r = n$ Equation (9) follows directly from the formula in [**FS**]. □

COROLLARY 4.2. *Let $p : F \to \pi$ be a 2-equivalence. Then p is an n-equivalence if and only if all Massey products in $H_1(\pi)$ of length less than n vanish.*

See also [**CGO**, Proposition 6.8].

PROOF. The "if" part follows directly from the above proposition. To prove the "only if" part we proceed by induction on n. The inductive step presents us with a map $\pi \to \pi/\pi_{n-1} \cong F/F_{n-1}$; consider the diagram

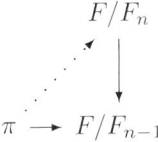

The obstruction to lifting this map is the pullback of the characteristic class in $H^2(F/F_{n-1}; F_{n-1}/F_n)$ of the central extension $F_{n-1}/F_n \to F/F_n \to F/F_{n-1}$. But Proposition 4.1 implies that $H^2(F/F_{n-1})$ is generated by Massey products of length $n-1$ and so the pullback is zero if and only if all Massey products of length $n-1$ vanish in $H^2(\pi)$. Thus, we can inductively lift the map to $\pi \to F/F_n$, thus to a map $\pi/\pi_n \to F/F_n$, which is still a 2-equivalence.

On the other hand, since p is a 2-equivalence, it induces an onto map $F/F_n \to \pi/\pi_n$, which is also a 2-equivalence. Composing with the map $\pi/\pi_n \to F/F_n$, we get an endomorphism of F/F_n which is a 2-equivalence. Stalling's theorem [**St**] implies that this endomorphism of F/F_n is an isomorphism which implies that the map $F/F_n \to \pi/\pi_n$ is one-to-one and thus an n-equivalence. $\square$

This proves the first assertion of Theorem 1.

COROLLARY 4.3. *Suppose $p : F \to \pi$ is an n-equivalence. Then there is an exact sequence:*
$$H_2(\pi) \xrightarrow{\hat{p}} F_n/F_{n+1} \xrightarrow{p_*} \pi_n/\pi_{n+1} \to 0,$$
where $\hat{p}$ is defined by the formula
$$\hat{p}(\alpha) = \sum_I (\langle u_I \rangle \cdot \alpha)\, t_I,$$
where $\alpha \in H_2(\pi)$, the summation is over I of length n, $\cdot : H^* \otimes H_* \to \mathbb{Z}$ is the evaluation map, and where the right hand side is asserted to lie in $\mathsf{L}_n(H) = F_n/F_{n+1} \subseteq \mathbb{Z}[\![t_1,\ldots,t_m]\!]$.

PROOF. Apply Stallings five-term exact sequence to the short exact sequence of groups $1 \to \pi_n \to \pi \to \pi/\pi_n \to 1$ to obtain
$$H_2(\pi) \to H_2(\pi/\pi_n) \to \pi_n/\pi_{n+1} \to 1.$$
Combining this with the map p gives the commutative diagram

$$\begin{array}{ccccccc} H_2(\pi) & \longrightarrow & H_2(\pi/\pi_n) & \longrightarrow & \pi_n/\pi_{n+1} & \longrightarrow & 1 \\ & & \cong \uparrow & & \uparrow p_* & & \\ & & H_2(F/F_n) & \xrightarrow{\cong} & F_n/F_{n+1} & & \end{array}$$

This diagram yields the exact sequence of the corollary, where $\hat{p}$ is defined as the composition
$$H_2(\pi) \to H_2(\pi/\pi_n) \cong H_2(F/F_n) \cong F_n/F_{n+1}$$
To prove the formula for $\hat{p}$ first note that, for any $\alpha \in H_2(F/F_n) \cong F_n/F_{n+1} = \mathsf{L}_n(H) \subseteq \mathbb{Z}[\![t_1,\ldots,t_m]\!]_n$ we have
$$\alpha = \sum_I (\langle u_I \rangle \cdot \alpha)\, t_I,$$
as follows directly from Proposition 4.1. But now the corollary follows from the definition of $\hat{p}$ and naturality. $\square$

This proves the second assertion of Theorem 1 for $K(\pi, 1)$ spaces.

REMARK 4.4. Given two choices p, p' of maps as in Corollary 4.3, we get a commutative diagram:

$$\begin{array}{ccccccc} H_2(\pi) & \longrightarrow & F_n/F_{n+1} & \longrightarrow & \pi_n/\pi_{n+1} & \longrightarrow & 0 \\ \downarrow & & \downarrow & & \downarrow & & \\ H_2(\pi) & \longrightarrow & F_n/F_{n+1} & \longrightarrow & \pi_n/\pi_{n+1} & \longrightarrow & 0 \end{array}$$

where the middle map is the automorphism of $F_n/F_{n+1} \cong H_2(F/F_n)$ defined by the composition of isomorphisms

$$H_2(F/F_n) \xrightarrow{p_*} H_2(\pi/\pi_n) \xleftarrow{p'_*} H_2(F/F_n).$$

In particular, if $p' \equiv p \mod F_2$ then the two exact rows are identical.

The third assertion of Theorem 1 for $K(\pi, 1)$ spaces follows from the following

PROPOSITION 4.5. *Let $p : F \to \pi$ be an n-equivalence and $\alpha_1, \ldots, \alpha_{n+1} \in H^1(\pi)$. Then we have:*

$$\alpha_1 \smile \langle \alpha_2, \ldots, \alpha_{n+1} \rangle = \langle \alpha_1, \ldots, \alpha_n \rangle \smile \alpha_{n+1}.$$

See also [**Kr**] for a related result.

PROOF. We will use Dwyer's formulation [**Dw**] of the Massey products. Choose cocycles a_i representing α_i, for $1 \leq i \leq n+1$. Since we are assuming all Massey products of length less than n are defined and vanish, we can choose cochains a_{ij}, for $1 \leq i < j \leq n+2$, with the exception of the three cases

$$i = 1, j = n+1 \qquad i = 2, j = n+2 \qquad i = 1, j = n+2$$

so that $a_{i,i+1} = a_i$ and $\delta a_{rs} = \sum_{r<i<s} a_{ri} \smile a_{is}$. For two of the three exceptional cases the cochains

$$b_{1,n+1} = \sum_{1<i<n+1} a_{1i} \smile a_{i,n+1} \quad \text{and} \quad b_{2,n+2} = \sum_{2<i<n+2} a_{2i} \smile a_{i,n+2}$$

the b_{ij} are cocycles but not necessarily coboundaries. In fact they represent the Massey products $\langle \alpha_1, \ldots, \alpha_n \rangle$ and $\langle \alpha_2, \ldots, \alpha_{n+1} \rangle$ respectively.

Thus $\langle \alpha_1, \ldots, \alpha_n \rangle \smile \alpha_{n+1}$ is represented by the cocycle $b_{1,n+1} \smile a_{n+1,n+2}$ and $\alpha_1 \smile \langle \alpha_2, \ldots, \alpha_{n+1} \rangle$ is represented by the cocycle $a_{12} \smile b_{2,n+2}$.

Now consider the cochain

$$c = \sum_{1<i<r<n+2} a_{1i} \smile a_{ir} \smile a_{r,n+2}$$

By grouping the terms in one way we see that

$$c = a_{12} \smile (\sum_{2<r<n+2} a_{2r} \smile a_{r,n+2}) + \sum_{2<i<r<n+2} (a_{1i} \smile a_{ir} \smile a_{r,n+2})$$

$$= a_{12} \smile b_{2,n+2} + \sum_{2<i<r<n+2} (a_{1i} \smile \delta a_{i,n+2})$$

Grouping the terms in another way gives

$$c = \sum_{1<i<r<n+1} (a_{1i} \smile a_{ir} \smile a_{r,n+2}) + (\sum_{1<i<n+1} a_{1i} \smile a_{i,n+1}) \smile a_{n+1,n+2}$$

$$= \sum_{1<r<n+1} (\delta a_{1r} \smile a_{r,n+2}) + b_{1,n+1} \smile a_{n+1,n+2}$$

Now subtracting these two formulae for c gives

$$a_{12} \smile b_{2,n+2} - b_{1,n+1} \smile a_{n+1,n+2} = \sum_{1<r<n+1} \delta a_{1r} \smile a_{r,n+2} - \sum_{2<i<n+2} a_{1i} \smile \delta a_{i,n+2}$$

$$= \delta(\sum_{2<i<n+1} a_{1i} \smile a_{i,n+2})$$

since a_{12} and $a_{n+1,n+2}$ are cocycles. Since the left side of this equation represents
$$\alpha_1 \smile \langle \alpha_2, \ldots, \alpha_{n+1}\rangle - \langle \alpha_1, \ldots, \alpha_n\rangle \smile \alpha_{n+1}$$
and the right side is a coboundary the proof is complete. □

This concludes the proof of Theorem 1 for $K(\pi, 1)$ spaces. The general case follows from the fact that the canonical map $X \to K(\pi, 1)$ induces an onto map $H_2(X) \to H_2(\pi)$. □

PROOF. (of Corollary 2.2) The first part is immediate from Theorem 1, using the Poincaré duality isomorphism $H^* = H_1(M, \mathbb{Q}) = H^2(M, \mathbb{Q})$. In local coordinates, it implies (see Corollary 4.3) that the Massey product
$$\mu_n(M) \in H \otimes \mathsf{L}_n(H) \cong \mathrm{Hom}(H^1(M), \mathsf{L}_n(M))$$
is given by
$$(10) \qquad \mu_n(M)(u) = \sum_I [M] \frown (u \smile \langle u_I\rangle)\, t_I,$$
where $u \in H^1(M)$, the summation is over I of length n, $\smile$ indicates cup product and $[M] \frown$ indicates cap product with the fundamental homology class of M. Let $[\cdot] : H \otimes \mathsf{L}(H) \to \mathsf{L}(H)$ be the Lie algebra bracket, defined by $[a \otimes b] = [a, b]$, for $a \in H, b \in \mathsf{L}(H)$. If we regard $\mathsf{L}(H) \subseteq \mathbb{Z}[\![t_1, \ldots, t_m]\!]$ then $[\cdot]$ can be expressed by the formula
$$[x_i \otimes c] = (1 + t_i)c - c(1 + t_i) = t_i c - c t_i.$$
Equation (10) implies that
$$\mu_n(M) = \sum_{i,I} x_i \otimes ([M] \frown (u_i \smile \langle u_I\rangle))\, t_I$$
and so
$$[\mu_n(M)] = \sum_{i,I} [M] \frown (u_i \smile \langle u_I\rangle)\, (t_i t_I - t_I t_i).$$

Thus Corollary 2.2 follows from the third assertion of Theorem 1 (or its coordinate version, Proposition 4.5). □

4.2. Realization results.

PROOF. (of Theorem 3) Given an element $\phi \in A_0(F/F_n)$ we construct maps $f^\pm : \Sigma_{g,1} \to K(F/F_n, 1)$, where $f^+_* : \pi_1(\Sigma_{g,1}) \to F/F_n$ corresponds to the canonical projection $p : F \to F/F_n$ under the identification of $\pi_1(\Sigma_{g,1})$ with F, and $f^-_* = \phi \circ p$. Since $\phi_*(\omega_g) = \omega_g$, we have $f^+|\partial\Sigma_{g,1} \simeq f^-|\partial\Sigma_{g,1}$ and so we can combine the two maps to define a map $f : \hat\Sigma_{g,1} \to K(F/F_n, 1)$, where $\hat\Sigma_{g,1}$ is the double of $\Sigma_{g,1}$. We would like to extend f to a map $\Phi : M \to K(F/F_n, 1)$, for some compact orientable 3-manifold with $\partial M = \hat\Sigma_{g,1}$ the obstruction to the existence of Φ is the element $\theta \in \Omega_2(F/F_n) \cong H_2(F/F_n)$ represented by f, where $\Omega_*(F/F_n)$ are the oriented bordism groups of F/F_n. Since $H_2(F/F_n) \neq 0$ we must be careful in our choices to assure that $\theta = 0$. Redo the construction of f^+, f^- and f but using $K(F/F_{n+1}, 1)$ instead of $K(F/F_n, 1)$ and using a lift of ϕ to an automorphism $\bar\phi$ of F/F_{n+1} instead of ϕ. Our restriction on ϕ assures that $\bar\phi(\omega_g) = \omega_g$ and so we obtain $\bar f : \hat\Sigma_{g,1} \to K(F/F_{n+1}, 1)$ and an obstruction element $\bar\theta \in \Omega_2(F/F_{n+1})$. Now this element may not be zero, but since the projection map $H_2(F/F_{n+1}) \to H_2(F/F_n)$

is zero, and clearly $\bar\theta$ maps to θ, we conclude that $\theta = 0$. Thus f extends to the desired $\Phi : M \to K(F/F_n, 1)$.

Let $i^\pm : \Sigma_{g,1} \to \partial M$ be the obvious diffeomorphisms onto the domains of $f^\pm$. It is clear that if M were a homology cylinder over Σ^+, then $\sigma_n(M) = \phi$. But this is not necessarily true and so we will perform surgery on the map Φ, adapting the arguments in [**KM**] to our situation. See also [**Tu**, Theorem 1] for similar surgery arguments which are used to show that any finite 3-dimensional Poincaré complex is homology equivalent to a closed 3-manifold.

LEMMA 4.6. *Suppose $\alpha \in \operatorname{Ker}\Phi_* : H_1(M) \to H$. Then there exists $\bar\alpha \in \pi_1(M)$ such that $\bar\alpha \in \operatorname{Ker}\Phi_* : \pi_1(M) \to F/F_n$ and $\bar\alpha$ represents α.*

PROOF. If $\bar\alpha \in \pi_1(M)$ is any representative of α, then $\Phi_*(\bar\alpha) \in F_2/F_n$. Choose an element $\beta \in \pi_1(\Sigma^+)_2$ so that $\Phi_*(\beta) = f_*^+(\beta) = \bar\alpha$. Then $\bar\alpha\beta^{-1} \in \operatorname{Ker}\Phi_*$ and $\bar\alpha\beta^{-1}$ represents α. $\square$

Thus for any $\alpha \in \operatorname{Ker}\Phi_* : H_1(M) \to H$ we can do surgery on a curve representing α and extend F over the trace of the surgery.

The first step in killing $\operatorname{Ker}\Phi_*$ will be to kill the torsion-free part. Note that $H_1(M) \cong H_1(\Sigma^+) \oplus \operatorname{Ker}\Phi_*$, since $\Phi_* \circ i_*^+$ is an isomorphism, and so, under the canonical map $H_1(M) \to H_1(M, \partial M)$, $\operatorname{Ker}\Phi_*$ maps onto $H_1(M, \partial M)$. Choose an element $\alpha \in \operatorname{Ker}\Phi_*$ which maps to a primitive element of $H_1(M, \partial M)$. Now surgery on a simple closed curve C representing α will produce a new manifold M' so that, if $\beta \in H_1(M')$ is the element represented by the meridian of C, then

$$H_1(M)/\langle\alpha\rangle \cong H_1(M')/\langle\beta\rangle \tag{11}$$

(see [**KM**]). Since α is primitive in $H_1(M, \partial M)$, there is a 2-cycle z in M whose intersection number with C is $+1$. Thus the intersection of z with M' is a 2-chain whose boundary is β. So, by Equation (11), $H_1(M') \cong H_1(M)/\langle\alpha\rangle$.

A sequence of such surgeries will kill the torsion-free part of $H_1(M, \partial M)$. But this implies that $\operatorname{Ker}\Phi_*$ is torsion by the following simple homology argument. Consider the exact sequence:

$$0 \to H_2(M) \to H_2(M, \partial M) \to H_1(\partial M) \to H_1(M) \to H_1(M, \partial M) \to 0$$

Since $\operatorname{rank} H_2(M) = \operatorname{rank} H_1(M, \partial M) = 0$ and $H_1(\Sigma^+)$ imbeds into $H_1(M)$, it follows that $\operatorname{rank} H_2(M, \partial M) = \operatorname{rank} H_1(M) \geq 2g$. But, since $\operatorname{rank} H_1(\partial M) = 4g$, we conclude that $\operatorname{rank} H_1(M) = 2g$. Therefore

$$2g = \operatorname{rank} H_1(M) = \operatorname{rank} H_1(\Sigma^+) + \operatorname{rank} \operatorname{Ker}\Phi_*$$

and so $\operatorname{rank}\operatorname{Ker}\Phi_* = 0$.

We now follow the argument in [**KM**] to kill the torsion group $T = \operatorname{Ker}\Phi_*$. The linking pairing $l : T \otimes T \to \mathbb{Q}/\mathbb{Z}$ is non-singular since $T = \operatorname{tors} H_1(M)$ maps isomorphically to $\operatorname{tors} H_1(M, \partial M) = H_1(M, \partial M)$. According to [**KM**, Lemma 6.3] if, for $\alpha \in T$, $l(\alpha, \alpha) \neq 0$, then we can choose the normal framing to any closed curve C representing α so that the element $\beta \in H_1(M')$ is of finite order smaller than the order of α. Thus the torsion subgroup of $H_1(M')$ is smaller than T. Continuing in this way we reach the point where all the self-linking numbers are 0. According to [**KM**, Lemma 6.5] this implies that T is a direct sum of copies of $\mathbb{Z}/2$. Now choose any non-zero element $\alpha \in T$. We will show that surgery on α reduces the rank of $H_1(M, \partial M; \mathbb{Z}/2)$. Denote by V the trace of the surgery and M' the result of the surgery. Then we have a diagram of homology groups (coefficients in $\mathbb{Z}/2$) with exact row:

$$H_2(V,\partial M') \longrightarrow H_2(V,M') \longrightarrow H_1(M',\partial M') \longrightarrow H_1(V,\partial M') \longrightarrow 0$$
$$\uparrow$$
$$H_2(M,\partial M)$$

Now $H_1(V,\partial M') \cong H_1(M,\partial M)/\langle\alpha\rangle$ and so has rank one less than $H_1(M,\partial M)$. Since $H_2(V,M')$ is generated by the transverse disk bounded by the meridian curve representing β, the dotted arrow can be interpreted as the ($\mathbb{Z}/2$) intersection number with α. By Poincaré duality this map is non-zero and so $H_1(M',\partial M') \cong H_1(V,\partial M')$, proving the claim. As in [**KM**] we can assume the normal framing chosen so that β has order 2 or ∞. Thus the possibilities for $H_1(M',\partial M';\mathbb{Z})$ are either $\mathbb{Z}\oplus(s-2)\mathbb{Z}/2$ or $\mathbb{Z}/4\oplus(s-2)\mathbb{Z}/2$, where $s = \operatorname{rank} H_1(M,\partial M)$. We can then do a surgery to kill the $\mathbb{Z}$ factor, in the first case, or reduce the order of $H_1(M',\partial M';\mathbb{Z})$, in the second case. Continuing this way we eventually kill $\operatorname{Ker}\Phi_*$, producing the desired (M,Φ). $\square$

PROOF. (of Proposition 2.5) Let $\mathsf{D}_n^a(H)$ denote the kernel of the natural projection $\mathsf{A}_0(F/F_{n+1}) \to \mathsf{A}_0(F/F_n)$. We first construct a map $D_n : \mathsf{D}_n^a(H) \to \mathsf{D}_n(H)$ as follows. If $h \in \mathsf{D}_n^a(H)$ we can write $h(a) = a\psi(a)$, where $\psi(a) \in F_n/F_{n+1} \cong \mathsf{L}_n(H)$. Then, we define $D_n(h)([a]) = \psi(a)$, where $[a] \in H$ and $a \in F/F_n$ is a lift of $[a]$. Using the isomorphism $\operatorname{Hom}(H,\mathsf{L}_n(H)) \cong H \otimes \mathsf{L}_n(H)$ this defines a map (denoted by the same name)

$$\mathsf{D}_n^a(H) \to H \otimes \mathsf{L}_n(H)$$

with corresponding description in local coordinates given by

$$D_n(h) = \sum_i x_i \otimes \psi(y_i) - y_i \otimes \psi(x_i) \in H \otimes \mathsf{L}_n(H).$$

If $h \in \mathsf{D}_n^a(H)$, as above, then $\prod_i[h(x_i),h(y_i)] \equiv \prod_i[x_i,y_i] \mod F_{n+2}$ and so

$$\prod_i[x_i\psi(x_i),y_i\psi(y_i)] \equiv \prod_i[x_i,y_i][\psi(x_i),y_i][x_i,\psi(y_i)] \mod F_{n+2}$$

Therefore $\prod_i[\psi(x_i),y_i][x_i,\psi(y_i)] \in F_{n+2}$, which implies that $D_n(h) \in \mathsf{D}_n(H)$.

It is clear that D_n is one-to-one. We now show that it is onto. Suppose we have an element $\theta = \sum_i(x_i\otimes\alpha_i - y_i\otimes\beta_i) \in \mathsf{D}_n(H)$. Lift α_i,β_i into F_n (denoted by the same symbols) and define an endomorphism h of F by

$$h(x_i) = x_i\alpha_i, \; h(y_i) = y_i\beta_i.$$

It follows by Stalling's theorem [**St**] that h induces an automorphism of F/F_{n+1} which restricts to the identity automorphism of F/F_n. We note that

$$h(\prod_i[x_i,y_i]) = \prod_i[x_i\alpha_i,y_i\beta_i] \equiv \prod_i[x_i,y_i][x_i,\alpha_i][\beta_i,y_i] \mod F_{n+2}$$

But $\prod_i[x_i,\alpha_i][\beta_i,y_i]$ represents the image of *theta* under the Lie bracket $H \otimes \mathsf{L}_n(H) \to \mathsf{L}_{n+1}(H)$, which vanishes since $\theta \in \mathsf{D}_n(H)$; thus $\prod_i[x_i,\alpha_i][\beta_i,y_i] \in F_{n+2}$. This shows that $h \in \mathsf{A}_0(F/F_{n+1})$ and clearly $D_n(h) = \theta$.

The fact that the projection $\mathsf{A}_0(F/F_{n+1}) \to \mathsf{A}_0(F/F_n)$ is onto follows immediately from Theorem 3. It is not hard, however, to give a direct argument; we leave this as an exercise for the reader. Finally, it is clear by the definitions that the diagram in Proposition 2.5 commutes, and that the sequence below it is exact. □

REMARK 4.7. The action of $A_0(F/F_n)$ on $\mathsf{D}_n^a(H) \cong \mathsf{D}_n(H)$ induced by conjugation by elements of $A_0(F/F_{n+1})$ coincides with the natural action of $A_0(F/F_2) \cong \mathrm{Sp}(2g,\mathbb{Z})$ on $\mathsf{D}_n(H)$, via the projection $A_0(F/F_n) \to A_0(F/F_2)$.

PROOF. (of Theorem 4) Let $M \in \mathcal{H}_{g,1}$ and define $S(M)^o = T_+ \cup M \cup T_-$, where $T_\pm$ are two copies of the solid handlebody T of genus g, which are attached to ∂M via the diffeomorphisms $i^\pm$ so that, referring to a basis $\{x_i, y_i\}$ of F corresponding to a symplectic basis of H, the $\{x_i\}$ are represented by the boundaries of meridian disks in T. Thus $\pi_1(T) = F'$, the free group generated by $\{y_i\}$ (or, more precisely, their images in $\pi_1(T)$). $S(M)^o$ is a 3-manifold with boundary S^2, which we can fill-in to obtain a closed 3-manifold $S(M)$. If $M \in \mathcal{H}_{g,1}[n]$, then the inclusion $T_+ \subseteq S(M)$ induces an isomorphism $p : F'/F_n' \cong \pi_1(S(M))/\pi_1(S(M))_n$ and we can consider $\mu_n(S(M),p) \in \mathsf{D}_n(H')$, where $H' = H_1(F')$. Suppose that $\sigma_{n+1}(M) = h \in A_0(F/F_{n+1})$. Set $a_i = \rho(h(x_i)) \in F_n'$, where $\rho : F \to F'$ is the projection defined by $\rho(x_i) = 1$. Then $\mu_n(S(M),p) = \sum_i [y_i] \otimes [a_i]$, where $[y_i] \in H', [a_i] \in \mathsf{L}_n(H')$ are the classes represented by y_i, a_i. This assertion is just the obvious generalization of Corollary 2.2 and the proof is the same.

Now let $\sum_i [y_i] \otimes [\lambda_i]$ be an arbitrary element in $\mathsf{D}_n(H')$, where $\lambda_i \in F_n'$, i.e. $\prod_i [y_i, \lambda_i] \in F_{n+2}$. We want to construct (N,p) such that $\mu_n(N,p) = \sum_i [y_i] \otimes [\lambda_i]$. Consider the endomorphism h of F defined by

$$(12) \quad \begin{aligned} h(x_i) &= x_i \lambda_i \\ h(y_i) &= \lambda_i^{-1} y_i \lambda_i. \end{aligned}$$

Denote also by h the induced automorphism of F/F_{n+1}. To see that $h \in A_0(F/F_{n+1})$, we compute

$$\prod_i [h(x_i), h(y_i)] = \prod_i (x_i y_i x_i^{-1} \lambda_i^{-1} y_i^{-1} \lambda_i) = \prod_i [x_i, y_i][y_i, \lambda_i^{-1}] \equiv \prod_i [x_i, y_i] \mod F_{n+2}.$$

Therefore, by Theorem 3, $h = \alpha_n(M)$ for some $M \in \mathcal{H}_{g,1}$. Since $\lambda_i \in F_n'$ we have $h \in \mathsf{D}_n^a(H)$ and so $M \in \mathcal{H}_{g,1}[n]$. By the discussion above, $\mu_n(N,p) = \sum_i [y_i] \otimes [\lambda_i]$. □

For completeness, we close this section by a sketch of a more direct proof of Theorem 4 using the results of [**O1, IO**]. Similar arguments can be found in [**CGO**].

LEMMA 4.8. *For every* $\alpha \in H_3(F/F_n)$ *there is a closed 3-manifold* M *and an n-equivalence* $p : F \to \pi \stackrel{def}{=} \pi_1(M)$ *such that* $p_*[M] = \alpha$.

REMARK 4.9. Since $\Omega_3(F/F_n) \cong H_3(F/F_n)$, it follows that every element $\alpha \in \Omega_3(F/F_n)$ is represented by some closed 3-manifold M and map $p : \pi \stackrel{def}{=} \pi_1(M) \to F/F_n$ so that $p_*[M] = \alpha$. The point is to arrange that $p_* : H_1(M) \cong H_1(F/F_n)$, which would imply that p is an n-equivalence.

PROOF. We apply the constructions and results of [**O1**]. Consider the mapping cone K_n of the natural map $K(F,1) \to K(F/F_n, 1)$ of Eilenberg-MacLane spaces. K_n is constructed from $K(F/F_n, 1)$ by adjoining 2-cells e_i^2 along the generators $x_i \in F \twoheadrightarrow F/F_n$. Then K_n is simply-connected and $H_i(K_n) \cong H_i(F/F_n)$ for $i \geq 2$. So we have the Hurewicz epimorphism $\rho : \pi_3(K_n) \twoheadrightarrow H_3(K_n) \cong H_3(F/F_n)$, where

π_3 denotes the third homotopy group. Suppose $\rho(\theta) = \alpha$. Then the Pontrjagin-Thom construction gives us a map $f : S^3 \to K_n$ representing θ, such that, if $x_i \in e_i^2$ is some interior point, then $f^{-1}(x_i) = L_i$, a zero-framed imbedded circle in S^3, see [**O1**]. Now let M^3 be the result of framed surgery on S^3 along the $\{L_i\}$. Then f induces a map $p : M \to K(F/F_n)$ and it is clear that $p_*[M] = \alpha$. Finally we note that $p_* : H_1(M) \cong H_1(F/F_n)$. $\square$

From another viewpoint, we have defined a homomorphism $\mu_n' : H_3(F/F_n) \to \mathsf{D}_n(H)$ by the formula

$$\mu_n'(\alpha) = \sum_{i,I} x_i \otimes (\alpha \frown (u_i \smile \langle u_I \rangle)) t_I \tag{13}$$

and so, by Corollaries 2.2 and 4.3 $\mu_n(M,p) = \mu_n'(p_*[M])$. It follows from Theorem 1 that the kernel of μ_n' is precisely the image of $H_3(F/F_{n+1}) \to H_3(F/F_n)$, inducing, therefore, an injection $\text{cok}(H_3(F/F_{n+1}) \to H_3(F/F_n)) \rightarrowtail \mathsf{D}_n(H)$. But it is shown in [**O1, IO**] that both sides are free finitely generated abelian groups of the same rank and so they are isomorphic. In fact K. Orr has pointed out to us that a straightforward examination of the spectral sequence of the group extension $F_n/F_{n+1} \to F/F_{n+1} \to F/F_n$ shows that the rank of $\text{cok}(H_3(F/F_{n+1}) \to H_3(F/F_n))$ is at least that of $\mathsf{D}_n(H)$. $\square$

4.3. Massey products and the Johnson homomorphism. In this section we will give a proof of Theorem 2. We first recall the definition of the Johnson homomorphism. Let $\mathsf{A}_0(F/F_n)$ be as in Section 2.4. We define the Johnson homomorphism

$$\tau_n : \mathsf{A}_0(F/F_n) \to \hom(H, \mathsf{L}_n(H))$$

where $H = H_1(F)$, as follows. If $\phi \in \mathsf{A}_0(F/F_n)$, then $\tau_n(\phi) \cdot [\alpha] = \alpha\phi(\alpha^{-1}) \in F_n/F_{n+1} \cong \mathsf{L}_n(H)$, where $[\alpha] \in H$ is the homology class of $\alpha \in F$. Let $K(\phi) \subseteq F$ be the normal closure of all elements of the form $\alpha\phi(\alpha^{-1})$ for $\alpha \in F$; note that $K(\phi) \subseteq F_n$ if $\phi \in \mathsf{A}_0(F/F_n)$. Let $\pi = F/K(\phi)$. Then the projection $F \to \pi$ induces an isomorphism $p_\phi : F/F_n \xrightarrow{\cong} \pi/\pi_n$ and the homomorphism $\hat{p}_\phi$ defined (as $\hat{p}$) in Corollary 4.3 is given by the composition

$$H_2(\pi) \cong K(\phi)/[F, K(\phi)] \to F_n/F_{n+1} \cong \mathsf{L}_n(H)$$

The first isomorphism is given by Hopf's theorem. Now consider the natural homomorphism $i : H \to H_2(\pi)$ defined by $i([\alpha]) = [\alpha\phi(\alpha)^{-1}]$, then

$$\tau_n(\phi) = \hat{p}_\phi \circ i. \tag{14}$$

Now let Σ be a compact orientable surface of genus g with one boundary component. Then $\pi_1(\Sigma) \cong F$, the free group on $2g$ generators. By Nielsen's theorem $\Gamma_{g,1} \subseteq \mathsf{A}(F)$ and so $\Gamma_{g,1}[n] = \Gamma_{g,1} \cap \mathsf{A}_0(F/F_n)$.

For $\phi \in \Gamma_g$ define an associated closed 3-manifold T_ϕ as follows. Let T_ϕ' be the mapping torus of ϕ, i.e. $I \times \Sigma$ with $1 \times \Sigma$ identified with $0 \times \Sigma$ by the homeomorphism $(1, x) \to (0, \phi(x))$ for every $x \in \Sigma$. Since $\phi|\partial\Sigma = \text{id}$ there is a canonical identification of $\partial T_\phi'$ with the torus $S^1 \times S^1$. T_ϕ is defined by pasting in $D^2 \times S^1$. Note that $\pi_1(T_\phi) \cong F/K(\phi)$, where $K(\phi)$ was defined above.

A theorem of the following sort was first suggested by Johnson [**Jo1**] and a proof (which we sketch, for completeness) was given in [**Ki**].

PROPOSITION 4.10. $\tau_n(\phi) = \mu_n(T_\phi)$, where $H_1(\Sigma)$ and $H^1(T_\phi)$ are identified by the string of isomorphisms

$$H^1(T_\phi) \xrightarrow{\cong} H^1(\Sigma) \cong H_1(\Sigma)$$

induced by the inclusion $\Sigma \subseteq T_\phi$ and Poincaré duality for Σ.

PROOF. It follows from the definitions that we need to establish the commutativity of the following diagram

$$\begin{array}{ccc} H_1(\Sigma) = H & \xrightarrow{i} & H_2(\pi) \\ \cong \downarrow & \swarrow j_* & \uparrow \theta \\ H^1(\Sigma) & \xleftarrow{j^*} H^1(T_\phi) \xleftarrow{\cong} & H_2(T_\phi) \end{array}$$

where i is defined in equation (14), $j : \Sigma \subseteq I \times 0 \subseteq I \times \Sigma \to T_\phi$ is the inclusion map inducing j^* and the Gysin homomorphism j_*, and θ is the Hopf map $H_2(X) \to H_2(\pi_1(X))$ when $X = T_\phi$.

Now suppose $\alpha \in \pi_1(\Sigma)$ is represented by a 1-cycle z in Σ. since $\phi(z)$ is homologous to z, there exists a 2-chain c in Σ such that $\partial c = \phi(z) - z$. Consider the chain $I \times z$ in $I \times \Sigma \to T_\phi$ with boundary $1 \times z - 0 \times z$. Since $0 \times z$ is identified with $1 \times \phi(z)$ in T_ϕ, the chain $\xi = I \times z + 1 \times c$ is a 2-cycle in T_ϕ—let β denote its homology class in $H_2(T_\phi)$.

It follows from the definition of θ that $\theta(\beta) = i([\alpha])$. On the other hand $j_*(\beta) = [\alpha]$ since the 2-cycle ξ intersects $\epsilon \times \Sigma$ transversely in the 1-cycle $\epsilon \times z$, if $0 < \epsilon < 1$. This establishes the desired commutativity. $\square$

Combining Proposition 4.10 and Corollary 4.3 we have

COROLLARY 4.11. If $\{x_i\}$ is a basis of $H = H_1(\Sigma) \cong H^1(T_\phi)$ and $\{u_i\}$ is the dual basis of $H_1(T_\phi)$, then

$$\tau_n(h) = \sum_{i,I} x_i \otimes ([T_\phi] \frown (u_i \smile \langle u_I \rangle)) t_I$$

using the inclusion $\mathsf{L}_n(H) \subseteq \mathbb{Z}[\![t_1, \ldots, t_{2g}]\!]_n$.

We now turn to the proof of Theorem 2. Fix an n-equivalence $p : F \to \pi_1(M)$, an imbedding $\iota : \Sigma \to M$ of a (closed) surface in a (closed) 3-manifold M, and an element $\phi \in \Gamma_g[n]$. The homomorphisms $f : \pi_1(M) \to F/F_{n+1}$ and $f_\phi : \pi_1(M_\phi) \to F/F_{n+1}$ induced by the n-equivalence p (in the discussion before the statement of the theorem) determine maps $f : M \to K(F/F_{n+1}, 1)$ and $f_\phi : M_\phi \to K(F/F_{n+1}, 1)$ in the Eilenberg-MacLane space $K(F/F_n, 1)$. We will construct a cobordism $F : V^4 \to K(F/F_{n+1})$ between f and f_ϕ.

First we push $\iota(\Sigma)$ in the positive and negative normal directions in M to obtain two copies of $\iota(\Sigma)$, Σ_p and Σ_n, respectively. Then we attach $I \times \iota(\Sigma)$ to M along its boundary by identifying $0 \times \iota(\Sigma)$ to Σ_n by the homeomorphism $(0, x) \to x$, for $x \in \iota(\Sigma)$, and $1 \times \iota(\Sigma)$ to Σ_p by $(1, x) \to \phi(x)$. We can now thicken up $I \times \iota(\Sigma)$ and M to obtain a manifold W, see Figure 2.

Note that the boundary of W consists of three components: M, M_ϕ and the mapping torus T_ϕ of ϕ on $\iota(\Sigma)$. We make one small modification to obtain V. Remove a disk D from $\iota(\Sigma)$ to obtain a surface Σ^o with boundary and lift ϕ to a

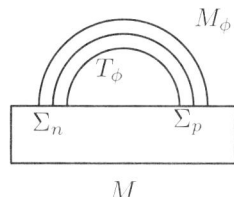

Figure 2. The manifold W. The handle shown represents a thickening of $\iota(\Sigma)$, shown as the core.

diffeomorphism ϕ^o of Σ^o. Then $S^1 \times D$ is naturally imbedded in T_ϕ, and so in the boundary of W. We attach $D^2 \times D$ to W along $S^1 \times D$ to obtain V.

The boundary of V is now given by $\partial V = M_\phi - T_{\phi^o} - M$. It is not difficult to check that the inclusions $M \to V \leftarrow M_\phi$ are $(n+1)$-equivalences since since $\phi \in \Gamma_g[n+1]$. Then f, f_ϕ extend to a map $F : V \to K(F/F_{n+1}, 1)$. In addition the inclusions $\Sigma^o \subseteq T_{\phi^o} \subseteq V$ induces isomorphisms

$$F/F_{n+1} \cong \pi_1(\Sigma^o)/\pi_1(\Sigma^o)_{n+1} \cong \pi_1(T_{\phi^o})/\pi_1(T_{\phi^o})_{n+1} \cong \pi_1(V)/\pi_1(V)_{n+1}$$

So the bordism (F, V) implies that $(f_\phi)_*[M_\phi] = f_*[M] + (p_\phi)_*[T_{\phi^o}] \in \Omega_3(F/F_{n+1}) \cong H_3(F/F_{n+1})$.

Applying the map of Equation (1) concludes the proof of Theorem 2. □

REMARK 4.12. Theorem 2 and Proposition 4.10 can be generalized easily to the context of homology cylinders, by essentially the same arguments. For any homology cylinder N there is an obvious notion of mapping torus T_N—then Proposition 4.10 can be rephrased, replacing T_ϕ by T_N, $\phi \in \Gamma_{g,1}[n]$ by $N \in \mathcal{H}_{g,1}[n]$ and τ_n by ς_n. For Theorem 2 we consider $N \in \mathcal{H}_{g,1}$ and an imbedding $\Sigma \subseteq M$, where Σ is a closed orientable surface of genus g. We can cut M open along Σ and paste in $\bar{N}$—where $\bar{N}$ is obtained from N by filling it in, in the obvious way, to get a homology cylinder over Σ—and so obtain a new manifold M_N. Then Theorem 2 can be rephrased, replacing M_ϕ by M_N.

REMARK 4.13. We can consider another filtration of $\mathcal{H}_{g,1}$. Let $\mathcal{H}_{g,1}(n)$ denote the set of all homology cylinders n-equivalent to $\Sigma_{g,1} \times I$ (see Section 2.5). Thus if $M \in \mathcal{H}_{g,1}(n)$ then $M - (\Sigma_{g,1} \times I) \in \mathcal{F}_n^Y(\Sigma_{g,1} \times I)$. It is easy to see that $\mathcal{H}_{g,1}(n)$ is a subsemigroup of $\mathcal{H}_{g,1}$. It is a natural conjecture that the quotient $\mathcal{H}_{g,1}/\mathcal{H}_{g,1}(n)$ is a group.

According to [**GGP**] we get the same filtration if we ask that M be obtained from $\Sigma_{g,1} \times I$ by cutting open along some imbedded closed orientable surface $\Sigma' \subseteq \Sigma_{g,1} \times I$ and reattaching by some element of $\mathcal{T}_n$, the n-th lower central series subgroup of the Torelli group $\mathcal{T}$ of Σ'. It is clear that $\mathcal{H}_{g,1}(n) \subseteq \mathcal{H}_{g,1}[n]$ since the effect of cutting and reattaching in M by an element of $\mathcal{T}_n$ does not change $\pi_1(M)/\pi_1(M)_n$. If $\mathcal{GH}_{g,1}(*)$ and $\mathcal{GH}_{g,1}[*]$ denote the associated graded groups of these filtrations then we have a natural map $\mathcal{GH}_{g,1}(*) \to \mathcal{GH}_{g,1}[*]$. Note also that the natural homomorphism $\Gamma_{g,1} \to \mathcal{H}_{g,1}$ induces maps $(\mathcal{T}_{g,1})_n \to \mathcal{H}_{g,1}(n)$ and so $\mathcal{G}(\mathcal{T}_{g,1})_* \to \mathcal{GH}_{g,1}(*)$.

Putting this, and some of the other maps constructed in this paper, all together, we have a commutative diagram:

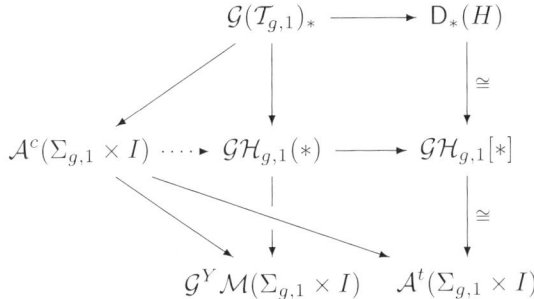

The dotted arrow denotes a conjectured lifting.

5. Questions

It is well known that there is a set of moves that generates (string) link concordance, [**Tr**]. These moves, together with the existence of the Kontsevich integral, were the key to the proof that the tree-level part of the Kontsevich integral of string-links is given by Milnor's invariants, or equivalently, by Massey products, see [**HM**].

QUESTION 1. Is there a set of local moves that generates homology cobordism of homology cylinders?

QUESTION 2. Does Theorem 6 generalize to homology cylinders, using our extension of the Johnson homomorphism? See Remark 4.13.

A positive answer to the above question would imply that the full tree-level part of the theory of finite type invariants on 3-manifolds is given by our extension of the Johnson homomorphism to homology cylinders.

QUESTION 3. If $(\Sigma_{g,1} \times I) - M \in \mathcal{F}_n^Y(\Sigma_{g,1} \times I)$, is $M \in \mathcal{H}_{g,1}(n) \otimes \mathbb{Q}$? See Remark 4.13. When $g = 0$ the answer is yes—see [**GL2**]. Are the μ-invariants of homology cylinders (see Remark 2.4) finite-type in the Goussarov-Habiro sense?

We now consider the group $\mathcal{H}_{g,1}^c$ of homology cobordism classes of homology cylinders defined in Remark 2.3. The subgroup $\mathcal{H}_{g,1}^c[2]$ (see Proposition 2.5) is the analogue of the Torelli group, which we denote $\mathcal{TH}_{g,1}^c$. We can consider the lower central series filtration $(\mathcal{TH}_{g,1}^c)_n$ and, just as in the case of the Torelli group (see [**Mo3**]), we have $(\mathcal{TH}_{g,1}^c)_n \subseteq \mathcal{H}_{g,1}^c[n+1]$. Recall also that Hain [**Ha**] proved that, over $\mathbb{Q}$, the associated graded Lie algebra of the lower central series filtration of the Torelli group maps onto the associated graded Lie algebra of the weight filtration of the Torelli group, and that these two filtrations are known to differ in degree 2 by a factor of $\mathbb{Q}$, [**Ha, Mo4**]. Whether they differ in degrees other than 2 is an interesting question.

QUESTION 4. What is the relation between the filtrations $(\mathcal{TH}_{g,1}^c)_n$ and $H_{g,1}^c[n]$?

Notice that the answer to above question is not known for the group of concordance classes of string-links, see [**HM**], but it is known (in the positive) for the group of homotopy classes of string-links, see [**HL**], and for the pure braid group, see [**Kh**].

QUESTION 5. We have $(\mathcal{T}_{g,1})_n \subseteq \mathcal{H}_{g,1}^c(n)$ (see Remark 4.13) and obviously $(\mathcal{T}_{g,1})_n \subseteq (\mathcal{T}\mathcal{H}_{g,1}^c)_n$. What is the relation between the filtrations $\mathcal{H}_{g,1}^c(n)$ and $(\mathcal{T}\mathcal{H}_{g,1}^c)_n$?

We now consider the center $\mathcal{Z}(\mathcal{H}_{g,1}^c)$ of $\mathcal{H}_{g,1}^c$. This contains, at least, the group θ^H of homology cobordism classes of homology 3-spheres (see Remark 2.3). Furthermore $\mathcal{Z}(\mathcal{H}_{g,1}^c)$ contains also the element τ_∂ defined by a Dehn twist about the boundary of $\Sigma_{g,1}$. For the mapping class group Γ_g it seems to be true (according to J. Birman and C. McMullen) that the center is trivial, at least for large g.

QUESTION 6. Determine $\mathcal{Z}(\mathcal{H}_{g,1}^c)$. Is it generated by θ^H and τ_∂?

We now consider the subgroup $\mathcal{H}_{g,1}^c[\infty] \subseteq \mathcal{H}_{g,1}^c$. This contains $\theta^H \times \mathcal{S}_g^c[\infty]$, where $\mathcal{S}_g^c[\infty]$ denotes the subgroup of the string-link concordance group $\mathcal{S}_g^c$ consisting of string links with vanishing μ-invariants (see Remarks 2.3 and 2.7). $\mathcal{S}_g^c[\infty]$ contains, for example, all boundary string links and, in particular, the knot concordance group, which is an abelian group of infinite rank (see[**Le1**]).

QUESTION 7. Determine $\mathcal{H}_{g,1}^c[\infty]$. Is it equal to $\theta^H \times \mathcal{S}_g^c[\infty]$?

QUESTION 8. Is $\mathcal{Z}(\mathcal{H}_{g,1}^c)/\mathcal{Z}(\mathcal{H}_{g,1}^c) \cap \mathcal{H}_{g,1}^c[\infty]$ the infinite cyclic group generated by τ_∂?

In [**Mo4**], Morita calculated the symplectic invariant part $\mathsf{D}(H)^{\mathfrak{sp}}$ of $\mathsf{D}(H)$ in terms of a beautiful space of chord diagrams. The group $\mathcal{Z}(\mathcal{H}_{g,1}^c)/(\mathcal{Z}(\mathcal{H}_{g,1}^c) \cap \mathcal{H}_{g,1}^c[\infty])$ is closely related to $\mathsf{D}(H)^{\mathfrak{sp}}$, since the image under the map σ_n of Theorem 3 of an element in $\mathcal{Z}(\mathcal{H}_{g,1}^c) \cap \mathcal{H}_{g,1}^c[n]$ lies in $\mathsf{D}_n(H)^{\mathfrak{sp}}$. Thus any element of $\mathcal{Z}(\mathcal{H}_{g,1}^c)/\mathcal{Z}(\mathcal{H}_{g,1}^c) \cap \mathcal{H}_{g,1}^c[\infty]$ provides a geometric construction of an element of $\mathsf{D}(H)^{\mathfrak{sp}}$.

The following question is important to the philosophical notion of finite type.

QUESTION 9. Is $\mathcal{H}_{g,1}^c$ finitely-generated? Is its abelianization finitely-generated? Note that both the mapping class group $\Gamma_{g,1}$ and the Torelli group $\mathcal{T}_{g,1}$ are finitely-generated. Note also that $\mathcal{S}_g^c$ and θ^H are infinitely-generated abelian (see [**F**]) and, as for the analogous question for string-links, since the knot concordance group has infinite rank, the abelianization of the string-link concordance group (on any number of strings) has infinite rank.

QUESTION 10. Let $\mathcal{H}_g$ denote the semigroup of homology cylinders over the *closed* surface Σ_g of genus g. The kernel of the obvious epimorphism $\mathcal{H}_{g,1} \to \mathcal{H}_g$ is related to concordance classes of framed proper arcs in $I \times \Sigma_g$. Describe this more explicitly and consider also the homology cobordism groups $\mathcal{H}_g^c$.

5.1. Note. The present paper was completed in 1999, and a related paper by the second author, which was completed later, appeared in [**Le3**].

References

[AMR1]   J. Andersen and J. Mattes and N. Reshetikhin, *The Poisson structure on the moduli space of flat connections and chord diagrams*, Topology **35** (1996) 1069–1083.

[AMR2]   J. Andersen and J. Mattes and N. Reshetikhin, *Quantization of the algebra of chord diagrams*, Math. Proc. Cambridge Phil. Soc. **124** (1998) 451–467.

[B-N]   D. Bar-Natan, *Vassiliev homotopy string-link invariants*, J. Knot Theory and its Ramifications **4** (1995) 13–32.

[CE]   H. Cartan, S. Eilenberg, *Homological Algebra*, Princeton University Press, 1956.

[CGO] T. Cochran, A. Gerges and K. Orr, *Dehn surgery equivalence relations on three-manifolds*, Math. Proc. Cambridge Philos. Soc. **131** (2001) 97–127.

[Dr]   V. G. Drinfeld, *On quasitriangular quasi Hopf algebras and a group closely connected with $Gal(\overline{\mathbb{Q}}/\mathbb{Q})$*, Leningrad Math. J. **2** (1991) 829–860.

[Dw]   W. Dwyer, *Homology, Massey products and maps between groups*, J. Pure and Applied Algebra **6** (1975) 177–190.

[FS]   R. Fenn, D. Sverje, *Massey products and lower central series of groups*, Canad. J. Math. **39** (1987) 322–337.

[F]   M. Furuta, *Homology cobordism group of homology 3-spheres*, Inventiones Math. **100** (1990) 339–355.

[GGP] S. Garoufalidis, M. Goussarov and M. Polyak, *Calculus of clovers and finite type invariants of 3-manifolds*, Geometry and Topology, **5** (2001) 75–108.

[GL1] S. Garoufalidis and J. Levine, *Finite type 3-manifold invariants, the mapping class group and blinks*, J. Diff. Geom. **47** (1997) 257–320.

[GL2] S. Garoufalidis and J. Levine, *Finite type 3-manifold invariants and the Torelli group I*, Inventiones Math. **131** (1998) 541–594.

[Gu1] M. Goussarov, *Finite type invariants and n-equivalence of 3-manifolds*, C. R. Acad. Sci. Paris Ser. I. Math. **329** (1999) 517–522.

[Gu2] M. Goussarov, *Variations of knotted graphs, geometric technique of n-equivalence*, St. Petersburg Math. J. **12** (2001).

[HL]   N. Habegger and X-S. Lin, *The classification of links up to homotopy*, Journal AMS **3** (1990) 389–419.

[HM]   _____ and G. Masbaum, *The Kontsevich Integral and Milnor's Invariants*, Topology **39** (2000) 1253–1289.

[Hb]   K. Habiro, *Clasper theory and finite type invariants of links*, Geometry and Topology **4** (2000), 1–83.

[Ha]   R. Hain, *Infinitesimal presentations of the Torelli groups*, Journal of AMS **10** (1997) 597–651.

[Ho]   H. Hopf, *Fundamentalgruppe und Zweite Bettische Gruppe*, Comment. Math. Helvetici **14** (1942) 257–309.

[IO]   K. Igusa, K. Orr, *Links, pictures and the homology of nilpotent groups*, Topology **40** (2001) 1125–1166.

[Ih]   Y. Ihara, *Some problems on three-point ramifications and associated large Galois representations*, Adv. Studies in Pure Math. **12** North-Holland (1987) 173–188.

[Jo1] D. Johnson, *An abelian quotient of the mapping class group*, Math. Ann. **249** (1980) 225-242.

[Jo2] D.Johnson, *A survey of the Torelli group*, Contemporary Math. **20** (1983) 163–179.

[KM]   M. Kervaire, J.W. Milnor, *Groups of homotopy spheres I*, Annals of Math. **77** (1963) 504–537.

[Ki]   T. Kitano, *Johnson's homomorphisms of subgroups of the mapping class group, the Magnus expansion and Massey higher products of mapping tori*, Topology and its Applic. **69** (1996) 165–172.

[Kh]   T. Kohno, *Série de Poincaré-Koszul associée aux groupes de Tresses Pures*, Inventiones Math. **82** (1985) 57–76.

[Ko1] M. Kontsevich, *Formal (non)-commutative symplectic geometry*, Gelfand Math. Seminars, 1990-92, Birkhauser, Boston, (1993) 173–188.

[Ko2] M. Kontsevich, *Feynmann diagrams and low-dimensional topology*, Proceedings of the first European Congress of Mathematicians, vol. 2, Progress in Math. **120** Birkhauser, Boston, (1994) 97–121.

[Ko3] M. Kontsevich, *Operads and motives in deformation quantization*, Lett. Math. Phys. **48** (1999) 35–72.

[Kr]   D. Kraines, *Massey higher products*, Transactions AMS **124** (1966) 431–449.

[LMO] T.T.Q. Le, J. Murakami, T. Ohtsuki, *A universal quantum invariant of 3-manifolds*, Topology **37** (1998) 539–574.

[La]   L. Lambe, *Two exact sequences in rational homotopy theory relating cup products and commutators*, Proc. AMS **96** (1986) 360–364.

[Le1] J. Levine, *Invariants of knot cobordism*, Inventiones Math. **8** (1969) 98–110.

[Le2]   J. Levine, *Link concordance and algebraic closure*, Comment. Math. Helvetici **64** (1989) 236–255.

[Le3]   _____, *Homology cylinders: an enlargement of the mapping class group*, Algebr. Geom. Topol. **1** (2001) 243–270.

[Mac]   S. MacLane, *Homology*, Springer-Verlag, 1963.

[MKS]   W. Magnus, A. Karras and D. Solitar, *Combinatorial group theory, presentations of groups in terms of generators and relations*, Wiley, New York, 1966.

[Ma]   W.S. Massey, *Higher order linking numbers*, J. Knot Theory and its Ramifications **7** (1998) 393–414.

[M]   S. V. Matveev, *Generalized surgery of three-dimensional manifolds and representations of homology spheres*, Math. Notices Acad. Sci. USSR, **42:2** (1987) 651–656.

[MM]   J.W. Milnor and J.C. Moore, *On the structure of Hopf algebras*, Ann. Math. **81** (1965) 211–264.

[Mo1]   S. Morita, *Casson's invariant for homology 3-spheres and characteristic classes of vector bundles I*, Topology **28** (1989) 305-323.

[Mo2]   S.Morita, *Abelian quotients of subgroups of the mapping class group of surfaces*, Duke Math. Journal **70** (1993) 699–726.

[Mo3]   S.Morita, *A linear representation of the mapping class group of orientable surfaces and characteristic classes of vector bundles*, Topology of Teichmüller spaces, S. Kojima et al editors, World Scientific, (1996) 159–186.

[Mo4]   S.Morita, *Structure of the mapping class group of surfaces: a survey and a prospect*, Proceedings of the Kirbyfest Geom. Topol.Monogr., **2** (1999) 349–406.

[Oh]   T. Ohtsuki, *A filtration of the set of integral homology 3-spheres*, Proceedings of the International Congress of Mathematicians, Vol. II, Berlin (1998) 473–482.

[O1]   K. Orr, *Homotopy invariants of links*, Inventiones Math. **95** (1989) 379–394.

[O2]   K.Orr, *Link concordance invariants and Massey products*, Topology **30** No. 4 (1991) 699-710.

[Qu]   D. Quillen, *On the associated graded ring of a group ring*, Journal of Algebra **10** (1968) 411–418.

[St]   J. Stallings, *Homology and central series of groups*, Journal of Algebra **2** (1965) 170–181.

[Su]   D. Sullivan, *On the intersection ring of compact 3-manifolds*, Topology **14** (1975) 275–277.

[Tr]   A. Tristram, *Some cobordism invariants for links*, Math. Proc. Cambridge Phil. Soc. **66** (1969) 251–264.

[Tu]   V. G. Turaev, *Nilpotent homotopy types of closed 3-manifolds*, Proceedings of Topology Conference in Leningrad, Lecture Notes in Mathematics **1060** (1982), Springer-Verlag, 355–366.

School of Mathematics, Georgia Institute of Technology, Atlanta, GA 30332-0160, USA
URL: http://www.math.gatech.edu/~stavros
E-mail address: stavros@math.gatech.edu

Department of Mathematics, Brandeis University, Waltham, MA 02454-9110, USA.
URL: http://www.brandeis.edu/~levine
E-mail address: levine@brandeis.edu

Multivalued Morse theory, asymptotic analysis and mirror symmetry

Kenji Fukaya

ABSTRACT. In this article we describe a series of conjectures which, if proven, will imply (*assuming* existence of a dual torus fibration) various predictions of the Mirror Symmetry Conjecture. We present evidence for our conjectures and prove some parts of them.

CONTENTS

1. Introduction
2. Semi-flat mirror symmetry: review
3. A-model: pseudoholomorphic curves
4. Including Lagrangian submanifolds
5. B-model: asymptotic analysis
6. Asymptotic analysis on the mirror bundle
References

1. Introduction

In this paper we are going to discuss how various conclusions of mirror symmetry are expected to follow in the case that the mirror manifold is obtained as a dual torus fibration, as was proposed by Strominger-Yau-Zaslow [**72**].

We start with a pair of manifolds $M, M^\vee$ which are compactifications of dual torus fibrations $M_0 \to B_0$, $M_0^\vee \to B_0$, whose fibers are Lagrangian submanifolds. In the case that they are globally a fiber bundle with torus fiber we can easily construct a complex structure on $M^\vee$ from the symplectic structure on M. However, it seems that only a torus is an example where such a fiber bundle structure exists globally. So, to study the case other than the torus, we must include singular fibers. The symplectic structure of M_0 is extended to the singular fiber, but the complex structure of $M_0^\vee$ does not extend. We need to include a "quantum correction" or "instanton correction" to the complex structure on $M_0^\vee$ so that it will be extended

---

2000 *Mathematics Subject Classification.* 53D40.
Key words and phrases. Floer homology, mirror symmetry, WKB analysis, Morse theory.

© 2005 American Mathematical Society

to the singular fibers. Such an "instanton correction" is expected to be described by the (pseudo)holomorphic curves in M.

This story is then expected to imply the most famous conclusion of mirror symmetry, that is the coincidence of Gromov-Witten potential and Yukawa coupling, since the former is a generating function of the number of holomorphic disks and the later is obtained from the "quantum correction" to the complex structures on $M_0^\vee$.

In this article, we study asymptotic behavior of the instanton effect when the diameter of the fiber is small.

The main idea of this article is to relate pseudoholomorphic curves on M, on the one hand, and the asymptotic behavior of the "quantum correction" to the complex structure on $M_0^\vee$, on the other hand, to a third object: Morse homotopy of a multivalued function on B. In this article we explain the procedure of how to reduce both problems to the Morse theory on the base space. However, there are several points which the author cannot make precise yet.

We also include the case when there is a Lagrangian submanifold L on M and a holomorphic vector bundle on $M^\vee$. In this case, we study the Homological Mirror Symmetry Conjecture due to Kontsevich [59]. (See [32, 33] for more precise statement on homological mirror symmetry.) Then, the story develops as follows. In the case that the restriction of the projection $M \to B$ to L is a covering map, the story is very much similar to the case that there are no singular fibers. Namely, we can construct a mirror bundle rather easily. (This part was worked out by the author in [32].) A new phenomenon occurs when $L \to B$ is not a covering map somewhere. Then, we need to study caustics or Lagrangian singularities which have been extensively studied in symplectic geometry. Let the caustics, $S(L) \subset B$, be the subset where $L \to B$ is not a covering map. In the mirror side, the mirror bundle $\mathcal{E}(L)$ with "classical" holomorphic structure can be defined on $\pi_{M^\vee}^{-1}(B \backslash S(L))$. This holomorphic structure does not extend to the preimage of the caustic, $\pi_{M^\vee}^{-1}(S(L))$. Hence, we need to include again an "instanton effect", which is described by a pseudoholomorphic disk in M with the boundary in L. Our idea in this article is to reduce both the asymptotic behavior of the holomorphic structure of the mirror bundle and holomorphic disks with boundary in our Lagrangian submanifold L to the Morse theory of multivalued functions on B.

Relation of Morse theory to pseudoholomorphic curves is classical and goes back to Floer's invention [22] of Floer homology of Lagrangian intersection. In [36] the author with Y.G. Oh generalized Floer's theory. Namely, Floer studied the case when there are two Lagrangian submanifolds (and one Morse function) and described the pseudoholomorphic strip in terms of gradient lines of the Morse function. In [36] we discussed the case when there are several Lagrangian submanifolds and several Morse functions and described pseudoholomorphic disks in terms of Morse homotopy. In [36], the case when M is a cotangent bundle was studied. The result of [36] was used by the author in [32] to study the homological mirror symmetry of complex torus. Recently Kontsevich and Soibelman [61] showed that the result of [36] can be generalized to the case when M is a torus T^{2n} (and the fiber is T^n instead of $\mathbb{R}^n$) and used it to give an alternative proof of a part of the results of [32]. (Kontsevich and Soibelman in [61] also proposed to use rigid analytic geometry to study mirror symmetry. We use in this paper asymptotic analysis, which is somewhat similar to their proposal.)

In both articles [36] and [61], the case when there are no caustics is discussed. The new point in this article is that we include caustics. We also study the closed strings and singular fibers. Then, we study a multivalued function on B, once in [36] we studied several functions simultaneously. Namely, instead of studying several functions we need to study several branches of a multivalued function. The roles of the branched loci and the critical points of the corresponding function are played respectively by caustics and singular loci.

The relation of Morse theory to asymptotic analysis is also classical. Especially, it was discussed in a famous paper by Witten [76]. In [76] Witten deformed the De Rham complex "by hand" by using a Morse function. In this article we consider Dolbault operator $\overline{\partial}$ on $M^{\vee}$ and study the asymptotic behavior of the solution of the Maurer-Cartan equation $\overline{\partial} B + B \wedge B = 0$ when the fiber shrinks to a point. For this purpose we develop B as a Fourier series along torus fibers. The idea to use Fourier series along fibers to study mirror symmetry appeared in Arinkin-Polishchuk [5] and was used by several authors (see for example by [61]). Then, in each Fourier mode, the equation $\overline{\partial} B = 0$ is equivalent to the deformed De Rham operator. Here the deformation is ruled by the same multivalued function which appears in the study of the pseudoholomorphic disks of the mirror manifold and is similar to Witten's Morse function. Thus, if we follow Witten's idea, the solution of the linearized equation $\overline{\partial} B = 0$ is described by the Morse theory of our multivalued function. We however need to study the nonlinear terms. The way we do so is actually the same as the one due to Feynman himself. Namely, we use Feynman diagrams to include the effect of the nonlinear terms.

We will define the multivalued function which plays the central role in our story in §2.1 and §3.1. Roughly speaking, a branch of the function at $x \in B$ corresponds to an element of the fundamental group of the fiber $M \to B$ at x. The pseudoholomorphic disk which is described by a branch is one which wraps a loop representing the element of the fundamental group corresponding to the branch.

On the other hand, a term of the Fourier expansion along the fiber of the dual fibration $M^{\vee} \to B$ also corresponds to the fundamental group of the fiber of $M \to B$. (This is a consequence of Pontrjyagin duality.)

Thus, in our argument, the cycle over which the closed string wraps corresponds to a Fourier mode in the mirror. This is exactly the idea of T-duality.

2. Semi-flat mirror symmetry: review

The content of this section is not new. There are many articles describing it. We refer, for example, [72, 64, 61, 50, 3, 15, 47, 70, 32, 35].

Let B_0 be an n dimensional manifold. We do not assume B_0 is compact. Let $\pi_\Lambda : \Lambda \to B_0$ be a bundle whose fiber is $\mathbb{Z}^n$ and whose structure group is $SL(n;\mathbb{Z})$. We put $\Lambda_x = \pi_\Lambda^{-1} x$. Let $E = \Lambda \otimes_{\mathbb{Z}} \mathbb{R}$, and let $\pi_E : E \to B_0$ be the projection. E is a real n dimensional vector bundle over B_0. We put $E_x = \pi_E^{-1} x$. We may regard Λ as a subbundle of E.

Since the structure group of E is reduced to a discrete group $SL(n;\mathbb{Z})$, it follows that E has a flat connection ∇^{GM}. Following [61], we call ∇^{GM} the *Gauss-Manin connection*. We put $M_0 = E/\Lambda$. M_0 is a total space of the fiber bundle $\pi_{M_0} : M_0 \to B_0$. We put $\pi_{M_0}^{-1}(x) = F_x = E_x/\Lambda_x$. The fiber F_x is diffeomorphic to an n dimensional torus T^n. The fiber bundle π_{M_0} has a flat connection induced from ∇^{GM}. We denote it also by ∇^{GM} and call it the Gauss-Manin connection.

The zero section $s_0 : B_0 \to E$ induces a section of π_{M_0}. We denote it by the same symbol $s_0 : B_0 \to M_0$.

Our manifold M_0 has locally a T^n action (whose orbits are the fibers). Because of the presence of the monodromy, this local action does not induce a global action of T^n. An appropriate language to describe this situation is called a T-structure and was introduced by Cheeger-Gromov [**17, 18**], to study collapsing Riemannian manifolds. We recall its definition here since the story of mirror symmetry is actually related to collapsing Riemannian manifolds at this point.

DEFINITION 2.1. Let M be a $2n$ dimensional manifold. A *pure polarized T-structure* of dimension n on M consists of the following data :
(1) An open covering $M = \cup U_i$.
(2) An effective action of T^n on U_i. We assume that the dimension of every orbit is n. We write the action as $(g, x) \mapsto g \cdot_i x$.
(3) If $U_i \cap U_j \neq \emptyset$ then $U_i \cap U_j$ is invariant of the both T^n actions (the ones on U_i and on U_j).
(4) An isomorphism $I_{i,j} : T^n \to T^n$ such that $g \cdot_i x = I_{i,j}(g) \cdot_j x$, for $x \in U_i \cap U_j$ and $g \in T^n$.

REMARK 2.1. The definition in Cheeger-Gromov [**17, 18**] uses more sophisticated language, the action of sheaf of groups. The definition here, which is equivalent to Cheeger-Gromov', is taken from [**25**].

Obviously, our manifold M_0 has a pure polarized T-structure of dimension n. One can define T-structure invariance of tensors in an obvious way.

2.1. Symplectic structure.

PROPOSITION 2.1. *There exists a bijection between the following two objects:*
(A) *The $\sim$ equivalence class of symplectic form ω on M_0 such that $\omega|_{F_x} \equiv 0$ and $\omega|_{s_0(B_0)} \equiv 0$. Here $\omega \sim \omega'$ if and only if there exists a diffeomorphism $f : M_0 \to M_0$ such that $\pi \circ f = \pi$ and $f^*\omega = \omega'$.*
(B) *An isomorphism $\varphi : TB_0 \cong E^*$ between tangent bundle of B_0 and the dual bundle of E, such that the dual connection ∇^{GM*} induces a **torsion free** connection on TB_0.*

PROOF. We first mention the following lemma.

LEMMA 2.1. *Any symplectic form ω as in (A) is $\sim$ equivalent to a unique symplectic form which is invariant under the T-structure on M_0.*

The proof is an immediate consequence of the existence of action-angle coordinates (see for example [**6**]) and is left to the reader. From now on we only consider symplectic forms which are invariant under our T-structure.

Using the Gauss-Manin connection, we have the following isomorphism (splitting into horizontal and vertical directions):
$$T_{s_0(x)}M_0 \cong T_xB_0 \oplus T_{s_0(x)}F_x \cong T_xB_0 \oplus E_x. \tag{1}$$
Let ω be a T-structure invariant nondegenerate 2-form on M_0 such that it vanishes on the fibers F_x and on $s_0(B_0)$. The restriction of ω to $T_xB_0 \otimes E_x$ induces a

homomorphism $\varphi_\omega : T_x B_0 \to E_x^*$. Since ω is nondegenerate and vanishes on the fibers F_x and on $s_0(B_0)$, it follows that φ_ω is an isomorphism.

Using the T-structure invariance of ω, it is easy to see that $\varphi_\omega = \varphi_{\omega'}$ implies $\omega = \omega'$. Moreover, any isomorphism $\varphi : T_x B_0 \to E_x^*$ is φ_ω for some ω. Thus, to complete the proof of Proposition 2.1, it suffices to show the following lemma.

LEMMA 2.2. $d\omega = 0$ if and only if $\varphi_\omega^* \nabla^{GM*}$ is torsion free.

PROOF. The problem is local with respect to B_0. So, we take a small open set U on B_0 and study $\pi_{M_0}^{-1} U$. We use the Gauss-Manin connection to split $\pi_{M_0}^{-1} U = U \times T^n$. Let (x^i) be coordinates of U and let y^i be affine coordinates of T^n. We may regard the basis of E_x as $\frac{\partial}{\partial y^i}$ $i = 1, \cdots, n$. Hence, dy^i $i = 1, \cdots, n$ may be regarded as a basis of E^*.

Now, we put $\varphi_\omega \left(\frac{\partial}{\partial x^i} \right) = \sum h_{ij}(x) dy^j$. Then, $\omega = \sum h_{ij}(x) dx^i \wedge dy^j$. Since ω is T-structure invariant it follows that h_{ij} depends only on $x \in B_0$ and is independent of y^j. It is straightforward calculation to check :

$$d\omega = 0 \quad \Leftrightarrow \quad \forall i \forall j \forall k \; \frac{\partial h_{ij}}{\partial x^k} = \frac{\partial h_{kj}}{\partial x^i} \quad \Leftrightarrow \quad T^{\varphi_\omega^* \nabla^{GM*}} = 0.$$

Here $T^{\varphi_\omega^* \nabla^{GM*}}$ is the torsion tensor of $\varphi_\omega^* \nabla^{GM*}$. The proof of Lemma 2.2 and hence of Proposition 2.1 is now complete. □

Let us consider the situation of Proposition 2.1. Then, we have a flat torsion free affine connection $\varphi_\omega^* \nabla^{GM*}$ on the tangent bundle. Hence, there exist affine coordinates (y^j) on B_0. Namely, $\frac{\partial}{\partial y^j}$ are parallel. Then, the function h_{ij} are constant. Hence, by linear change of the affine coordinates (y^j) we can assume $h_{ij} = \delta_{ij}$. It follows that $\omega = \sum dx^i \wedge dy^i$. Namely, (x^i, y^j) $i, j = 1, \cdots, n$, are the Darboux coordinates of our symplectic manifold M_0. In this case we call x^j *symplectic coordinates* of B_0 and (x^j, y^j) symplectic coordinates of M.

2.2. Complex structure.

PROPOSITION 2.2. *There exists a bijection between the following two objects:*
(A) *A complex structure J on M_0 which is invariant under the T-structure and which satisfies $J(T_x B_0) \subseteq T_{s(x)} F_x$, $J(T_x F_x) \subseteq T_{s(x)} B_0$. (Here we use decomposition (1) to regard $T_x B_0 \subset T_{s_0(x)} M_0$.)*
(B) *An isomorphism $\varphi : TB_0 \cong E$ between the tangent bundle of B_0 and E such that $\varphi^* \nabla^{GM}$ is a **torsion free** connection on TB_0.*

REMARK 2.2. The condition $J(T_x B) \subseteq T_{s_0(x)} F_x$, $J(T_{s_0(x)} F_x) \subseteq T_x B$ we assumed in Proposition 2.2 (A) looks rather restrictive. To remove it we need to consider the B field in the mirror side (the symplectic structure side). We do not discuss it in this article, see [**64, 15, 3, 32, 35**].

PROOF. Let J be an almost complex structure which satisfies Proposition 2.2 (B) except integrability. The restriction of J to $T_x B_0 \subset T_{s_0(x)} M_0$ defines a homomorphism $\varphi_J : TB_0 \to E$. It is easy to see that φ_J is an isomorphism. On the other hand, any φ as in Proposition 2.2 (B) is obtained as φ_J. Therefore, to prove Proposition 2.2, it suffices to show the following.

LEMMA 2.3. *J is integrable if and only if $\varphi_J^* \nabla^{GM}$ is torsion free.*

PROOF. Let (x^i, y^j) be local coordinates of M_0 as in the proof of Lemma 2.2. We put $J\left(\frac{\partial}{\partial x^i}\right) = \sum g_i^j(x)\frac{\partial}{\partial y^j}$. It follows that $\varphi_J\left(\frac{\partial}{\partial x^i}\right) = \sum g_i^j(x)\frac{\partial}{\partial y^j}$. We can check by a simple calculation that

$$\overline{\partial}_J \circ \overline{\partial}_J = 0 \quad \Leftrightarrow \quad \forall i \forall j \forall k \ \frac{\partial g_i^j}{\partial x^k} = \frac{\partial g_k^j}{\partial x^i} \quad \Leftrightarrow \quad T^{\varphi_J^* \nabla^{GM}} = 0.$$

Here $\overline{\partial}_J$ is the Dolbaut operator associated to the almost complex structure J. Lemma 2.3 and Proposition 2.2 follow. □

We take affine local coordinates x^i on B_0 with respect to the flat affine connection $\varphi^* \nabla^{GM}$. We can choose them so that $J\left(\frac{\partial}{\partial x^i}\right) = \frac{\partial}{\partial y^i}$. Then, $z^i = x^i + \sqrt{-1}y^i$ are complex coordinates of M_0. In this case we say the coordinates x^i are *complex coordinates* (of B_0).

2.3. Kähler and Calabi-Yau structures. Let $B_0 = \cup U_a$ be an open covering such that each of U_a is contractible.

PROPOSITION 2.3. *There exists a bijection between the following two objects:*
(A) *A Kähler structure J, ω on M_0 such that J satisfies Proposition 2.2 (A) and that ω satisfies Proposition 2.1 (A).*
(B) *A pair of an isomorphism $\varphi : TB_0 \cong E^*$ and a functions $h_a : U_a \to \mathbb{R}$ such that $\varphi^* \nabla^{GM}$ is a **torsion free** connection on TB_0 and $g_{ij} = \frac{\partial^2 h_a}{\partial x^i \partial x^j}$ is a Riemannian metric g_a on U_a. (Here (x^i) are affine coordinates with respect to the flat affine connection $\varphi^* \nabla^{GM}$.) Moreover, g_a is independent of the choice of a neighborhood $U_a \ni x$.*

PROOF. Let J be as in Proposition 2.2 (A) and ω be as in Proposition 2.1 (A). We assume that (M_0, ω, J) is Kähler. Namely, we assume that $g(V, W) = \omega(V, JW)$ is a Riemannian metric. We take symplectic coordinates (x^i, y^j) and put $J\left(\frac{\partial}{\partial x^i}\right) = \sum g_{ij}(x)\frac{\partial}{\partial y^j}$. Note $\omega = \sum dx^i \wedge dy^i$. Hence,

$$g\left(\frac{\partial}{\partial x^i}, \frac{\partial}{\partial x^j}\right) = \omega\left(\frac{\partial}{\partial x^i}, J\frac{\partial}{\partial x^j}\right) = g_{ij}.$$

Hence, $\sum g_{ij}(x) dx^i dx^j$ is a Riemannian metric on B_0.

As we established in §1.1, the integrability of J is equivalent to $\frac{\partial g_{ij}}{\partial x^k} = \frac{\partial g_{kj}}{\partial x^i}$. This condition is equivalent to $d(\sum_j g_{ij} dx^j) = 0$. Hence, locally, there exist smooth functions h^i such that $\sum_j g_{ij} dx^j = dh_i$. Since $g_{ij} = g_{ji}$, it follows that $\frac{\partial h_i}{\partial x^j} = \frac{\partial h_j}{\partial x^i}$. Therefore, there exists, locally, a smooth function h such that $h_i = \frac{\partial h}{\partial x^i}$. Thus, we have :

(2) $$g_{ij} = \frac{\partial^2 h}{\partial x^i \partial x^j}.$$

We thus show that an element of (A) gives an element of (B). The proof of the converse is similar. □

REMARK 2.3. An affine flat manifold with a Riemannian metric satisfying (2) is called a *Hessian manifold*. It has been studied for a long time in differential geometry. (According to [**71**], it first appeared in [**62**], see also [**1, 19**].) Its relation to mirror symmetry is observed by [**50**].

Let us calculate the complex coordinates of B_0 in terms of h and symplectic coordinates. We put

$$\check{x}^i = \frac{\partial h}{\partial x^i}. \tag{3}$$

Since g_{ij} is nondegenerate, it follows from (3) that $(\check{x}^i)$ are local coordinates. We have

$$J\left(\frac{\partial}{\partial y^j}\right) = \sum g^{ij}(x)\frac{\partial}{\partial x^i} = \sum g^{ij}(x)\frac{\partial^2 h}{\partial x^i \partial x^k}\frac{\partial}{\partial \check{x}^k} = \frac{\partial}{\partial \check{x}^j}.$$

Namely, $\check{x}^j$ are complex coordinates.

We remark that (3) is a classical Legendre transformation with respect to the convex function h (see [**6**]). Thus, we obtain:

PROPOSITION 2.4. *In the situation of Proposition 2.3 complex coordinates $(\check{x}^j)$ are Legendre transformations of symplectic coordinates (x^j).*

As for Calabi-Yau structure we have the following.

LEMMA 2.4. *In the situation of Proposition 2.3, the Kähler metric induced from J and ω is Ricci flat if and only if*

$$\det\left(\frac{\partial^2 h}{\partial x^i \partial x^j}\right) = 1.$$

We omit the proof, see [**19, 64, 61, 47**].

2.4. Mirror symmetry without quantum effect.

The discussion of §§1.1, 1.2 imply a simple version of mirror symmetry as follows. Let M_0, E, Λ be as above. Let E^* be the dual bundle to E. It has a dual (flat) connection ∇^{GM*}. We put $\Lambda_x^* = \{\gamma \in E_x^* | \gamma(\Lambda_x) \subseteq \mathbb{Z}\}$, which is the dual lattice. Its union $\Lambda^* = \cup_x \Lambda_x^* \subseteq E^*$ is a $\mathbb{Z}^n$ bundle over B_0, and $\Lambda^* \otimes_{\mathbb{R}} \mathbb{R} = E^*$. We put $M_0^{\vee} = E^*/\Lambda^*$, and call $M_0^{\vee}$ a *mirror* of M_0.

PROPOSITION 2.5. *The set of the symplectic structures ω on M_0 satisfying Proposition 2.1 (A) corresponds one-to-one to the set of complex structures J on $M_0^{\vee}$ satisfying Proposition 2.2 (A).*

PROOF. They both correspond to an isomorphism $\varphi : TB_0 \to E$ such that the pull back $\varphi^*\nabla^{GM*}$ is torsion free, by Propositions 2.1 and 2.2. $\square$

We remark moreover that the symplectic coordinates for M_0 will become complex coordinates for $M_0^{\vee}$. We may exchange the role of M_0 and $M_0^{\vee}$ in Proposition 2.5. Then, a complex structure on M_0 corresponds to a symplectic structure on $M_0^{\vee}$. Based on the discussion of §1.4 we can consider the situation where M_0 is both symplectic and complex (that is Kähler), and also the situation when M_0 is a Calabi-Yau manifold. We omit it (see [**64, 61, 47**]).

Proposition 2.5 is a mirror symmetry in the sense that it exchanges the role of symplectic and complex structures. To study more interesting part of mirror symmetry, we need to consider the case when M is not globally a fiber bundle but has singular fibers. Then, the complex structure J is not invariant under the T-structure (see §4). We also need to take into account the "quantum effect". Studying these points is the main purpose of this article. The purpose of this section is to provide a background to study the "quantum effect", which is the most interesting part of the story of mirror symmetry.

In this article, we start with a manifold M which admits a fibration $\pi_M : M \to B$. Such a fibration is conjectured to exist in the case when the complex structure of M is in a neighborhood of large complex structure limit. (However, we need to allow M to have a singular fiber unless it is a torus.) Several authors [47, 61] have proposed and/or partially proved the existence of such a singular fibration. To prove the existence of torus fibration, the theory of collapsing Riemannian manifolds may be useful, see [47, 61].

Actually, the following has been proved:

THEOREM 2.1. ([23, 24]) *Let M_i be a sequence of Riemannian manifolds. We assume that the sectional curvature of the M_i is uniformly bounded in absolute value. We also assume that M_i converges to B, another (smooth) Riemannian manifold, with respect to the Gromov-Hausdorff convergence [42]. Then, for sufficiently large i, there exists a fiber bundle $M_i \to B$. The fiber is an infra nilmanifold and the structure group is reduced to the group of affine diffeomorphisms.*

Here, a nilmanifold is a quotient of a nilpotent Lie group by its cocompact discrete subgroup. It has an equivariant affine structure. An infra-nilmanifold is the quotient of a nilmanifold by an affine action of a finite group.

REMARK 2.4. Yamaguchi [78] has proved the first half of the conclusion under the weaker assumption that the sectional curvature is uniformly bounded from below.

REMARK 2.5. If ϵ is the diameter of the fiber and if the volume of the M_i is of order ϵ^k where $k = \dim M_i - \dim B$ in the situation of Theorem 2.1, then it follows from Corollary A1.8 of [16] that the fibers are the quotient of a torus by a finite group.

In our situation, our Riemannian manifolds are of bounded *Ricci* curvature but not of bounded *sectional* curvature. So, Theorem 2.1 does not directly apply. In the situation of bounded *Ricci* curvature Bando-Kasue-Nakajima [12] (see also [2]) proved the following.

THEOREM 2.2. *Let M_i be a sequence of Einstein-Kähler manifolds of complex dimension 2. We assume that the Ricci curvature of the M_i are 0 or ± 1. We also assume that the volume of the M_i are greater that $\epsilon > 0$ independent of i and the diameter of the M_i is bounded. Then, the limit of M_i with respect to the Gromov-Hausdorff convergence is an orbifold M with A, D, E type singularities. Moreover, for large i, the M_i are diffeomorphic to the minimal resolution of M.*

Let (M, J_i) be a maximal degenerate family of Calabi-Yau manifolds (see [65] for its definition.) Let us take the Ricci flat Kähler metrics g_i on M whose existence was established by Yau's proof of the Calabi conjecture. We normalize the metrics so that the diameter of (M, g_i) is equal to 1. It follows from a famous result by Gromov ([42], Theorem 5.3) that there exists a subsequence of the (M, g_i) which converges to some metric space. (However, it is not known whether it converges without taking a subsequence.) In this situation, let us propose the following conjecture, which is a variant of one in [61, 47].

CONJECTURE 2.1. *The sequence (M, g_i) converges in the Gromov-Hausdorff topology to a space B with the following properties:*
(1) *B is homeomorphic to an n-dimensional simplicial complex.*

(2) There exists a codimension ≥ 2 subcomplex $S(B)$ such that $B\backslash S(B)$ is a smooth Riemannian manifold.
(3) Let $\varphi_i : (M, g_i) \to B$ be Haudsorff ϵ_i-approximations with $\lim_{i\to\infty} \epsilon_i = 0$, in the sense of [**25**], Definition 2.1. Then, for each δ, the sectional curvature of the (M, g_i) on $\varphi_i^{-1}(B\backslash U_\delta(S(B)))$ is uniformly bounded for large i. (Here $U_\delta(S(B))$ is the δ neighborhood of $S(B)$.)

A result of Gross-Wilson [**47**] implies Conjecture 2.1 in the case of an elliptic K3 surface.

We remark that Conjecture 2.1, if proved, together with Theorem 2.1 will imply the existence of a fiber bundle structure over $B\backslash S(B)$. Remark 2.5 may imply that the fibers are tori.

However, estimating the size of the set where the sectional curvature goes to infinity seems to be an extremely hard problem. So, the author feels that it is too difficult to attack this problem using the technology of global Riemannian geometry established at this stage. We will explain a gauge theory analogue of Conjecture 2.1 in §5 (Conjecture 6.5), which seems more tractable.

3. A-model: pseudoholomorphic curves

3.1. Multivalued function. Now, we introduce the main new player in our story, a multivalued function.

Let $\pi_{M_0} : M_0 \to B_0$, $M_0 = E/\Lambda$ be as in §1.

ASSUMPTION 3.1. (1) $\pi_M : M \to B$ is a map from a $2n$ dimensional manifold M to an n dimensional manifold B. We assume that $\dim(B\backslash B_0) \leq \dim B - 2$.
(2) We may identify M_0 and B_0 with dense open subsets of M and B, respectively.
(3) We assume that $\pi_M^{-1}(B_0) = M_0$ and that the restriction of π_M to M_0 coincides with π_{M_0}.
(4) There exists a symplectic structure on M which extends one on M_0 satisfying Proposition 2.1 (A).

We put $S(B) = B\backslash B_0$ and call it the *singular locus*.

REMARK 3.1. In Assumption 3.1 (1), we did not say so much about the singular locus $S(B)$ and how the map π_M can be singular there. Actually, the author does not know which conditions we need to put on $S(B)$ and on the singular fibers in order to make the argument below work. In that sense, Assumption 3.1 is a working hypothesis rather than a definition. We add more assumptions on the singular fibers in §2.4. The study of special Lagrangian fibrations may suggest adequate conditions to impose. Much progress in this direction is underway, especially by M. Gross, see [**44, 46, 45, 41, 79, 43, 70, 53, 54**] for example.

Under this assumption we will define a *multivalued* function on B_0 as follows.

DEFINITION 3.1. We put
$$\tilde{B}_0 = \left\{ (x, [\varphi]) \;\middle|\; \begin{array}{l} x \in B_0, \; \varphi : (D^2, \partial D^2) \to (M, F_x), \\ [\varphi] \in \pi_2(M, F_x), [\partial\varphi] \neq 0 \in \pi_1(F_x). \end{array} \right\}$$
There exists a covering $\tilde{B}_0 \to B_0 : (x, [\varphi]) \mapsto x$.

DEFINITION 3.2. We define a function f on $\tilde{B}_0$ by $f(x,[\varphi]) = \int_{D^2} \varphi^*\omega$. Here ω is the symplectic form of M.

Since f is a function on a covering space of B_0, we may regard it as a multivalued function on B_0 and still call it f.

We need the following variant of f in the next section. Let
$$\tilde{B}_0^+ = \{(x,[\gamma]) | x \in B_0, [\gamma] \in \pi_1(F_x)\}.$$
We define a map $\tilde{B}_0 \to \tilde{B}_0^+$ by $(x,[\varphi]) \mapsto (x,[\varphi|_{\partial D^2}])$.

LEMMA 3.1. *There exists a closed 1-form θ on $\tilde{B}_0^+$ whose pull back to $\tilde{B}_0$ is df.*

PROOF. We fix x_0. For each $(x,[\gamma]) \in \tilde{B}_0^+$ we take a path $\ell : [0,1] \to B_0$ joining x to x_0. We lift it to $\tilde{\ell} : [0,1] \to \tilde{B}_0^+$ such that $\tilde{\ell}(0) = (x,[\gamma])$. We put $\tilde{\ell}(\tau) = (\ell(\tau),[\gamma_\tau])$ where $\gamma_\tau : S^1 \to F_{\ell(\tau)}$ depends continuously on τ. We put $u(\tau,t) = \gamma_\tau(t)$ and $f^+(x,[\gamma]) = \int_{[0,1]\times S^1} u^*\omega$. We remark that up to constant f^+ is independent of the choice of x_0 and the path ℓ. Hence, $\theta = df^+$ is well defined. It is easy to see that the pull back of θ is df. $\square$

Hereafter, we write df in place of θ by abuse of notation (f then is locally well defined as a function on $\tilde{B}_0^+$ up to an additive constant).

3.2. Free field. By free fields in A-model, we mean a pseudoholomorphic cylinder. The purpose of this section is to describe the pseudoholomorphic cylinder in terms of the gradient vector fields of our multivalued function f. For this purpose, we first need to choose an almost complex structure on M_0. We choose one which is T-structure invariant and satisfies Proposition 2.2 (A). (We, however, do not assume J is integrable.)

Tensors ω and J induce a Riemannian metric g on B_0 as follows. We identify $T_x B_0$ with the subspace of $T_{s_0(x)} M_0$ consisting of horizontal vectors (here we use the Gauss-Manin connection). Then, we put $g(V,W) = \omega(V, JW)$, for $V, W \in T_x B_0 \subseteq T_{s_0(x)} M$. Using the symplectic coordinates (x^i, y^j) we have
$$g = \sum g_{ij}(x) dx^i dx^j, \quad \text{where} \quad J\left(\frac{\partial}{\partial x^i}\right) = \sum g_{ij}(x) \frac{\partial}{\partial y^j}.$$

We are going to study J-holomorphic map $u : [a,b] \times S^1 \to M$. We use τ for the coordinate on the interval $[a,b]$ and t for the coordinate on the circle $S^1 = \mathbb{R}/\mathbb{Z}$. We use the complex structure $J\left(\frac{\partial}{\partial \tau}\right) = \frac{\partial}{\partial t}$ on $[a,b] \times S^1$. Our main result in this section is the following Proposition 3.1.

Let $\ell : [a,b] \to B_0$ be a smooth curve and $\tilde{\ell} : [a,b] \to \tilde{B}_0^+$ be its lift. We put $\tilde{\ell}(\tau) = (\ell(\tau),[\gamma_\tau])$, where $\gamma_\tau \in \pi_1(F_{\ell(\tau)})$ and $\gamma_\tau : S^1 \to F_{\ell(\tau)}$. We can choose γ_τ so that it is an affine geodesic and $\gamma_\tau(0)$ is contained in the zero section $s_0(B_0)$. We put

(4) $$u_{\tilde{\ell}}(\tau,t) = \gamma_\tau(t), \quad [a,b] \times S^1 \to M_0.$$

The Riemannian metric g on B_0 induces a Riemannian metric on $\tilde{B}_0^+$. Hence, our closed 1-form df determines a vector field $\operatorname{grad} f$ on $\tilde{B}_0^+$.

PROPOSITION 3.1. *$u_{\tilde{\ell}}$ is J-holomorphic if and only if $\frac{d\tilde{\ell}}{d\tau}(\tau) = \operatorname{grad} f$.*

PROOF. The problem is local. We use symplectic coordinates (x^i, y^j). We can take $\gamma_{\tau_0} = (\gamma_1, \cdots, \gamma_n) \in \mathbb{Z}^n$ with respect to y^j-coordinates of T^n. (We identify $F_{\ell(\tau)} \cong T^n \cong \mathbb{R}^n/\mathbb{Z}^n$.) We put $\ell(\tau) = (\ell_1(\tau), \cdots, \ell_n(\tau))$. Then, $\frac{\partial u_{\tilde{\ell}}}{\partial t} = \sum_{j=1}^n \gamma_j \frac{\partial}{\partial y^j}$. Hence, $u_{\tilde{\ell}}$ is J-holomorphic if and only if

$$(5) \quad \sum_{j=1}^n \gamma_j \frac{\partial}{\partial y^j} = J\left(\sum_i \frac{d\ell_i}{d\tau} \frac{\partial}{\partial x^i}\right) = \sum_{i,j} g_{ij}(x) \frac{d\ell_i}{d\tau} \frac{\partial}{\partial y^j}.$$

On the other hand, we find easily that $f(x^1, \cdots, x^n) = \sum x^i \gamma_i + \text{const}$. Hence, $\mathrm{grad} f(x, [\gamma]) = \sum_{i,j} \gamma_i g^{ij} \frac{\partial}{\partial x^j}$. The proposition follows. $\square$

3.3. Interaction and/or scattering.
In this subsection, we discuss mainly the case of $n = 2$. The case $n > 2$ is discussed briefly at the end of this subsection. Interaction in our sense occurs when two gradient lines (which appeared in Proposition 3.1) intersect each other. Hence, it is a local phenomenon in B_0. So, we consider the case $M_0 = \mathbb{R}^2 \times T^2$. Let (x^1, x^2) be coordinates of $\mathbb{R}^2$ and (y^1, y^2) be coordinates of T^2. We take them so that $\omega = dx^1 \wedge dy^1 + dx^2 \wedge dy^2$, $J\left(\frac{\partial}{\partial x^i}\right) = \frac{\partial}{\partial y^i}$. We can choose such a coordinate system without losing generality since the problem is local in B_0.

Let $\gamma = (\gamma_1, \gamma_2) \in \mathbb{Z}^2 = \pi_1(T^2)$, $\mu = (\mu_1, \mu_2) \in \mathbb{Z}^2 = \pi_1(T^2)$, and $w = (w_1, w_2) \in T^2$. We put

$$(6) \quad u_{\gamma,w}(\tau, t) = (\tau\gamma_1, \tau\gamma_2, t\gamma_1 + w_1, t\gamma_2 + w_2) \in \mathbb{R}^2 \times T^2.$$

$u_{\gamma,w} : \mathbb{R} \times S^1$ is J-holomorphic by Proposition 3.1. We define $u_{\mu,w}$ and $u_{\gamma+\mu,w}$ in a similar way.

We consider $\Sigma = \mathbb{P}^1 \setminus \{0, 1, \infty\}$. It has three ends which we write End_0, End_1, End_∞. We conformally identify End_0, End_1 with $(-\infty, 0] \times S^1$ and End_∞ with $[0, \infty) \times S^1$. We *fix* sufficiently large positive number τ_0 and consider the following boundary conditions :

CONDITION 3.1. (1) There exist t_0, C and ϵ, such that for each $(\tau, t) \in \mathrm{End}_0 \cong (-\infty, 0] \times S^1$ we have :

$$\|u(\tau, t) - u_{\gamma,\omega_0}(\tau + \tau_0, t - t_0)\| < C \exp(\epsilon\tau).$$

We denote this condition by $u \sim_{0;\tau_0} u_{\gamma,\omega_0}$.

(2) There exists t_1, C and ϵ, such that for each $(\tau, t) \in \mathrm{End}_1 \cong (-\infty, 0] \times S^1$ we have :

$$\|u(\tau, t) - u_{\mu,\omega_1}(\tau + \tau_0, t - t_1)\| < C \exp(\epsilon\tau).$$

We denote this condition by $u \sim_{1;T} u_{\mu,\omega_1}$.

(3) There exists τ_∞, t_∞, C and ϵ, such that for each $(\tau, t) \in \mathrm{End}_\infty \cong [0, \infty) \times S^1$ we have :

$$\|u(\tau, t) - u_{\mu,w_\infty}(\tau - \tau_\infty, t - t_\infty)\| < C \exp(-\epsilon\tau).$$

We denote this condition by $u \sim_\infty u_{\gamma+\mu, w_\infty}$.

In case (3) is satisfied for some ω_∞, we write $u \sim_\infty u_{\gamma+\mu,*}$.

PROPOSITION 3.2. *Let τ_0 be a sufficiently large positive number and we fix $\omega_0, \omega_1 \in T^2$. We assume that the greatest common divisor of four numbers γ_1, γ_2, μ_1, μ_2 is 1. Then, the number of J-holomorphic maps $u : \Sigma \to \mathbb{R}^2 \times T^2$, such that $u \sim_{0;\tau_0} u_{\gamma,\omega_0}$, $u \sim_{1;\tau_0} u_{\mu,\omega_1}$, $u \sim_\infty u_{\gamma+\mu,*}$ is $|\gamma_1\mu_2 - \gamma_2\mu_1|$.*

PROOF. We define an isomorphism $\mathfrak{J} : \mathbb{R}^2 \times T^2 \to \mathbb{C}^2_\times$ by

$$\mathfrak{J}(x_1, x_2, y_1, y_2) = (\exp(2\pi(x_1 + \sqrt{-1}y_1)), \exp(2\pi(x_2 + \sqrt{-1}y_2))).$$

(Here $\mathbb{C}_\times = \mathbb{C}\setminus\{0\}$.) We define $u_{\alpha,\beta}$ by

(7) $$\mathfrak{J} \circ u_{\alpha,\beta}(z) = (\alpha z^{\gamma_1}(z-1)^{\mu_1}, \beta z^{\gamma_2}(z-1)^{\mu_2}).$$

LEMMA 3.2. *There exists w_0, w_1, w_∞ (depending on α, β) such that $u_{\alpha,\beta} \sim_0 u_{\gamma,w_0}$, $u_{\alpha,\beta} \sim_1 u_{\mu,w_1}$, and $u_{\alpha,\beta} \sim_\infty u_{\gamma+\mu,w_\infty}$.*

Moreover, any J-holomorphic curve u such that $u \sim_0 u_{\gamma,w_0}$, to $u \sim_1 u_{\mu,w_1}$, and to $u \sim_\infty u_{\gamma+\mu,w_\infty}$ for some w_0, w_1, w_∞, is equal to $u_{\alpha,\beta}$ for some α, β.

PROOF. The first half of the lemma is obvious. Let u be as in the second half of the lemma. We consider the composition $\mathfrak{J} \circ u$ and regard it as $\mathbb{P}^1 \setminus \{0, 1, \infty\} \to \mathbb{C}^2_\times$. Our boundary condition implies that it extends to a holomorphic map : $\mathbb{P}^1 \to \mathbb{P}^2$. Moreover, the intersection number of it with the divisors $z_1 = 0$, $z_2 = 0$, and $\mathbb{P}^2 \setminus \mathbb{C}^2$ are the same as $u_{\alpha,\beta}$. The lemma follows easily. □

Let us write $w_0(\alpha, \beta), w_1(\alpha, \beta)$ for w_0, w_1 as determined in Lemma 3.2.

Our assumption on the greatest common divisor of four numbers $\gamma_1, \gamma_2, \mu_1, \mu_2$ implies that if $(\alpha, \beta) \neq (\alpha', \beta')$ then $u_{\alpha,\beta} \neq u_{\alpha',\beta'}$.

We remark that changing absolute value of α, β corresponds to the shift along τ direction in each end $\cong (-\infty, 0] \times S^1$ (or $[0, \infty) \times S^1$). Since we fixed τ_0 in Condition 3.1, there is a unique choice of the absolute value of α, β. Hence, it suffices to study the argument of α, β. We remark $w_0(u\alpha, v\beta) = \left(\frac{\text{Arg}\, u}{2\pi}, \frac{\text{Arg}\, v}{2\pi}\right) + w_0(\alpha, \beta)$, $w_1(u\alpha, v\beta) = \left(\frac{\text{Arg}\, u}{2\pi}, \frac{\text{Arg}\, v}{2\pi}\right) + w_1(\alpha, \beta)$. (Here, we regard $w_0(\alpha, \beta), w_1(\alpha, \beta) \in T^2 = \mathbb{R}^2/\mathbb{Z}^2$.) We remark also that we have a freedom to choose t_0, t_1 in Condition 3.1. This means that u is asymptotic to $u_{\gamma,w}$ at 0 if and only if it is asymptotic to $u_{\gamma,w+r\gamma}$ at 0, where r is an arbitrary real number. Therefore, $u_{\alpha,\beta}$ satisfies Condition 3.1 if and only if $\left(\frac{\text{Arg}\, \alpha}{2\pi}, \frac{\text{Arg}\, \beta}{2\pi}\right) \in \gamma \cap \mu$. (Here, γ, μ in the right-hand side is the closed geodesic representing the homotopy class γ, μ and containing the origin.)

Therefore, the number of solutions is equal to the intersection number $|\gamma \cdot \mu|$ which is $|\gamma_1\mu_2 - \gamma_2\mu_1|$. The proof of Proposition 3.2 is complete. □

REMARK 3.2. It is easy to see from the proof above that our solution u is transversal if we work out the Fredholm theory of our J-holomorphic curve equation.

REMARK 3.3. From the proof above we find that our solution in Proposition 3.2 is asymptotic to $u_{\gamma+\mu,w}$ where w runs on $\gamma \cap \mu$.

REMARK 3.4. Let k, m be natural numbers whose greatest common divisor is 1. We may consider the holomorphic cylinders $u_{k\gamma}$ and $u_{m\mu}$ as initial data in Proposition 3.2. We then obtain solutions which are asymptotic to $u_{k\gamma+m\mu,w}$. So, starting from a pair of "closed strings", interaction produces an infinite number of "closed strings". This remark will become important when we compare Propositions 3.2 to corresponding results in B-model (see §4).

REMARK 3.5. We remark that in Proposition 3.2 we consider only the case when two strings interact. We can find a similar solution related to the interaction of several closed strings. Namely, we can generalize (7) to

$$\mathfrak{I} \circ u(z) = \left(\alpha \prod_{i=1}^{k}(z-a_i)^{\gamma_i}, \beta \prod_{i=1}^{k}(z-b_i)^{\mu_i} \right). \tag{8}$$

here a_i, b_i are points on $\mathbb{C}$ disjoint to each other. The holomorphic map (8) is parametrized by the configuration space on $\mathbb{C}P^1$ and describes interaction of k closed strings.

Now, we proceed to the case when $n > 2$. We will study J-holomorphic curves $u : \Sigma \to \mathbb{R}^n \times T^n$. Let x^i be coordinates of $\mathbb{R}^n$ and (y^i) be coordinates of T^n. We use complex structure $J\left(\frac{\partial}{\partial x^i}\right) = \frac{\partial}{\partial y^i}$ and a symplectic structure $\omega = \sum dx^i \wedge dy^i$. Let $\gamma = (\gamma_1, \cdots, \gamma_n) \in \mathbb{Z}^n \cong \pi_1 T^n$, $\mu = (\mu_1, \cdots, \mu_n) \in \mathbb{Z}^n = \pi_1 T^n$, $w = (w_1, \cdots, w_n) \in T^n = \mathbb{R}^n / \mathbb{Z}^n$. We put

$$u_{\gamma,w}(\tau, t) = (\tau\gamma, t\gamma + w) \in \mathbb{R}^n \times T^n.$$

$u_{\gamma,w} : \mathbb{R} \times S^1 \to \mathbb{R}^n \times T^n$ is J-holomorphic by Proposition 3.1.

We can define asymptotic boundary conditions, $u \sim_{0;\tau_0} u_{\gamma,w_0}$, $u \sim_{1;\tau_0} u_{\mu,w_1}$, $u \sim_\infty u_{\gamma+\mu,w_\infty}$, $u \sim_\infty u_{\gamma+\mu,*}$ in the same way as Condition 3.1, where $\gamma, \mu \in \mathbb{Z}^2$, $\omega_0, \omega_1, \omega_\infty \in T^n$, and τ_0 is a sufficiently large positive number.

Let $\gamma \in \mathbb{Z}^n$. It determines an S^1 action on T^n by

$$t \cdot [t_1, \cdots, t_n] = [t_1 + t\gamma_1, \cdots, t_n + t\gamma_n]. \tag{9}$$

We write the quotient space of this action by T^n/S_γ^1. (Note $T^n/S_\gamma^1 \cong T^n$). We also write S_γ^1 to show that this S^1 acts by (9). For a cycle A_γ in T^n/S_γ^1, we write its pull back to T^n by $S_\gamma^1 A_\gamma$.

Now, we set the boundary condition as follows. Let A_γ, A_μ be cycles in T^n/S_γ^1, T^n/S_μ^1. We assume that $S_\gamma^1 A_\gamma$ is transversal to $S_\mu^1 A_\mu$ and $S_\gamma^1 A_\gamma \cap S_\mu^1 A_\mu$ is transversal to orbits of $S_{\gamma+\mu}^1$. We consider J-holomorphic map $u : \mathbb{P}^1 \setminus \{0, 1, \infty\} \to \mathbb{R}^n \times T^n$ such that

$$\begin{cases} u \sim_{0;\tau_0} u_{\gamma,w_0} & \text{for some } w_0 \in S_\gamma^1 A_\gamma, \\ u \sim_{1;\tau_0} u_{\mu,w_1} & \text{for some } w_1 \in S_\mu^1 A_\mu, \\ u \sim_\infty u_{\gamma+\mu,*}. \end{cases} \tag{10}$$

We denote the moduli space of such u by $\mathcal{M}(\gamma, \mu, A_\gamma, A_\mu)$.

PROPOSITION 3.3. $\mathcal{M}(\gamma, \mu, A_\gamma, A_\mu)$ is diffeomorphic to $S_\gamma^1 A_\gamma \cap S_\mu^1 A_\mu$.

Let $u \in \mathcal{M}(\gamma, \mu, A_\gamma, A_\mu)$. We assume $\gamma + \mu$ is prime in $\mathbb{Z}^n$. The set of ω_∞ such that $u \sim_\infty u_{\gamma+\mu,\omega_\infty}$ is an orbit of $S_{\gamma+\mu}^1$. Hence, we have an evaluation map $ev_\infty : \mathcal{M}(\gamma, \mu, A_\gamma, A_\mu) \to T^n/S_{\gamma+\mu}^1$.

PROPOSITION 3.4. $ev_{\infty*}[\mathcal{M}(\gamma, \mu, A_\gamma, A_\mu)] = \pi_*[S_\gamma^1 A_\gamma \cap S_\mu^1 A_\mu] \in H(T^n/S_{\gamma+\mu}^1)$, where $\pi : T^n \to T^n/S_{\gamma+\mu}^1$.

The proofs of Propositions 3.3 and 3.4 are straightforward analogues of the proof of Proposition 3.2 and are omitted.

3.4. Singular locus. As we have mentioned already, the author is not yet able to impose correct conditions on the singular fiber. However, we state a conjecture (in case $2n = 4$) which we expect to hold under appropriate conditions on the singular fiber. Also, we can check it in the case of simplest singular fiber.

Let $x_0 \in S(B)$. We choose a sufficiently small open neighborhood U of x_0. Let $x \in U \backslash S(B)$. The fiber $\pi_M^{-1}(x)$ is a torus T^n which we write F_x.

DEFINITION 3.3. An element $(x, [u]) \in \tilde{B}_0$ is said to be a *vanishing cycle* if we can take a representative u of $[u] \in \pi_2(M, F_x)$ such that $u(D^2) \subset \pi_M^{-1} U$.

We say $[u] \in \pi_2(M, F_x)$, $[\partial u] \in \pi_1(F_x)$ is a vanishing cycle, too.

EXAMPLE 3.1. Let $\pi : M \to B$ be an elliptic fibration in the sense of Kodaira [57]. (We do not need to assume M is compact.) The fibers of π are complex submanifolds. We assume M is hyperkähler. Then, by the hyperkähler twist (see [51]) we find another Kähler structure such that the fibers of π are Lagrangian submanifolds with respect to ω. (It is actually a special Lagrangian submanifold.) (Namely hyperkähler manifold has a family of kähler structure parametrized by S^2 and complex submanifold of one complex structure will become special Lagrangian submanifold with respect to the other complex structure.) Thus, any singular fiber in Kodaira's list appears as a singular fiber of our fibration satisfying Assumption 3.1.

The simplest among them is type I singular fiber. It is an immersed S^2 with transverse self-intersections. The inverse image $\pi^{-1}(U)$ is homotopy equivalent to the image of S^2. Let $F_x = T^2$ be a nearby fiber. Then, the vanishing cycles are $k\ell \in \mathbb{Z}^2$, where ℓ is a cycle which "shrinks" to the singular point of the singular fiber.

The open neighborhood U in the base space B is identified with a disk whose origin is the singular locus. $D^2 \backslash \{0\}$ has a flat affine structure induced by the Gauss-Manin connection. This affine manifold is described as follows. Let us consider a connected manifold Ω together with a map $\pi : \Omega \to D^2 \backslash \{0\}$ such that π is a diffeomorphism on the domain $\{(x, y) \in D^2 \,|\, y > (1 + \epsilon)x\} \cup \{(x, y) \in D^2 \,|\, y < -\epsilon x\} \cup \{(x, y) \in D^2 | x < 0\}$ and is two to one on $\{(x, y) \in D^2 \,|\, (1+\epsilon)x > y > -\epsilon x\}$. We consider a diffeomorphism $(x, y) \mapsto (x, y+x)$. from $\{(x,y) \in D^2 \,|\, \epsilon x > y > -\epsilon x\}$ to $\{(x, y) \in D^2 \,|\, (1 + \epsilon)x > y > (1 - \epsilon)x\}$. We use it to glue two different sheets of Ω to get a flat affine manifold. It is isomorphic to $U \backslash S(B) \cong B \backslash \{0\}$ as an affine manifold. To see this, it suffices to recall that the monodromy of a type I singular fiber is

$$\begin{pmatrix} 1 & 1 \\ 0 & 1 \end{pmatrix}$$

We can also easily calculate our multivalued function corresponding to the vanishing cycles and find $f([x, y], k\ell) = ckx$. Note the right-hand side is independent of the gluing map $(x, y) \mapsto (x, y + x)$, and hence defines a function on B_0. Here $c > 0$ depends on the normalization of our Kähler form.

Let us take an n fold covering $D^2 \backslash \{0\} \to D^2 \backslash \{0\}$ and pull back our M, and then obtain a type I_n singular fiber. Our function f will become $f([x,y], k\ell) = ck \operatorname{Re}(x + \sqrt{-1}y)^n$.

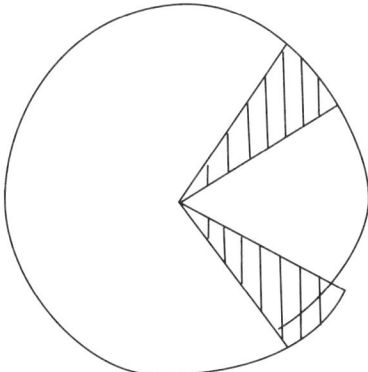

FIGURE 3.1.

In §2.2, 2.3 we found a relation of the gradient line of our multivalued function to holomorphic annuli. Our conjecture in this section is a relation of the pseudoholomorphic disk representing the vanishing cycle to the gradient line of f which starts at singular locus.

We take and fix a Riemannian metric g on B. It induces a Riemannian metric on $\tilde{B}_0$.

DEFINITION 3.4. We define *unstable manifold* $\mathrm{Unst}(U; f; g)$ of f as the set of $(x, [u]) \in \tilde{B}_0$ such that:
(1) $x \in U$;
(2) $[u]$ is a vanishing cycle;
(3) We have

$$\lim_{\tau \to -\infty} \pi \exp(\tau \,\mathrm{grad} f)(x) \in S(B), \tag{11}$$

where $\mathrm{grad} f$ is a gradient vector field of f with respect to the metric g, $\exp(\tau \,\mathrm{grad} f_\gamma)$ is the one-parameter group of transformations generated by it, and $\pi : \tilde{B}_0 \to B_0$ is the projection.

We next define a moduli space of pseudoholomorphic disks which bound the fiber F_x. In this subsection, we only need the case when the disk is contained in $\pi^{-1}(U)$, a small neighborhood of the singular fiber. But, since we study a more general case in the next subsection, we define it in general.

DEFINITION 3.5. We consider a pair (x, u) of $u : D^2 \to M$ and $x \in B_0$ such that (1) u is J-holomorphic; (2) $u(\partial D^2) \subseteq F_x$. We take the quotient of the set of all such (x, u) by an obvious action of $PSL(2; \mathbb{R}) \cong \mathrm{Aut}(D^2, j_{D^2})$. (Here j_{D^2} is the standard complex structure of the 2-disk D^2.) We denote the quotient space by $\mathcal{M}(M; J; D^2)$.

We define a map $\pi : \mathcal{M}(M; J; D^2) \to \tilde{B}_0$ by $(x, u) \mapsto (x, [u])$. (Here $[u]$ is the homotopy class of u in $\pi_2(M, F_x)$.)

We need the following results:

THEOREM 3.1. *There exists $\eta : \pi_0(\tilde{B}_0) \to \mathbb{Z}$ and a space $\mathcal{CM}(M; J; D^2)$ containing $\mathcal{M}(M; J; D^2)$ such that :*

(1) $\mathcal{CM}(M; J; D^2)$ has a Kuranishi structure with corners of dimension
$$\eta(\beta) + 2n - 3;$$
(2) $\pi : \mathcal{M}(M; J; D^2) \to \tilde{B}_0$ extends to $\pi : \mathcal{CM}(M; J; D^2) \to \tilde{B}_0$;
(3) Let $E < \infty$ and $K \subset B_0$ be a compact set, and let
$$\tilde{B}_0(K, E) = f^{-1}((-\infty, E]) \cap \pi^{-1}(K) \subset \tilde{B}_0.$$
Then, the inverse image $\pi^{-1}(\tilde{B}_0(K, E)) \subset \mathcal{CM}(M; J; D^2)$ is compact.

The space $\mathcal{CM}(M; J; D^2)$ is the moduli space of stable maps from a Riemann surface with boundary (see [**37**] §3 for its definition, which is a minor modification of the one in [**60**].)

The homomorphism $\eta(\beta)$ is called the *Maslov index*, see [**7**] for its definition.

We do not define the notion of Kuranishi structure with corners (see [**38**].) The reader who is not familiar with this notion may read the statement (1) as follows: "After a generic perturbation, $\mathcal{CM}(M; J; D^2)$ becomes a smooth manifold with corners of dimension $\eta(\beta) + 2n - 3$". Note that Theorem 3.1(3) is a consequence of Gromov compactness. Namely, the moduli space of pseudoholomorphic disks with bounded energy and compact family of boundary values is compact.

The proof of Theorem 3.1 is almost the same as the ome of Proposition 17.1 in [**37**] (preprint version of December 2000). We only need to remark the following point. In the definition of our moduli space we allow x to move in B_0. In [**37**], we considered the case when x (or F_x) is fixed. In that case the dimension is $\eta(\beta) + n - 3$. The freedom to move x increases the dimension by n.

ASSUMPTION 3.2. We assume $\eta : \tilde{B}_0 \to \mathbb{Z}$ is zero.

REMARK 3.6. It is known that the Maslov index is zero for special Lagrangian submanifolds in Calabi-Yau manifold. Hence, Assumption 3.2 holds for special Lagrangian fibrations.

REMARK 3.7. In case when M is simply connected, we can easily show the following: If Maslov index of vanishing cycles are all zero then Assumption 3.2 holds. Hence, Assumption 3.2 may be regarded as a condition on singular fibers.

By Assumption 3.2, the dimension of $\mathcal{CM}(M; J; D^2)$ is $2n - 3$. Let us assume $n = 2$. Then, $\mathcal{CM}(M; J; D^2)$ is one dimensional. Hence, it is a curve in $\tilde{B}_0$, after appropriate perturbation. We remark that the dimension of a space with Kuranishi structure is, by definition, its virtual dimension (index of linearized operator). Hence, we need to take a perturbation to obtain a space whose actual dimension is equal to the virtual dimension. Existence of such perturbation is proved in general in [**38**]. Hereafter, we write $\mathcal{M}(M; J; D^2)$ for its perturbation representing its virtual fundamental chain constructed in [**38**] by abuse of notation.

Now, we restrict ourselves to a neighborhood U of a singular fiber F_{x_0}. We consider $\tilde{U}_{\text{vanish}} = \{(x, [u]) \in \tilde{B}_0 | x \in U, [u]$ is a vanishing cycle.$\}$ We put $\mathcal{M}_{\text{vanish}}(U) = \pi^{-1}\tilde{U}_{\text{vanish}} \cap \mathcal{M}(M; J; D^2)$. Now, we can state our conjecture. In this conjecture we assume $[u] \in \pi_2(M; F_x)$ is primitive (i.e., it is not of the form $k[u']$ with $k > 1$). We need this assumption since the author does not know the correct way to count multiply covered disks. (In physics literature it is conjectured that d^{-2} is the contribution of d-fold covered disks. However, there is no mathematical argument to justify this choice.)

CONJECTURE 3.1. *If $n = 2$, then $\mathcal{M}_{\text{vanish}}(U)$ is isotopic to $\text{Unst}(U; f; g)$ if we take appropriate choice of g, J and perturbation. In case the fiber is a multiple fiber with multiplicity d we need to take d copies of $\text{Unst}(U; f; g)$.*

THEOREM 3.2. *Conjecture 3.1 is true in case of type I_n singular fiber in Example 3.1.*

PROOF. As we discussed in Example 3.1, the submanifold $\text{Unst}(U; f; g)$ is an axis $x \geq 0, y = 0$ for $\gamma = \ell$ and $x \leq 0, y = 0$ for $\gamma = -\ell$. It is proved in [37], Chapter 7 that the same holds for $\mathcal{M}_{\text{vanish}}(U)$.

The case of type I_n singular fiber can be proved by taking n fold cover of the type I case. □

REMARK 3.8. In case $n \geq 3$ we need to be more careful to state the conjecture, since then $\dim \mathcal{M}_{\text{vanish}}(U) \geq n$. We need to cut it by using cycles in the fiber T^n. (Compare [35], §5, where the author explains how to use $\mathcal{M}_{\text{vanish}}(U)$ to construct an object of derived category of the category of coherent sheaves in the mirror manifold.) The idea to cut moduli space by cycles is naturally related to Propositions 3.3, 3.4. However, at the time of writing this article the author has not yet found the best way to generalize Conjecture 3.1 to higher dimensions.

We next show another evidence of Conjecture 3.1. But it is not strictly an example of Conjecture 3.1, since M is singular there.

EXAMPLE 3.2. Let us consider $T^4 = \mathbb{C}^2/\mathbb{Z}^4$ the quotient of quaternion by its lattice. It is a hyperkähler manifold. We divide it by $\{\pm 1\}$ to obtain M, which is an orbifold. (If we blow up M, it will become a K3 surface. But we do not blow it up here.) There is a projection $\pi : M \to T^2/\{\pm 1\} = S^2$. The general fibers are T^2 and there are 4 singular fibers which are S^2. The fibers are complex submanifolds with respect to the Kähler structure. We take hyperkhäler twist and the fibers become special Lagrangian submanifolds. Let us take a neighborhood of a singular locus $p \in S^2$. We can identify it with $\mathbb{R}^2/\{\pm 1\}$ where $p = 0$. All the elements of $\pi_1(T^2) = \mathbb{Z}^2$ are vanishing cycles. We can calculate easily that $f((x_1, x_2); (k, \ell)) = c(kx_1 + \ell x_2)$. Here (x_1, x_2) are the coordinates of $\mathbb{R}^2$. The multivalued function $c(kx + \ell y)$ induces a multivalued function on $\mathbb{R}^2/\{\pm 1\}$. The unstable manifolds are unions of all lines of rational slope.

We can find pseudoholomorphic disks corresponding to each slope as follows. Our M in a neighborhood of our singular fiber is obtained by dividing $\mathbb{R}^2 \times \mathbb{R}^2$ by G. Where $1 \to \mathbb{Z}^2 \to G \to \{\pm 1\} \to 1$. Here $\mathbb{Z}^2$ acts as translation of the second factor of $\mathbb{R}^2 \times \mathbb{R}^2$ and $\{-1\}$ acts by $(x_1, x_2, y_1, y_2) \mapsto (-x_1, -x_2, -y_1, -y_2)$. We consider $\varphi(u, v) = (ku + \delta_1, \ell v + \delta_2, ku, \ell v) : \mathbb{C} \to \mathbb{R}^2 \times \mathbb{R}^2$ where $\delta_i = 0$ or $1/2$. φ is a pseudoholomorphic map. It induces a map $\mathbb{C}/G \to M$. It is easy to see that if we restrict $\{(u, v) | |v| < r\}$ then φ defines a pseudoholomorphic disk which bounds $F_{(kr, \ell r)}$. By changing the choices of δ_i we have 2 pseudoholomorphic disks. (There are 4 choices for δ_i. But each two of them define the same map.) Thus, an analogue of Conjecture 3.1 in case when $M = T^2/\{\pm 1\}$ holds.

To have more evidence of Conjecture 3.1, we introduce its analogue by including Lagrangian submanifolds and studying its caustics, in the next section.

3.5. Feynman diagram description of pseudoholomorphic disks: closed string version. Combining the pictures developed in §§2.1 - 2.4, we obtain a description of the moduli space $\mathcal{M}(M; J; D^2)$. In this subsection, we only consider the case when $n = 2$, since we can formulate Conjecture 3.1 precisely only in this case. However, the author believes that a similar idea works in higher dimensions, too.

We consider $\pi_M : M \to B^2$ which satisfies Assumptions 3.1, 3.2. We consider the following data, $\Gamma, l, \tilde{\psi}, m$. We will explain later how it corresponds to J-holomorphic disks.

DEFINITION 3.6. A *finite oriented graph* Γ consists of the following data:
(1) A finite set Vertex(Γ), the set of vertices;
(2) A finite set Edge(Γ), the set of edges;
(3) Maps $\partial_{\text{source}} : \text{Edge}(\Gamma) \to \text{Vertex}(\Gamma)$, $\partial_{\text{target}} : \text{Edge}(\Gamma) \to \text{Vertex}(\Gamma)$.

We take copies of intervals $[0,1]_e$ corresponding to each element of $e \in \text{Edge}(\Gamma)$ and copies of points v corresponding to each element $v \in \text{Vertex}(\Gamma)$. We identify $\{0\} \in [0,1]_e$ with $\partial_{\text{source}}(e)$ and $\{1\} \in [0,1]_e$ with $\partial_{\text{target}}(e)$. We thus obtain a one-dimensional complex, which we call $|\Gamma|$. We put $|e| = [0,1]_e \subset |\gamma|$. For a vertex v we let $|v| \in |\gamma|$ a point corresponding to it.

We say that Γ is a *tree* if $|\Gamma|$ is connected and simply connected.

We consider oriented trees Γ which satisfy the following conditions:

CONDITION 3.2. (1) Vertex(Γ) is decomposed into a disjoint union of Vertex$_{\text{int}}$(Γ) and Vertex$_{\text{ext}}$(Γ);
(2) If $v \in \text{Vertex}_{\text{int}}(\Gamma)$ then $|\partial_{\text{target}}^{-1}(v)| = 2$, $|\partial_{\text{source}}^{-1}(v)| = 1$;
(3) There exists $v_{\text{last}} \in \text{Vertex}_{\text{ext}}(\Gamma)$, the last vertex, such that $|\partial_{\text{target}}^{-1}(v)| = 1$, $|\partial_{\text{source}}^{-1}(v)| = 0$;
(4) If $v \in \text{Vertex}_{\text{ext}}(\Gamma) \setminus \{v_{\text{last}}\}$ then $|\partial_{\text{target}}^{-1}(v)| = 0$, $|\partial_{\text{source}}^{-1}(v)| = 1$.

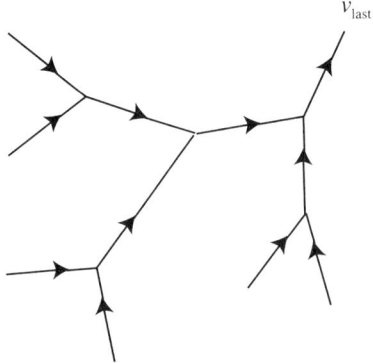

FIGURE 3.2.

We say an element of Vertex$_{\text{int}}(\Gamma)$ an *interior vertex* and an element of Vertex$_{\text{ext}}(\Gamma)$ an *exterior vertex*. An edge e is called an *exterior edge* if $\{\partial_{\text{target}}(e), \partial_{\text{source}}(e)\} \cap \text{Vertex}_{\text{ext}}(\Gamma) \neq \emptyset$. Otherwise it is called an *interior edge*. Hereafter, we put Vertex$_{\text{sta}}(\Gamma) = \text{Vertex}_{\text{ext}}(\Gamma) \setminus \{v_{\text{last}}\}$

We will now describe some more data. Let Γ be an oriented tree satisfying Condition 3.2. A *metric* on Γ is a map $l : \text{Edge}(\Gamma) \to \mathbb{R}_+ \cup \{\infty\}$. We assume one of the following conditions:

CONDITION 3.3. (a) $l(e) = \infty$ if and only if $e \in \text{Vertex}_{\text{sta}}(\Gamma)$;
(b) $l(e) = \infty$ if and only if $e \in \text{Vertex}_{\text{ext}}(\Gamma)$.

Let l satisfies Condition 3.3 (a). Let us consider $|\Gamma|$ minus the union of $|v|$ for $v \in \text{Vertex}_{\text{sta}}(\Gamma)$. Then, we define a metric on it by identifying the interior edge $|e|$ with the arc of length $l(e)$ and identify $|e|$ minus $|v|$ for $v = \partial_{\text{source}}(e) \in \text{Vertex}_{\text{sta}}(\Gamma)$ with $(-\infty, 0]$. We denote it by $|(\Gamma, l)|$.

In case l satisfies Condition 3.3 (b), we identify $|e| \cong [0, \infty)$ if $\partial_{\text{target}} e = v_{\text{last}}$ and define $|(\Gamma, l)|$ in a similar way. (Compare [**36**].)

The third data $\tilde{\psi}$ is a continuous map from $|(\Gamma, l)| \backslash \{\text{vertices}\}$ to $\tilde{B}_0$ such that the composition $\pi \circ \tilde{\psi}$ extends to a continuous map $|(\Gamma, l)| \to B$.

The last data m is a multiplicity. Namely, $m : \text{Edge}(\Gamma) \to \mathbb{Z}_+$.

Now, our assumption for $\tilde{\psi}$ is as follows. We fix a Riemannian metric g on B. It induces a Riemannian metric on $\tilde{B}_0$. Hence, we have a gradient vector field $\text{grad} f$ of f on $\tilde{B}_0$.

CONDITION 3.4. The restriction of $\tilde{\psi}$ to $\text{Int}|e| \cong (-m(e), 0)$ satisfies the differential equation

$$\frac{d\psi}{d\tau} = \text{grad} f_{\beta(e)}.$$

We assume moreover the following conditions 3.5. Let e be an edge. We identify it with $[0, l(e)]$. We put

$$\tilde{\psi}(e; \text{target}) = \lim_{t \to l(e)} \tilde{\psi}(t), \quad \tilde{\psi}(e; \text{source}) = \lim_{t \to 0} \tilde{\psi}(t).$$

We define β by :

$$\tilde{\psi}(e; \text{target}) = (|\partial_{\text{target}}(e)|, \beta_{\text{target}}(e)),$$
$$\tilde{\psi}(e; \text{source}) = (|\partial_{\text{source}}(e)|, \beta_{\text{source}}(e)).$$

CONDITION 3.5. (1) If $\partial_{\text{source}}(e) \in \text{Vertex}_{\text{sta}}(\Gamma)$ then $\lim_{t \to -\infty} \psi(t)$ converges to a point of the singular locus $S(B)$. And $\beta_{\text{source}}(e)$ is a vanishing cycle. (2) Let $v \in \text{Vertex}_{\text{int}}(\Gamma)$. Let $e_1(v), e_2(v), e_0(v)$ be three edges such that $\partial_{\text{source}}(e_1(v)) = \partial_{\text{source}}(e_2(v)) = \partial_{\text{target}}(e_0(v)) = v$. Then, $m(e_1(v))/m(e_0(v))$, $m(e_2(v))/m(e_0(v))$ are integers and we have

$$m(e_1(v))\beta_{\text{target}}(e_1(v)) + m(e_2(v))\beta_{\text{target}}(e_2(v)) = m(e_0(v))\beta_{\text{source}}(e_0(v))$$

in $\pi_2(M, F_{\psi(|v|)})$. (3) If $\partial_{\text{target}}(e) = v_{\text{last}}$ then $m(e) = 1$.

DEFINITION 3.7. The moduli space $\mathcal{M}_{\text{Morse}}(M; g)$ is a set of all $(\Gamma, l, \tilde{\psi}, m)$ satisfying Conditions 3.3(b), 3.4, 3.5.

We define a map $\text{ev}_{\text{last}} : \mathcal{M}_{\text{Morse}}(M; g) \to \tilde{B}_0$ by $\text{ev}_{\text{last}}(\Gamma, l, \tilde{\psi}, m) = \tilde{\psi}(e; \text{target})$ where e is the edge such that $\partial_{\text{target}} e = v_{\text{last}}$.

We can prove transversally properties of $\mathcal{M}_{\text{Morse}}(M; g)$ in the same way as [**36**]. Our conjecture is a relation of $\mathcal{M}_{\text{Morse}}(M; g)$ to $\mathcal{M}(M; J; D^2)$ introduced in §2.4. To compare $\mathcal{M}_{\text{Morse}}(M; g)$ with $\mathcal{M}(M; J; D^2)$ we need to modify it by putting in the multiplicity appearing in Proposition 3.2. Here we only consider the case $m \equiv 1$. The case $m \neq 1$ is more complicated and the author has not yet obtained an answer in the case $m \neq 1$ at the time of writing this article.

DEFINITION 3.8. Let $m \equiv 1$. We define multiplicity multi$(\Gamma, l, \tilde{\psi}, m)$ of $(\Gamma, l, \tilde{\psi}, m)$
$\in \mathcal{M}_{\mathrm{Morse}}(M; g)$ by

$$\mathrm{multi}(\Gamma, l, \tilde{\psi}, m) = \prod_{v \in \mathrm{Vertex}_{\mathrm{int}}(\Gamma)} |\beta_{\mathrm{target}}(e_1(v)) \cap \beta_{\mathrm{target}}(e_2(v))|$$

where $e_1(v)$, $e_2(v)$ are as in Condition 3.5.

We take multi$(\Gamma, l, \tilde{\psi}, m)$ copies of each $(\Gamma, l, \tilde{\psi}, m)$ and obtain a moduli space $\mathcal{M}_{\mathrm{Morse}}(M; g)$

To state our conjecture we need to specify an almost complex structure. Namely, we consider a family of almost complex structures J_ϵ defined by:

$$(12) \qquad J_\epsilon\left(\frac{\partial}{\partial x^i}\right) = \epsilon^{-1} \sum_j g_{ij}(x^1, \cdots, x^n) \frac{\partial}{\partial y^j}$$

outside a small neighborhood of $\pi_M^{-1}(S(B))$. Here (x^i, y^j) are symplectic coordinates as in §1.1 and g_{ij} is a Riemannian metric on B independent of the number ϵ.

CONJECTURE 3.2. *For small ϵ, the moduli space $\mathcal{M}_{\mathrm{Morse}}(M; g)$ is isotopic to $\mathcal{M}(M; J; D^2)$ as submanifolds in $\tilde{B}_0$.*

In the same way as [36], we can reduce Conjecture 3.1 to Conjecture 3.2, Propositions 3.1, 3.2 and a gluing argument which now has become standard in geometric analysis, as we explain below. It proves Conjecture 3.1 in the case when $m \equiv 1$ and all the singular fibers are of type I_n for some n.

Suppose (Γ, l) satisfies Condition 3.2 and Definition 3.7. We construct a family of Riemann surfaces $D(\Gamma, l; \epsilon; \theta)$ as follows. For each edge e of Γ we take $[-l(e)/\epsilon, 0] \times S^1$. For each vertex $v \in \mathrm{Vertex}_{\mathrm{int}}(\Gamma)$ we take a pair of pants Σ_v. We cut Σ_v by a union of three arcs as in Figure 3.3 below into three pieces. We identify them with $[-l(e)/\epsilon, 0] \times S^1$ corresponding to the three edges $e \in \partial_{\mathrm{target}}^{-1}(v) \cup \partial_{\mathrm{source}}^{-1}(v)$. We thus obtain a Riemann surface with $|\mathrm{Vertex}_{\mathrm{ext}}(\Gamma) \setminus \{v_{\mathrm{last}}\}|$ punctures and one boundary component. We conformally fill the punctures and obtain a Riemann surface which is isomorphic to a disk. We remark that there is a freedom to glue a pair of pants with $[-l(e)/\epsilon, 0] \times S^1$. We denote the resulting disk by $D(\Gamma, l; \epsilon; \theta)$, where θ parametrizes the way of glueing ($\theta \in (S^1)^{\text{number of edges}}$).

Now, given $\tilde{\psi}$ and m, we obtain a map $D(\Gamma, l; \epsilon; \theta) \to M$ as follows. (We fix a choice of θ while we construct a map.) Let e be an edge. The restriction of $\tilde{\psi}$ to $|e|$ defines a map $u_{\tilde{\psi}} : (-l(e)/\epsilon, 0) \times S^1 \to M$ by (6). It is J_ϵ-holomorphic by Proposition 3.1

We next consider Σ_v. Let e_1, e_2, e_0 be as in Condition 3.5 (2). We apply Proposition 3.2 to $\gamma = \beta_{\mathrm{target}}(e_1(v))$, $\mu = \beta_{\mathrm{target}}(e_2(v))$ to obtain $u_v : \Sigma_v \to \mathbb{R}^2 \times T^2$. We identify a neighborhood of $\pi^{-1}(\psi(|v|))$ with $\mathbb{R}^2 \times T^2$. Hence, we have $u_v : \Sigma_v \to M$. We remark that Condition 3.5 (2) implies $\beta(e_0) = \beta(e_1) + \beta(e_2) = \gamma + \mu$. We use it, Proposition 3.2 and the definitions of u_v, u_e to see that u_v and u_e can be glued modulo small difference. We can adjust this difference by using partition of unity. (We have a choice of θ such that this gluing works.) The number of approximate solutions is the multiplicity multi$(\Gamma, l, \tilde{\psi}, m)$ by Proposition 3.2.

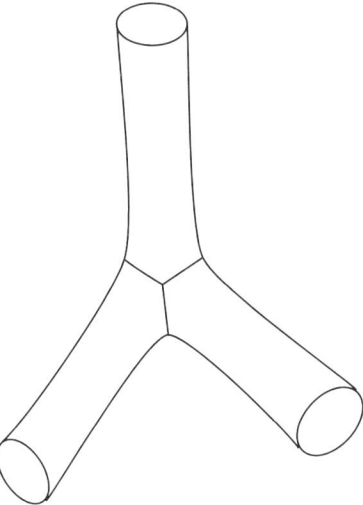

Figure 3.3.

We finally apply Conjecture 3.1 (which we have proved in case of type I_n singular fibers), to construct J_ϵ-holomorphic map in a neighborhood of the punctures corresponding to elements of $\text{Vertex}_{\text{sta}}(\Gamma)$.

Thus, we obtain an approximate solution of the J_ϵ-holomorphic curve equation for $D(\Gamma, l; \epsilon; \theta) \to M$. The final step is a standard Implicit Function Theorem similar to [**36**].

We need careful analysis to show that we obtain all the pseudoholomorphic curves in this way. The author did not work it out yet, at the time of writing this article.

REMARK 3.9. Actually, we did not use at all the fact that Γ is a tree in the argument above. In case when Γ is not a tree, $D(\Gamma, l; \epsilon; \theta)$ is not necessarily a disk but is a Riemann surface of higher genus with boundary. We thus can describe the holomorphic map from the modular space of Riemann surfaces of higher genus with (or without) boundary to $\mathcal{M}_{\text{Morse}}(M; g)$ (compare [**27**] and Remark 5.4).

4. Including Lagrangian submanifolds

As we mentioned in §2.4, to handle singular fibers in arbitrary dimension is harder. To have more examples to apply the ideas presented in the last section, we include Lagrangian submanifolds into the story. Then, we are attacking the Homological Mirror Symmetry Conjecture by Kontsevich [**59**, **58**]. Let $\pi_M : M \to B$ satisfy Assumption 3.1. Let $L \subset M$ be a Lagrangian submanifold.

DEFINITION 4.1. A point $x \in B$ is said to be in the *caustic* of L if F_x is not transversal to L. We denote the caustic by $S(L)$.

We remark that the restriction of π_M to L is an immersion over $M \backslash S(L)$. Caustics have been studied extensively in symplectic geometry especially by Arnol'd and his school [**8**, **9**].

4.1. Multivalued function.

We define a multivalued function f_L on $B\backslash S(L)$ as follows. We first define a covering space $\tilde{B}_L$ over $B\backslash S(L)$.

We fix some notations. We consider $(A; B_1, \cdots, B_n; p_1, \cdots, p_m)$ such that $B_i \subset A$ and $p_i \in A$.

A map $\varphi : (A; B_1, \cdots, B_n; p_1, \cdots, p_m) \to (A'; B'_1, \cdots, B'_n; p'_1, \cdots, p'_m)$ is by definition a continuous map $\varphi : A \to A'$ such that $\varphi(B_i) \subseteq B'_i$, $\varphi(p_i) = p'_i$. Let

$$\Pi((A; B_1, \cdots, B_n; p_1, \cdots, p_m), (A'; B'_1, \cdots, B'_n; p'_1, \cdots, p'_m))$$

be the set of homotopy classes of such maps. We put

$$\partial_+ D^2 = \{z \in \partial D^2 | \operatorname{Im} z > 0\}, \quad \partial_- D^2 = \{z \in \partial D^2 | \operatorname{Im} z < 0\}.$$

Let $L_1, L_2 \subset M$ and $p, q \in L_1 \cap L_2$. We put

$$\pi_2(M; L_1, L_2; p, q) = \Pi((D^2; \partial_+ D^2, \partial_- D^2; -1, +1), (M; L_1, L_2; p, q)).$$

Let $p, q \in X$. We put

$$\pi_1(X; p, q) = \Pi(([0,1]; 0, 1), (X; p, q)).$$

There exist obvious products

$$\pi_2(M; L_1, L_2; p, q) \times \pi_2(M; L_1, L_2; q, r) \to \pi_2(M; L_1, L_2; p, r)$$
$$\pi_1(X; p, q) \times \pi_1(X; q, r) \to \pi_1(X; p, r)$$

which are associative.
In other words $\pi_2(M; L_1, L_2; p, q)$ and $\pi_1(X; p, q)$ are groupoids.

Moreover, there exist boundary homomorphisms

$$\partial_+ : \pi_2(M; L_1, L_2; p, q) \to \pi_1(L_1; p, q),$$
$$\partial_- : \pi_2(M; L_1, L_2; p, q) \to \pi_1(L_2; p, q)$$

which are groupoid homomorphisms.

DEFINITION 4.2. We denote by $\tilde{B}_L$ the set of all $(x; p, q; [u])$ where $x \in B\backslash S(L)$, $p, q \in F_x \cap L$, and $[u] \in \pi_2(M; L, F_x; p, q)$.

There exists an obvious map $\pi_{\tilde{B}_L} : \tilde{B}_L \to B\backslash S(L)$ which is a covering over each connected component of $B\backslash S(L)$. (The order of the inverse image may depend on the component.)

We define another covering space $\tilde{B}_L^+$ as follows. $\tilde{B}_L^+$ is the set of all $(x; p, q; [\gamma])$, such that $p, q \in F_x \cap L$, $[\gamma] \in \pi_1(F_x; p, q)$. The map $\pi_{\tilde{B}_L^+} : \tilde{B}_L^+ \to B\backslash S(L)$ is defined in an obvious way. It is a covering over each connected component of $B\backslash S(L)$. We also have a map $\partial_+ : \tilde{B}_L \to \tilde{B}_L^+$.

We now define

$$(13) \qquad f_L(x; p, q; [u]) = \int_{D^2} u^*\omega.$$

We remark that the right-hand side of (13) is independent of the representative u since L and F_x are Lagrangian submanifolds.

LEMMA 4.1. *There exists a closed 1-form θ on $\tilde{B}_L^+$ which pulls back to df_L on $\tilde{B}_L$.*

The proof is the same as the proof of Lemma 3.1 and is omitted. From now on we write df_L in place of θ by abuse of notation. We can define a multivalued function by using several Lagrangian submanifolds, in a similar way. We do not explain this generalization which is rather straightforward.

EXAMPLE 4.1. Let $h(y^1, \cdots, y^n) : \mathbb{R}^n \to \mathbb{R}$ be a function. We assume $dh(0) = 0$. We put

$$L_h = \left\{ \left(\frac{\partial h}{\partial y^1}, \cdots, \frac{\partial h}{\partial y^n}, y^1, \cdots, y^n \right) \middle| (y^1, \cdots, y^n) \in \mathbb{R}^n \right\}.$$

Then, L_h is a Lagrangian submanifold of $\mathbb{R}^{2n}$. We divide the y direction by $\mathbb{Z}^n$ and obtain $M = \mathbb{R}^n \times T^n$. We consider the projection $\pi_M : M \to \mathbb{R}^n$ defined by $\pi(x^1, \cdots, x^n, y^1, \cdots, y^n) = (x^1, \cdots, x^n)$.

We remark that $x \in S(L_h)$ if and only if $x = dh(y)$ for some critical point y of h.

Suppose $\pi^{-1}(x)$ intersects L_h at two different points $p = (x, y)$ and $q = (x, y')$, where $dh(y) = dh(y') = x$. We identify D^2 with $[0,1]^2$ and define u by

$$u(s_1, s_2) = ((1 - s_1)x + s_1 dh(s_2 y + (1 - s_2) y'), s_2 y + (1 - s_2) y').$$

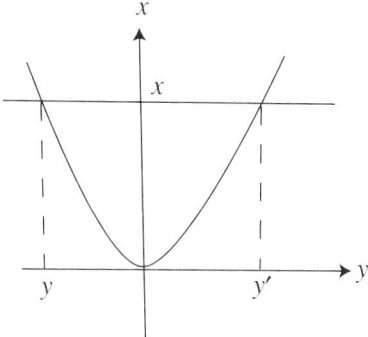

FIGURE 4.1.

Clearly $(x; p, q, [u])$ is an element of $\tilde{B}_{L_h}$. We can easily calculate

$$f_L(x; p, q; [u]) = \frac{1}{2}(h(y') - h(y)).$$

We will use in §5.6 a similar but a bit different multivalued function, which we define here. Let L_1, L_2 be two Lagrangian submanifolds of M. We assume that they intersect each other transversally. Let $B_{00} = B_0 \backslash (S(L_1) \cup S(L_2))$. We put $\chi = \exp(2\pi\sqrt{-1}/3)$. Let $\partial_i D^2$ be the part of ∂D^2 between χ^i and χ^{i+1} ($i = 0, 1, 2$).

DEFINITION 4.3. $\tilde{B}_{L_1, L_2}$ is the set of all $= (x; [u]; p, q; x_0)$ such that :
(1) $x \in B_{00}$,
(2) $p \in F_x \cap L_1$,
(3) $q \in F_x \cap L_2$, $x_0 \in L_1 \cap L_2$,
(4) $[u] \in \Pi((D^2; \partial_0, \partial_1, \partial_2; \chi^0, \chi^1, \chi^2), (M; F_x, L_1, L_2; p, x_0, q))$.

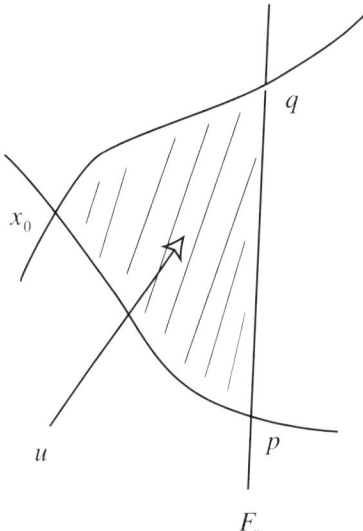

Figure 4.2.

There exists an obvious immersion $\tilde{B}_{L_1,L_2} \to B_{00}$. We define

(14) $$f_{L_1,L_2}(x;[u];p,q;x_0) = \int_{D^2} u^*\omega.$$

We also define

(15) $$\tilde{B}^+_{L_1,L_2} = \{(x;[\gamma];p,q) | x \in B_{00}, p \in F_x \cap L_1,$$
$$q \in F_x \cap L_2, [\gamma] \in \pi_1(F_x;p,q)\}.$$

There exists an obvious immersion $\tilde{B}_{L_1,L_2} \to \tilde{B}^+_{L_1,L_2}$. In the same way as Lemma 4.1, we can find a closed 1-form θ on $\tilde{B}^+_{L_1,L_2}$ which pulls back to df_{L_1,L_2}. We write $\theta = df_{L_1,L_2}$ by abuse of notation. (Then, f_{L_1,L_2} is well defined locally up to constant difference on $\tilde{B}^+_{L_1,L_2}$.)

4.2. Free field. We discuss an open string analogue of the discussion of §2.2. Actually, as we mentioned in introduction, this case is classical and is due to Floer [**22**].

In this section, we consider the case when $S(B) = S(L) = \emptyset$. (We do not assume B is complete. So, in other words, we are working on the complement of $S(B) \cup S(L)$.)

We fix a Riemannian metric g on B. It defines a family of almost complex structures J_ϵ by formula (12) in §2.4. We are going to study J_ϵ holomorphic map $u(\tau,t) : (a,b) \times [0,1] \to M$ such that $u(\tau,0), u(\tau,1) \in L$.

Let $\ell(\tau) : [a,b] \to B$ be a path with bounded C^1 norm. We take its lift $\tilde{\ell} = (\ell(\tau); p(\tau), q(\tau); \gamma_\tau) : [a,b] \to \tilde{B}^+_L$. We define $u^\epsilon_{\tilde{\ell}} : (a/\epsilon, b/\epsilon) \times [0,1] \to M$ as follows. We remark that we have a canonical flat affine structure on the fibers F_x. Namely, we take (y^i) as affine coordinates. (Here y^i are coordinates such that $\omega = \sum dx^i \wedge dy^i$, see the end of §1.1.) In other words (y^i) are angle coordinate.

We take the representative $\gamma_\tau : [0,1] \to F_{\ell(\tau)}$ as the unique affine geodesic joining $p(\tau)$ with $q(\tau)$ in the homotopy class $\in \pi_1(F_{\ell(\tau)}; p(\tau), q(\tau))$. We now put

(16) $$u_{\tilde{\ell}}^\epsilon(\tau, t) = \gamma_{\epsilon\tau}(t).$$

The boundary condition $u(\tau, 0), u(\tau, 1) \in L$ is obviously satisfied.

We remark that J_ϵ and our symplectic structure ω determine a Riemannian metric g_ϵ on M. It then is easy to see

(17) $$\left\|\frac{\partial u_{\tilde{\ell}}^\epsilon}{\partial \tau}\right\|_{g_\epsilon} \sim C\epsilon, \quad \left\|\frac{\partial u_{\tilde{\ell}}^\epsilon}{\partial t}\right\|_{g_\epsilon} \sim C\epsilon.$$

We also remark that g induces a Riemannian metric on $\tilde{B}_L^+$. Hence, we have a gradient vector field $\operatorname{grad} f_L$ on $\tilde{B}_L^+$.

PROPOSITION 4.1. *If we assume*

(18) $$\frac{d\tilde{\ell}}{d\tau} = \operatorname{grad} f_L,$$

then we have

(19) $$\left\|\frac{\partial u_{\tilde{\ell}}^\epsilon}{\partial \tau} + J_\epsilon\left(\frac{\partial u_{\tilde{\ell}}^\epsilon}{\partial t}\right)\right\|_{g_\epsilon} \leq C\epsilon^2.$$

In view of (17), Formula (19) means that $u_{\tilde{\ell}}^\epsilon$ is an approximate solution of the J_ϵ holomorphic curve equation. The proof of Proposition 4.1 is a straightforward calculation and is omitted.

REMARK 4.1. We remark that Proposition 3.1 says that the gradient line of the multivalued function f corresponds to the exact solution of the pseudoholomorphic curve equation. On the other hand, Proposition 4.1 only gives an approximate solution. The difference is that we put here an extra datum L which can be quite arbitrary. So, it is in general impossible to get an explicit solution of our J_ϵ holomorphic curve equation together with L as a boundary value.

4.3. Open string vs open string scattering: review. This section is a review of [36] with only a minor modification. We study the interaction of two open strings. The problem is local in B so we may consider the following situation. $B = \mathbb{R}^n$, $M = \mathbb{R}^n \times T^n$. Moreover, we may consider only a Lagrangian submanifold L which is a union of "horizontal" ones. Namely, we take $\alpha_i \in T^n$, and put $L_{\alpha_i} = \mathbb{R}^n \times \{\alpha_i\}$. Then, we study the case $L = L_{\alpha_1} \cup L_{\alpha_2} \cup L_{\alpha_3}$.

Let $x \in \mathbb{R}^n$. We put $p_i(x) = (x, \alpha_i) \in F_x$. The element of $\pi_1(F_x; p_i(x), p_j(x))$ corresponds to the element $\gamma_{ij} \in \mathbb{R}^n$ such that $\gamma_{ij} \equiv \alpha_j - \alpha_i \mod \mathbb{Z}^n$. We put

(20) $$\tilde{\ell}_{ij}(\tau) = (\tau\gamma_{ij}; p_i(\tau\gamma_{ij}), p_j(\tau\gamma_{ij}); \gamma_{ij}) \in \tilde{B}_L^+.$$

The curve $\tilde{\ell}_{ij} : \mathbb{R} \to \tilde{B}_L^+$ is a gradient line of df_L. In this case, we can easily show that $u_{\tilde{\ell}_{ij}}^\epsilon$ defined by (16) is J_ϵ holomorphic. (Here we take standard Euclidean metric on $\mathbb{R}^n$ to define J_ϵ.) Since the discussion of this subsection is same for all ϵ, we omit ϵ hereafter in this subsection.

Proposition 4.2 below asserts that we have a holomorphic disk which is asymptotic to $u_{\tilde{\ell}_{12}}$, $u_{\tilde{\ell}_{23}}$, $u_{\tilde{\ell}_{13}}$ at their three ends.

Let us introduce some notations to state it. We put $\chi = e^{2\pi\sqrt{-1}/3}$. Let Σ be a Riemann surface with boundary conformal to $D^2 \setminus \{1, \chi, \chi^2\}$. Σ has three ends corresponding to $1, \chi, \chi^2$. We write them End_1, End_χ, End_{χ^2}. We take and fix a conformal isomorphism $\mathrm{End}_\chi \cong \mathrm{End}_{\chi^2} \cong (-\infty, 0] \times [0,1]$, $\mathrm{End}_1 \cong [0,\infty) \times [0,1]$. We use the symbol τ for the first factor and t for the second factor.

PROPOSITION 4.2. ([**36**]) *There exists a J_ϵ holomorphic map $u : \Sigma \to \mathbb{R}^n \times T^n$ such that $u(\partial\Sigma) \subset L$ and that*

(21a) $$|u(\tau, t) - u_{\tilde{\ell}_{12}}(\tau + \tau_1, t)| < Ce^\tau \quad \text{on } \mathrm{End}_{\chi^2},$$

(21b) $$|u(\tau, t) - u_{\tilde{\ell}_{23}}(\tau + \tau_2, t)| < Ce^\tau \quad \text{on } \mathrm{End}_\chi,$$

(21c) $$|u(\tau, t) - u_{\tilde{\ell}_{23}}(\tau - \tau_3, t)| < Ce^{-\tau} \quad \text{on } \mathrm{End}_1.$$

PROOF. We can identify $\Sigma = \{z \in \mathbb{C} \mid \mathrm{Im}\, z \geq 1, z \neq \pm 1\}$, such that ends End_χ, End_{χ^2}, End_1 correspond to $-1, +1\, \infty$ respectively. We next identify $\mathbb{R}^n \times T^n \cong \mathbb{C}_*^n$. We can identify L_1, L_2, L_3 as $L_1 = \mathbb{R}_+^n$, $L_2 = \{(z_1, \cdots, z_n) \mid \arg z_i = \pi \alpha_i\}$, $L_3 = \{(z_1, \cdots, z_n) \mid \arg z_i = \pi \beta_i\}$. We now put

$$u(z) = ((z+1)^{\alpha_1}(z-1)^{\beta_1 - \alpha_1}, \cdots, (z+1)^{\alpha_n}(z-1)^{\beta_n - \alpha_n}).$$

Here, we choose branch of z^α such that $z^\alpha \in \mathbb{R}_+$ if $z \in \mathbb{R}_+$. It is easy to check (21). (Note $u(\{z \in \mathbb{R} \mid z > 1\}) \subseteq L_1$, $u(\{z \in \mathbb{R} \mid 1 > z > -1\}) \subseteq L_2$, $u(\{z \in \mathbb{R} \mid z < -1\}) \subseteq L_3$.) $\square$

REMARK 4.2. We remark that there is one difference between our case and the case of [**36**]. Namely, here we are studying torus bundles and in [**36**] the case when fibers are a vector space is studied. We specify γ_{ij} to handle this point. Then, the two cases are almost the same as far as the argument of this section concerns except the following point. Let us consider the case $\alpha_1 = \alpha_3$. In this case we obtain a solution u which corresponds to the interactions of many open strings. Let us describe it.

Let us consider Σ conformal to D^2 minus $k+1$ points on the boundary. Σ has $k+1$ ends End_i, $i = 0, \cdots, k$. We identify $\mathrm{End}_i \cong (-\infty, 0] \times [0,1]$ for $i \neq 0$ and $\mathrm{End}_i \cong [0,\infty) \times [0,1]$.

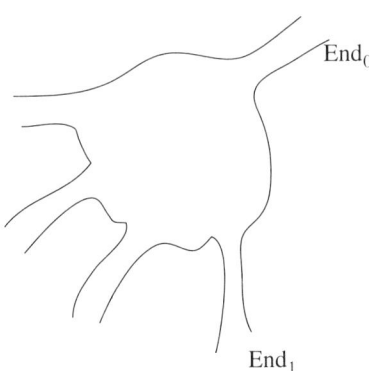

FIGURE 4.3.

We put $\gamma'_{12} = (k-1)(\gamma_{12} + \gamma_{21})/2 + \gamma_{12}$ if k is odd and $\gamma'_{11} = k(\gamma_{12} + \gamma_{21})/2$ if k is even. We then put

$$\ell'_{12,k} = (\tau\gamma'_{12}; p_1(\tau\gamma'_{12}), p_2(\tau\gamma'_{12}); \gamma'_{12}), \quad \text{if } k \text{ is odd},$$
$$\ell'_{11,k} = (\tau\gamma'_{11}; p_1(\tau\gamma'_{11}), p_2(\tau\gamma'_{11}); \gamma'_{11}), \quad \text{if } k \text{ is even}.$$

Then, there exists a J-holomorphic $u : \Sigma \to M$ such that $u(\partial\Sigma) \subset L = L_1 \cup L_2$ and

(22a) $\quad |u(\tau,t) - u^\epsilon_{\tilde{\ell}_{12}}(\tau - \tau_i, t)| < Ce^\tau \quad$ on End_i with i odd,

(22b) $\quad |u(\tau,t) - u^\epsilon_{\tilde{\ell}_{21}}(\tau - \tau_i, t)| < Ce^\tau \quad$ on End_i with i even,

(22c) $\quad |u(\tau,t) - u^\epsilon_{\tilde{\ell}'_{12}}(\tau - \tau_0, t)| < Ce^{-\tau} \quad$ on End_0 if k is odd,

(22d) $\quad |u(\tau,t) - u^\epsilon_{\tilde{\ell}'_{11}}(\tau - \tau_0, t)| < Ce^{-\tau} \quad$ on End_0 if k is even.

Such u is actually constructed also in [**36**] where the solution which gives the interaction of arbitrary many open strings is constructed.

The interaction described by this solution u will become important when we discuss its B model analogue in §4.6.

We also point out that the gluing procedure given by our u here is the same one as studied by Hutchings and Li [**52**] in the case of closed orbit of ordinary differential equation which appears in Novikov homology [**67**].

4.4. Open string vs closed string scattering. We next discuss a mixture of the two kinds of scattering. Let us consider the same situation as the last section. ($M = \mathbb{R}^n \times T^n$ etc.) We suppose we have two horizontal Lagrangian submanifolds $L_i = \mathbb{R}^n \times \{\alpha_i\}$, and put $L = L_1 \cup L_2$.

Let $\gamma_{12} \in \mathbb{R}^n$ such that $\gamma_{12} \equiv \alpha_2 - \alpha_1 \mod \mathbb{Z}^n$. We have a J-holomorphic curve $u_{\tilde{\ell}_{12}}$ defined by (16).

We next take $\gamma \in \mathbb{Z}^n = \pi_1(F_x)$ and $w \in T^n$. Then, we put

$$u_{\gamma,w}(\tau,t) = (\tau\gamma, t\gamma + w) \in \mathbb{R}^n \times T^n.$$

$u_{\gamma,w} : \mathbb{R} \times S^1 \to \mathbb{R}^n \times T^n$ is J-holomorphic by Proposition 3.1.

We put $\gamma'_{12} = \gamma_{12} + \gamma$ and use it in place of γ_{12} to obtain $u_{\tilde{\ell}'_{12}}$ by (16).

Now, the main result of this subsection is about the J-holomorphic curve which is asymptotic to $u_{\tilde{\ell}_{12}}$, $u_{\gamma,w}$, $u_{\tilde{\ell}'_{12}}$ in its three ends. Let us define some notations to state it.

For $s \in (-1,1)$, let Σ_s be a Riemann surface with boundary which is conformal to $D^2 \setminus \{s\sqrt{-1}, -1, +1\}$. Σ_s has three ends $\text{End}_{s\sqrt{-1}}$, $\text{End}_{\pm 1}$ corresponding to the points removed. We fix a conformal diffeomorphisms $\text{End}_{s\sqrt{-1}} \cong (-\infty, 0] \times S^1$, $\text{End}_{-1} \cong (-\infty, 0] \times [0,1]$, $\text{End}_{+1} \cong [0,\infty) \times [0,1]$. Let A be a codimension-one cycle of T^n/S^1_γ. (Here we are using the notation of §2.3.) We take a representative $\gamma_{12} \in \pi_1(T^n; \alpha_1, \alpha_2)$ which is an affine geodesic. We assume that $AS^1_\gamma \cap \{\alpha_1, \alpha_2\} = \emptyset$ and γ_{12} intersects transversally with AS^1_γ.

PROPOSITION 4.3. *We consider the set of pairs* (u,s) *of* $s \in (-1,1)$ *and J-holomorphic maps* $u : \Sigma_s \to \mathbb{R}^n \times T^n$ *with the following boundary conditions. There* exists $\tau_0, \tau_{\pm 1}$ and $w \in AS^1_\gamma$ such that :

(23a) $\qquad |u(\tau,t) - u_{\gamma,w}(\tau+\tau_0,t)| < Ce^{+C\tau} \quad$ on $\mathrm{End}_{s\sqrt{-1}}$,

(23b) $\qquad |u(\tau,t) - u_{\tilde{\ell}_{12}}(\tau+\tau_{-1},t)| < Ce^{+C\tau} \quad$ on End_{-1},

(23c) $\qquad |u(\tau,t) - u_{\tilde{\ell}'_{12}}(\tau-\tau_{+1},t)| < Ce^{-C\tau} \quad$ on End_{+1}.

Then, the number of such solutions is exactly equal to the intersection number $|AS^1_\gamma \cap \gamma_{12}|$.

PROOF. We prove only the case $n = 2$. The general case is similar. We identify $\Sigma_s \cong \{z \in \mathbb{C} | \mathrm{Im}\, z \geq 0,\ z \neq s' + \sqrt{-1}, 0\}$. Then, s' moves on $\mathbb{R}$ while s moves on $(-1,1)$. Here, 0, $s' + \sqrt{-1}$, ∞ corresponds the ends End_{-1}, $\mathrm{End}_{s\sqrt{-1}}$, End_{+1} respectively. We also identify $\mathbb{R}^2 \times T^2 \cong \mathbb{C}^2_*$. It suffices to consider the case when $L_1 = \mathbb{R}^2_+ \subset \mathbb{C}^2_*$. $L_2 = \{(z_1,z_2) \in \mathbb{C}^2_* | \mathrm{Arg}\, z_1 = \pi\alpha_1, \mathrm{Arg}\, z_2 = \pi\alpha_2\}$ for $i = 1,2$. We can also take $\gamma = (\gamma_1, \gamma_2)$ and A is one point. Hence, $AS^1_\gamma = \{a+t\gamma | t \in \mathbb{R}\} \subset \mathbb{R}^2/\mathbb{Z}^2$, $a = (a_1, a_2)$.

An intersection point in $AS^1_\gamma \cap \gamma_{12}$ then corresponds to a pair of $\rho \in [0,1]$, $\theta \in [0,1]$ such that

(24) $\qquad (\rho\alpha_1, \rho\alpha_2) = 2\pi(a_1 + \theta\gamma_1, a_2 + \theta\gamma_2).$

It follows that, for each point of $AS^1_\gamma \cap \gamma_{12}$, we can take s' such that

$$\mathrm{Arg}(s'+\sqrt{-1})^{\alpha_1} = 2\pi(a_1+\theta_1\gamma_1), \quad \mathrm{Arg}(s'+\sqrt{-1})^{\alpha_2} = 2\pi(a_2+\theta_2\gamma_2).$$

Here, we take the branch of z^{α_i} so that it is real on $\mathbb{R}_+$. (Note $\alpha_i \in (0, 2\pi)$ and γ_i are integers.) We now put

(25) $\qquad \begin{aligned} u(z) = \big(&z^{\alpha_1}(z-s'-\sqrt{-1})^{\gamma_1}(z-s'+\sqrt{-1})^{\gamma_1}, \\ &z^{\alpha_2}(z-s'-\sqrt{-1})^{\gamma_2}(z-s'+\sqrt{-1})^{\gamma_2}\big). \end{aligned}$

It is easy to check (23). It is also easy to show that all solution are obtained in this way. $\square$

We can combine constructions of §§2.3, 3.3, 3.4 to obtain a holomorphic curve in $\mathbb{R}^n \times T^n$ which describes interactions between arbitrarily many open and closed strings.

4.5. Caustics, elementary catastrophes, and bifurcations of pseudo-holomorphic disks. We now explain a conjecture which is an open string version of Conjecture 3.1. We are going to study a neighborhood of the point $x \in M_0$ where L is tangent to the fiber. So, we consider $\mathbb{R}^n$ bundle instead of T^n bundle. We define an analogue of $U(f_\gamma; g)$. Let $x_0 \in S(L)$. We take its neighborhood U in B. Let $x \in U \setminus S(L)$ and $(x; p, q; [u]) \in \tilde{B}_L$.

DEFINITION 4.4. We say that $[u] \in \pi_2(M; L, F_x; p, q)$ is a *vanishing disk* if we can choose a representative in the class $[u]$ which is contained in $\pi^{-1}(U)$. We say $[\partial_+ u] \in \pi_1(F_x; p, q)$ is a *vanishing arc* if $[u]$ is a vanishing disk.

We remark that an analogy of Assumption 3.2 is not appropriate in our situation, as we can see from examples we will show later. So, we define $\mathrm{Unst}(U)$ as a subset of $\tilde{B}_L$ as follows. We use a pull back of a Riemannian metric on B to define a Riemannian metric on $\tilde{B}_L$. Then, a gradient vector field $\mathrm{grad} f_L$ is induced on

$\tilde{B}_L$. Let $\text{ext}(\tau\,\text{grad}\,f_L)$ be the one parameter group of transformations generated by it.

DEFINITION 4.5. We define $\text{Unst}(U)$ as the set of all $(x;p,q;[u]) \in \tilde{B}_L$, such that :
(1) $x \in U$. u is a vanishing disk.
(2) $\pi\left(\text{ext}(-\tau\text{grad}f_L)(x;p,q;[u])\right) \in U$ for all $\tau > 0$.
(3) $\lim_{\tau\to\infty} \pi\left(\text{ext}(-\tau\text{grad}f_L)(x;p,q;[u])\right) \in S(L)$.

We define moduli space of pseudoholomorphic disks as follows:

DEFINITION 4.6. We let $\tilde{\mathcal{M}}((M,J_\epsilon);L)$ be the set of all $(x;p,q;u)$ such that $(x;p,q;[u]) \in \tilde{B}_L$, where $[u]$ is a homotopy class of u, and that u is J_ϵ holomorphic. It has a natural action of $\mathbb{R} = \text{Aut}(D^2;\pm 1)$. We denote by $\mathcal{M}((M,J_\epsilon);L)$ the quotient space by this action.

If U is a small neighborhood of $x_0 \in S(L)$, we define $\mathcal{M}_{\text{vanish}}(U;(M,J_\epsilon);L)$ as the set of all $[x;p,q;u] \in \mathcal{M}((M,J_\epsilon);L)$ such that $x \in U$ and $u(\pi(D^2)) \subset U$. (In other words, u is a vanishing disk.)

There is a natural map $\mathcal{M}((M,J_\epsilon);L) \to \tilde{B}_L$.

CONJECTURE 4.1. *$\mathcal{M}_{\text{vanish}}(U;(M,J_\epsilon);L)$ is isotopic to $\text{Unst}(U)$ as a submanifold of $\tilde{B}_L$ after appropriate perturbation.*

Let us give some examples and arguments that justify the conjecture.

EXAMPLE 4.2. The simplest caustic is a fold. Namely, we take $n = 1$ and let L is given by $x = 3y^2$ where $x = \text{const}$ is the fiber. This is the case when we put $h(y) = y^3$ in Example 4.1. Hence, $f_L(x) = x^{1/3}$, which is defined only on the domain $x > 0$. In this case $\text{Unst}(U)$ is the part $x > 0$ (and corresponds to the unique branch of the covering $\tilde{B}_L \to \{x|x > 0\}$). It is easy to see that $\mathcal{M}_{\text{vanish}}(U;(M,J_\epsilon);L)$ is the same.

EXAMPLE 4.3. The next case to study is a cusp. We take $h(y_1,y_2) = y_1^4 + y_1 y_2^2$ in Example 4.1. Then, L is defined by $(x_1,x_2) = \Psi(y_1,y_2) = (4y_1^3 + y_2^2, 2y_1 y_2)$. The caustics $S(L)$ is given by $\Psi(\{(y_1,y_2) \mid 6y_1^3 = y_2^2\})$ and is given by $\ell(t) = (7t^8, 2\sqrt{6}t^5)$. Let us decompose $\mathbb{R}^2 \backslash S(L) = \Omega_0 \cup \Omega_1$ where Ω_0 contains the domain $x_1 < 0$. If $x \in \Omega_0$ then F_x intersects with L at one point. If $x \in \Omega_1$ then F_x intersects with L at three points. Let $p_i(x)$ $i = 1,2,3$ be those three points. We choose them so that $|p_1(x) - p_2(x)| \to 0$ as x converges to $S(L)_- = S(L) \cap \{x_2 < 0\}$ and $|p_3(x) - p_2(x)| \to 0$ as x converges to $S(L)_+ = S(L) \cap \{x_2 > 0\}$. There is one vanishing arc for each of $S(L)_-$ and $S(L)_+$, the ones joining $p_1(x)$ to $p_2(x)$ and $p_3(x)$ to $p_2(x)$, respectively. The gradient lines of the corresponding branch of our function is obtained by Example 3.2 above. The gradient line starting $S(L)_-$ dies when it meets with $S(L)_+$ (since there will no longer be a corresponding branch of f_L.) There are two other branches of f_L which correspond to the vanishing arcs joining p_1 to p_3 and p_3 to p_1. The gradient line of these branches starting origin do not exit.

EXAMPLE 4.4. Let us take $M = \mathbb{C}^2$, and let (x,y) be the complex coordinates of it. We regard M as a hyperkähler manifold and take hyperkähler twist so that $(x,y) \mapsto x$ will have a Lagrangian submanifold as fibers. We put $x = x_1 + \sqrt{-1}x_2$, $y = y_1 + \sqrt{-1}y_2$. (Note that the symplectic form is $dx_1 \wedge dy_1 + dx_2 \wedge dy_2$.) Let L is given by $x = y^2$. It is a complex submanifold with respect to the original complex structure. Hence, after hyperkähler twist, L will become a (special) Lagrangian submanifold. The intersection $L \cap F_x$ consists of two points $(x, \pm x^{1/2})$. Hence, there are exactly two vanishing arcs, one from $(x, -x^{1/2})$ to $(x, x^{1/2})$, the other from $(x, x^{1/2})$ to $(x, -x^{1/2})$. (These two branch interchange when we go around origin once.)

We put $x = re^{\sqrt{-1}\theta}$ and parametrize L as $(r, \theta) \mapsto (re^{2\sqrt{-1}\theta}, \sqrt{r}e^{\sqrt{-1}\theta})$. Then, $(s, t) \mapsto (sre^{2\sqrt{-1}t}, t\sqrt{sr}e^{\sqrt{-1}r})$, $(s, t) \in [0, 1] \times [-1, 1]$ is a parameterization of a vanishing disk. We use it to calculate our multivalued function f_L and obtain

$$f_L(x) = \pm \frac{4}{3} \operatorname{Re} x^{3/2}.$$

Therefore, there exist exactly three gradient lines starting from origin.

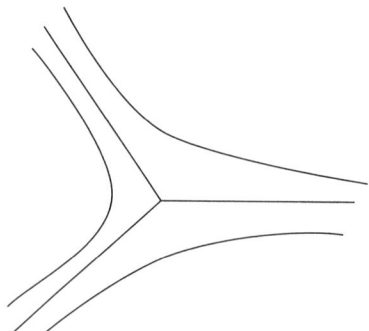

FIGURE 4.4.

Let us explain an argument which suggests Figure 4.4 also appears as the moduli space of J-holomorphic discs. This argument may work in more general situations. So, we discuss the case of Example 4.1 in general. Namely, our Lagrangian submanifold L is obtained as a graph of a 1-form dh where h is a function of $y_1, \cdots, y_n$. (We remark that in [36] we consider the case when our Lagrangian submanifold is a graph of 1-form which is an exterior derivative of a function of x. In that case, the Lagrangian submanifold is transveral to the fibers F_x. Here we are considering the opposite extreme namely the case when the Lagrangian submanifold is tangent to the fiber at origin.) We put

$$h_{x_1,\cdots,x_n}(y_1, \cdots, y_n) = h(y_1, \cdots, y_n) - \sum x_i y_i.$$

It is easy to see that $(x, y) \in L \cap F_x$ if and only if y is a critical point of h_x. We now define another moduli space as follows:

DEFINITION 4.7. We define an moduli space $\mathcal{M}_{\text{Morse,vanish}}(M; L)$ as the set of all $(x; p_1(x), p_2(x); [\ell])$'s such that $p_i = (x, y_i) \in L \cap F_x$ and $\ell : \mathbb{R} \to \mathbb{R}^n$ such that $d\ell/d\tau = \operatorname{grad} h_x$ and $\lim_{\tau \to -\infty} \ell(\tau) = y_1$, $\lim_{\tau \to +\infty} \ell(\tau) = y_2$. Here $[\ell]$ is the $\sim$ equivalence class, and we define $\sim$ by $\ell(\tau) \sim \ell(\tau + \tau_0)$.

There is an obvious map $\mathcal{M}_{\text{Morse,vanish}}(M; L) \to \tilde{B}_L$. Now, we can divide Conjecture 4.1 into two pieces.

CONJECTURE 4.2. $\mathcal{M}_{\text{vanish}}(U;(M,J_\epsilon);L)$ *is isotopic to* $\mathcal{M}_{\text{Morse,vanish}}(M;L)$ *as a submanifold of* $\tilde{B}_L$ *after appropriate perturbation.*

CONJECTURE 4.3. $\mathcal{M}_{\text{Morse,vanish}}(M;L)$ *is isotopic to* $\text{Unst}(U)$ *as a submanifolds of* $\tilde{B}_L$.

REMARK 4.3. In Conjecture 4.1, we consider the general case and here we consider the case of Example 4.1. However, we can divide B into two directions, one is reduced to Example 4.1 and the other direction is transversal to L. In that way, to study Conjecture 4.1, it suffices to consider the case of Example 4.1.

Conjecture 4.2 asserts that the moduli space of gradient lines is isotopic to the moduli space of J_ϵ holomorphic disks. These kinds of results are established in various settings, but not yet proved in the particular setting of Conjecture 4.2. The first result of this kind is due to Floer [22], that is the case of free field we discussed in §3.2. The author together with Y.G. Oh generalized it in [36] to the case when interaction exists.

Floer used in [22] another result (Proposition 5.1 in [22]) which is closely related to the case $h(y) = y^3$ of Conjecture 4.2, to prove invariance of Floer homology under Hamiltonian diffeomorphisms. The argument in [22], §5 is rather sketchy. The argument of [22], §19 is one using the center manifold in the case appearing there, though the name "center manifold" is not used in [22]. The idea to use the center manifold is used later by Morga-Mrowka-Ruberman in a related context of gauge theory. However, their argument is also rather sketchy. Giving a more detailed argument in [22], §5 should be a way to solve Conjecture 4.2. In a recent paper [63], Y.J. Lee justified a similar degenerate gluing in the situation of Floer homology of periodic Hamiltonian system. The author is sorry that he used this kind of degenerate gluing in his paper [30] without writing a detailed proof.

Conjecture 4.3 is one on finite dimensional Morse theory. However, it is not trivial one. The author can not prove it in general at the time of writing this article. Instead, let us check it in the case of Examples 4.3 and 4.4.

We first consider the case of Examples 4.3. We have $h_{x_1,x_2}(y_1,y_2) = y_1^4 + y_1 y_2^2 - x_1 y_1 - x_2 y_2$. If $x \in \Omega_1$, then $h_{x_1,x_2}(y_1,y_2)$ has three critical points $y^1(x), y^2(x), y^3(x)$ $\in \mathbb{R}^2$ such that $p_i(x) = (x, y^i(x))$. By direct calculation we find that $y^2(x)$ is a saddle point and $y^1(x), y^3(x)$ are local minimum. There exists a gradient lines of h_x from $y^2(x)$ to $y^1(x)$ and from $y^2(x)$ to $y^3(x)$. However, there is no gradient line from $y^1(x)$ to $y^3(x)$ or from $y^3(x)$ to $y^1(x)$, for any x.

This is consistent with the description of $\mathcal{M}_{\text{vanish}}(U;(M,J_\epsilon);L)$ we gave before. (We remark that $\mathcal{M}_{\text{vanish}}(U;(M,J_\epsilon);L)$ is related to the gradient line of f_L which is *different* from h_{x_1,x_2}.) Thus, Conjecture 4.3 holds in this case.

We next consider the case of Example 4.4. We put $h(y_1,y_2) = \text{Re}\, y^3/3$. It is easy to check that the graph of dh is our Lagrangian submanifold L. For $x \neq 0$, L intersects with the fiber F_x at two points $p_\pm(x) = (x, \pm\sqrt{x})$ These two points are critical points $h_x(y) = h(y) - x_1 y_1 - x_2 y_2$. We remark that they both are saddle points and have the same Morse index ($=1$). Hence, the virtual dimension of the moduli space of gradient lines joining them is -1. It means that for generic x there exists no such gradient line, but there may exist such a gradient line when x is contained in a codimension-one subset of B.

We can study gradient vector field of it drawn below and calculate $\mathcal{M}_{\text{Morse,vanish}}(M;L)$ explicitly, and we find that it is exactly the same as the space $\text{Unst}(U)$ before. Thus, Conjecture 4.3 holds in this case also.

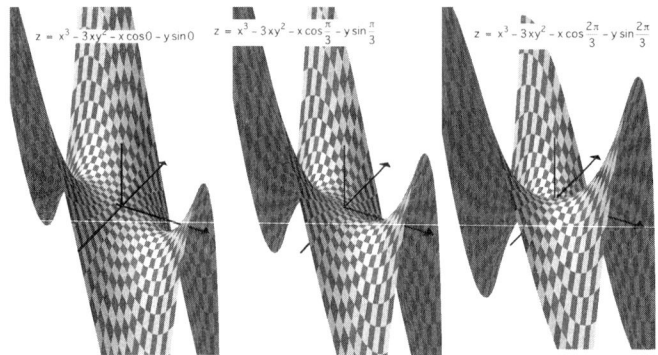

FIGURE 4.5.

More generally, let us consider the case when $L = \{(y^n, y) | y \in \mathbb{C}\}$. Here, we identify $\mathbb{R}^4 = \mathbb{C}^2$ with complex coordinate (x, y). We apply hyperkähler twist, then L will become a special Lagrangian submanifold. It is easy to see that L is obtained from $h(y_1, y_2) = \frac{1}{n+1} \operatorname{Re} y^{n+1}$ as in Example 4.1. In this case, we can check $f_L(x) = \frac{n}{n+1} \operatorname{Re}(\exp(2\pi\sqrt{-1}k/n) - \exp(2\pi\sqrt{-1}k'/n)x^{(n+1)/n})$. One might check Conjecture 4.3 in this case also in a similar way by direct calculation. However, the calculation looks rather cumbersome and the author did not complete it at the time of writing this article.

REMARK 4.4. The bifurcation of a pseudoholomorphic disk shown in Figure 4.4 was presented by the author in his talk at the conference "Integrable Systems in Differential Geometry" which was held in July 2000 at University of Tokyo. After his talk, Yousuke Ohyama pointed him out that Figure 4.4 looks quite similar to Stokes line which appears in exact WKB calculus (see [**75**, **4**]). Since lines in Figure 4.4 are related to asymptotic analysis, as we will explain in the next two sections, it is natural to expect that it is related to WKB calculus. However, the author does not yet know the precise relation of it to Stokes line. The difficulty for him to do so is that in our case the fibers are totally real while in exact WKB calculus the fibers are complex submanifolds.

The author outlined the article in his lecture at the Tokyo Metropolitan University in 2002 January. After his lecture, Akishi Kato pointed him out that the story of string junction (see [**39**]) is similar to one discussed in this subsection. The essential difference is that in the story of string junction, the fibers are complex tori and a Lagrangian submanifold of the total space is reduced to a graph of the base space. In our case, fibers are Lagrangian submanifolds and a pseudoholomorphic curve of the total space is reduced to a graph in the base space. The story of string junction may be more directly related to exact WKB calculus. In the case that the dimension is $4 = 2 \times 2$, these two cases interchange by hyperkähler twist.

REMARK 4.5. The Lagrangian submanifold in Example 4.4 is not generic in the sense of Lagrangian singularity. (In the case $n = 2$, only cusps and folds appear as generic singularities.) However, it is generic among special Lagrangian submanifolds. It seems interesting to classify generic Lagrangian singularity for special Lagrangian submanifold in higher dimension, especially in the case $n = 3$.

During the conference "D-branes and Mirror Symmetry" held at RIMS in 2002 February, D. Joyce pointed out to the author that in the case that the dimension is $6 = 3 \times 2$, the fold also appears as a Lagrangian singularity of a *special* Lagrangian submanifold. He also pointed out that to have a good restriction of Lagrangian singularity, one needs to consider the limiting behavior of the special Lagrangian submanifold as the diameter of the fiber goes to zero.

We finally give another piece of evidence for Conjecture 4.1. Namely, we show its analogue in the case of an immersed Lagrangian submanifold.

EXAMPLE 4.5. ([**36, 32**]) Let $q_1, q_2 : \mathbb{R}^n \to \mathbb{R}$ be quadratic forms. Then, the graph of its exterior derivatives are (linear) Lagrangian submanifolds in $T^*\mathbb{R}^n = \mathbb{R}^{2n}$. We denote them by L_1, L_2. They intersect each other at $0 \in \mathbb{R}^{2n}$. We assume that $q_2(x) - q_1(x)$ is nondegenerate. Then, L_1 and L_2 intersect transversally. We put $L = L_1 \cup L_2$. So, L is an immersed Lagrangian submanifold. The analogy of the caustics $S(L)$ in our situation is the image of the singular point 0. Namely, $S(L) = \{0\}$.

Let $x \in \mathbb{R}^n$. $p_i(x) = (x, dq_i(x)) \in L_i$ is a unique point in $L_i \cap T_x^*(\mathbb{R}^n)$. Let $u_x : D^2 \to T^*\mathbb{R}^n$ be the map onto the affine triangle whose vertices are $0, p_1, p_2$. The analogy of f_L in our case is

$$f(x) = \int_{D^2} u_x^* \omega = q_2(x) - q_1(x).$$

Hence, the analogy of $\mathrm{Unst}(U)$ is the unstable manifold of the gradient vector field of $q_2(x) - q_1(x)$. It is then easy to see that the unstable manifold is the linear subspace spanned by the negative eigenvectors of $q_2(x) - q_1(x)$.

We next consider the analogue of $\mathcal{M}_{\mathrm{vanish}}(U; (M, J_\epsilon); L)$ in our situation. Let x^i be the standard coordinate of $\mathbb{R}^n$. We take coordinates (x^i, y^i) of $\mathbb{R}^{2n}$ so that $\omega = \sum dx^i \wedge dy^i$. We define an almost complex structure J_ϵ by $J_\epsilon(\partial/\partial x^i) = \frac{1}{\epsilon} \partial/\partial y^i$. Let us put $\chi = \exp(2\pi\sqrt{-1}/3)$. Let us define $\partial_i D^2$ $i = 1, 2, 0$ so that $\cup_{i=0}^3 \partial_i D^2 = \partial D^2$ and $\partial_i D^2 \cap \partial_{i+1} D^2 = \chi^{i+1}$ for $i = 0, 1, 2$. ($\partial_3 D^2 = \partial_0 D^2$ by convention.)

We denote by $\mathcal{M}_{\mathrm{vanish}}(\mathbb{R}^n; (\mathbb{R}^{2n}, J_\epsilon); L)$ the set of all (x, u) where $x \in \mathbb{R}^n$, $u : (D^2; \partial_0 D^2, \partial_1 D^2, \partial_2 D^2; 1, \chi, \chi^2) \to (T^*\mathbb{R}^n; T_x^*\mathbb{R}^n, L_1, L_2; 0, p_1, p_2)$ and u is J_ϵ holomorphic. We define $\mathrm{ev} : \mathcal{M}_{\mathrm{vanish}}(\mathbb{R}^n; (\mathbb{R}^{2n}, J_\epsilon); L) \to \mathbb{R}^n$ by $(x, u) \mapsto x$.

It is proved in [**36**] that the image of $\mathrm{ev} : \mathcal{M}_{\mathrm{vanish}}(\mathbb{R}^n; (\mathbb{R}^{2n}, J_\epsilon); L) \to \mathbb{R}^n$ converges to the unstable manifold of $q_2(x) - q_1(x)$ as ϵ goes to zero. Thus, analogy of Conjecture 4.1 holds. (Note that the result of [**36**] implies a similar result also in case when q_i are not necessarily quadratic.)

4.6. Feynman diagram description of pseudoholomorphic disks: open string version. Now, combining the discussions thus far we can describe the J_ϵ holomorphic disk which bounds L (assuming Conjecture 4.1).

In this section, we describe the case when $S(B) = \emptyset$, that is the case when M is a torus. In the general case, we need both closed string vs. open string and open string vs. open string scattering. We will not discuss the general case, since there are some points which are not yet clear to the author. (We will mention it at the end of this section.)

There are two cases we can study this way. One is the case of pseudoholomorphic disk $u : D^2 \to M$ such that $u(\partial_- D) \subset L$ and $u(\partial_+ D) \subset F_x$, the other is the case of pseudoholomorphic disk $u : D^2 \to M$ with $u(\partial D) \subset L$.

REMARK 4.6. In the case of closed strings, we did not discuss the case of J_ϵ-holomorphic sphere (that is, a closed string analogue of $u : (D^2, \partial) \to (M, L)$.) The reason we did not do so is that, in the case $n = 2$, (that is the only case we discussed there), the virtual dimension of the pseudoholomorphic sphere in a Calabi-Yau 2-fold (K3 surface or torus) is -1. Hence, in the generic situation there is no such a sphere. If we consider the case $n = 3$, or if we consider a family of Calabi-Yau 2-folds, then it is possible to analyze the J_ϵ-holomorphic sphere in a similar way. (We need to clarify what happens in the neighborhood of the singular fiber to do so.)

We use the notations of §2.5. Let Γ be a finite oriented graph.

DEFINITION 4.8. A *ribbon structure* of Γ is a choice of cyclic order of the set $\partial^{-1}_{\text{target}}(v) \cup \partial^{-1}_{\text{source}}(v)$ for each vertex v.

A finite oriented graph equipped with a ribbon structure is called a *finite oriented ribbon graph*.

If there exists an embedding of Γ into an oriented surface, then it induces a ribbon structure on Γ. In case Γ is a tree then the ribbon structure determines an embedding of Γ into $\mathbb{R}^2$ up to isotopy.

We now consider the moduli space of data $\Gamma, l, \tilde{\psi}$ similar to ones in §2.7. Let us describe it precisely.

CONDITION 4.1. Γ is a finite oriented *ribbon* graph which satisfies Conditions 3.2 (1),(3),(4), and the following condition:

(2′) If $v \in \text{Vertex}_{\text{int}}(\Gamma)$ then $|\partial^{-1}_{\text{target}}(v)| \geq 2$, $|\partial^{-1}_{\text{source}}(v)| = 1$.

We assume either Condition 3.3 (a) or (b).

We next describe the data $\tilde{\psi}$. It is a continuous map $\psi : |\Gamma| \to B$ together with a lift $\tilde{\psi} : |\Gamma|\setminus\{\text{vertices}\} \to \tilde{B}_L$ which satisfies the following Condition 4.2. To state the condition we need some notations. For $\tau \in |\Gamma|\setminus\{\text{vertices}\}$, we put $\tilde{\psi}(\tau) = (\psi(\tau); p(\tau), q(\tau); \beta(\tau))$, where $\beta(\tau)$ is an element of $\pi_2(M; F_{\psi(\tau)}; p(\tau), q(\tau))$. Let v be an interior vertex of Γ. We number $\partial^{-1}_{\text{target}}(v) \cup \partial^{-1}_{\text{source}}(v)$ as follows. There exists a unique edge e such that $\partial_{\text{source}}(e) = v$. We denote it by $e_0(v)$. Now, using the cyclic order induced by the ribbon structure, we order the edges in $\partial^{-1}_{\text{target}}(v)$ and denote them by $e_1(v), \cdots, e_{v(m)}(v)$.

CONDITION 4.2. (1) If we restrict $\tilde{\psi}$ to a edge $|e|$ and if we identify $|e| \cong [0, l(e)]$ then

(26) $$\frac{d\psi}{d\tau} = \text{grad} f_L,$$

where we use τ as a coordinate of $[0, l(e)]$ and the branch of our multivalued function f_L is determined by $\beta(\tau)$.

(2) Let $\tau_{i,j} \in |e_i(v)|$ such that $\lim_{j\to\infty} \tau_{i,j} = |v|$ then
$$\lim_{j\to\infty} p(\tau_{i+1,j}) = \lim_{j\to\infty} q(\tau_{i,j}) \in F_{\psi(|v|)}, \qquad i = 1, \cdots, m(v) - 1,$$
$$\lim_{j\to\infty} q(\tau_{m(v),j}) = \lim_{j\to\infty} q(\tau_{0,j}) \in F_{\psi(|v|)},$$
$$\lim_{j\to\infty} p(\tau_{1,j}) = \lim_{j\to\infty} p(\tau_{0,j}) \in F_{\psi(|v|)}.$$

We also have
$$\sum_{i=1}^{m(v)} \lim_{j\to\infty} \beta(\tau_{i,j}) = \lim_{j\to\infty} \beta(\tau_{0,j}).$$

Note we use the obvious sum $\pi_2(M, F_x; p_1, p_2) \times \pi_2(M, F_x; p_2, p_3) \to \pi_2(M, F_x; p_1, p_3)$ in the left-hand side. They both are elements of the homotopy set : $\pi_2(M; F_{\psi(|v|)}; \lim_{j\to\infty} p(\tau_{0,j}), \lim_{j\to\infty} q(\tau_{0,j}))$.

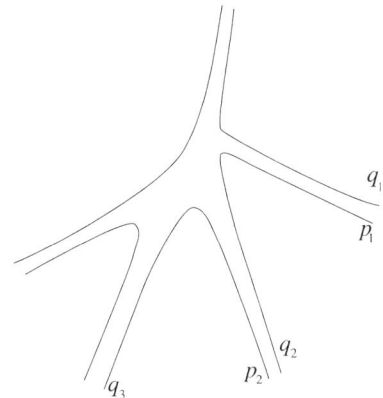

FIGURE 4.6.

(3) Let $v \in \text{Vertex}_{\text{sta}}(\Gamma)$ and $\partial_{\text{source}}(e) = v$. we identify $|e| \cong (-\infty, 0]$ and use τ as a coordinate. Then, $\beta(\tau)$ is a vanishing disk and $\lim_{\tau \to -\infty} \psi(\tau) \in S(L)$.

(4) The case Condition 3.3(b) is satisfied : Let e_{last} be the edge such that $\partial_{\text{target}}(e_{\text{last}}) = v_{\text{last}}$. We identify $|e_{\text{last}}| \cong [0, \infty)$ and use τ as a coordinate. Then, $-\partial_+ \beta(\tau)$ is a vanishing disk and $\lim_{\tau \to \infty} \psi(\tau) \in S(L)$.

The case Condition 3.3(a) is satisfied : We identify $|e| \cong [0, |l(e_{\text{last}})|]$ (Note $l(e_{\text{last}}) < \infty$ in this case.) We define
$$\widetilde{\text{ev}}(\Gamma, l, \tilde{\psi}) = \tilde{\psi}(v_{\text{last}}), \quad \text{ev}(\Gamma, l, \tilde{\psi}) = \psi(v_{\text{last}}).$$

Let $\alpha \in \pi_2(M, L)$. We consider also the following Condition 4.3 in case Condition 3.3(b) is satisfied. We use the notation of Condition 4.2(4). We choose a disk u bounding $-\partial_+ \beta(t)$ and its image is in $\pi^{-1}(U)$ where U is a small neighborhood of $\lim_{t\to -\infty} \psi(t) \in S(L)$. (Such u exists by (4)).

CONDITION 4.3. $\alpha = [\beta(t) \cup u]$.

DEFINITION 4.9. We denote by $\mathcal{M}_{\text{Morse}}(M, L; \alpha)$ the set of all $(\Gamma, l, \tilde{\psi})$ satisfying Conditions 3.3(b), 4.1, 4.2, 4.3.

Let denote by $\mathcal{M}_{\text{Morse}}(M, L)$ the set of all $(\Gamma, l, \tilde{\psi})$ satisfying Conditions 3.3(a), 4.1, 4.2. We define $\widetilde{\text{ev}} : \mathcal{M}_{\text{Morse}}(M, L) \to \tilde{B}_L$ and $\text{ev} : \mathcal{M}_{\text{Morse}}(M, L) \to B$ as in Condition 4.2(4).

Now, we can state an analogue of Conjecture 3.1

CONJECTURE 4.4. *We assume $S(B) = \emptyset$. Then, the moduli space of J_ϵ-holomorphic disks $u : D^2 \to M$ such that $u(\partial D^2) \subset L$ and $[u] = \alpha$ is cobordant to the moduli space $\mathcal{M}_{\text{Morse}}(M, L; \alpha)$.*

We consider the moduli space of (u, x) of pairs of point $x \in B$ and a pseudoholomorphic map $u : (D^2; \partial_- D^2, \partial_+ D^2) \to (M; L, F_x)$. We consider a map $(u, x) \mapsto x$. Then, it is bordant to the moduli space $\mathcal{M}_{\text{Morse}}(M, L)$ together with ev.

In a way similar to the discussion after Conjecture 3.1 one can reduce Conjecture 4.4 to the gluing argument in [**36**], results of §3.3 and Conjectures in §3.5.

REMARK 4.7. We do not need to restrict ourselves to the case of pseudoholomorphic disks in Conjecture 4.4. Namely, we can generalize it to the case of Riemann surface of higher genus with boundary. In the case when $M = T^*N$ where N is a closed 3 manifold, it is related to Witten's paper [**77**]. After [**77**] was distributed (in 1992), the author pointed out in [**29**] (which was written and submitted for publication in 1993) its relation to Morse theory. This point of view is further amplified by [**27**], [**36**], [**28**]. In those papers, the case when fibers are vector spaces is considered. Recently Kontsevich-Soibelman [**61**] showed that the same idea can be applied to the case when fibers are tori. (They, however, consider the case $S(L) = B(L) = \emptyset$.)

REMARK 4.8. After reading this and the last sections, it is easy to guess that there is a version which involves both open and closed string, using open string vs. closed string scattering discussed in §3.4. We expect it describes the moduli space of pseudoholomorphic disks $(D^2, \partial D^2) \to (M, L)$ in the case when $S(B) \neq \emptyset$. The reason we did not state a conjecture in this general case is as follows. In §3.4, the open string vs. closed string scattering is described by the intersection numbers of the bounding arcs $u(\partial_+ u)$ and $v(\partial D^2)$. Here $u : (D^2; \partial_+ D^2, \partial_- D^2; -1, +1) \to (M; L, F_x; p, q)$ is an open string and $v : (D^2, \partial D^2) \to (M, F_x)$ is a closed string. The trouble is that the intersection number $u(\partial_+ u) \cap v(\partial D^2)$ is not an invariant of the homology (or homotopy) class $[v(\partial D^2)] \in H_1(F_x)$. Actually, if we move v continuously then as it passes through one of the points p, q the intersection number changes. So, we need to study more to find a correct formulation of the conjecture. We leave it to the future research.

5. B-model: asymptotic analysis

Let us start with closed string version. Let $M_0 \to B \backslash S(B)$ be as in Assumption 3.1. We consider its mirror $M_0^\vee$ constructed in §1. The symplectic structure ω on M_0 induces a complex structure J_0 on $M_0^\vee$. We can extend ω to the compactification M. So, we put it as a part of Assumption 3.1. On the other hand, the complex structure J_0 does *not* extend to M. In physics literature, there are various arguments which suggest that the contribution of pseudoholomorphic disks bounding the fibers F_x modifies the complex structure J_0 so that it can be extended to the singular fibers. There are explicit formulae which describe modified complex structure locally in a neighborhood of some singular fibers. In Ooguri-Vafa [**68**] such a formula is given in the case of type I singular fiber. Recently, it was used by [**47**] to study collapsing of a K3 surface in a neighborhood of large complex structure limit. More precisely, in [**47**], Gross-Willson used the metric constructed by [**68**] in each neighborhood of singular fiber (type I singular fiber), glue it with a flat Calabi-Yau metric (which is a Kähler metric with respect to J_0) by partition of unity, and use functional analysis to show that there exists a Calabi-Yau metric which is close to the metric in a neighborhood of singular fiber and to a flat metric outside a neighborhood of singular fiber. The approximate solution they start with is fairly close to the actual solution. Namely, the difference between the approximate solution and the actual solution is of order $e^{-C/\epsilon}$ where ϵ is the diameter of the fiber. It is known in physics literature that the "instanton effect" (which is the contribution of pseudoholomorphic disks in the situation we are discussing now) is of exponentially small order. However, for various purposes we need a more precise information on the metric or on the complex structure.

One of the important consequences of mirror symmetry is the coincidence of the product structure of sheaf cohomology (Yukawa coupling) with the Gromov-Witten invariant of its mirror. When one tries to prove mirror symmetry using (singular) torus fibration picture we are discussing here, one needs to study the "instanton effect" to the complex structure. So, we need to study this exponentially small part of the complex structure. In asymptotic analysis, it is known that the tunnel effect (see §4.2 for its definition in a situation which is closely related to one we are interested in) is also a term of exponentially small order. The main purpose of this section is to explain that Yukawa coupling is regarded as a kind of tunnel effect (see §4.2) and show a relation of it to the Morse theory picture of pseudoholomorphic disks described in §2,3. However, our discussion is incomplete by various reasons which are explained in §4.3. Since the case of instanton effect to J_0 is harder, in the next section we study a simpler case of vector bundles, which is related to homological mirror symmetry. In that case, we can present the story in more detail.

5.1. Free field: Fourier expansion of complex structure. This section is influenced much by Arinkin-Polishchuk's paper [**5**] which discusses the relation between the path space $\pi_1(F_x; F_x \cap L)$ and Fourier expansion. Here, we consider "closed string case" but the basic idea is the same. The open string version (which is more directly an analogue of Arinkin-Polishchuk's paper), will be discussed in the next section.

Let us consider $M_0 \to B \backslash S(B)$ be as Assumption 3.1. We consider its mirror $M_0^\vee$ which has a T-structure invariant complex structure J_0 induced by the symplectic structure of M_0, by Proposition 1.5. Namely, if we take symplectic coordinates (x^i, y^i) of M_0 (such that $\omega = \sum dx^i \wedge dy^i$), then the complex coordinates of the mirror are $z_\epsilon^i = x^i + \sqrt{-1}\epsilon y^{i*}$. Here, (y^{i*}) are dual coordinates in the following sense: There exists an open set U of B such that $\pi_{M_0}^{-1}(U) = U \times T^n$. Then, (x^i) are coordinates of U and (y^i) are coordinates of $T^n = V/\Gamma$ such that $V \cong \mathbb{R}^n$, $\Gamma = \mathbb{Z}^n$. The corresponding open set of $M_0^\vee$ is $U \times T^{n*}$ where $T^{n*} = V^*/\Gamma^\vee$. Here, V is a dual vector space to V and $\Gamma^\vee$ is a dual lattice of Γ. Then, y^{i*} is the dual basis of y^i.

We denote by $\overline{\partial}_{0,\epsilon}$ the Dolbaut differential associated to $J_{0,\epsilon}$, where

$$J_{0,\epsilon}\left(\frac{\partial}{\partial x^i}\right) = \frac{1}{\epsilon}\left(\frac{\partial}{\partial y^i}\right).$$

We consider another almost complex structure on M_0 and let $\overline{\partial}$ is its Dolbaut differential. We put

(27) $$\overline{\partial} = \overline{\partial}_{0,\epsilon} + \mathfrak{B} \circ \partial_{0,\epsilon}$$

where $\mathfrak{B} \in \Gamma(M_0; TM_0 \otimes \Lambda^{0,1})$. According to Kodaira-Spencer and Newlander-Nirenberg, the condition for $\overline{\partial}$ to be integrable is

(28) $$\overline{\partial}_{0,\epsilon}\mathfrak{B} + \frac{1}{2}[\mathfrak{B},\mathfrak{B}] = 0.$$

We are going to study equation (28) using fiberwise Fourier expansion.

LEMMA 5.1. *There exists a canonical isomorphism*

(29) $$\Gamma(B\backslash S(B); TB \otimes \Lambda^k(B)) \cong \Gamma(M_0^\vee; M_0^\vee \otimes \Lambda^{0,k}(M_0^\vee))^T.$$

Here the right-hand side is the set of T-structure invariant sections.

PROOF. We use local complex coordinates $z_\epsilon^i = x^i + \sqrt{-1}\epsilon y^{i*}$ as above. A T-structure invariant section locally can be written as

$$s = \sum s^i_J(x^1,\cdots,x^n)\frac{\partial}{\partial z_\epsilon^i} \otimes d\overline{z}_\epsilon^J,$$

where $J = (j_1,\cdots,j_k)$ is a multi index. Here the T-structure invariance of s is equivalent to the fact that s^i_J depends only on x^k and is independent of y^{k*}. We put

$$\overline{s} = \sum s^i_J(x^1,\cdots,x^n)\frac{\partial}{\partial x^i} \otimes dx^J.$$

It is easy to see that $s \mapsto \overline{s}$ gives the required isomorphism. □

Hereafter, we denotes the isomorphism (29) by $\mathfrak{I}_\epsilon$.

Let $x \in B\backslash S(B)$. We remark that $F_x = \pi^{-1}_{M_0^\vee}$ is the dual torus of $F_x^\vee = \pi^{-1}_{M_0}$. Hence, by Pontrjyagin duality we have a canonical isomorphism

$$\pi_1(F_x) = Hom(F_x^\vee, U(1)).$$

For $\gamma \in \pi_1(F_x)$, we denote by $\exp\gamma^*$ the corresponding element of $Hom(F_x^\vee, U(1))$. We then have immediately the following:

LEMMA 5.2. *Let U be an contractible open neighborhood of x in $B\backslash S(B)$. Then, any element s of $\Gamma(\pi^{-1}_{M_0^\vee}(U); M_0^\vee \otimes \Lambda^{0,k}(M_0^\vee))$ can be uniquely expanded as*

(30) $$s = \sum_{\gamma \in \pi_1(F_x^\vee)} (\exp\gamma^*) s_\gamma$$

such that s_γ is invariant of T-structure.

Now, let $\mathfrak{B}$ be as in (27). We expand $\mathfrak{B}$ as

(31) $$\mathfrak{B} = \sum_{\gamma \in \pi_1(F_x^\vee)} \exp\gamma^* \, \mathfrak{I}_\epsilon(\mathfrak{B}_\gamma)$$

locally. We assume $\mathfrak{B}_0 = 0$. Namely, the T-structure invariant part of $\mathfrak{B}$ is zero. (We recall that the T-structure invariant part can be included to $J_{0,\epsilon}$. Hence, we can assume it is zero without loss of generality.)

In this section we discuss Free field. Namely, we study an equation

$$\overline{\partial}_{0,\epsilon}\mathfrak{B} = 0, \tag{32}$$

which is a linearization of (20).

PROPOSITION 5.1. $\mathfrak{B}$ satisfies (32) if and only if

$$d(e^{2\pi f_\gamma/\epsilon}\mathfrak{B}_\gamma) = 0. \tag{33}$$

Here, f_γ is a branch of the multivalued function f, defined in §2.1, corresponding to $[u] \in \pi_2(N; F_x)$ such that $[\partial u] = [\gamma]$ and d is the exterior derivative of TB valued 1-form $\mathfrak{B}_\gamma$. We define exterior derivative of TB valued forms by using flat affine structure of $B \backslash S(B)$.

PROOF. The problem is local so we may use local coordinates (x^i, y^{i*}). Let $(\gamma_1, \cdots, \gamma_n) \in \mathbb{Z}^n \cong \pi_1(F_x)$. As we see before $f_\gamma = \sum x^i \gamma_i + c$ where c is a constant. On the other hand, $\exp \gamma^* = \exp\left(2\pi\sqrt{-1}\sum \gamma_i y^{i*}\right)$. Hence,

$$\overline{\partial}_{0,\epsilon}\mathfrak{B} = \frac{1}{2}\sum \left(\frac{\partial}{\partial x^i} + \frac{2\pi}{\epsilon}\gamma_i\right) \exp \gamma^* d\overline{z}_\epsilon^i \wedge \mathfrak{I}_\epsilon(\mathfrak{B}_\gamma).$$

The proposition follows. □

Thus, the linearization (32) of (20) is equivalent to a system of differential equations (33). Equation (33) is a multivalued version of the deformed De Rham complex studied by a celebrated paper of Witten [**76**], which we review in the next subsection.

5.2. Supersymmetry and Morse theory, revisited. In this section we review [**76, 49**] and propose some conjectures which generalize it. The generalization we discuss in this section is based on Morse homotopy [**29, 36**] (see also [**14**]) and is closely related to [**77**], §8 of [**27**] and also to [**61**].

Let X be a Riemannian manifold and f be a Morse function on it. (In our situation of multivalued Morse theory, X will correspond to B and f to our multivalued function.) We consider the De Rham complex $(\Gamma(X; \Lambda(X)), d)$. Let $* : \Gamma(X; \Lambda^k(X)) \to \Gamma(X; \Lambda^{n-k}(X))$ be the Hodge star operator (where $n = \dim X$). Witten considered the differential operator

$$d_{f,\epsilon} = \exp(-f/\epsilon) \circ d \circ \exp(f/\epsilon) \tag{34}$$

and

$$\Delta_{f,\epsilon} = d_{f,\epsilon} \circ d_{f,\epsilon}^* + d_{f,\epsilon}^* \circ d_{f,\epsilon}.$$

Here $d_{f,\epsilon}^*$ is the L^2 dual to $d_{f,\epsilon}$ and is

$$d_{f,\epsilon}^* = -\exp(-f/\epsilon) \circ * \circ d \circ * \circ \exp(f/\epsilon).$$

Obviously, $(\Gamma(X; \Lambda(X)), d_{f,\epsilon})$ is a cochain complex and its cohomology is isomorphic to the cohomology of $(\Gamma(X; \Lambda(X)), d)$, the De Rham cohomology. We remark that operator (33) is similar to the operator (34).

Witten observed the following Theorem 5.1 which is proved rigorously, for example, in [**69**]. Let $Cr(f)$ be the set of all critical points of f. Let $Cr_k(f) \subseteq Cr(f)$

be the subset of critical points of Morse index $n - k$. Hereafter, we denote the Morse index of p by $n - \eta(p)$. For $\delta > 0$ we put

$$\mathcal{H}_\delta^k(X; \Delta_{f,\epsilon}) = \{u \in \Gamma(X; \Lambda^k(X)) \mid \Delta_{f,\epsilon} u = \lambda u, \ \lambda < \delta\}.$$

THEOREM 5.1. *For sufficiently small ϵ, δ, the dimension of the vector space $\mathcal{H}_\delta^k(X; \Delta_{f,\epsilon})$ is equal to the order of $Cr_k(f)$. Moreover, there exists $u_p \in \mathcal{H}_\delta^k(X; \Delta_{f,\epsilon})$ for each $p \in Cr_k(f)$ such that*

$$|u_p| < C \exp(-c/\epsilon) \|u_p\|_{L^2}$$

outside a small neighborhood of p.

Hereafter, we normalize so that $\|u_p\| = 1$.

For our purpose we need more detailed information. Namely, we need to study tunnel amplitude and also interactions (that is k point function with $k \geq 3$). We start with tunnel amplitude. Let $p \in Cr_k(f)$, $q \in Cr_{k+1}(f)$. We are going to study $\langle d_{f,\epsilon} u_p, u_q \rangle_{L^2}$. An observation by Witten, which is proved rigorously by B.Helffer and J.Sjöstrand in [**49**], is that this number is described by the number of gradient lines joining p to q. To state it precisely we need some notations.

We assume that $\text{grad} f$ is Morse-Smale. Namely, its stable manifolds and unstable manifolds intersect transversally. Let us consider the set of all $\ell : \mathbb{R} \to X$ such that $d\ell/d\tau = \text{grad}_{\ell(\tau)} f$ and $\lim_{\tau \to -\infty} \ell(\tau) = p$, $\lim_{\tau \to +\infty} \ell(\tau) = q$. We say $\ell \sim \ell'$ if there exists τ_0 such that $\ell'(\tau) = \ell(\tau + \tau_0)$. We denote by $\mathcal{M}(X; f; p, q)$ the set of all $\sim$ equivalence classes of such ℓ. Since we assumed $\text{grad} f$ to be Morse-Smale it follows that $\mathcal{M}(X; f; p, q)$ is a smooth manifold of dimension $\mu(q) - \mu(p) - 1$.

To fix orientation on $\mathcal{M}(X; f; p, q)$ we need to fix the following extra data (see [**38**] the beginning of page 1037 for detail). We fix an orientation of $T_p W^u(f; p)$ where $W^u(f; p) = \{x \in X \mid \lim_{\tau \to -\infty} \exp(\tau \text{grad} f)(x) = p\}$ is the unstable manifold of the gradient flow. (Here $\exp(\tau \text{grad} f)$ is the one parameter group of transformations associated to $\text{grad} f$.) We also fix an orientation of $T_q W^u(f; q)$ As explained in [**38**] it induces an orientation of $\mathcal{M}(X; f; p, q)$. As we remarked in [**38**] the orientation induced on $\mathcal{M}(X; f; p, q)$ is independent of the orientation of X. (Actually, we do not need to assume that X is orientable.)

We also remark that after normalization there are two choices u_p. (Namely, we can replace u_p by $-u_p$. To fix this choice we need to orient $W^u(f; p)$, see [**69**], page 166. The eigenform e_0 there depends on the *oriented* orthonormal basis (dx^i).)

Now, we have :

THEOREM 5.2. (Helffer-Sjöstrand)

$$\langle d_{f,\epsilon} u_p, u_q \rangle_{L^2} \sim e^{(f(p)-f(q))/\epsilon} [\mathcal{M}(X; f; p, q)].$$

Here, $[\mathcal{M}(X; f; p, q)] \in \mathbb{Z}$ is the order counted with sign and $\sim$ means that the difference C_ϵ between left and right-hand side is estimated as

$$\lim_{\epsilon \to 0} e^{(f(q)-f(p))/\epsilon} |C_\epsilon| = 0.$$

Theorem 5.1 means that u_p is concentrated on a neighborhood of p. Theorem 5.2 means that u_p propagates along the gradient line of f.

For our purpose, we need even more detailed information which is related to the product structure. Namely, we need to find a relation between a number similar to $\langle d_{f,\epsilon} u_p, u_q \rangle$ and Morse homotopy ([**29, 36**]). For this purpose we need to take several Morse functions as follows. Let f_i $i = 0, 1, 2$ be functions. We assume $f_i - f_j$ $i \ne j$ are Morse functions. We take $p_1 \in Cr(f_1 - f_0)$, $p_2 \in Cr(f_2 - f_1)$ and $q \in Cr(f_2 - f_0)$. We assume $\mu(p_1) + \mu(p_2) = \mu(q)$ where $n - \mu$ is the Morse index. We want to study asymptotic behavior of $\langle u_{p_1} \wedge u_{p_2}, u_q \rangle$. We let $\mathcal{M}(X; f_1, f_2, f_3; p_1, p_2, q)$ be the set of all $x \in X$ such that

$$\lim_{\tau \to -\infty} \exp(\tau(\mathrm{grad} f_1 - \mathrm{grad} f_0))(x) = p_1,$$
$$\lim_{\tau \to -\infty} \exp(\tau(\mathrm{grad} f_2 - \mathrm{grad} f_1))(x) = p_2,$$
$$\lim_{\tau \to +\infty} \exp(\tau(\mathrm{grad} f_2 - \mathrm{grad} f_0))(x) = q.$$

$\mathcal{M}(X; f_0, f_1, f_2; p_1, p_2, q)$ is an intersection of two unstable manifolds and a stable manifold. For generic f_1, f_2, f_3, it is a smooth manifold of dimension $\mu(p_1) + \mu(p_2) - \mu(q) = 0$. So, we assume that $\mathcal{M}(X; f_0, f_1, f_2; p_1, p_2, q)$ consists of finitely many points. The orientations of $T_{p_1} W^u(f_1 - f_0; p_1)$, $T_{p_2} W^u(f_2 - f_1; p_2)$, $T_q W^u(f_2 - f_0; q)$ induce an orientation of the moduli space $\mathcal{M}(X; f_0, f_1, f_2; p_1, p_2, q)$.

CONJECTURE 5.1. *Using notations similar to Theorem 5.1 we have :*

$$\langle u_{p_1} \wedge u_{p_2}, u_q \rangle$$
$$\sim e^{-(f_1(p_1) - f_0(p_1) + f_2(p_2) - f_1(p_2) + f_0(q) - f_2(q))/\epsilon}[\mathcal{M}(X; f_0, f_1, f_2; p_1, p_2, q)].$$

The idea behind Conjecture 5.1 is as follows. As we mentioned before, u_{p_i} are concentrated in a neighborhood of p_i. Hence, $u_{p_1} \wedge u_{p_2}$ is zero up to exponentially small terms. To study those small terms, we need to see how they propagate. Theorem 5.1 suggests that they propagate along the gradient lines of $f_{i+1} - f_i$. Hence, $u_{p_1} \wedge u_{p_2}$ will be nonzero at the point where two gradient lines, one of $f_1 - f_0$ the other of $f_2 - f_1$, intersect each other. From the intersection points, it propagates along the gradient line of $f_2 - f_1$. Hence, the left-hand side can be described by the moduli space $\mathcal{M}(X; f_0, f_1, f_2; p_1, p_2, q)$.

Conjecture 5.1 is concerned with the three points function. We need to further generalize it to four points functions or more. Let us first discuss the generalization of the left-hand side. (Compare [61] §6.4 and [37] §A6.)

Let Γ be a ribbon tree with $k + 1$ exterior vertices. We assume Condition 3.2 (1),(2),(3),(4) and $|\Gamma|$ be the corresponding one dimensional complex (see Definition 2.6). There exists an embedding $|\Gamma| \subset D^2$ which is compatible with ribbon structure and such that the exterior vertices are on ∂D^2. We let $v_{\mathrm{last}} = v_0$ (the last exterior vertex) and number the other exterior vertices $v_1, \cdots, v_k$ so that $v_0, v_1, \cdots, v_k$ respects the cyclic order of ∂D^2. The domain $D^2 \setminus |\Gamma|$ has exactly $k + 1$ connected components D_i, $i = 0, \cdots, k$. We number them so that $v_i \in \overline{D}_i \cap \overline{D}_{i-1}$ for $i = 1, \cdots, k$ and $v_0 \in \overline{D}_0 \cap \overline{D}_k$. For each edge e of Γ there exists exactly two of D_i's $D_{\mathrm{right}(e)}$ and $D_{\mathrm{left}(e)}$, such that $|e| \subset \overline{D}_{\mathrm{right}(e)} \cap \overline{D}_{\mathrm{left}(e)}$ and that $\mathrm{left}(e) > \mathrm{right}(e)$. We take smooth functions f_i, $i = 0, \cdots, k$ such that $f_i - f_j$ are Morse functions.

We put $f_e = f_{\mathrm{left}(e)} - f_{\mathrm{right}(e)}$ and consider the operators $\Delta_{f_e, \epsilon}$, $d_{f_e, \epsilon}$ in §4.1. Let $\Pi^k_{f_e, \epsilon, \delta} : L^2(X; \Lambda^k(X)) \to \mathcal{H}^k_\delta(X; \Delta_{f_e, \epsilon})$ be the orthonormal projection. The following lemma can be proved in the same way as usual Harmonic analysis.

LEMMA 5.3. *There exists a pseudo differential operator*

$$Q^k_{f_e, \epsilon, \delta} : \Gamma(X; \Lambda^k(X)) \to \Gamma(X; \Lambda^k(X))$$

of degree -2 such that
$$\Delta_{f_e,\epsilon} \circ Q^k_{f_e,\epsilon,\delta} - Q^k_{f_e,\epsilon,\delta} \circ \Delta^k_{f_e,\epsilon,\delta} = 1 - \Pi^k_{f_e,\epsilon,\delta}.$$

DEFINITION 5.1. We define the *propagator*
$$G^k_{f_e,\epsilon,\delta} : \Gamma(X; \Lambda^k(X)) \to \Gamma(X; \Lambda^{k+1}(X))$$
by
$$G^k_{f_e,\epsilon,\delta} = d^*_{f_e,\epsilon} \circ Q^k_{f_e,\epsilon,\delta}.$$

We can easily prove that

(35) $$d_{f_e,\epsilon} \circ G^k_{f_e,\epsilon,\delta} + G^{k-1}_{f_e,\epsilon,\delta} \circ d_{f_e,\epsilon} = 1 - \Pi^k_{f_e,\epsilon,\delta}.$$

REMARK 5.1. We consider two chain maps id : $(\mathcal{H}^*_\delta(X; \Delta_{f,\epsilon}); d_{f,\epsilon}) \to (\Gamma(X; \Lambda(X)); d_{f_e,\epsilon})$ and $\Pi_{f_e,\epsilon,\delta} : (\Gamma(X; \Lambda(X)); d_{f_e,\epsilon}) \to (\mathcal{H}^*_\delta(X; \Delta_{f_e,\epsilon}); d_{f_e,\epsilon})$. The composition $\Pi_{f_e,\epsilon,\delta} \circ$ id is obviously an identity. On the other hand, (35) implies that $G_{f_e,\epsilon,\delta}$ is a chain homotopy from identity to id $\circ \Pi_{f_e,\epsilon,\delta}$.

Now, using propagator $G_{f_e,\epsilon,\delta}$ we define operators
$$\mathfrak{m}_\Gamma : \mathcal{H}^*_\delta(X; \Delta_{f_1-f_0,\epsilon}) \otimes \cdots \otimes \mathcal{H}^*_\delta(X; \Delta_{f_k-f_{k-1},\epsilon}) \to \mathcal{H}^*_\delta(X; \Delta_{f_k-f_0,\epsilon})$$
$$\varphi_\Gamma : \mathcal{H}^*_\delta(X; \Delta_{f_1-f_0,\epsilon}) \otimes \cdots \otimes \mathcal{H}^*_\delta(X; \Delta_{f_k-f_{k-1},\epsilon}) \to \Gamma(X; \Lambda(X))$$

as follows. The definition is by induction on k, the number of exterior vertices plus one.

In the case $k = 0$, there exists only one Γ, say Γ_0. We put
$$\mathfrak{m}_{\Gamma_0} = d_{f_e,\epsilon}, \qquad \varphi_{\Gamma_0} = \mathrm{id}.$$

We assume that we have constructed $\mathfrak{m}_\Gamma$ up to $k-1$. Let v_{last} be the last vertex of Γ. There exists a unique edge e_{last} which has v_{last} as a vertex. We remove $|e_{\text{last}}|$ from $|\Gamma|$. Then, $|\Gamma| \setminus |e_{\text{last}}|$ is a union of two ribbon trees Γ_1, Γ_2. We now put

(36) $$\varphi_\Gamma = G_{f_{e_{\text{last}}},\epsilon,\delta} \circ \wedge \circ (\varphi_{\Gamma_1} \otimes \varphi_{\Gamma_2})$$

(37) $$\mathfrak{m}_\Gamma = \Pi_{f_{e_{\text{last}}},\epsilon,\delta} \circ \wedge \circ (\varphi_{\Gamma_1} \otimes \varphi_{\Gamma_2}).$$

The operator $\mathfrak{m}_\Gamma$ defines an A_∞ structure and φ_Γ defines a homotopy equivalence from it to the De Rham complex of X. More precisely we have the following Theorem 5.3.

We put $\mathfrak{m}_k = \sum_\Gamma \mathfrak{m}_\Gamma, \varphi_k = \sum_\Gamma \varphi_\Gamma$, where the sum is taken over all ribbon trees Γ which satisfies Condition 3.2 (1),(2),(3),(4) and has $k+1$ vertices.

THEOREM 5.3. $\mathfrak{m}_k$ *satisfies A_∞ formulae. Namely, for any elements* $x_i \in \mathcal{H}^*_\delta(X; \Delta_{f_i-f_{i-1},\epsilon})$, *we have* :

(38) $$\sum_{k+\ell=n+1} \sum_{i=1}^{n+1-\ell} (-1)^* \mathfrak{m}_k(x_1, \cdots, x_{i-1}, \mathfrak{m}_\ell(x_i, \cdots, x_{i+\ell-1}), x_{i+\ell}, \cdots, x_n) = 0.$$

Here the sign is defined by : $* = \deg x_1 + \cdots + \deg x_{i-1} + i - 1$.

The homomorphisms $\varphi_k : \mathcal{H}^*_\delta(X; \Delta_{f_1-f_0,\epsilon}) \otimes \cdots \otimes \mathcal{H}^*_\delta(X; \Delta_{f_k-f_{k-1},\epsilon}) \to \Gamma(X; \Lambda(X))$ *define an A_∞ functor.*

We do not prove Theorem 5.3, since we do not use it. We omit the definition of A_∞ functor. See [**56, 40, 34, 33, 37**] for more detail on homological algebra of A_∞ algebra and of A_∞ category.

REMARK 5.2. In the case when $f_j - f_i$ are perfect Morse functions (that is the case when $\mathcal{H}^*_\delta(X; \Delta_{f_k-f_{k-1},\epsilon})$ is isomorphic to the De Rham cohomology) Theorem 5.3 implies that there exists an A_∞ structure on cohomology group which is homotopy equivalent to the De Rham complex (and an A_∞ algebra). (Actually, we need to say A_∞ category instead of A_∞ algebra, since $\mathfrak{m}_k$ depends on the choice of f_i.) This statement was first proved by Kadeishvili in [**55**] see also [**48, 66, 61, 33**]. We can prove Theorem 5.3 in the same way. The relation of this construction to Feyman diagram is found in [**61**] and in [**37**].

Now, let us consider $p_i \in Cr(f_i - f_{i-1})$, $q \in Cr(f_k - f_0)$. We assume $\sum \mu(p_i) - \mu(q) = k - 2$. It is easy to see that the degree of the operator $\mathfrak{m}_k$ is $2 - k$. So, the number

$$\langle \mathfrak{m}_\Gamma(u_{p_1} \otimes \cdots \otimes u_{p_k}), u_q \rangle$$

is well defined. It is a generalization of the left-hand side of Conjecture 5.1.

We next describe the generalization of the right-hand side of Conjecture 5.1. It is quite similar to one we discussed in §3.6. Let Γ, f_i, p_i, q be as above. We consider the pair (ℓ, ψ) where $\ell : \text{Edge} \to \mathbb{R}_+ \cup \{\infty\}$, and $\psi : |\Gamma| \setminus \{|v_0|, \cdots, |v_k|\} \to X$ is a continuous map which is smooth on each edge. We require the following condition.

CONDITION 5.1. (1) $\ell(e)$ is ∞ if e is exterior. $\ell(e)$ is finite if e is interior. (2) For each interior edge e, we identify $|e| \cong [0, \ell(e)]$. If $\partial_{\text{source}}(e) \in \text{Vertex}_{\text{ext}}(\Gamma)$ we identify $|e| \cong (-\infty, 0]$. It $\partial_{\text{target}}(e) = v_{\text{last}}$, we identify $|e| \cong [0, \infty)$. Now, we have

$$\frac{d\psi}{d\tau} = \text{grad} f_e$$

on each edge e. (3) If $\partial_{\text{source}}(e) = v_i \in \text{Vertex}_{\text{ext}}(\Gamma)$ and if $\tau_i \in (-\infty, 0] \cong |e|$ with $\lim_i \tau_i = -\infty$, then we have $\lim_i \psi(\tau_i) = p_i$. (4) If $\partial_{\text{target}}(e) = v_{\text{last}}$, $\tau_i \in [0, +\infty, 0) \cong |e|$ with $\lim_i \tau_i = \infty$, then we have $\lim_i \psi(\tau_i) = q$.

DEFINITION 5.2. $\mathcal{M}(X; \Gamma; f_0, \cdots, f_k; p_1, \cdots, p_k, q)$ is the set of all (ℓ, ψ) satisfying Condition 5.1.

The following lemma is proved in [**36**].

LEMMA 5.4. *For generic f_i the moduli space $\mathcal{M}(X; \Gamma; f_0, \cdots, f_k; p_1, \cdots, p_k, q)$ is a smooth manifold of dimension $\sum \mu(p_i) - \mu(q) - k + 2$.*

Thus, by our assumption, $\mathcal{M}(X; \Gamma; f_0, \cdots, f_k; p_1, \cdots, p_k, q)$ consists of finitely many points. We can now state our conjecture. We put

$$\lambda = f_k(q) - f_1(q) - \sum_i (f_i(p_i) - f_{i-1}(p_i)).$$

CONJECTURE 5.2. *Orientations of unstable manifolds determine an orientation of $\mathcal{M}(X; \Gamma; f_0, \cdots, f_k; p_1, \cdots, p_k, q)$, so that the following asymptotic expansion holds.*

$$\langle \mathfrak{m}_\Gamma(u_{p_1} \otimes \cdots \otimes u_{p_k}), u_q \rangle \sim e^{-\lambda/\epsilon}[\mathcal{M}(X; \Gamma; f_0, \cdots, f_k; p_1, \cdots, p_k, q)].$$

To explain the meaning of Conjecture 5.2 we recall the following Theorem 5.4 proved in [**26, 36**]. We define the Morse complex $MC(f_i - f_{i-1})$ by

$$MC(f_i - f_{i-1})^k \cong \bigoplus_{p \in Cr_k(f_i - f_{i-1})} \mathbb{R}[p]$$

$$d[p] = \sum_{q \in Cr_{k+1}(f_i - f_{i-1})} [\mathcal{M}(X; f; p, q)][q].$$

We define A_∞ operation by using the moduli space $\mathcal{M}(X; \Gamma; f_0, \cdots, f_k; p_1, \cdots, p_k, q)$ as follows.

$$\mathfrak{m}_\Gamma([p_1], \cdots, [p_k]) = \sum [\mathcal{M}(X; \Gamma; f_0, \cdots, f_k; p_1, \cdots, p_k, q)][q],$$
$$\mathfrak{m}_k = \sum \mathfrak{m}_\Gamma \quad k \geq 2,$$
$$\mathfrak{m}_1 = d.$$

THEOREM 5.4. ([**36**]) $\mathfrak{m}_k$ satisfies the A_∞ formula (38).

Now, we consider an isomorphism $I : MC(f_i - f_{i-1}) \cong \mathcal{H}_\delta(X; \Delta_{f_1 - f_0, \epsilon})$ by

$$I([p]) = e^{f_i(p) - f_{i-1}(p)} u_p.$$

Then, Conjecture 5.2 and Theorem 5.2 will imply that I is an isomorphism of the A_∞ category. On the other hand, Theorem 5.3 implies that $(\mathcal{H}_\delta(X; \Delta_{f_1 - f_0, \epsilon}), \mathfrak{m}_k)$ is homotopy equivalent to the De Rham complex.

We remark that it is proved in [**61**] that $MC(f_i - f_{i-1})$ is homotopy equivalent to the De Rham complex. However, Conjecture 5.2 itself is not discussed there. One can try to generalize the argument of Helffer-Sjöstrand [**49**] to prove Conjectures 5.1,5.2 though the author had not yet done so at the time of writing this article.

5.3. Quantum correction to complex structure: a plan. We will now pursue the analogy of the results and conjectures in §4.2 to the study of (28). Proposition 5.1 shows that the operator $\bar{\partial}_{0,\epsilon}$ reduces to the operator $d_{f,\epsilon}$ in each Fourier mode. (Here f is our multivalued function.)

Let us next study the solution of (28) itself (not its linearization). In this section we state a conjecture of the asymptotic behavior of $\mathfrak{B}$ solving (28).

To state our conjecture, we need to assume that $\dim_\mathbb{C} M = 2$, since otherwise the closed string - closed string scattering is more complicated to describe. In other words, we consider the case when M is a $K3$ surface.

Let us consider the covering space $\tilde{B}_0$ of $B_0 = B \backslash S(B)$ introduced in §2.1. Let $\tilde{\mathfrak{B}} \in \Gamma(\tilde{B}_0^+, T\tilde{B}_0 \otimes \Lambda^1 \tilde{B}_0)$. We can associate $\mathfrak{B} \in \Gamma(M_0^\vee, T_\mathbb{C} M_0^\vee \otimes \Lambda^{0,1} M_0^\vee)$ as follows.

Let $p \in B_0$. For each $[u] \in \pi_2(M_0, F_x)$ we have $\tilde{\mathfrak{B}}(p, [u])$ which can be identified as an element of of the fiber of $TB \otimes \Lambda^1 B$ at p. We take it as $\mathfrak{B}_{[u]}$ at p. Then, we put

(39) $$\mathfrak{B} = \sum_{[u] \in \pi_2(M_0, F_x)} e^{-2\pi f_u/\epsilon} [\partial u]^* \otimes \mathfrak{I}_\epsilon(\mathfrak{B}_{[u]}),$$

as in (31). Here, f_u is the branch of our multivalued function f specified by $[u] \in \pi_2(M_0, F_x)$, and $[\partial u] \in \pi_1(F_x)$. (Actually, the author does not know whether the right-hand converges. So, to be precise, we need to regard the right hand side as an asymptotic expansion, see Conjecture 5.3).

Thus, $\mathfrak{B}_{[u]}$ in (39) and $\mathfrak{B}_\gamma$ in (31) is related by the formula

$$\mathfrak{B}_\gamma = \sum_{[u] \in \pi_2(M_0, F_x), \partial[u] = \gamma} e^{-2\pi f_u/\epsilon} \mathfrak{B}_{[u]}.$$

$\overline{\partial}\mathfrak{B} = 0$ then becomes $d\mathfrak{B}_{[u]} = 0$.

Now, we take $\mathrm{ev}_{\mathrm{last}} : \mathcal{M}_{\mathrm{Morse}}(M; g) \to \tilde{B}_0$ as in Definitions 3.7.3.8. (We remark that the multiplicity of the moduli space $\mathcal{M}_{\mathrm{Morse}}(M; g)$ was left as an open problem in §2.5 in the case of multiple covered disk. So, the definition of $\mathcal{M}_{\mathrm{Morse}}(M; g)$ was not rigorous in that case.)

It is easy to check that the dimension of $\mathcal{M}_{\mathrm{Morse}}(M; g)$ is one in our case. We regard it as a distribution valued 1-form on it. (1 is the codimension of $\mathcal{M}_{\mathrm{Morse}}(M; g)$ in $\tilde{B}_0$. Note $\dim_{\mathbb{C}} M^\vee = 2$.) To add $T\tilde{B}_0^+ \cong \pi^* TB$ factor, we proceed as follows. We consider the canonical isomorphism

(40) $$H_1(F_x) \cong H^1(F_x) \cong T^*_{s_0(x)} F_x \cong T_x(B_0).$$

Here, the first isomorphism is the Poincaré duality. (Note we are considering the case $\dim F_x = 2$.) There exists a second isomorphism since F_x is a torus equipped with an affine structure. The third isomorphism is induced by the symplectic form on M_0. We remark that all the isomorphisms in (40) are canonical. By (40) element $[\gamma] \in H_1(F_x)$ corresponds to an element of $T_x(B_0)$ which we write $d_{[\gamma]}$.

Let U_x be a small open neighborhood of x in B_0. We consider an open subset $\pi_{M_0^\vee}^{-1}(U_x)$ of $M_0^\vee$. Our vector field $d_{[\gamma]}$ determines a T-structure invariant section of $\Lambda^{0,1} M_0^\vee$ on $\pi_{M_0^\vee}^{-1}(U_x)$ as before. Namely, using complex coordinates (x^i, y^{i*}), we send $\partial/\partial x^i$ to $\partial/\partial \bar{z}_\epsilon^i$, where $z^i = x^i + \sqrt{-1}\epsilon y^{i*}$. We write it as $\partial_{[\gamma], \epsilon}$.

Now, let $x \in B_0$. For each $[u] \in \pi_2(M, F_x)$ with nonzero $[\partial u] \in \pi_1(F_x)$, we have a neighborhood $U_{(x,[u])}$ of $(x, [u]) \in \tilde{B}_0$. We restrict the distribution valued 1-form $\mathrm{ev}_{\mathrm{last}\,*}\,\mathcal{M}_{\mathrm{Morse}}(M; g)$ to $U_{(x,[u])}$ and write it as $\mathrm{ev}_{\mathrm{last}\,*}\,\mathcal{M}_{\mathrm{Morse}}(M; g)|_{U_{(x,[\gamma])}}$. It induces a section of $\Lambda^{0,1} M_0^\vee$ as before. (Namely, we send dx^i to $d\bar{z}_\epsilon^i$). We write it by the same symbol $\mathrm{ev}_{\mathrm{last}\,*}\,\mathcal{M}_{\mathrm{Morse}}(M; g)|_{U_{(x,[u])}}$ by abuse of notation.

Now, on a neighborhood U_x of x in B_0, we "define" $\mathfrak{B}$ by

(41)
$$\mathfrak{B} \text{ "=" } \sum_{\gamma \in \pi_1(F_x)} \sum_{[u] \in \pi_2(M, F_x), \partial[u] = \gamma} e^{-2\pi f_u/\epsilon} \exp \gamma^*$$
$$\partial_{[\gamma], \epsilon} \otimes \mathrm{ev}_{\mathrm{last}\,*}\,\mathcal{M}_{\mathrm{Morse}}(M; g)|_{U_{(x,[u])}}$$
$$\text{" } \in \text{ "} \quad W^{-\infty}(\pi_{M_0^\vee}^{-1}(U_x), T_{\mathbb{C}} M_0^\vee \otimes \Lambda^{0,1} M_0^\vee).$$

Here, $W^{-\infty}$ denotes the space of distribution valued sections. We put the equality in (41) in the quote because of the convergence problem mentioned before. To go around the convergence problem we take (41) as an asymptotic expansion. (It is proposed in [**61**] to use rigid analytic geometry to go around convergence problem in a related context. To use asymptotic expansion is similar to their proposal.) To state our conjecture, we also need to take smoothening of distribution valued forms. Let us discuss those points below.

We consider a family of smoothening of the distribution valued 1-form $\text{ev}_{\text{last}\,*}\,\mathcal{M}_{\text{Morse}}(M;g)$ and denote it by
$$\mathfrak{M}_\delta\left(\text{ev}_{\text{last}\,*}\,\mathcal{M}_{\text{Morse}}(M;g)\right) \in \Gamma(\tilde{B}_0, T\tilde{B}_0 \otimes \Lambda^1 \tilde{B}_0).$$
Here, we say that $\mathfrak{M}_\delta(T)$ is a smoothening of a distribution T if the support of $\mathfrak{M}_\delta(T)$ is in the δ neighborhood of T, if $\lim_{\delta \to 0} \mathfrak{M}_\delta(T) = T$ as distribution, and if $\mathfrak{M}_\delta(T)$ is smooth.

We next take a large positive number λ. We consider the set $\tilde{B}_{0,\lambda}$ of all $(x,[u])$ such that $f(x,[u]) < \lambda$. The restrictions to $\tilde{B}_{0,\lambda}$ of the distribution valued section $\text{ev}_{\text{last}\,*}\,\mathcal{M}_{\text{Morse}}(M;g)$ and of its smoothening $\mathfrak{M}_\delta\left(\text{ev}_{\text{last}\,*}\,\mathcal{M}_{\text{Morse}}(M;g)\right)$, have compact support, by a Morse theory analogue of Gromov compactness (which is easy to prove). We denote the restriction of $\mathfrak{M}_\delta\left(\text{ev}_{\text{last}\,*}\,\mathcal{M}_{\text{Morse}}(M;g)\right)$ to $\tilde{B}_{0,\lambda}$ by the symbol
$\mathfrak{M}_{\delta,\lambda}\left(\text{ev}_{\text{last}\,*}\,\mathcal{M}_{\text{Morse}}(M;g)\right)$.

Now, instead of (41), we consider
$$\mathfrak{B}(\epsilon,\delta,\lambda) = \sum_{\gamma \in \pi_1(F_x)} \sum_{[u] \in \pi_2(M,F_x), \partial[u]=\gamma} e^{-2\pi f_u/\epsilon} \exp \gamma^*$$
(42)
$$\partial_{[\gamma],\epsilon} \otimes \mathfrak{M}_{\delta,\lambda}\left(\text{ev}_{\text{last}\,*}\,\mathcal{M}_{\text{Morse}}(M;g)\right)|_{U_{(x,[u])}}$$
$$\in \Gamma(\pi_{M_0^\vee}^{-1}(U_x), T_\mathbb{C} M_0^\vee \otimes \Lambda^{0,1} M_0^\vee).$$

We remark that (42) is a rigorously defined smooth section if we specify the choice of the smoothening $\mathfrak{M}_\delta$ and the multiplicity we put for multiple covered disks. We also remark that (42) is patched together and gives a section on $M_0^\vee$.

Now, we are ready to state the conjecture on the asymptotic behavior of the complex structure of the mirror.

CONJECTURE 5.3. *Let M be a K3 surface with Lagrangian torus fibration and $M^\vee$ be its mirror. Then, for sufficiently small ϵ,δ and sufficiently large λ, there exists a section $\tilde{\mathfrak{B}}(\epsilon,\delta,\lambda) \in \Gamma(M_0^\vee, T_\mathbb{C} M_0^\vee \otimes \Lambda^{0,1} M_0^\vee)$, the choice of the smoothening $\mathfrak{M}_\delta$ and of the multiplicity of multiple covered disks. They have the following properties.*

(1) *$\tilde{\mathfrak{B}}(\epsilon,\delta,\lambda)$ is a solution of*
$$\overline{\partial}_{0,\epsilon}\tilde{\mathfrak{B}}(\epsilon,\delta,\lambda) + \frac{1}{2}[\tilde{\mathfrak{B}}(\epsilon,\delta,\lambda), \tilde{\mathfrak{B}}(\epsilon,\delta,\lambda)] = 0.$$

In other words, $\overline{\partial}_{0,\epsilon} + \tilde{\mathfrak{B}}(\epsilon,\delta,\lambda) \circ \partial_{0,\epsilon}$ is an (integrable) complex structure on $M_0^\vee$.

(2) *The complex structure $\overline{\partial}_{0,\epsilon} + \tilde{\mathfrak{B}}(\epsilon,\delta,\lambda) \circ \partial_{0,\epsilon}$ extends to $M^\vee$.*

(3) *There exists $C(\lambda)$ independent of ϵ such that*
$$\|\tilde{\mathfrak{B}}(\epsilon,\delta,\lambda) - \mathfrak{B}(\epsilon,\delta,\lambda)\| < C(\lambda)e^{-\lambda/\epsilon},$$
where $\mathfrak{B}(\epsilon,\delta,\lambda)$ is as in (42).

Compared to the discussion of §2, the following two points are missing in the discussion of this section.

(1) The nonlinear term of (28) is the mixture of the bracket $[\,,\,]$ and wedge product. On the other hand, we consider the usual wedge product of differential forms in Conjecture 5.2. Hence, we need to discuss interaction or scattering of equation

(28) which is different from the one in the last section. We conjecture that it coincides with closed string - closed string scattering we described in §2.3.

(2) In the case of our multivalued function f, its critical point set is in $S(B)$. In other words, $S(B)$ plays the role of $Cr(f)$ in the last section. So, we need to study the behavior of $\mathfrak{B}$ in a neighborhood of singular locus and compare it to the unstable manifold of f (which we discussed in §2.4), in order to study Conjecture 5.3. The behavior of $\mathfrak{B}$ in a neighborhood of the singular fiber should be determined by (2) of Conjecture 5.3. In the case when the singular locus is of type I, Ooguri-Vafa [**68**] found a complex structure (in fact they found a hyperkähler metric) on a neighborhood of the singular locus. They found that it is related to the counting of holomorphic disks. (They count the multiplicity of dth multiple covered disk as d^{-2}.)

The author could not resolve these issues at the time of writing this article. So, in the next section, we will study a slightly simpler case, that is the case of a holomorphic structure on the mirror bundle and will provide a more detailed treatment of the above points, in their open string version.

REMARK 5.3. Our conjecture 5.3 is on the complex structure. The symplectic structure of $M^{\vee}$ is not included in the story. One may try to go further and to formulate a conjecture on the asymptotic behavior of Calabi-Yau metric in a similar way. The trouble to do so is that the Einstein-Kähler equation is more complicated than the equation $\overline{\partial} \circ \overline{\partial} = 0$. Namely, we need one more equation to solve. The equation $\overline{\partial} \circ \overline{\partial} = 0$ is not elliptic hence its solution is far from being unique. We need gauge fixing to have an elliptic equation. Requiring Einstein-Kähler condition is one (the ideal) way to do so. The author does not know yet appropriate Feynman rule for the Einstein-Kähler equation. He does not know either how it is related to the holomorphic sphere in the mirror.

REMARK 5.4. In [**35**], the author discussed the quantum correction to the complex structure in a different way. Namely, we study the family of Floer homologies and the quantum correction to the complex structure of the D-Brane moduli space (the moduli space of the pair of Lagrangian submanifold and a flat $U(1)$ bundle on it) is described so that the family of Floer homology will become a holomorphic family. Then, the quantum effect is related to the family version of the obstruction theory of well definedness of Floer homology in [**37**].

The two points of view (one of [**35**] and one of this article) actually coincide with each other. As we discussed in [**35**], the family of Floer homology will not give a holomorphic family with respect to the complex structure $J_{0,\epsilon}$ because of the wall crossing of Floer homology. On the other hand, Conjecture 3.2 implies that wall crossing occurs exactly on $\mathrm{ev}_{\mathrm{last}\,*}\,\mathcal{M}_{\mathrm{Morse}}(M;g)$.

REMARK 5.5. It seems likely that the discussion of this section can be generalized to the case of higher loops (namely, we including graphs which not trees). Then, if everything works ideally, we might prove the coincidence of the Quantum Kodaira Spencer theory [**13**] with the Gromov-Witten invariant of higher genus.

In the case when M is the cotangent bundle and the open string version, this possibility was suggested already in [**27, 36**]. Namely, it was conjectured that higher genus Morse homotopy coincides with the invariant of Chern-Simons Perturbation theory [**10, 11**].

If we continue our heuristic argument, we can give an idea which might imply the coincidence of Gromov-Witten potential of M with Yukawa coupling of $M^\vee$. We do not try to do so in this article since the author is unable to realize this idea yet. We present its open string analogue in the next section. In that case, we can state at least a conjecture in a precise way.

6. Asymptotic analysis on the mirror bundle

6.1. Semiflat homological mirror symmetry: review.

In this subsection, we study semiflat mirror symmetry in the case when there is a Lagrangian submanifold. The construction of this subsection is not new. It first appeared in 1998 July in [32] and was used by [5]. Several other authors found the same construction independently, see for example [3, 15, 61, 64]. The content of this subsection can be regarded as a real analogue of Fourier-Mukai transformation. (In other words, the *classical part* of homological mirror symmetry is a Fourier-Mukai transformation.)

Let $\pi_{M_0} : M_0 \to B_0$ be as in §1. Let M satisfy Assumption 3.1. We consider a Lagrangian submanifold L of M. Let $S(L) \subset B$ be its caustics. We put $B_{00} = B_0 \backslash S(L)$. We assume the following for simplicity:

ASSUMPTION 6.1. (1) $\dim B - \dim S(L) \geq 2$;
(2) The intersections between F_x and L are positive for $x \in B_{00}$ (i.e., each of them contributes $+1$ to the intersection number).

We remark that Assumption 6.1 (1) does *not* hold for generic Lagrangian submanifold L since the generic Lagrangian singularity of a codimension-one that is a fold. However, Assumption 6.1 holds for generic special Lagrangian submanifold if $\dim_{\mathbb{C}} M = 2$. We remark that Assumption 6.1 (2) is not generic either. It holds if M is hyperkähler and if L is obtained from a complex submanifold by hyperkähler twist. We may remove Assumption 6.1. We need some more discussions to do so, which we do not mention in this article. As for as the argument of this subsection concerns, we do not need to use Assumption 6.1.

We put $M_{00}^\vee = \pi_{M^\vee}^{-1}(B_{00}) \subset M_0^\vee$. We put the complex structure $J_{0,\epsilon}$ on it. We put

(43) $\qquad \tilde{B}_L^0 = \{(x,p) \mid x \in B_0, \; p \in F_x \cap L, \; F_x \text{ is transversal to } L \text{ at } p\}.$

Note $\tilde{B}_L^0$ is different from $\tilde{B}_L$ defined in Definition 4.2. We put $\tilde{B}_L^{00} = \tilde{B}_L^0 \cap \pi^{-1}(B_{00})$. By Assumption 6.1, the order of $F_x \cap L$ is independent of $x \in B_{00}$, hence $(x,p) \mapsto x$ defines a covering space $\pi_{\tilde{B}_L^{00}} : \tilde{B}_L^{00} \to B_{00}$. We put $r = \sharp(F_x \cap L)$, which will be the rank of the mirror bundle.

Before going further let us mention the plan of this section. We defined a family of complex structures J_ϵ on M_0 that is $J_\epsilon\left(\frac{\partial}{\partial x_i}\right) = \frac{1}{\epsilon}\frac{\partial}{\partial y_i^*}$. In this subsection, we are going to construct a holomorphic vector bundle $(\mathcal{E}(L)_0, \overline{\partial}_{0,\epsilon})$. The holomorphic structure $\overline{\partial}_{0,\epsilon}$ we define in this subsection is a "classical" one. Namely, we do not include an instanton effect which is caused by pseudoholomorphic disks. Actually, we need to include instanton effect to obtain the mirror bundle. (We will discuss it in §5.2 ∼ §5.5. Also, we need a twist by a local system $\rho : \pi_1(\tilde{B}_L^0) \to \{\pm 1\}$. ρ is expected to be related to orientation problem of the moduli space of pseudo holomorphic disks. (See [37] §25.6 for family version of orientation problem of pseudoholomorphic disks.) We call it orientation twist (see §5.5 and 5.6 for its definition). It modifies $\mathcal{E}(L)_0$ to $\mathcal{E}(L)_0^\rho$. Holomorphic structure $\overline{\partial}_{0,\epsilon}$ induces one on $\mathcal{E}(L)_0^\rho$ which we denote by the same symbol. The mirror bundle is $(\mathcal{E}(L)_0^\rho, \overline{\partial}_{0,\epsilon} + \mathfrak{B})$, where $\mathfrak{B}$ is a section of $End(\mathcal{E}(L)_0^\rho) \otimes \Lambda^{0,1}$ satisfying the equation $\overline{\partial}_{0,\epsilon} \mathfrak{B} + \mathfrak{B} \wedge \mathfrak{B} = 0$.

Let us start the construction of mirror bundle $\mathcal{E}(L)_0^\rho$. We recall that there exists a flat $SL(n,\mathbb{Z})$ bundle $E \to M_0$ and its lattice $\Lambda \subset E$ such that $M_0 = E/\Lambda$. The mirror manifold is the dual fibration $M_0^\vee = E^*/\Lambda^\vee$ (see §1). We first define a bundle $\tilde{\mathcal{E}}(L)_0^\rho$ on E^*.

Let $\rho : \pi_1(\tilde{B}_L^0) \to \{\pm 1\}$ be a homomorphism which we will define later. Let $\mathbb{C}_\rho \to \tilde{B}_L^0$ be the flat $U(1)$ bundle associated to ρ. We define

$$(44) \qquad \tilde{\mathcal{E}}(L)_0^\rho = \pi_{E^*}^* (\pi_{\tilde{B}_L^{00}}!(\mathbb{C}_\rho)).$$

We next lift the $\Lambda^\vee$ action on E^* to $\tilde{\mathcal{E}}(L)_0^\rho$. Let $(x, y^*) \in E_x^*$ and $(x, y) \in \tilde{B}_L^0$. (Here, $x \in B_{00}$). Let $v \in (\mathbb{C}_\rho)_{\tilde{B}_L^0}$ (the fiber of the flat bundle $\mathbb{C}_\rho$ at $\tilde{B}_L'$). v determines an element of the fiber of $\tilde{\mathcal{E}}(L)^\rho$ at (x, y^*) which we write $[x, y^*; y; v]$. Let $(x, \alpha) \in \Lambda_x^\vee$. We remark that $\langle \alpha, y^* \rangle$ is well defined as an element of $\mathbb{R}/\mathbb{Z}$. Hence, $\exp(2\pi\sqrt{-1}\langle \alpha, y^* \rangle) \in U(1)$ is well defined.

DEFINITION 6.1. We put :

$$(45) \qquad \alpha \cdot [x, y^*; y; v] = \exp(2\pi\sqrt{-1}\langle \alpha, y^* \rangle)[x, y^* + \alpha; y; v].$$

We define $\mathcal{E}(L)_0^\rho \to M_0^\vee$ to be the quotient bundle of $\tilde{\mathcal{E}}(L)_0^\rho$ by this $\Lambda^\vee$ action.

Hereafter, we omit y^* from the notation $[x, y^*; y; v]$ and will write $[x; y; v]$.

We next define a holomorphic structure on $\mathcal{E}(L)_0^\rho$. For this purpose we first define a local holomorphic frame of $\tilde{\mathcal{E}}(L)_0^\rho$. Let $s_0 : B_0 \to M_0$ be the zero section. We put

$$(46) \qquad \tilde{B}_{L_0, L} = \{(x, p, [\ell]) \mid (x, p) \in \tilde{B}_L^{00}, [\ell] \in \pi_1(F_x; s_0(x), p)\}.$$

In other words, $[\ell]$ is the homotopy class of an arc in F_x joining $s_0(x) \in F_x \cap L_0$ with p. The map $(x, p, [\ell]) \mapsto x$ defines a covering map $\pi_{\tilde{B}_{L_0,L}} : \tilde{B}_{L_0,L} \to B_{00}$. There exists also another covering map $\pi_{\tilde{B}_{L_0,L}, \tilde{B}_L^{00}} : \tilde{B}_{L_0,L} \to \tilde{B}_L^{00}$.

Now, let U be a contractible open subset of B_{00}. We choose a section $\pi_{\tilde{B}_L^{00}}^{-1}(U) \to \pi_{\tilde{B}_{L_0,L}}^{-1}(U)$ to $\pi_{\tilde{B}_{L_0,L}, \tilde{B}_L^{00}}$. Such a section exists since U is contractible. We denote it by $(x, p) \mapsto (x, p, [\ell_U(x, p)])$. In other words, we choose a homotopy class of a path $\ell_U(x, p)$ joining $s_0(x)$ to $p \in F_x \cap L$.

We remark that $\pi_{\tilde{B}_L^{00}}^{-1}(U)$ is a disjoint union of r copies of U. In other words, there exist (continuous) sections $p_i : U \to \pi_{\tilde{B}_L^{00}}^{-1}(U)$, $i = 1, \cdots, r$ such that the union of their images is $\pi_{\tilde{B}_L'}^{-1}(U)$. We put $\ell_i(x) = \ell_U(x, p_i(x))$. We also fix a trivialization of $\mathbb{C}_\rho$ on $\pi_{\tilde{B}_L^{00}}^{-1}(U)$.

Now, let $(x, y^*) \in E_x$, $x \in U$. The basis of the fiber of $\tilde{\mathcal{E}}(L)_0^\rho$ is $[x; p_i(x); 1]$. (Here, 1 is a canonical basis of the fiber of $\tilde{\mathcal{E}}(L)_0^\rho$ at $(x, p_i(x))$ induced by its trivialization.) We lift $\ell_i(x)$ to a path $\tilde{\ell}_i(x)(t)$ in E_x such that $\tilde{\ell}_i(x)(0) = 0$. We put $\tilde{p}_i(x) = \tilde{\ell}_i(x)(1) \in E_x$. (Note, $\tilde{p}_i(x) \equiv p_i(x) \mod \Lambda_x$.) We put

$$(47) \qquad H_i(x, y^*) = \langle y^*, \tilde{p}_i(x) \rangle \in \mathbb{R}.$$

We next fix a base point x_U in U. Let $x \in U$. We joint x_U with x by a path m in U. We can assume that

$$\varphi_{i,x}(s,t) = \ell_i(m(s))(t) \in F_{m(s)}$$

is a smooth map $\varphi_{i,x} : [0,1]^2 \to M_0$. (We remark that the image of $\varphi_{i,x}$ is a disk which bounds $F_{x_U} \cup F_x \cup L \cup s_0(U)$ as in Figure 6.1.)

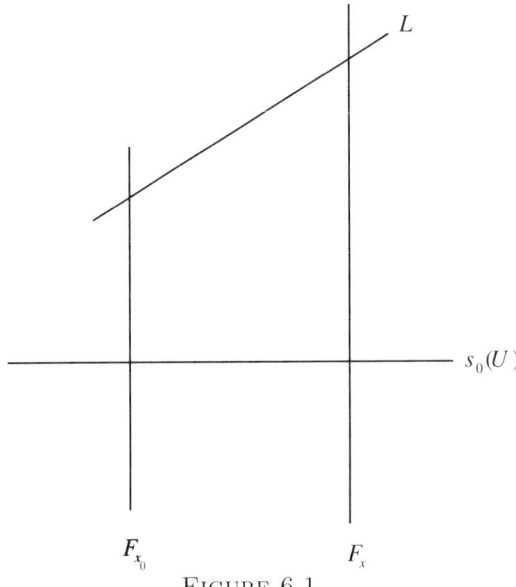

FIGURE 6.1.

We put

(48) $$Q_i(x) = \int_{[0,1]^2} \varphi_{i,x}^* \omega.$$

We now define local frame $s_{i,\epsilon}$ by

$$s_{i,\epsilon}(x, y^*) = e^{-2\pi\left(\frac{1}{\epsilon} Q_i(x) + \sqrt{-1} H_i(x, y^*)\right)} [x; p_i(x); 1].$$

Here, $[x; p_i(x); 1]$, $i = 1, \cdots, r$, is a basis of the fiber of $\tilde{\mathcal{E}}(L)_0^\rho$ at (x, y^*) we introduced before.

Using symplectic coordinates x^j we have $H_i = \sum_{j=1}^n \frac{\partial Q_i}{\partial x^j} y^{j*}$.

LEMMA 6.1. *There exists a unique holomorphic structure of the bundle $\mathcal{E}(L)_0^\rho$ on the complex manifold $(M_0^\vee, J_\epsilon)$ such that $s_{i,\epsilon}$ is holomorphic with respect to the induced holomorphic structure on $\tilde{\mathcal{E}}(L)_0^\rho$.*

PROOF. We need to show that the holomorphic structure of $\tilde{\mathcal{E}}(L)_0^\rho$ defined by $s_{i,\epsilon}$ can be patched on the intersection of different U and also to show that it is invariant of the Λ action.

To prove the first point we only need to show that the holomorphic structure on $\tilde{\mathcal{E}}(L)_0^\rho$ is independence of $\ell_i(x)$ and of the base point p_U. The later is immediate since by changing p_U the function Q_i changes by a constant and H_i does not change. To show the independence of $\ell_i(x)$ we consider alternative choice of ℓ_i say ℓ_i'. We write $\tilde{p}_i'(x)$, Q_i', H_i' the ones obtained from ℓ_i'. We remark that $p_i(x) - p_i'(x) =$

$q(x) \in \Lambda_x$. Since we are working in a small open set, we can take symplectic coordinates (x^j, y^j). We may then regard $q(x) \in \mathbb{R}^n$ (by using y-coordinate.) Now, it is easy to see from the definition that

$$Q_i(x) - Q_i'(x) = \langle x - x_U, q(x) \rangle$$

here the right-hand side is the standard inner product of $\mathbb{R}^n$. On the other hand, we have

$$H_i(x, y^*) - H_i'(x, y^*) = \langle y^*, q(x) \rangle.$$

We remark that $x^j + \sqrt{-1}\epsilon y^{j*}$ are the complex coordinates of $(M_0^\vee, J_\epsilon)$. Hence, $\frac{1}{\epsilon}(Q_i(x) - Q_i'(x)) + \sqrt{-1}(H_i(x, y^*) - H_i'(x, y^*))$ is a holomorphic function of (x, y^*).

The proof of Λ invariance is easy and is omitted. The proof of Lemma 6.1 is now complete. $\square$

We thus have defined a holomorphic structure $\overline{\partial}_{0,\epsilon}$ on $\tilde{\mathcal{E}}(L)_0^\rho$. In later sections, we use a section of $End(\mathcal{E}(L)_0^\rho) \otimes \Lambda^{0,1}$ to modify the holomorphic structure $\overline{\partial}_{0,\epsilon}$. So, let us describe a local holomorphic frame of $End(\mathcal{E}(L)_0^\rho)$.

Actually, later on we will need to consider a more general case of $Hom(\mathcal{E}(L_2)_0^{\rho_2}, \mathcal{E}(L_1)_0^{\rho_1})$, where L_1, L_2 are Lagrangian submanifolds. So, we discuss this case here. We put $B_{00} = B_0 \backslash S(L_1) \cup S(L_2)$ and consider the fiber product $\tilde{B}_{L_1}^{00} \times_{B_{00}} \tilde{B}_{L_2}^{00}$. Its elements are triples $(x; p_{1,i}(x), p_{2,j}(x))$ where $p_{a,i} : U \to \tilde{B}_{L_a}^{00}$, $i = 1, \cdots, r_a$, $a = 1, 2$ are sections as before. ($r_a = \sharp(F_a \cap L_a) = \text{rank}\,\mathcal{E}(L_a)_0^{\rho_a}$.) By using the trivialization of $\mathbb{C}_{\rho_a}$ we introduced before, a basis of a fiber of $Hom(\mathcal{E}(L_2)_0^{\rho_2}, \mathcal{E}(L_1)_0^{\rho_1})$ at (x, y^*) is $[x; p_{1,i}(x), p_{2,j}(x); 1]$ where $i = 1, \cdots, r_1$, $j = 1, \cdots, r_2$. Then, for each i, j we have a local holomorphic frame $s_{i,j;\epsilon}$ of $Hom(\mathcal{E}(L_2)_0^{\rho_2}, \mathcal{E}(L_1)_0^{\rho_1})$ defined by

$$(49) \qquad s_{i,j;\epsilon}(x, y^*) = e^{-2\pi\left(\frac{1}{\epsilon}Q_{i,j}(x) + \sqrt{-1}H_{i,j}(x,y^*)\right)}[x; p_{1,i}(x), p_{2,j}(x); 1],$$

Here, $Q_{i,j}$, $H_{i,j}$ are defined as follows. Let us consider $Q_{1,i}, H_{1,i}$ which is defined by using Lagrangian submanifold L_1 as before. We define $Q_{2,j}, H_{2,j}$ using L_2 in a similar way. We now put $Q_{i,j} = Q_{2,i} - Q_{1,j}$, $H_{i,j} = H_{2,i} - H_{1,j}$. Actually, $Q_{i,j}$ is a branch of the function f_{L_1, L_2} defined at the end of §3.1. In particular, in case $L_1 = L_2 = L$, we have

$$(50) \qquad Q_{i,j}(x) = f_L(x; p_i(x), p_j(x); [\gamma]) + \text{const.}$$

where the right-hand side is as in (13), $(x; p_i(x), p_j(x); [\gamma]) \in \tilde{B}_L^+$ and γ is chosen as follows. We recall that we took lifts $\tilde{p}_i(x), \tilde{p}_j(x) \in E_x$. We put $\tilde{\gamma}(t) = (1-t)\tilde{p}_i(x) + t\tilde{p}_j(x)$. It is a path in E_x and induces a path γ in F_x. (We write $[\gamma] \in \pi_1(F_x; p_i(x), p_j(x))$ in (50) since f_L depends only on $(x; p_i(x), p_j(x); [\gamma])$ and is independent of $[u]$ with $[\partial_+ u] = [\gamma]$ up to constant difference.) Formula (50) is immediate from the definition of $Q_{i,j}$ and f_L.

6.2. Free field: the case of vector bundle. We consider sections $\mathfrak{B} \in End(\mathcal{E}(L)_0^\rho) \otimes \Lambda^{0,k}$, $s \in \mathcal{E}(L)_0^\rho \otimes \Lambda^{0,k}$ and study linear equations

$$(51) \qquad \overline{\partial}_{0,\epsilon} \mathfrak{B} = 0, \qquad \overline{\partial}_{0,\epsilon} \mathfrak{s} = 0.$$

We recall that the condition that $\overline{\partial}_{0,\epsilon} + \mathfrak{B}$ defines a holomorphic structure on $\mathcal{E}(L)_0^\rho$ is

$$(52) \qquad \overline{\partial}_{0,\epsilon} \mathfrak{B} + \mathfrak{B} \wedge \mathfrak{B} = 0$$

($k = 1$) and $\mathfrak{s}$ is a holomorphic section of $\overline{\partial}_{0,\epsilon} + \mathfrak{B}$ is

(53) $$\overline{\partial}_{0,\epsilon}\mathfrak{s} + \mathfrak{B} \wedge \mathfrak{s} = 0.$$

Hence, the equations (51) are their linearizations. We are studying (51) by using Fourier expansion along fiber. (As we mentioned before this idea is similar to one in [**5**] and is also quite natural from the point of view of T-duality.) To describe Fourier expansion, we proceed as follows. Let $x \in B_{00}$. We consider $(x, y), (x, y_1), (x, y_2) \in F_x \cap L$ and let $(x, 0) = s_0(x)$. We then consider the relative homotopy groups $\pi_1(F_x; (x, y_1), (x, y_2))$ and $\pi_1(F_x; (x, 0), (x, y))$. For elements $[\ell] \in \pi_1(F_x; (x, y_1), (x, y_2))$,
$[\gamma] \in \pi_1(F_x; (x, 0), (x, y))$, and $(x, y^*) \in F_x^\vee$, we define

$$\langle [\ell], y^* \rangle \in \mathbb{R}/\mathbb{Z}, \quad \langle [\gamma], y^* \rangle \in \mathbb{R}/\mathbb{Z}$$

as follows. We lift the path ℓ to $\tilde{\ell} : [0, 1] \to E_x$. Then, $\tilde{\ell}(1) - \tilde{\ell}(0) \in E_x$ is independent of the choice of the lift. So, if we lift y^* to an element $\tilde{y}^*$ of E_x^* then we have $\langle \tilde{\ell}(1) - \tilde{\ell}(0), \tilde{y}^* \rangle \in \mathbb{R}$. This number is independent of the lift $\tilde{y}^*$ modulo $\mathbb{Z}$. Hence, we obtain $\langle [\ell], y^* \rangle \in \mathbb{R}/\mathbb{Z}$. The definition of $\langle [\gamma], y^* \rangle \in \mathbb{R}/\mathbb{Z}$ is similar.

We next identify $\pi^*_{M^\vee} \Lambda^k B \cong \Lambda^{0,k} M_0^\vee$ by sending dx^i to $d\bar{z}^i_\epsilon$. Here, x^i are symplectic coordinates of $M_0 \to B_0$ (hence, they are complex coordinate of $M_0^\vee \to B_0$), and $z^i_\epsilon = x^i + \epsilon\sqrt{-1} y^{i*}$. We write this isomorphism as $\mathfrak{I}_\epsilon : \pi^*_{M^\vee} \Lambda^k B \cong \Lambda^{0,k} M_0^\vee$.

Now, we expand

(54) $$\mathfrak{B}(x, y^*) = \sum_{p_1, p_2, \ell} e^{-2\pi\sqrt{-1}\langle [\ell], y^* \rangle} [x; p_1, p_2; 1] \otimes \mathfrak{I}_\epsilon(\mathfrak{B}_{p_1, p_2, \ell}(x)).$$

Here, we remark that $[x; p_1, p_2; 1]$ is a basis of the fiber of $End(\mathcal{E}(L)_0^\rho)$ at each (x, y^*). $\mathfrak{B}_{p_1, p_2, \ell}$ are 1 forms on B. (Then, $\mathfrak{I}_\epsilon(\mathfrak{B}_{p_1, p_2, \ell})$ is a T-structure invariant $(0, 1)$ form on M_0.)

In a similar way we have

(55) $$\mathfrak{s}(x, y^*) = \sum_{p, \gamma} e^{-2\pi\sqrt{-1}\langle [\gamma], y^* \rangle} [x; p; 1] \otimes \mathfrak{I}_\epsilon(\mathfrak{s}_{p, \gamma}(x)).$$

Now, the following lemma is obtained in a way similar to Proposition 5.1 and the definition of the holomorphic structure in the last subsection. To state it we recall some notations. Let $[\gamma] \in \pi_1(F_x; p_1, p_2)$. Then, $(x; p_1, p_2; [\gamma])$ is an element of $\tilde{B}_L^+$. Hence, $f_L(x; p_1, p_2; [\ell])$ is well defined up to constant by Lemma 4.1.

LEMMA 6.2. *The first equation of* (51) *is equivalent to*

(56) $$d(e^{\frac{2\pi}{\epsilon} f_L(x; p_1, p_2; [\gamma])} \mathfrak{B}_{p_1, p_2, \ell}(x)) = 0.$$

Let us denote by $L_{\rm st}$ the image of $s_0 : B \to M$, the zero section. (It will correspond to the structure sheaf of the mirror manifold, see [**32**].)

We remark that

(57) $$f_{L_{\rm st}, L}(x, p_i(x); [\gamma]) = Q_i(x) + {\rm const}$$

where Q_i is as in the last subsection, and $[\gamma]$ is chosen as follows. We define a lift $\tilde{p}_i(x) \in E_x$ in the last subsection. Then, $\gamma(t) \equiv t\tilde{p}_i(x) \mod \Lambda_x$.

LEMMA 6.3. *The second equation of* (51) *is equivalent to*

(58) $$d(e^{\frac{2\pi}{\epsilon} f_{L_{\rm st}, L}(x; p; [\gamma])} \mathfrak{s}_{p, \gamma}(x)) = 0.$$

Let us now explain how the solutions of (56), (58) are related to the Morse theory of our multivalued functions f_L, $f_{L_{\text{st}},L}$.

Here we discuss the case of $\mathfrak{B}$. We first remark that we may regard $\mathfrak{B}_{p_1,p_2,\ell}$ as a differential form on $\tilde{B}_L^+$ rather than one on B. In other words, we consider a map

$$\tilde{\mathfrak{I}}_\epsilon : \Gamma_c(\tilde{B}_L^+; \Lambda^k) \to \Gamma(M_0^\vee; End(\mathcal{E}(L)_0^\rho) \otimes \Lambda^{0,k})$$

as follows. (Here, Γ_c denotes the space of sections of compact support.) Let $\mathfrak{A} \in \Gamma_c(\tilde{B}_L^+; \Lambda^k)$ and $x \in B_0$. There exist finitely many $(p,q;[\gamma])$ in $\pi_{\tilde{B}_L^+}^{-1}(x)$ such that $\mathfrak{A}(x; p_1, p_2; [\gamma])$ is nonzero. We identify $\Lambda_{x;p,q;[\gamma]}^k \tilde{B}_L^+ \cong \Lambda_x^k B$. We then put

$$(59) \quad \tilde{\mathfrak{I}}_\epsilon(\mathfrak{A})(x, y^*) = \sum_{p_1, p_2, [\gamma]} e^{-2\pi\sqrt{-1}\langle [\ell], y^* \rangle} [x; p_1, p_2; 1] \otimes \mathfrak{I}_\epsilon(\mathfrak{A}(x; p_1, p_2; [\gamma])).$$

(59) is a generalization of (54). We have an obvious analogue of Lemma 6.2

We recall that the function f_L is well defined on $\tilde{B}_L^+$ up to constant difference. Hence, the gradient vector field $\text{grad} f_L$ on $\tilde{B}_L^+$ is well defined. Here we used the Riemannian metric on $\tilde{B}_L^+$ induced by one on B_0. We remark that this Riemannian metric is not complete since at the caustics the map $\tilde{B}_L^+ \to B_{00}$ is not a covering map. Let $\mathfrak{C}\tilde{B}_L^+$ be the completion of $\tilde{B}_L^+$. Note however the vector field $\text{grad} f_L$ is complete, since the caustics coincide with the set of critical points of f_L.

Let $\tau \mapsto e^{\tau \text{grad} f_L}$ be the one parameter group of transformations on $\tilde{B}_L^+$ generated by $\text{grad} f_L$. For $\mathfrak{A} \in \Gamma_c(\tilde{B}_L^+; \Lambda^k)$ we put

$$\psi_\tau \mathfrak{A} = \left(e^{-\tau \text{grad} f_L}\right)^* \mathfrak{A},$$

$$G_{\tau_1, \tau_2} \mathfrak{A} = \int_{\tau_1}^{\tau_2} i_{\text{grad} f_L}(\psi_\tau \mathfrak{A}) d\tau$$

LEMMA 6.4. *We have*

$$d \circ G_{\tau_1, \tau_2} + G_{\tau_1, \tau_2} \circ d = \psi_{\tau_2} - \psi_{\tau_1}.$$

PROOF. By $d \circ i_X + i_X \circ d = L_X$, $L_{\text{grad} f_L}(\psi_\tau \mathfrak{A}) = \frac{d}{d\tau}(\psi_\tau \mathfrak{A})$, we have

$$(d \circ G_{\tau_1, \tau_2} + G_{\tau_1, \tau_2} \circ d)(\mathfrak{A}) = \int_{\tau_1}^{\tau_2} \frac{d}{d\tau}(\psi_\tau \mathfrak{A}) d\tau = (\psi_{\tau_2} - \psi_{\tau_1})(\mathfrak{A}).$$

$\square$

We set

$$U(f_L) = \{(x; p_1, p_2; [\gamma]) \in \tilde{B}_L^+ \mid \lim_{\tau \to \infty} e^{\tau \text{grad} f_L}(x; p_1, p_2; [\gamma])$$
$$\text{is bounded in } \mathfrak{C}\tilde{B}_L^+\}.$$

Let $\mathbf{x} = (x; p_1, p_2; [\gamma]) \in \tilde{B}_L^+$. We put $\ell(\tau) = e^{\tau \text{grad} f_L}(\mathbf{x})$ and

$$(60) \quad f_{L,\tau_0}(\mathbf{x}) = \int_0^{\tau_0} \ell^* df_L(e^{\tau \text{grad} f_L}(\mathbf{x})) d\tau,$$

$$f'_{L,\tau_0}(\mathbf{x}) = f_{L,\tau_0}(e^{-\tau_0 \text{grad} f_L} \mathbf{x}).$$

We remark that f_L is well defined only up to constant. But the right-hand side is well defined. Hence, we define the left-hand side by (60). The following lemma is easy to see.

LEMMA 6.5. *If* $\mathbf{x} \notin U(f_L)$ *then* $\lim_{\tau \to +\infty} f_{L,\tau}(\mathbf{x}) = +\infty$.

We next put

$$(61) \qquad \mathfrak{G}_{\tau_1,\tau_2}\mathfrak{A} = \int_{\tau_1}^{\tau_2} \tilde{\mathfrak{I}}_\epsilon \left(e^{-2\pi f'_{L,\tau}} i_{\mathrm{grad} f_L}(\psi_\tau \mathfrak{A}) \right) d\tau,$$

and

$$(62) \qquad \Psi_\tau \mathfrak{A} = \tilde{\mathfrak{I}}_\epsilon(e^{-2\pi f'_{L,\tau}} \psi_\tau \mathfrak{A}).$$

Note $\Psi_0 \mathfrak{A} = \tilde{\mathfrak{I}}_\epsilon(\mathfrak{A})$. Combining Lemmata 6.2, 6.4, we have

$$(63) \qquad \overline{\partial} \circ \mathfrak{G}_{\tau_1,\tau_2} + \mathfrak{G}_{\tau_1,\tau_2} \circ d = \Psi_{\tau_2} - \Psi_{\tau_1}.$$

Lemma 6.5 implies that

$$\lim_{\tau \to +\infty} \Psi_\tau \mathfrak{A} = 0$$

if the support of $\mathfrak{A}$ is disjoint from $U(f_L)$. This implies that $\lim_{\tau \to +\infty} \mathfrak{G}_{\tau,0}$ plays the role of a propagator. Namely, we have the following:

PROPOSITION 6.1. *If* $d\mathfrak{A} = 0$ *and the support of* $\mathfrak{A}$ *is disjoint with* $U(L)$, *then* $\lim_{\tau \to +\infty} \mathfrak{G}_{\tau,0}(\mathfrak{A})$ *converges and we have*

$$-\overline{\partial} \left(\lim_{\tau \to +\infty} \mathfrak{G}_{\tau,0}(\mathfrak{A}) \right) = \tilde{\mathfrak{I}}_\epsilon(\mathfrak{A}).$$

Moreover, we can choose a generator of the $\overline{\partial}$ cohomology so that its support is near $U(f_L)$. This is related to Witten's argument we discussed in §4.2.

Proposition 6.1 implies that we can take a solution of (51) so that its Fourier component propagates along the gradient vector field of f_L. We can prove a similar proposition for $\mathfrak{s}$.

6.3. Interaction. In §5.2, we studied the linearized equation (51). As we discussed in §5.2 we are studying the solution which propagates along the gradient vector field of f_L. The effect of the nonlinear term appears when two of such gradient lines intersect each other (in B). The problem is thus local in B. Hence, we consider the following case. $M = \mathbb{R}^n \times T^n$ and L is a disjoint union of Lagrangian submanifolds obtained by $\mathbb{R}^n \times \{p_i\}$, $i = 1, \cdots, k$. We take a symplectic coordinate x^i, y^i of M. We take the metric $\sum dx^i \otimes dx^i$ on $B = \mathbb{R}^n$. The mirror manifold then is $M^\vee = \mathbb{R}^n \times T^n$ whose coordinates are (x^i) and (y^{i*}) with complex structure $J_\epsilon(\frac{\partial}{\partial x^i}) = \frac{1}{\epsilon} \frac{\partial}{\partial y^{i*}}$. The mirror bundle $\mathcal{E}_0$ is a trivial bundle with frame $[x; p_i]$. A holomorphic frame is

$$(64) \qquad e_{i;\epsilon}(x, y^*) = e^{-2\pi(\frac{1}{\epsilon}\langle p_i, x\rangle + \sqrt{-1}\langle p_i, y^*\rangle)}[x; p_i].$$

Here, $p_i, x, y^* \in \mathbb{R}^n$ and $\langle \cdot, \cdot \rangle$ is the standard inner product of $\mathbb{R}^n$.

We first explain the simplest case to illustrate the problem and then proceed to more general case. We first consider the case $k = 3$, $n = 2$. Let $w_{i,j} \in \mathbb{R}^2_{>0}$. We assume $w_{i,j} \equiv p_j - p_i \mod \mathbb{Z}^2$. Let $\Pi_{i,j}$ be the projection

$$(65) \qquad \Pi_{i,j}(x) = x - \frac{\langle w_{i,j}, x \rangle}{\langle w_{i,j}, w_{i,j} \rangle} w_{i,j}$$

to the plain orthogonal to $w_{i,j}$. We identify $\Pi_{i,j}(\mathbb{R}^2) \cong \mathbb{R}$. Let $\mathfrak{b}_\delta$ be a 1 form on $\mathbb{R}$ such that the support of $\mathfrak{b}_\delta$ is on $[-\delta, \delta]$ and that $\int \mathfrak{b}_\delta = 1$. (In other words $\mathfrak{b}_\delta$ is a smooth approximation of delta current.) We consider

$$\mathfrak{B}_{i,j;\epsilon;\delta} = e^{-2\pi(c_0 + \frac{1}{\epsilon}\langle w_{i,j}, x\rangle + \sqrt{-1}\langle w_{i,j}, y^*\rangle)}$$

$$\mathfrak{I}_\epsilon(\Pi_{i,j}^*(\mathfrak{b}_\delta)) \otimes [x; p_j, p_i] \in \Gamma(\mathbb{R}^2 \times T^2; End(\mathcal{E}_0) \otimes \Lambda^{0,1}).$$

Here, c_0 is a positive constant. Here, and hereafter, we put $[x; p_j, p_i] = [x; p_j] \otimes [x; p_i]^*$.

It follows from the construction of the last subsection that $\overline{\partial}_{0,\epsilon} \mathfrak{B}_{i,j;\epsilon;\delta} = 0$. Moreover, since $\mathfrak{B}_{i,j;\epsilon;\delta} \wedge \mathfrak{B}_{i,j;\epsilon;\delta} = 0$ (for $i \neq j$) it follows that $\overline{\partial}_{0,\epsilon} + \mathfrak{B}_{i,j;\epsilon;\delta}$ is another holomorphic structure of $\mathcal{E}_0$. However, if we consider $\mathfrak{B}_{1,2;\epsilon;\delta} + \mathfrak{B}_{2,3;\epsilon;\delta}$, it no longer satisfies the Maurer-Cartan equation, since $\mathfrak{B}_{2,3;\epsilon;\delta} \wedge \mathfrak{B}_{1,2;\epsilon;\delta}$ is nonzero in a neighborhood of origin. We thus need to study the following scattering problem. We assume that $w_{1,2}$ and $w_{2,3}$ are linearly independent.

PROBLEM 6.1. Find an element $\mathfrak{B} \in \Gamma(\mathbb{R}^2 \times T^2; End(\mathcal{E}_0) \otimes \Lambda^{0,1})$ with the following properties.
(1) $\overline{\partial}_{0,\epsilon}\mathfrak{B} + \mathfrak{B} \wedge \mathfrak{B} = 0$.
(2) There exists $C > 0$ such that if $x^1 + x^2 < -C$ then $\mathfrak{B} = \mathfrak{B}_{1,2;\epsilon;\delta} + \mathfrak{B}_{2,3;\epsilon;\delta}$.

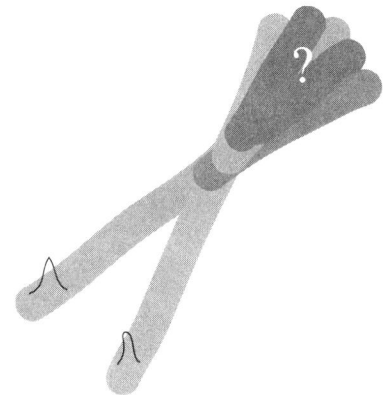

FIGURE 6.2.

To solve this problem it suffices to find $\mathfrak{B}'_{1,3} = B_{1,3}[x; p_3, p_1]$ such that

$$\overline{\partial}\mathfrak{B}'_{1,3} = \mathfrak{B}_{2,3;\epsilon;\delta} \wedge \mathfrak{B}_{1,2;\epsilon;\delta}$$

and that the support of $\mathfrak{B}'_{1,3}$ is away from $x^1 + x^2 < -C$. If we put $\mathfrak{B}'_{1,3} = B_{1,3}[x; p_3, p_1]$ then $\mathfrak{B}'_{1,3} \wedge \mathfrak{B}_{1,2;\epsilon;\delta}$, $\mathfrak{B}'_{1,3} \wedge \mathfrak{B}_{2,3;\epsilon;\delta}$, $\mathfrak{B}_{1,2;\epsilon;\delta} \wedge \mathfrak{B}'_{1,3}$, $\mathfrak{B}_{2,3;\epsilon;\delta} \wedge \mathfrak{B}'_{1,3}$ are all zero. This means that once two particles (corresponding to 1, 2 and 2, 3 components) interact, new particle (corresponding to 1, 3 component) are born. But, they never interact again. So, it suffices to find $\mathfrak{B}'_{1,3}$ in order to solve Problem 6.1. Such $\mathfrak{B}'_{1,3}$ can be obtained by applying Proposition 6.1. Let us describe the answer below.

To find $\mathfrak{B}'_{1,3}$ we consider the 2-form $\Pi^*_{2,3}(\mathfrak{b}_\delta) \wedge \Pi^*_{1,2}(\mathfrak{b}_\delta) = \xi(x^1, x^2)dx^1 \wedge dx^2$, and define

$$\rho(x) = \int_{t > 0} \xi(x - tw_{1,3})dt.$$

Then,
$$d(\rho(x)i_{w_{1,3}}(dx^1 \wedge dx^2)) = -\Pi_{2,3}^*(\mathfrak{b}_\delta) \wedge \Pi_{1,2}^*(\mathfrak{b}_\delta).$$
(Here, we regard $w_{1,3}$ as a vector field on $\mathbb{R}^2$.) Now, we put
$$B_{1,3} = e^{-2\pi(2c_0 + \frac{1}{\epsilon}\langle w_{1,3}, x\rangle + \sqrt{-1}\langle w_{1,3}, y^*\rangle)}\rho(x)\mathfrak{I}_\epsilon(i_{w_{1,3}}(dx^1 \wedge dx^2)).$$
Thus, we have :

PROPOSITION 6.2. $\mathfrak{B} = \mathfrak{B}_{1,2;\epsilon;\delta} + \mathfrak{B}_{2,3;\epsilon;\delta} + B_{1,3}[x; p_3, p_1]$ *satisfies* (1),(2) *of Problem* 6.1.

The next case to study is the case when $n = 2$ and $k = 2$ and we start with $\mathfrak{B}_{1,2}$ and $\mathfrak{B}_{2,1}$. (We recall $w_{1,2}, w_{2,1}$ are both in $\mathbb{R}^2_{>0}$.)

We define $\mathfrak{B}_{1,2;\epsilon;\delta}$ and $\mathfrak{B}_{2,1;\epsilon;\delta}$ in the same way. Now, our problem is:

PROBLEM 6.2. Find an element $\mathfrak{B} \in \Gamma(\mathbb{R}^2 \times T^2; End(\mathcal{E}_0) \otimes \Lambda^{0,1})$ with the following properties:
(1) $\overline{\partial}_{0,\epsilon}\mathfrak{B} + \mathfrak{B} \wedge \mathfrak{B} = 0$.
(2) There exists $C > 0$ such that if $x^1 + x^2 < -C$ then $\mathfrak{B} = \mathfrak{B}_{1,2;\epsilon;\delta} + \mathfrak{B}_{2,1;\epsilon;\delta}$.

This problem looks quite similar to Problem 6.1, but its answer is much more complicated. The reason is as follows. We first find $\mathfrak{B}'_{1,1}$ and $\mathfrak{B}'_{2,2}$ such that
$$\overline{\partial}_{0,\epsilon}\mathfrak{B}'_{1,1} = -\mathfrak{B}_{2,1;\epsilon;\delta} \wedge \mathfrak{B}_{1,2;\epsilon;\delta}$$
$$\overline{\partial}_{0,\epsilon}\mathfrak{B}'_{2,2} = -\mathfrak{B}_{1,2;\epsilon;\delta} \wedge \mathfrak{B}_{2,1;\epsilon;\delta}.$$
Contrary to the case of Proposition 6.1, $\mathfrak{B}_{2,1;\epsilon;\delta} \wedge \mathfrak{B}'_{1,1}$ etc. are again nonzero. Hence, we add more terms whose $\overline{\partial}$ boundaries are $\mathfrak{B}_{2,1;\epsilon;\delta} \wedge \mathfrak{B}'_{1,1}$ etc. In this way we need to add infinitely many terms.

REMARK 6.1. This phenomenon is the B-model analogue of the phenomenon we discussed in Remark 4.2.

To describe those infinitely many terms, we use Feynman diagram as follows. We consider a graph Γ which satisfies Condition 3.2. Let l be a metric on it. We endow it with the following additional structure. Let $I : |\Gamma| \to D^2$ be an embedding which preserves ribbon structure and such that $I^{-1}(\partial D^2)$ consists of the exterior vertices. Let $\alpha : \pi_0(D^2 \backslash \{I(|\Gamma|)\}) \to \{0, 1\}$ be a map satisfying the following condition.

For each $e \in \text{Edge}(\Gamma)$, let $\text{left}(e) \in \pi_0(D^2 \backslash I(|\Gamma|))$ [resp. $\text{right}(e) \in \pi_0(D^2 \backslash I(|\Gamma|))$] be the component which is in the left [resp. right] hand side of e. (Note e is oriented.) Let $v_1, \cdots, v_k$ be the set of all exterior vertices minus v_{last}. We assume that $v_{\text{last}}, v_1, \cdots, v_k$ respects the cyclic order. Let e_i be the unique edge such that $\partial_{\text{source}} e_i = v_i$. We assume that $\alpha(\text{left}(e_i)) \neq \alpha(\text{right}(e_i))$ for each i. (We remark that $\alpha(\text{left}(e)) \neq \alpha(\text{right}(e))$ may happen for interior edge e and for the last edge.)

We define a form $B_{(\Gamma, l, \alpha)}$ on $\mathbb{R}^2$ by induction on the number of vertices of Γ.

To start the induction, we consider the case of Γ_{-1} and Γ_0. Here Γ_{-1} is the tree without interior vertex and Γ_0 has three edges and one interior vertex. There are two choice of α for Γ_{-1}. The first one, $\alpha_{1,2}$, is such that $\alpha_{1,2}(\text{right}(e)) = 1$, $\alpha_{1,2}(\text{left}(e)) = 2$, where e is the unique edge of Γ_{-1}. Let $\alpha_{2,1}$ be the other choice of α. The metric l is trivial. We now put
$$B_{(\Gamma_{-1}, \cdot, \alpha_{1,2})} = \Pi_{1,2}^*(\mathfrak{b}_\delta)[x; p_2, p_1].$$

We define $B_{(\Gamma_{-1},\cdot,\alpha_{2,1})}$ in the same way by exchanging 1 and 2.

We next consider the case of Γ_0. There is only one edge e_{last} of Γ_0 with finite length (e_{last} is the edge which has v_{last} as a vertex). Let $l(e_{\text{last}})$ be its length. There are two choices of α. One of them, α_1, is such that $\alpha(D) = 1$ for two of the domains D. The other, α_2, is such that $\alpha(D) = 2$ for two of the domains D (see Figure 6.3 below).

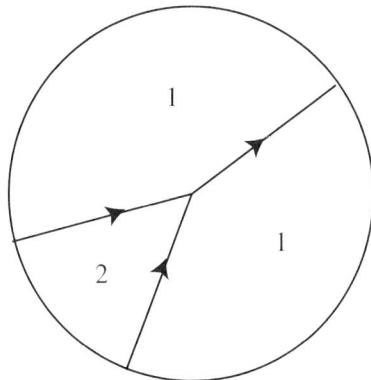

FIGURE 6.3.

We then put
$$B_{(\Gamma_0,l,\alpha_1)} = i_{w_{1,2}+w_{2,1}} \exp(-l(e_{\text{last}})(w_{1,2}+w_{2,1}))^* \\ B_{(\Gamma_{-1},\cdot,\alpha_{2,1})} \wedge B_{(\Gamma_{-1},\cdot,\alpha_{1,2})}.$$

Let us denote by $Met(\Gamma)$ the set of all metrics l on Γ satisfying Condition 3.3 (a). Clearly $Met(\Gamma_0) = (0,\infty)$. The following lemma can be proved in the same way as Lemma 6.4.

LEMMA 6.6.
$$d\left(\int_{l=0}^{\infty} B_{(\Gamma_0,l,\alpha_1)}\right) = -B_{(\Gamma_{-1},\cdot,\alpha_{2,1})} \wedge B_{(\Gamma_{-1},\cdot,\alpha_{1,2})}.$$

Now, we describe the inductive step of the construction of $B_{(\Gamma,l,\alpha)}$. We assume that $B_{(\Gamma,l,\alpha)}$ is constructed in the case when the number of interior vertices of Γ is smaller than k. Let us consider Γ with k interior vertices. We take the last vertex v_{last} and the last edge e_{last}. We remove it from Γ. Then, we get a union of two trees, which we call Γ_1 and Γ_2. (We use the ribbon structure at the vertex $\partial_{\text{source}} e_{\text{last}}$ to number Γ_1, Γ_2 so that $e_{\text{last}}, \Gamma_1, \Gamma_2$ respects the cyclic order.)

The map α for Γ induces maps α_1, α_2 for Γ_1, Γ_2 respectively. There exists an obvious map $\pi : Met(\Gamma) \to Met(\Gamma_1) \times Met(\Gamma_2)$, whose fiber is $(0,\infty)$, parametrizing the length of e_{last}. For $l_i \in Met(\Gamma_i)$ let $l = l(l_1, l_2; s)$ be the element of $Met(\Gamma)$ such that $\pi(l) = (l_1, l_2)$ and $l(e_{\text{last}}) = s$.

To define $B_{(\Gamma,l(l_1,l_2,s),\alpha)}$ we define $w(e)$ for each edge e. Let e,e' be edges of Γ. We say $e < e'$ if any path joining e to v_{last} intersects with e'. $<$ is a partial order. We define $w(e)$ by induction on this order.

If e is minimal then e is an exterior vertex. We put $a = \alpha(\text{right}(e))$, $b = \alpha(\text{left}(e))$ and $w(e) = w_{a,b}$. (Note $a \neq b$ by assumption. Hence, $w(e) = w_{1,2}$ or

$v_{2,1}$.) Suppose $w(e)$ is defined for all edges smaller than e. Let $\partial_{\text{source}} e = v$. The three edges which have v as a vertex are denoted by e, e_1, e_2. Obviously $e_1, e_2 < e$. Hence, $w(e_i)$ are defined. We put $w(e) = w(e_1) + w(e_2)$. We thus have defined $w(e)$.

We now put :

$$\tag{66} B_{(\Gamma, l(l_1, l_2, s), \alpha)} = i_{w(e_{\text{last}})} \exp(-s w(e_{\text{last}}))^* \left(B_{(\Gamma_1, l_1, \alpha_1)} \wedge B_{(\Gamma_2, l_2, \alpha_2)} \right).$$

The following lemma is easily proved in a way similar to the proof of Lemma 6.6.

LEMMA 6.7.
$$d \left(\int_{s=0}^{\infty} B_{(\Gamma, l(l_1, l_2, s), \alpha)} \right) = -B_{(\Gamma_1, l_1, \alpha_1)} \wedge B_{(\Gamma_2, l_2, \alpha_2)}.$$

Now, we consider the infinite sum

$$B = \sum_{\Gamma} e^{-c_0 k(\Gamma)} \int_{l \in Met(\Gamma)} B_{(\Gamma, l, \alpha)}$$

where $k(\Gamma) + 1$ is the number of exterior edges. (We put an obvious measure on $Met(\Gamma)$.) We can easily show that B converges if we take c_0 large.

Lemmas 6.6,6.7 imply the following :

PROPOSITION 6.3.
$$dB + B \wedge B = 0.$$

We now modify $B_{(\Gamma, l, \alpha)}$ to obtain $\mathfrak{B}_{(\Gamma, l, \alpha)}$ so that a statement similar to Proposition 6.3 holds for $\overline{\partial}$ in place of d. In order to do so, we define a function f similar to f_L. For this purpose, we first define $\exp_{(\Gamma, l, \alpha)} : \mathbb{R}^2 \times |(\Gamma, l)| \to \mathbb{R}^2$. (The definition below is the same as one used in [36] §16.) Here $|(\Gamma, l)|$ is as in §2. We define $\exp_{(\Gamma, l, \alpha)}$ on $\mathbb{R}^2 \times |e|$ for edges e by opposite induction with respect to the order $<$ on edges. The biggest edge with respect to $<$ is e_{last}. We identify $|e_{\text{last}}| = [-l(e_{\text{last}}), 0]$. We put

$$\exp_{(\Gamma, l, \alpha)}(x, \tau) = x + \tau w(e_{\text{last}}).$$

Let us assume that $\exp_{(\Gamma, l, \alpha)}$ is defined on $\mathbb{R}^2 \times |e|$ for e bigger than e_0. We put $\partial_{\text{target}} e_0 = v$. By the inductive hypothesis, $\exp_{(\Gamma, l, \alpha)}$ is defined on $\mathbb{R}^2 \times \{|v|\}$. We identify $|e_0|$ with $[-l(e_0), 0]$. (In case when $l(e_0) = \infty$ we identify $|e_0| = (-\infty, 0]$.) We now put

$$\exp_{(\Gamma, l, \alpha)}(x, \tau) = x + \tau w(e_0).$$

Now, let $v_1, \cdots, v_k, v_{\text{last}}$ be all the exterior edges of Γ. Let e_i be the exterior edges such that $\partial_{\text{source}} e_i = v_i$. We put $v_i' = \partial_{\text{target}} e_i$. Let $C_{1,f}(\Gamma)$ be the set of all edges of finite length. (Namely, the set of edges other than e_i.) We define $w(\alpha, i) = w_{a,b}$ where $a = \alpha(\text{right}(e))$, $b = \alpha(\text{left}(e))$. We put

$$f_{(\Gamma, l, \alpha)}(x) = \sum_i \langle w(\alpha, i), \exp_{(\Gamma, l, \alpha)}(x, |v_i'|) \rangle + \sum_{e \in C_{1,f}(\Gamma)} l(e) \|w(e)\|^2.$$

We now define

$$\mathfrak{B}_{(\Gamma,l,\alpha)} = e^{-2\pi(kc_0 + \frac{1}{\epsilon}f_{(\Gamma,l,\alpha)}(x) + \sqrt{-1}\langle y^*, w(e_{\text{last}})\rangle)} \mathfrak{I}_\epsilon(B_{(\Gamma,l,\alpha)}),$$

$$\mathfrak{B} = \sum_{\Gamma,\alpha} \int_{l \in Met(\Gamma)} \mathfrak{B}_{(\Gamma,l,\alpha)}.$$

$\mathfrak{B}$ converges if c_0 is large. The following theorem is a consequence of Proposition 6.3 and the argument of the last subsection.

THEOREM 6.1. $\mathfrak{B}$ *satisfies* (1), (2) *of Problem* 6.2.

We remark again that the procedure we have used to construct a solution of Problem 6.2 exactly corresponds to the one described in Remark 4.2.

In this subsection, we considered only two cases. They are generic in (real) dimension 4. The other cases in dimension 4 can be treated in a similar way. The difference between 4 dimensional case and higher dimensional case is as follows. Here we can start with the gauge field propagating along union of two lines. In higher dimension we need to study the gauge field propagating on submanifolds (foliated by lines). The discussion, however, is basically the same and can be organized by a similar Feynman diagram. We leave the precise study of it to the future research.

6.4. Monodromy around caustics and orientation twist.

We next study the Maurer-Cartan equation $\overline{\partial}_{0,\epsilon}\mathfrak{B} + \mathfrak{B} \wedge \mathfrak{B} = 0$ in a neighborhood of the caustics. Thus, our discussion is local. For the sake of preciseness, in this subsection we only consider the following example (though the author believes that the story can be generalized to the caustics of generic Lagrangian submanifolds of arbitrary dimension).

Let us consider $\mathbb{R}^4 = \mathbb{C}^2$. Let (x^1, x^2, y^1, y^2) be the coordinates such that $\omega = \sum dx^i \wedge dy^i$ and $J\left(\frac{\partial}{\partial x^i}\right) = \frac{\partial}{\partial y^i}$. Let L be a special Lagrangian submanifold in $\mathbb{R}^4$ which is tangent to the fiber $x = 0$. L is a complex submanifold with respect to the other complex structure $I\left(\frac{\partial}{\partial x^1}\right) = \frac{\partial}{\partial x^2}$, $I\left(\frac{\partial}{\partial y^1}\right) = \frac{\partial}{\partial y^2}$. We assume that the restriction of $\pi : (x, y) \mapsto x$ to L is a branched double covering and 0 is the branch point. Namely, with respect to the complex structure I, the restriction of π to L is holomorphic which is conjugate to $z \mapsto z^2$ at 0.

The mirror manifold is $\mathbb{R}^4$ with coordinates $(x^1, x^2, y^{1*}, y^{2*})$. The complex structure is $J_\epsilon\left(\frac{\partial}{\partial x^i}\right) = \frac{1}{\epsilon}\frac{\partial}{\partial y^{i*}}$.

The mirror bundle $\mathcal{E}_0$ without quantum correction is given as follows. For each $x \neq 0$, $F_x \cap L$ consists of two points $p_1(x) = (x, y_1(x))$, $p_2(x) = (x, y_2(x))$. (We remark however that $p_1(x)$ and $p_2(x)$ exchange when we go around origin. So, $p_i(x)$ is not globally well defined.) The fiber of $\mathcal{E}_0$ at (x, y^*) is of rank 2 and its basis consists of $[x; p_1(x)]$ and $[x; p_2(x)]$. However, this is not a holomorphic basis and we need to put some weight function to obtain a holomorphic basis (as in §5.1). Let us apply the definition of §5.1 in this case. There is a multivalued (2-valued) closed 1-form θ such that L is a graph is θ. We can find a 2-valued function f such $\theta = df$ and $f(0) = 0$. Here f is a function on the double cover of $\mathbb{R}^2 \setminus \{0\}$. If we regard it as a 2-valued function on $\mathbb{R}^2 \setminus \{0\}$, it changes the sign as we go a round the origin once. We have: $f(x) = \pm\frac{2}{3}\operatorname{Re}(x^1 + \sqrt{-1}x^2)^{3/2}$ (see Example 4.4).

If $p_i(x) = (x, df(x))$ then we put
$$Q_i(x) = \frac{1}{2}\int_0^1 df = \frac{1}{2}f(x), \quad H_i(x, y^*) = \langle df(x), y^*\rangle = y^{1*}\frac{\partial f}{\partial x^1} + y^{2*}\frac{\partial f}{\partial x^2}.$$

Then, $e^{2\pi(\frac{1}{\epsilon}Q_i(x) + \sqrt{-1}H_i(x,y^*))}[x; p_i(x)]$ is the holomorphic local frame we defined in §5.1. (Note that $Q_2 = -Q_1, H_2 = -H_1$ in our case.)

The holomorphic local frame of $End(\mathcal{E}_{0,\epsilon})$ is

(67)
$$\begin{aligned} e^{4\pi(\frac{1}{\epsilon}Q_i(x) + \sqrt{-1}H_i(x,y^*))}[x; p_j(x), p_i(x)] & \quad \text{if } i \neq j, \\ [x; p_i(x), p_i(x)] & \quad i = 1, 2. \end{aligned}$$

We next define the orientation twist $\rho : \pi_1(L\setminus\{(0,0)\}) \to \{\pm 1\}$. Note $\pi_1(L\setminus\{(0,0)\}) = \mathbb{Z}$. ρ in our case is the unique nontrivial homomorphism $\mathbb{Z} \to \{\pm 1\}$. The reason we need it becomes clearer later in this subsection.

We thus obtain a holomorphic vector bundle $\mathcal{E}_0^\rho$. To describe it precisely we proceed as follows. We cut $\mathbb{R}^2\setminus\{0\}$ by the half line $\{(x^1, x^2)|x^2 = 0, x^1 < 0\}$. We denote it by $\mathbb{R}^2_{\text{cut}}$. The two branches f_1, f_2 of f is well defined globally. We choose f_i so that $f_1 \geq 0$ for $x = (x^1, 0)$ with $x^1 \geq 0$.

On $\mathbb{R}^2_{\text{cut}} \times \mathbb{R}^2$, we define holomorphic vector bundle by taking the local holomorphic frame $e^{2\pi(\frac{1}{\epsilon}Q_i(x) + \sqrt{-1}H_i(x,y^*))}[x; p_i(x)]$. Let $(x^1, 0)$ be the point on the cut. We consider $(x^1, +0)$ [resp. $(x^1, -0)$] which are limit of (x^1, x^2) when x^2 goes to zero from above [resp. below]. We now put

(68)
$$\begin{aligned} [(x^1, +0); p_1(x)] &= -[(x^1, -0); p_2(x)], \\ [(x^1, +0); p_2(x)] &= [(x^1, -0); p_1(x)]. \end{aligned}$$

Gluing the trivial bundle by this isomorphism we obtain $\mathcal{E}_0^\rho$.

Now, an important observation is that neither $\mathcal{E}_0$ nor $\mathcal{E}_0^\rho$ can be extended to the fiber at origin $\{0\} \times \mathbb{R}^2$. This is because there is a nontrivial monodromy of our bundle. This phenomenon is "gauge theory analogue" of the phenomenon we met in §4, that is the T-structure invariant complex structure can not be extended to the singular fiber. The idea we explained in §4 (which in case of type I singular fiber is already established by [**68, 47**]) is that the instanton effect gives a "quantum correction" to the complex structure and then the complex structure with "quantum correction" can be extended to the singular fiber. The "quantum correction" in the case of complex structure of manifold itself is a family version of the obstruction class $\mathfrak{m}_0$ of the well-definedness of Floer homology of Lagrangian submanifold (see [**37, 35**]). Namely, it is obtained by using the moduli space of pseudoholomorphic disks whose boundary belong to the fiber F_x of the fibration $M_0 \to B_0$. In our case of the holomorphic structure of a vector bundle, the "quantum correction" is obtained by counting pseudoholomorphic strips, whose boundary belong to the two Lagrangian submanifolds L and the fiber F_x. We discussed the moduli space of such pseudoholomorphic disks in §3, especially in subsection 3.5. Conjecture 4.1 there asserts that this moduli space is described by the gradient lines of the function f_L, which is nothing but the function f we have used above. (In §3.5 we verified, in our case, half of the conjecture, namely Conjecture 4.2 was proven there.) In the rest of this subsection, we use the unstable manifold of f to modify the complex structure of $\mathcal{E}_0^\rho$ so that it can be extended to $\{0\} \times \mathbb{R}^2$.

REMARK 6.2. We remark that our situation, where the holomorphic bundle on $\mathbb{R}^2\setminus\{0\} \times T^2$ does not extend, is a bit unusual. Usually in complex geometry, we meet a situation where a coherent sheaf is singular on a complex subvariety. In our case, $\{0\} \times T^2$ is totally real. Hence, the bundle $\mathcal{E}_0^\rho$, $\mathcal{E}_0$ can not be extended even as a coherent sheaf, without including the quantum effect.

We first recall that the unstable manifold of our multivalued function f is a union of the three half lines

$$(0, \infty) \to \mathbb{R}^2, \quad s \mapsto (s, 0), (s \cos \pm 2\pi/3, s \sin \pm 2\pi/3).$$

Let $w_0 = (1, 0), w_\pm = (s \cos \pm 2\pi/3, s \sin \pm 2\pi/3)$ be their tangent vectors. We define $\Pi_0, \Pi_\pm : \mathbb{R}^2 \to \mathbb{R}$ using $w_0, w_\pm$ in the same way as (65). We take a sufficiently small δ and a 1-form $\mathfrak{b}_\delta$ whose support is in $[-\delta, \delta]$ and $\int \mathfrak{b}_\delta = 1$, as in §5.3. We now define

(69)
$$\begin{cases} B_0 = -\operatorname{Hev}(\langle x, w_0 \rangle) \Pi_0^*(\mathfrak{b}_\delta)[x; p_2(x), p_1(x)], \\ B_+ = \operatorname{Hev}(\langle x, w_+ \rangle) \Pi_+^*(\mathfrak{b}_\delta)[x; p_1(x), p_2(x)], \\ B_- = \operatorname{Hev}(\langle x, w_- \rangle) \Pi_-^*(\mathfrak{b}_\delta)[x; p_1(x), p_2(x)]. \end{cases}$$

Here, Hev is the Heaviside function.

We regard $B_0, B_\pm$ as smooth forms on $\mathbb{R}^2 \setminus D^2$ with coefficient in the bundle $End(\pi!\mathbb{C}_\rho)$, where $\pi : L \to \mathbb{R}^2$ is the projection and $\mathbb{C}_\rho$ is a flat line bundle on $L \setminus (\{0\} \times \mathbb{R}^2)$ with monodromy ρ (note that $\mathcal{E}_0^\rho$ is a pull back of $\pi!\mathbb{C}_\rho$). These forms are closed and their supports (restricted to $\mathbb{R}^2 \setminus D^2$) are disjoint. Hence, they define a flat connection on $\pi!\mathbb{C}_\rho$ which we write $d_{0,\rho} + B$. (We remark that the rank two bundle $\pi!\mathbb{C}_\rho$ has a flat connection $d_{0,\rho}$ induced by the flat connection on $\mathbb{C}_\rho$.) Now, we have:

LEMMA 6.8. *The bundle $\pi!\mathbb{C}_\rho$ on $\mathbb{R}^2 \setminus D^2$ equipped with the flat connection $d_{0,\rho} + B$ can be extended to a flat bundle on $\mathbb{R}^2$.*

PROOF. B_0, B_+, B_- gives the monodromy $\begin{pmatrix} 1 & -1 \\ 0 & 1 \end{pmatrix}, \begin{pmatrix} 1 & 0 \\ 1 & 1 \end{pmatrix}, \begin{pmatrix} 1 & 0 \\ 1 & 1 \end{pmatrix}$ respectively. We have

(70)
$$\begin{pmatrix} 1 & 0 \\ 1 & 1 \end{pmatrix} \begin{pmatrix} 1 & -1 \\ 0 & 1 \end{pmatrix} \begin{pmatrix} 1 & 0 \\ 1 & 1 \end{pmatrix} = \begin{pmatrix} 0 & -1 \\ 1 & 0 \end{pmatrix}$$

which cancels the monodromy (68). □

REMARK 6.3. The cancellation of the holonomy as in (70) looks rather accidental. However, if we use the fact that the gradient lines here coincide with the wall of the wall crossing phenomenon of the Floer homology as is discussed in [**37, 35**], then one finds that it should occur in general.

REMARK 6.4. The reason we need the orientation twist ρ and the minus sign in the definition of B_0 (69) can be seen from the proof of Lemma 6.8 above. However, this reason is rather computational. It seems likely that there is a better explanation based on the orientation problem of the pseudoholomorphic disks.

Now, we put weight to $B_0, B_\pm$ in a way similar to §5.2 and will prove a complex analogue of Lemma 6.8. Namely, we put

(71)
$$\begin{cases} \mathfrak{B}_{0,\epsilon} = -e^{4\pi(\frac{1}{\epsilon}Q_1(x)+\sqrt{-1}H_1(x))}\operatorname{Hev}(\langle x, w_0\rangle) \\ \qquad\qquad \mathfrak{I}_\epsilon \Pi_0^*(\mathfrak{b}_\delta)[x; p_2(x), p_1(x)], \\ \mathfrak{B}_{+,\epsilon} = e^{4\pi(\frac{1}{\epsilon}Q_2(x)+\sqrt{-1}H_2(x))}\operatorname{Hev}(\langle x, w_+\rangle) \\ \qquad\qquad \mathfrak{I}_\epsilon \Pi_+^*(\mathfrak{b}_\delta)[x; p_1(x), p_2(x)], \\ \mathfrak{B}_{-,\epsilon} = e^{4\pi(\frac{1}{\epsilon}Q_2(x)+\sqrt{-1}H_2(x))}\operatorname{Hev}(\langle x, w_-\rangle) \\ \qquad\qquad \mathfrak{I}_\epsilon \Pi_-^*(\mathfrak{b}_\delta)[x; p_1(x), p_2(x)]. \end{cases}$$

We then set $\mathfrak{B}_\epsilon = \mathfrak{B}_{0,\epsilon} + \mathfrak{B}_{+,\epsilon} + \mathfrak{B}_{-,\epsilon}$. We can easily show

(72) $$\overline{\partial}_{0,\epsilon}\mathfrak{B}_\epsilon + \mathfrak{B}_\epsilon \wedge \mathfrak{B}_\epsilon = 0.$$

Now, we have :

PROPOSITION 6.4. *The bundle $\mathcal{E}_{0,\rho,\epsilon}$ with holomorphic structure $\overline{\partial}_{0,\epsilon} + \mathfrak{B}_\epsilon$ can be extended to a holomorphic vector bundle on $\mathbb{R}^2 \times T^2$.*

PROOF. Let we take sections $\mathbf{s}_i(x)$ $(i=1,2)$ of $\pi!\mathbb{C}_\rho$ such that $(d+B)\mathbf{s}_i = 0$ and are linearly independent. (Lemma 6.8 implies the existence of such sections). We write them as

$$\mathbf{s}_i(x) = s_{i,1}(x)[x; p_1(x)] + s_{i,2}(x)[x; p_2(x)].$$

Then, we define $\mathbf{e}_i(x, y^*)$ by

$$\mathbf{e}_i(x, y^*) = e^{-2\pi(\frac{1}{\epsilon}Q_1(x)+\sqrt{-1}H_1(x,y^*))} s_{i,1}(x)[x; p_1(x)]$$
$$+ e^{-2\pi(\frac{1}{\epsilon}Q_2(x)+\sqrt{-1}H_2(x,y^*))} s_{i,2}(x)[x; p_2(x)].$$

By a straightforward calculation, we can check

$$(\overline{\partial}_{0,\epsilon} + \mathfrak{B}_\epsilon)\mathbf{e}_i = 0.$$

The proposition follows easily. □

6.5. Feynman diagram description of holomorphic structure of vector bundles. Using the results of subsections 5.2, 5.3, 5.4, we can define holomorphic structures on the mirror bundle (together with quantum effect) at least in the asymptotic sense. The result (which we describe in this subsection) is parallel to the description on the pseudoholomorphic disks we gave in §3.6.

We consider the following situation. Let us take $L \subseteq T^4$ a complex submanifold (Riemann surface) with respect to the complex structure I. It then becomes a (special) Lagrangian submanifold of T^4 (with respect to the symplectic structure ω which is a Kähler form of the complex structure J.) We consider Lagrangian fibration $T^4 \to T^2$. We restrict it to L. Then, with respect to the complex structure I, this map $L \to T^2$ is holomorphic. Hence, it is a branched covering. We assume that it is locally conjugate to $z \mapsto z^2$ at caustics. (This holds for generic L.)

REMARK 6.5. The author thanks to N. Leung who showed this example L to the author and asked what is its mirror bundle, after the author's talk at RIMS during the conference "Algebraic geometry and Integrable system" in 2000 June.

Let (x^1, x^2), (y^1, y^2) be the (symplectic) coordinates of T^4. Let T^4 be the mirror torus with complex structure J_ϵ (so that $z^i_\epsilon = x^i + \epsilon\sqrt{-1}y^{i*}$ are its complex coordinate).

Let $S(L) \subset T^2$ be the caustics. We put $B_0 = T^2 \backslash S(L)$. We put
$$L_0 = \{p \in L \mid d\pi : T_pL \to T_{\pi p}T^2 \text{ is zero.}\}$$
We define $\rho : \pi_1(L_0) \to \{\pm 1\}$ so that holonomy around each point in $L \backslash L_0$ is nontrivial.

Using these data, we defined (in subsection 5.1) a holomorphic vector bundle $\mathcal{E}(L)_0^\rho$ on $B_0 \times T^2$. We denote its holomorphic structure by $\overline{\partial}_{0,\epsilon}$. The main result of this section is the following Theorem 6.2. We consider the following assumption. We use the notation of §3.6 (Definition 4.9).

ASSUMPTION 6.2. λ is a positive number. The moduli space $\mathcal{M}_{\mathrm{Morse}}(M, L; \alpha)$ is empty if $\alpha \cap \omega \leq \lambda$.

THEOREM 6.2. *We assume Assumption 6.2, Then, for each ϵ, there exists a section $\mathfrak{B}_{\epsilon;\lambda} \in \Gamma(B_0 \times T^2; End(\mathcal{E}(L)_0^\rho) \otimes \Lambda^{0,1})$ with the following properties.*
(1) *There exists $C_{\lambda,k}$ depending on λ but is independent of ϵ such that*
$$\left\| \left(\overline{\partial}_{0,\epsilon} + \mathfrak{B}_{\epsilon;\lambda} \right)^2 \right\|_{C^k} \leq C_{\lambda,k} e^{-\lambda/\epsilon}.$$
Here, the left-hand side is the C^k norm.
(2) *The complex vector bundle $\mathcal{E}(L)_0^\rho$ together with holomorphic structure $\partial_{0,\epsilon} + \mathfrak{B}_{\epsilon;\lambda}$ extends to one on T^4.*

REMARK 6.6. If Conjecture 4.4 is correct, Assumption 6.2 is equivalent to the following : There are no pseudoholomorphic disks $\varphi : (D^2, \partial D^2) \to (T^4, L)$ such that $\int \varphi^* \omega \leq \lambda$.

Hence, Assumption 6.2 is related to the vanishing of the obstruction to define Floer homology of Lagrangian submanifolds, introduced in [**37**].

We remark that Assumption 6.2 is satisfied for generic L in our case when dimension is 4 and the Maslov index of L vanishes.

PROOF. We will prove more, namely, we will give an explicit description of $\mathfrak{B}_{\epsilon;\lambda}$ based on Feynman diagram.

We first take sufficiently small c (which depends on λ and will be specified later) and identify c neighborhood $B_c(q)$ of each point of the caustics q with a ball in $\mathbb{R}^2$ of large radius. The restriction of the bundle $\mathcal{E}(L)_0^\rho$ to $(B_c(q) \backslash \{q\}) \times T^2$ can be divided as a direct sum of the two bundles $\mathcal{E}(q)_1 \oplus \mathcal{E}(q)_2$. $\mathcal{E}(q)_1$ is the rank two bundle whose basis is identified to the two points in $F_p \cap L$ which are boundary points of the vanishing arc. We rescale the form $\mathcal{B}_\epsilon$ constructed in the last section and regard it as the section of $End(\mathcal{E}(q)_1) \otimes \Lambda^{0,1}$. We denote it by $\mathfrak{B}_\epsilon(q)$. Then, $\partial_{0,\epsilon} + \sum_q \mathfrak{B}_\epsilon(q)$ is a holomorphic structure defined in the union of $(B_c(q) \backslash \{q\}) \times T^2$ and it extends to the union of $B_c(q) \times T^2$.

We are going to extend it globally so that Theorem 6.2 (1) is satisfied.

To do so, we need to define a moduli space which is a kind of smoothening of one in §3.6. In §3.6 we considered the gradient lines which start exactly at caustics. We relax this condition a bit as follows.

Let $\ell_{i,q} : (-\infty, 0] \to B_c(q)$, $i = 1, 2, 3$ be the three gradient lines starting the caustics and corresponding to the vanishing cycles. There exist small transversals

$u_{i,q} : [-1,1] \to B_c(q)$ and a smooth section $\mathfrak{b} : [-1,1] \to u_{i,q}^* \Lambda^1$ of compact support, such that $\mathfrak{B}_\epsilon(q)$ in a neighborhood of the image $u_{i,q}$ is

$$
(73) \qquad e^{-4\pi(\frac{1}{\epsilon}f_L + \sqrt{-1}\sum_i \frac{\partial f_L}{\partial x^i} y^{*i})} \mathfrak{I}_\epsilon B
$$

where

$$
(74) \qquad B = \int_{-1}^1 dt \int_{-\delta}^\delta ds \exp(-s \operatorname{grad} f_L)^* \mathfrak{b}(t)
$$
$$
\otimes [u_{i,q}(v); p_b(u_{i,q}(v)), p_a(u_{i,q}(v))].
$$

Here, f_L is a branch of our multivalued function and $p_a(x), p_b(x)$ are two intersection points of $F_x \cap L$ which are two boundary points of the vanishing arc.

The existence of such $u_{i,q}$ and $\mathfrak{b}$ is immediate from the construction of the last subsection.

Now, we consider ribbon graph Γ satisfying Condition 4.1. Let l be a metric on it. We assume here that *all the edges are of finite length*. Let $\psi : |\Gamma| \to B$ be a continuous map and $\tilde\psi : |\Gamma|\backslash\{\text{vertices}\} \to \tilde B_L$ be its lift. We assume Condition 4.2 (1),(2). In place of Condition 4.2 (3), we assume the following.

CONDITION 6.1. Let $v \in \text{Vertex}_{\text{sta}}(\Gamma)$ and $\partial_{\text{source}}(e) = v$. We identify $|e| = [0, l(e)]$. Then, $\psi(0) = u_{i,q}(t)$ where $u_{i,q}$ is as above and $t \in [-1,1]$. We put $\tilde\psi(s) = (\psi(s); p(s), q(s); \beta(s))$ where $\beta(s) \in \pi_1(F_{\psi(s)}; p(s), q(s))$. Then, $\beta(0)$ is the vanishing arc at q.

DEFINITION 6.2. The set of all $(\Gamma, l, \psi, \tilde\psi)$ such that Conditions 4.2 and 4.4 (1)(2), 6.1 are satisfied and that l is finite, is denoted by $\mathcal{M}_c(L)$.

We define a function E on $\mathcal{M}_c(L)$ by

$$
(75) \qquad E(\Gamma, l, \psi, \tilde\psi) = \int_{\beta(l(e_{\text{last}}))} \omega.
$$

We set

$$
(76) \qquad ev(\Gamma, l, \psi, \tilde\psi) = \psi(v_{\text{last}}).
$$

We will define 1-form $\mathfrak{b}(\Gamma, l, \psi, \tilde\psi)$ at $ev(\Gamma, l, \psi, \tilde\psi)$ as follows. We define $\mathfrak{b}(v)$ for each vertex v by the induction on the partial order $<$. Here we say $v < v'$ if every arc joining v with v_{last} contains v'. First, for $v \in \text{Vertex}_{\text{sta}}(\Gamma)$ we have $\psi(0) = u_{i,q}(t)$. Then, we put $\mathfrak{b}(v) = \mathfrak{b}(t)$, where $\mathfrak{b}(t)$ is as in (74). Let us assume that $\mathfrak{b}(v)$ is defined for any v with $v < v_0$. Let us define $\mathfrak{b}(v_0)$. We consider two edges e_1, e_2 such that $\partial_{\text{target}} e_i = v_0$. Let $v_i = \partial_{\text{source}} e_i$. We now put

$$
(77) \qquad \mathfrak{b}(v_0) = i_{\operatorname{grad} f_0}(-\exp(l(e_2)\operatorname{grad} f_2)^* \mathfrak{b}(v_2)
$$
$$
\wedge \exp(-l(e_1)\operatorname{grad} f_1)^* \mathfrak{b}(v_1))
$$

Here, f_1, f_2, f_0 are three branches of f_L which correspond to the lift $\tilde\psi$ on e_1, e_2 and e_0 respectively. Here, e_0 is the edge such that $\partial_{\text{source}} e_0 = v_0$.

We then define

$$
\mathfrak{b}(\Gamma, l, \psi, \tilde\psi) = \mathfrak{b}(v_{\text{last}}).
$$

We put

$$
\mathcal{M}_c(L; \lambda) = \{(\Gamma, l, \psi, \tilde\psi) \in \mathcal{M}_c(L) | E(\Gamma, l, \psi, \tilde\psi) \leq \lambda\}
$$

The following is easy to prove:

LEMMA 6.9. *Under Assumption 6.2, we can choose the positive number c so small that $\mathcal{M}_c(L;\lambda)$ is compact.*

For $(\Gamma, l, \psi, \tilde{\psi}) \in \mathcal{M}_c(L)$, we can fix a branch of f_L at $ev(\Gamma, l, \psi, \tilde{\psi})$. We use this branch to define

$$e^{-2\pi(\frac{1}{\epsilon}f_L + \sqrt{-1}\sum_i \frac{\partial f_L}{\partial x^i} y^{*i})} \circ ev.$$

Hereafter, we omit ev for simplicity. The choice of branch determines a basis of $End(\mathcal{E}_{0,\epsilon}^\rho)$ also. We write it as $e(\Gamma, l, \psi, \tilde{\psi})$. We choose $\delta > 0$ small such that Assumption 6.2 holds for $\lambda + \delta$ in place of λ. Let $\chi : \mathbb{R} \to [0,1]$ be a smooth function such that $\chi = 0$ on $[\lambda + \delta, \infty)$ and $\chi = 1$ on $[0, \lambda]$. Now, we define :

DEFINITION 6.3.
$$\mathfrak{B}_{\epsilon,\lambda} = \int_{\mathcal{M}_c(L;\lambda+\delta)} \chi(E(\Gamma,l,\psi,\tilde{\psi})) e^{-2\pi(\frac{1}{\epsilon}f_L + \sqrt{-1}\sum_i \frac{\partial f_L}{\partial x^i} y^{*i})}$$
$$\mathfrak{I}_\epsilon(\mathfrak{b}(\Gamma,l,\psi,\tilde{\psi})) \otimes e(\Gamma,l,\psi,\tilde{\psi}).$$

Theorem 6.2 now follows from the constructions of the last three subsections. □

Theorem 6.2 gives only an asymptotic solution. It seems likely that we can use the Implicit Function Theorem to find an actual solution nearby. Since the author did not work it out at the time of writing this article, he states it as a conjecture.

CONJECTURE 6.1. *We assume Assumption 6.2 for all λ. Then, for each λ, there exists $\epsilon(\lambda, k)$ such that there exists $\mathfrak{B}'_{\epsilon,\lambda}$ for each $\epsilon < \epsilon(\lambda, k)$ with the following properties.*
(1) $\bar{\partial}_{0,\epsilon}\mathfrak{B}'_{\epsilon,\lambda} + \mathfrak{B}'_{\epsilon,\lambda} \wedge \mathfrak{B}'_{\epsilon,\lambda} = 0$.
(2) $\left\|\mathfrak{B}_{\epsilon;\lambda} - \mathfrak{B}'_{\epsilon;\lambda}\right\|_{C^k} \leq C_{\lambda,k} e^{-\lambda/\epsilon}$.

6.6. Homoloical mirror symmetry. In §5.5, we described a procedure to construct a mirror bundle from an unobstructed Lagrangian submanifold in the 4-dimensional case. This is a part of the Homological Mirror Symmetry Conjecture by Kontsevich [**58, 59**]. (See [**32, 33**] for precise formulation.) In this section, we explain how the other parts of the Homological Mirror Symmetry Conjecture may follow from a similar argument. We recall that the first part of the Homological Mirror Symmetry Conjecture is $HF(L_1, L_2) \cong \text{Ext}(L_1, L_2)$, where $HF(L_1, L_2)$ is the Floer cohomology and $\text{Ext}(L_1, L_2)$ is the extension. We remark that $\text{Ext}(L_1, L_2)$ is isomorphic to the $\bar{\partial}$ cohomology group $H_{\bar{\partial}}(M^\vee; Hom(\mathcal{E}(L_1)_\epsilon, \mathcal{E}(L_2)_\epsilon))$.

Let us fix a hermitian metric on $\mathcal{E}(L_a)_\epsilon$ $a = 1, 2$ and let $\Delta_{\bar{\partial},\epsilon}$ be the Laplace operator acting on $\Gamma(M^\vee; Hom(\mathcal{E}(L_1)_\epsilon, \mathcal{E}(L_2)_\epsilon) \otimes \Lambda^{0,k})$. The following may be regarded as an analogue of Witten's statement Theorem 5.1 in §4.2. We assume that L_1 is transversal to L_2.

CONJECTURE 6.2. *There exists $\delta > 0$ such that if ϵ is sufficiently small then the rank of the vector space*

$$\mathfrak{H}_\epsilon^k(\mathcal{E}(L_1)_\epsilon, \mathcal{E}(L_2)_\epsilon)$$
$$= \{u \in \Gamma(M^\vee; Hom(\mathcal{E}(L_1)_\epsilon, \mathcal{E}(L_2)_\epsilon) \otimes \Lambda^{0,k}) | \Delta_{\bar{\partial},\epsilon} u = \lambda u, \quad \lambda < e^{-\frac{\delta}{\epsilon}}\}$$

is equal to the number of the points in the intersection $L_1 \cap L_2$ whose Maslov index is k. Moreover, element u_q corresponding to $(q,p) = L_1 \cap L_2$ is concentrated in a neighborhood of q in the following sense. For any c there exists C such that

$$|u_q| < C e^{-1/C\epsilon} \|u_q\|_{L^2}$$

outside $\pi_{M^\vee}^{-1} B_c(q)$.

Then, the next thing to study is the tunnel amplitude and interactions. Our plan is to reduce both of them to the Morse homotopy of multivalued functions. In the A-model side (pseudoholomorphic disk), we discussed it already in §3. (We discussed there the case when there is one Lagrangian submanifold only. But to generalize it to our situation is fairly straightforward.) We start with B-model side. Our discussion here is very close to §4.2.

We first consider

(78) $$\overline{\partial} : \mathfrak{H}_\epsilon^k(\mathcal{E}(L_1)_\epsilon, \mathcal{E}(L_2)_\epsilon) \to \mathfrak{H}_\epsilon^{k+1}(\mathcal{E}(L_1)_\epsilon, \mathcal{E}(L_2)_\epsilon)$$

and study $\langle \overline{\partial}(u_p), u_q \rangle$. In case there is no quantum correction to the holomorphic structure of $\mathcal{E}(L_a)_\epsilon$ its asymptotic behavior is described by the gradient line of f_{L_1,L_2} in the same way as Theorem 5.2. Namely, the asymptotic behavior of $\overline{\partial}_{0,\epsilon}\mathfrak{s} = 0$ can be analyzed by the Laplace operator deformed by f_{L_1,L_2} in each Fourier mode. However, since there is a quantum correction to the holomorphic structure on $\mathcal{E}(L_a)_\epsilon$, we need to study the equation

(79) $$\overline{\partial}_{0,\epsilon}\mathfrak{s} + \mathfrak{B}^2 \wedge \mathfrak{s} + \mathfrak{s} \wedge \mathfrak{B}^1 = 0.$$

Here, $\overline{\partial}_{0,\epsilon} + \mathfrak{B}^a$ is the holomorphic structure of $\mathcal{E}(L_a)_\epsilon$.

Before stating a conjecture describing the asymptotic behavior of the solution of (79) we generalize the story. Namely, we study more general operations than (78). Let us consider several Lagrangian submanifolds L_a, $a = 0, \cdots, A$. We assume that they are mutually transversal. Let

$$\Pi_{L_a,L_b} : L^2(M^\vee; Hom(E(L_a)_\epsilon, E(L_b)_\epsilon)) \to \mathfrak{H}_\epsilon^k(\mathcal{E}(L_a)_\epsilon, \mathcal{E}(L_b)_\epsilon)$$

be the orthonormal projection. We denote by Δ_{L_a,L_b} the $\overline{\partial}$ Laplace operator on $\Gamma(M^\vee; Hom(\mathcal{E}(L_a)_\epsilon, \mathcal{E}(L_b)_\epsilon))$. In a way similar to §4.2 we can construct a propagator G_{L_a,L_b} such that

$$G_{L_a,L_b} \circ \overline{\partial} + \overline{\partial} \circ G_{L_a,L_b} = \mathrm{id} - \Pi_{L_a,L_b}.$$

Using it we will define

$$\mathfrak{m}_{\Gamma,\alpha} : \bigotimes_{j=1}^{r} \mathfrak{H}_\epsilon(\mathcal{E}(L_{\alpha(j-1)})_\epsilon, \mathcal{E}(L_{\alpha(j)})_\epsilon) \to \mathfrak{H}_\epsilon(\mathcal{E}(L_{\alpha(0)})_\epsilon, \mathcal{E}(L_{\alpha(r)})_\epsilon)$$

in a way similar to §4.2. Here Γ, α are as follows. Γ is a ribbon tree satisfying Condition 3.2. We embed it as $|\Gamma| \subset D^2$ such that the inverse image of ∂D^2 is the set of exterior vertices. α is a map $\alpha : \pi_0(D^2 \backslash |\Gamma|) \to \{1, \cdots, A\}$. In other words, we assign a Lagrangian submanifold to each connected component of $D^2 \backslash |\Gamma|$. Let $r+1$ be the order of $\pi_0(D^2 \backslash |\Gamma|)$. We number the element of $\pi_0(D^2 \backslash |\Gamma|)$ clockwise. We write them $D_0, \cdots, D_r$ respectively and put $\alpha(a) = \alpha(D_a)$.

We define $\mathfrak{m}_{\Gamma,\alpha}$ together with $\varphi_{\Gamma,\alpha}$ by induction on the number of vertices of Γ. In case $\Gamma = \Gamma_{-1}$ has no interior vertex, $\mathfrak{m}_{\Gamma_{-1},\alpha}$ is the operator (78) and $\varphi_{\Gamma_{-1},\alpha}$ is the identity.

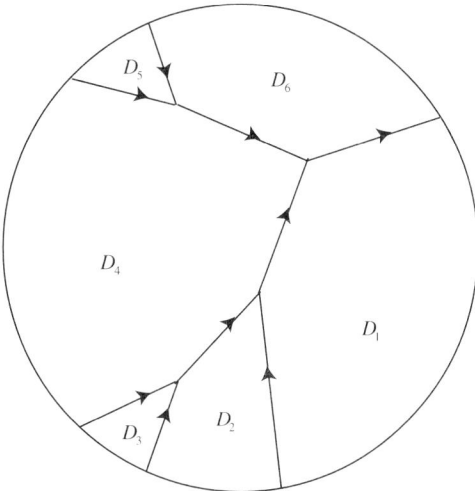

FIGURE 6.4.

We assume that $\mathfrak{m}_{\Gamma,\alpha}$ and $\varphi_{\Gamma,\alpha}$ are defined for the graphs Γ with less than v interior vertices. Let Γ be the ribbon tree with v interior vertices. We remove the last edge from it in the same way as in §4.2. We then get two ribbon trees Γ_1, Γ_2. α also determines α_1, α_2. Now, we put

(80) $$\begin{cases} \mathfrak{m}_{\Gamma,\alpha} = \Pi_{L_{\alpha(0),\alpha(r)}} \circ (\mathfrak{m}_{\Gamma_1,\alpha_1} \wedge \mathfrak{m}_{\Gamma_2,\alpha_2}) \\ \varphi_{\Gamma,\alpha} = G_{L_{\alpha(0),\alpha(r)}} \circ (\mathfrak{m}_{\Gamma_1,\alpha_1} \wedge \mathfrak{m}_{\Gamma_2,\alpha_2}). \end{cases}$$

We then define

$$\mathfrak{m}_k = \sum_{\Gamma,\alpha} \mathfrak{m}_{\Gamma,\alpha}, \quad \varphi_k = \sum_{\Gamma,\alpha} \varphi_{\Gamma,\alpha},$$

where the sum is taken over all Γ with $k+1$ exterior vertices. In a way similar to the proof of Theorem 5.3 we can prove that $\mathfrak{m}_k$ satisfies the A_∞ relation and hence defines an A_∞ category and φ_k defines an A_∞ functor.

We are now going to describe the asymptotic behavior of the operator $\mathfrak{m}_{\Gamma,\alpha}$ by Morse homotopy of several multivalued functions f_{L_a,L_b}. Because of the presence of the quantum correction to the holomorphic structures of $\mathcal{E}(L_a)$ we need to use a bit more complicated Feynman diagram to do so.

ASSUMPTION 6.3. We consider a ribbon tree Γ_+ which satisfies Condition 3.2 (1),(3),(4) and Condition 4.1(2'). We assume that the set of edges of Γ_+ is split into a disjoint union $\mathrm{Edge}(\Gamma_+) = \mathrm{Edge}(\Gamma)_0 \cup \mathrm{Edge}(\Gamma)_+$.

(1) We assume that after removing the edges in $\mathrm{Edge}(\Gamma)_+$ we still have a ribbon tree which we write $\mathrm{Red}(\Gamma_+) = \Gamma$. We assume Γ satisfies Condition 3.2 (1),(2),(3),(4).
(2) Let $|\Gamma_+| \subset D^2$ be the embedding as before. $\alpha : \pi_0(D^2 \setminus |\Gamma|) \to \{1, \cdots, A\}$ is a map as before.
(3) We divide the difference $|\Gamma_+| \setminus |\Gamma|$ into the connected components $\Gamma_+^{(i)}$. We assume the closures of $\Gamma_+^{(i)}$ are disjoint from each other. We also assume that each of $\Gamma_+^{(i)}$ satisfies Condition 3.2 (1),(3),(4) and Condition 4.1 (2').

(4) Let v be the vertex on $|\Gamma_+^{(i)}| \cap |\Gamma|$. (We remark that $|\Gamma_+^{(i)}| \cap |\Gamma|$ consists of one point.) We assume that v is contained in the interior of the edge of Γ. We also assume that v is the last vertex of $\Gamma_+^{(i)}$.

We draw the Figure of $|\Gamma_+|$ below. In Figure 6.5 we draw the edge of Γ by a real line and the edge in $\mathrm{Edge}(\Gamma)_+$ by dotted line.

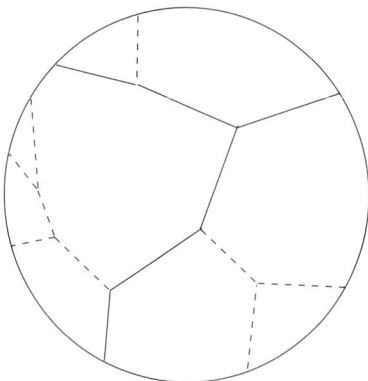

FIGURE 6.5.

We next discuss some other data we need. l is a metric on Γ_+. We assume Condition 3.3 (b). It induces a metric on Γ and $\Gamma_+^{(i)}$.

$\psi : |\Gamma_+| \to B$ is a continuous map. $\tilde\psi$ is a lift of it at each edges such that :

CONDITION 6.2. Let $e \in \mathrm{Edge}(\Gamma)_+$. $|e|$ is contained in a component $D(e)$ of $D^2 \setminus |\Gamma|$. Then, the restriction of $\tilde\psi$ to $|e|$ is a map to $\tilde{B}^+_{L_{\alpha(D(e))}}$.

Let $e \in \mathrm{Edge}(\Gamma)$. We have two components $\mathrm{left}(e)$, [resp. $\mathrm{right}(e)$] which is in the left [resp. right] hand side of e. Then, the restriction of $\tilde\psi$ to $|e|$ is a map to $\tilde{B}^+_{L_{\alpha(\mathrm{left}(e))}, L_{\alpha(\mathrm{right}(e))}}$.

We assume the lift $\tilde\psi$ is a gradient line. We specify the branch as follows. We write $\tilde\psi = (\psi; p, q; \beta)$.

ASSUMPTION 6.4. (1) Let v be an exterior vertex of Γ_+. We choose $\tau_i \in |\Gamma_+|$ converging to v. Let e be the unique edge such that $v \in \partial e$.
We first consider the case $v \in \Gamma$. Then,
$$\lim_{i \to \infty} \psi(\tau_i) \in \pi_M \left(L_{\alpha(\mathrm{left}(e))} \cap L_{\alpha(\mathrm{right}(e))} \right).$$
We assume moreover that $\beta(\tau_i)$
(which is in $\pi_1(F_x; F_x \cap L_{\alpha(\mathrm{left}(e))}, F_x \cap L_{\alpha(\mathrm{right}(e))})$) is a vanishing arc.
(2) We next consider the case when $v \in \Gamma_+^{(i)}$. We can assume $|\Gamma_+^{(i)}| \subset D_a$. Then,
$$\lim_{i \to \infty} \psi(\tau_i) \in S(L_{\alpha(D_a)}).$$
We also assume that $\beta(\tau_i)$ is a vanishing arc.
(3) Let v be an interior vertex of Γ_+. We first assume that v is an interior vertex of Γ. Let e_1, e_2, e_0 be the three edges of Γ such that $\partial_{\mathrm{target}} e_1 = \partial_{\mathrm{target}} e_2 = v =$

$\partial_{\text{source}} e_0$. We assume e_1, e_2, e_0 respects the cyclic order of the ribbon structure at v. We choose $\tau_{i,j}$ on e_i converging to v as $j \to \infty$. Then,

$$\lim_{j \to \infty} \beta(\tau_{1,j}) + \lim_{j \to \infty} \beta(\tau_{2,j}) = \lim_{j \to \infty} \beta(\tau_{0,j}).$$

Here the left-hand side is the product $\pi_1(F_x; F_x \cap L_{\text{left}(e_1)}, F_x \cap L_{\text{right}(e_1)}) \times \pi_1(F_x; F_x \cap L_{\text{left}(e_2)}, F_x \cap L_{\text{right}(e_2)}) \to \pi_1(F_x; F_x \cap L_{\text{left}(e_1)}, F_x \cap L_{\text{right}(e_2)})$.

(4) We consider each of $\Gamma_+^{(i)}$. Then, we assume that the restriction of $\tilde{\psi}$ there satisfies Condition 4.2 (2).

(5) Let v be the vertex in $|\Gamma| \cap |\Gamma_+^{(i)}|$. Let us take $\tau_j \in |\Gamma_+^{(i)}|$ converging to v. Let $v = \tau_0 \in |e|$ where e is an edge of Γ. (Compare Assumption 6.3 (4).) Suppose $|\Gamma_+^{(i)}| \in D_a$, $a = \text{right}(e)$. We take points $\tau \pm 1/j$ on $|e|$ converging to v. We then have

$$\lim_{j \to \infty} \beta(\tau_j) + \lim_{j \to \infty} \beta(\tau - 1/j) = \lim_{j \to \infty} \beta(\tau + 1/j).$$

Here, the sum in the left-hand side is $\pi_1(F_x; F_x \cap L_a, F_x \cap L_a) \times \pi_1(F_x; F_x \cap L_{\text{left}(e)}, F_x \cap L_{\text{right}(e)}) \to \pi_1(F_x; F_x \cap L_{\text{left}(e)}, F_x \cap L_{\text{right}(e)})$. In case the $a = \text{left}(e)$, we have

$$\lim_{j \to \infty} \beta(\tau - 1/j) + \lim_{j \to \infty} \beta(\tau_j) = \lim_{j \to \infty} \beta(\tau + 1/j).$$

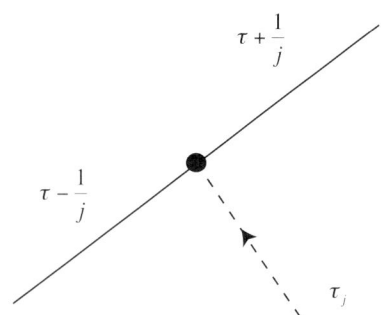

FIGURE 6.6.

Our main assumption is that $\tilde{\psi}$ is a gradient line:

ASSUMPTION 6.5. On $|e|$ with $e \in \text{Edge}(\Gamma)_0$, we have

$$\frac{\partial \tilde{\psi}}{\partial \tau} = \text{grad} f_{L_{\text{right}(e)}, L_{\text{left}(e)}}.$$

We also assume that the restriction of $\tilde{\psi}$ to each of $\Gamma_+^{(i)}$ satisfies Condition 4.2 (1).

DEFINITION 6.4. Let $\mathcal{M}_{\text{Morse}}(M; L_0, \cdots, L_A; \Gamma_+, \alpha)$ be the set of all $(l, \tilde{\psi})$ satisfying Assumptions 6.3, 6.4, 6.5 and Condition 6.2.

We define an energy E of an element $(l, \tilde{\psi})$ of $\mathcal{M}_{\text{Morse}}(M; L_0, \cdots, L_A; \Gamma_+, \alpha)$ as follows.

(81) $$E(l, \tilde{\psi}) = \sum_{e \in \text{Edge}(\Gamma_+)} \int_{\tau \in |e|} \left\| \frac{\partial \psi}{\partial \tau} \right\| d\tau.$$

Let us fix (Γ, α). Let us number the components of $D^2 \setminus |\Gamma|$ clockwise and denote the components by $D_0, \cdots, D_r$. Let $v_1, \cdots, v_r$ be exterior vertices such that $v_{\text{last}}, v_1, \cdots, v_r$ respects the cyclic order. We take edges e_i such that $\partial_{\text{source}} e_i = v_i$. Let us take Γ_+ such that $\text{Red}(\Gamma_+) = \Gamma$. We assume that $\text{right}(e_1) = D_0$, $\text{left}(e_r) = D_r$. Let $(l, \tilde{\psi}) \in \mathcal{M}_{\text{Morse}}(M; L_0, \cdots, L_A; \Gamma_+, \alpha)$. For each $v_i \in \text{Vertex}(\Gamma)$, the map $\tilde{\psi}$ determines an element of $L_{i-1} \cap L_i$. We denote it by $\text{ev}_i(l, \tilde{\psi})$.

For $p_i \in L_{i-1} \cap L_i$ we have an element u_{p_i} of $\mathfrak{H}_\epsilon(\mathcal{E}(L_{i-1})_\epsilon, \mathcal{E}(L_i)_\epsilon)$ (if we assume Conjecture 6.2.) Now, we can state the following.

CONJECTURE 6.3.
$$\tag{82} \langle \mathfrak{m}_{\Gamma, \alpha}(u_{p_1}, \cdots, u_{p_r}), u_{p_0} \rangle \sim \sum \pm \text{ext}(-E(l, \tilde{\psi}))$$
where the sum is taken over all Γ_+ with $\text{Red}(\Gamma_+) = \Gamma$ and $(l, \tilde{\psi}) \in \mathcal{M}_{\text{Morse}}(M; L_0, \cdots, L_A; \Gamma_+, \alpha)$ such that $\text{ev}_i(l, \tilde{\psi}) = p_i$.

The sign $\pm$ is determined by the orientation of the moduli space $\mathcal{M}_{\text{Morse}}(M; L_0, \cdots, L_A; \Gamma_+, \alpha)$.

It might be possible to prove Conjecture 6.3 by using an argument similar to those given in this section. In fact, the conditions we put on the Feynman diagram for $\mathcal{M}_{\text{Morse}}(M; L_0, \cdots, L_A; \Gamma_+, \alpha)$ are the Feynman rules for equation (79).

An argument similar to the one in §3 may imply that the number in the right-hand side of (82) coincides with the matrix element of the product structure of the Floer homology. In fact, this is correct only in the case when there are no holomorphic disks whose boundary belong to L_a. In case such holomorphic disks exist we need to include the correction terms as in [**34, 37**]. It seems likely that one can modify the argument of this section to include those correction terms. But we do not discuss it here since the formula will become very complicated and since we assumed that there is no holomorphic disk bounding L in Theorem 6.2. So, our conjecture is :

CONJECTURE 6.4. *Suppose that there are no holomorphic disks whose boundary belong to L_a. Then, the A_∞ structure on Floer homology defined in [**34**] is*

$$\langle \mathfrak{m}_k(p_1, \cdots, p_r), p_0 \rangle = \sum_\Gamma \sum \pm \text{ext}(-E(l, \tilde{\psi}))$$

where the sum in the right-hand side is taken in the same way as Conjecture 6.3.

Conjectures 6.3 and 6.4, if proved, will imply the Homological Mirror Symmetry Conjecture. Actually, the way it was proved in the case of affine Lagrangian submanifolds in a complex torus in [**32**] is similar to the one we are discussing here.

We finally state another conjecture which is related to the construction to the opposite direction. Namely, the construction of Lagrangian submanifold out of holomorphic vector bundles. As we mentioned at the beginning of §1.1, this is a gauge theory analogue of the theory of collapsing Riemannian manifolds. Let $\pi_{M^\vee} : M^\vee \to B$ be as in §1. Namely, $\pi_{M^\vee}$ is a singular fibration such that the general fibers are Lagrangian tori. We assume that there exists a family of complex structures J_ϵ on $M^\vee$ and a family of Calabi-Yau metrics g_ϵ on $M^\vee$ such that J_ϵ satisfies the assumption of §1.2 Proposition 2.2 and the diameter of the fibers with respect to g_ϵ are $\sim \epsilon$. We also suppose that there exists a family of holomorphic vector bundles $\mathcal{E}_\epsilon$ on $(M^\vee, J_\epsilon)$. We assume that they are stable. Then, by [**20, 74**] there exists a family of Einstein-Hermitian connections ∇_ϵ on $\mathcal{E}_\epsilon$.

CONJECTURE 6.5. *There exists a rectifiable subset $S(\mathcal{E}) \subseteq B$ of codimension ≥ 2 such that for each δ the curvature of the restriction of ∇_ϵ to $M^\vee \backslash \pi_{M^\vee}^{-1}(U_\delta(S(\mathcal{E})))$ is uniformly bounded. Here $U_\delta(S(\mathcal{E}))$ is the δ neighborhood of $S(\mathcal{E})$.*

For $x \notin S(\mathcal{E})$, the restriction of ∇_ϵ to the fiber $F_x^\vee$ converges to a flat connection as $\epsilon \to 0$. Hence, it determines an element of $S_k(F_x)$. Here, S_k is the k-th symmetric power and k is the rank of the bundle $\mathcal{E}$. We denote this element by $s(x)$. The set $\{(x,y) | y \in s(x)\}$ is a submanifold of M the mirror manifold. Its closure is a Lagrangian submanifold. We write it $L(\mathcal{E})$.

There exists a family of special Lagrangian submanifolds L_ϵ on the mirror manifold M_ϵ such that L_ϵ converges to L.

REMARK 6.7. In the case when the dimension of M is 4, the Einstein-Hermitian connections ∇_ϵ consist a family of self dual connections. Hence, we have a family of self dual connections on a family of 4 manifolds which collapses to a 2 manifold. This is the situation studied extensively in the study of Atiyah-Floer conjecture. So, in this situation, by using the method of [**21, 31**], one may solve Conjecture 6.5.

REMARK 6.8. Let us consider the case when the dimension of M is 6. Then, the situation is harder. However, we still may prove the conjecture in the following way. We may use the result by Tian [**73**] to conclude that the set where the curvature of ∇_ϵ diverges is a complex subvariety of complex codimension 2. Namely, it is a holomorphic curve. As we studied in section 2, the holomorphic curve (with finite energy) on $M^\vee$ is in a small neighborhood of $\pi_{M^\vee}^{-1}(\Gamma)$ where Γ is a graph in B. Thus, we may take $S(\mathcal{E}) = \Gamma$.

Thus, we have described a series of conjectures, which, if proven, will confirm most of the predictions of mirror symmetry. Some of the conjectures may be proved in the near future. However, some of them seem to be very hard, and so there is yet a long way to go.

References

[1] S. Amari. *Differential-geometrical methods in statistics*. Springer-Verlag, New York, 1985.

[2] M. Anderson. *Convergence and rigidity of manifolds under Ricci curvature bounds*. Invent. Math., **102 (2)** (1990), 429–445.

[3] B. Andreas, G. Curio, D.H. Ruiperez, and S.-T. Yau. *Fourier-Mukai transform and mirror symmetry for D-Branes on elliptic Calabi-Yau*. Preprint, math.AG/0012196.

[4] T. Aoki, T. Kawai, and Y. Takei. *Algebraic analysis of singular perturbations—on exact WKB analysis*. Sugaku Expositions, **8(2)** (1995), 217–240. Sugaku Expositions.

[5] D. Arinkin and A. Polishchuk. *Fukaya category and Fourier transform*. Preprint, math.AG/9811023.

[6] V. I. Arnol'd. *Mathematical methods of classical mechanics*. Springer-Verlag, New York, second edition, 1989.

[7] V.I. Arnol'd and A. Givental. *Symplectic geometry*. In: Dynamical systems. IV, ed. S. P. Novikov, 4–136, Springer-Verlag, Berlin, 1990.

[8] V.I. Arnol'd, S.M. Guseĭn-Zade, and A.N. Varchenko. *Singularities of differentiable maps, Vol. I*. Birkhäuser Boston Inc., Boston, MA, 1985.

[9] V.I. Arnol'd, S.M. Guseĭn-Zade, and A.N. Varchenko. *Singularities of differentiable maps, Vol. II*. Birkhäuser Boston Inc., Boston, MA, 1988.

[10] S. Axerlod and I. Singer. *Chern-Simons perturbatoin theory I*. In: Proc. XXth International Conference on Differential Theoretic method in Theoretical Physics, eds. S. Catto and A. Rocha, 3 – 45. World Scientifique, 1991.

[11] S. Axerlod and I. Singer. *Chern-Simons perturbatoin theory II*. J. Diff. Geom., **39** (1991), 173–213.
[12] S. Bando, A. Kasue, and H. Nakajima. *On a construction of coordinates at infinity on manifolds with fast curvature decay and maximal volume growth*. Invent. Math., **97(2)** (1989), 313–349.
[13] M. Bershadsky, M. Ceccoti, S. Ooguri, and C. Vafa. *Kodaira-Spencer theory of gravity and exact results for quantum string amplitude*. Commun. Math. Phys., **165** (1994), 311–427, 1994.
[14] M. Betz, and R. Cohen. *Graph moduli spaces and cohomology operations*. Turkish J. Math., **1**, (1994), 23–41.
[15] U. Bruzzo, G. Marelli, and F. Pioli. *A Fourier transform for sheaves on Lagrangian families of real tori*. preprint, math.DG/0105196.
[16] J. Cheeger, K. Fukaya, and M. Gromov. *Nilpotent structures and invariant metrics on collapsed manifolds*. J. Amer. Math. Soc., **5(2)** (1992), 327–372.
[17] J. Cheeger and M. Gromov. *Collapsing Riemannian manifolds while keeping their curvature bounded. I*. J. Differential Geom., **23(3)** (1986), 309–346.
[18] J. Cheeger and M. Gromov. *Collapsing Riemannian manifolds while keeping their curvature bounded. II*. J. Differential Geom., **32(1)** (1990), 269–298.
[19] S.Y. Cheng and S.T. Yau. *The real Monge-Ampère equation and affine flat structures*. In "Proceedings of the 1980 Beijing Symposium on Differential Geometry and Differential Equations, Vol. 1, 2, 3 (Beijing, 1980)", 339–370, Science Press, Beijing, 1982.
[20] S. Donaldson. *Infinite determinants, stable bundles and curvature*. Duke Math. J., **54** (1987), 231–247.
[21] S. Dostoglou, and D. Salamon. *Self-dual instantons and holomorphic curves*. Ann. of Math., **139** (1994), 581–640.
[22] A. Floer. *Morse theory for Lagrangian intersectoins*. J. Diff. Geom., **28** (1988), 513–547.
[23] K. Fukaya. *Collapsing Riemannian manifolds to ones of lower dimensions*. J. Differential Geom., **25(1)** (1987), 139–156.
[24] K. Fukaya. *Collapsing Riemannian manifolds to ones of lower dimensions II*. J. Math. Soc. Japan, **41** (1989), 333–356.
[25] K. Fukaya. *Hausdorff convergence of Riemannian manifolds and its applications*. In "Recent topics in differential and analytic geometry", 143–238. Academic Press, Boston, MA, 1990.
[26] K. Fukaya. *Morse homotopy, A^∞-categories, and Floer homologies*. In H. J. Kim, editor, "Proc. of the 1993 Garc Workshop on Geometry and Topology", Lecture Notes series **18**, Seoul Nat. Univ., 1–102. 1993. http://www.kusm.kyoto-u.ac.jp/ ~fukaya/ fukaya.html.
[27] K. Fukaya. *Morse homotopy and Chern-Simons perturbation theory*. Comm. Math. Phys., **181(1)** (1996), 37–90.
[28] K. Fukaya. *Informal note on topology, geometry and topological field theory*. In "Geometry from the Pacific Rim (Singapore, 1994)", 99–116. de Gruyter, Berlin, 1997.
[29] K. Fukaya. *Morse homotopy and its quantization*. In "Geometric topology (Athens, GA, 1993)", 409–440. Amer. Math. Soc., Providence, RI, 1997.
[30] K. Fukaya. *The symplectic s-cobordism conjecture: a summary*. In "Geometry and physics (Aarhus, 1995)", pages 209–219. Dekker, New York, 1997.
[31] K. Fukaya. *Anti-self-dual equation on 4-manifolds with degenerate metric*. Geom. Funct. Anal., **8** (1998), 466–528.
[32] K. Fukaya. *Mirror symmetry of abelian variety and multi theta functions*. to appear in J. Alg. Geom., 1998.
[33] K. Fukaya. *Deformation theory, homological algebra, and mirror symmetry*. preprint, 2001.
[34] K. Fukaya. *Floer homology and mirror symmetry II*. to appear in Adv. Studies in Pure Math., 2001.
[35] K. Fukaya. *Floer homology for families - report of a project in progress -*. to appear in Contemporary Math., 2001.
[36] K. Fukaya and Y.G. Oh. *Zero-loop open strings in the cotangent bundle and Morse homotopy*. Asian J. Math., **1** (1997), 99–180.
[37] K. Fukaya, Y.G. Oh, H.Ohta, and K.Ono. *Langrangian intersection Floer theory -anomaly and obstruction-*. preprint, http://www.kusm.kyoto-u.ac.jp/~fukaya/ fukaya.html, 2000.
[38] K. Fukaya and K. Ono. *Arnold conjecture and Gromov-Witten invariants*. Topology, **38** (1999), 933–1048.

[39] M. Gaberdiel and B. Zwiebach. *Exceptional groups from open string.* Nucl. Phys. B, **151**, (1998).

[40] E. Getzler and J. Jones. A_∞-*algebras and the cyclic bar complex.* Illinois J. Math., **34(2)** (1990), 256–283.

[41] E. Goldstein. *Calibrated fibrations on complete manifolds via torus action.* preprint, math.DG/0002097.

[42] M. Gromov. *Structures métriques pour les variétés riemanniennes.* CEDIC, Paris, 1981. Edited by J. Lafontaine and P. Pansu.

[43] M. Gross. *Examples of special Lagrangian fibrations.* preprint, math.AG/0012002.

[44] M. Gross. *Special Lagrangian fibrations. I. Topology.* In "Integrable systems and algebraic geometry (Kobe/Kyoto, 1997)", 156–193. World Sci. Publishing, River Edge, NJ, 1998.

[45] M. Gross. *Special Lagrangian fibrations. II. Geometry. A survey of techniques in the study of special Lagrangian fibrations.* In "Surveys in differential geometry: differential geometry inspired by string theory", 341–403. Int. Press, Boston, MA, 1999.

[46] M. Gross. *Topological mirror symmetry.* Invent. Math., **144(1)** (2001), 75–137.

[47] M. Gross and P. Wilson. *Large complex structure limits of K3 surfaces.* preprint, math.DG/0008018.

[48] V. Gugenheim and J. Stasheff. *On perturbations and A_∞-structures.* Bull. Soc. Math. Belg. Sér. A, **38** (1987), 237–246.

[49] B. Helffer and J. Sjöstrand. *Puits multiples en mécanique semi-classique. IV. Étude du complexe de Witten.* Comm. Partial Differential Equations, **10(3)** (1985), 245–340.

[50] N. Hitchin. *The moduli space of special Lagrangian submanifolds.* Ann. Scuola Norm. Sup. Pisa Cl. Sci. (4), **25(3-4)** (1998), 503–515. Dedicated to Ennio De Giorgi.

[51] N. Hitchin, A. Karlhede, U. Lindström, and M. Roček. *Hyper-Kähler metrics and supersymmetry.* Comm. Math. Phys., **108(4)** (1987), 535–589.

[52] M. Hutchings and Y.J. Lee. *Circle-valued Morse theory and Reidemeister torsion.* Geom. Topol., **3** (1999), 369–396 (electronic).

[53] D. Joyce. *U(1)-invariant special Lagrangian 3-folds I. nonsingular solutions.* preprint, math.DG/0111324.

[54] D. Joyce. *U(1)-invariant special Lagrangian 3-folds II. existence of singular solutions.* preprint, math.DG/0111326.

[55] T. V. Kadeishvili. *The algebraic structure in the homology of an $A(\infty)$-algebra.* Soobshch. Akad. Nauk Gruzin. SSR, **108(2)** (1983), 249–252.

[56] B. Keller. *In troduction to A infinity algebra and modules.* preprint, math.RA/9910179, 1999.

[57] K. Kodaira. *On compact analytic surfaces II.* Ann. Math. **77** (1963), 523–626.

[58] M. Kontsevich. A_∞ *algebras in mirror symmetry.* MPI Arbeitstagung talk, 1993.

[59] M. Kontsevich. *Homological algebra of mirror symmetry.* In "'Proceedings of the International Congress of Mathematicians, Zürich", volume I, pages 120–139. Birkhäuser, 1995.

[60] M. Kontsevich. *Enumeration of rational curves via torus actions.* In "The moduli space of curves", 335–368. Birkhäuser, 1995.

[61] M. Kontsevich and Y. Soibelman. *Homological mirror symmetry and torus fibrations.* In "Symplectic Geometry and Mirror Symmetry". World Sci. Press, 2001.

[62] J.L. Koszul. *Domaines bornés homogènes et orbites de groupes de transformations affines.* Bull. Soc. Math. France, **89** (1961), 515–533.

[63] Y. J. Lee. *Reidemeister torsion in symplectic Floer theory and counting pseudo-holomorphic tori.* preprint, math.DG/0111313.

[64] N. C. Leung. *Mirror symmetry without corrections.* preprint, math.DG/0009235.

[65] B.H. Lian, A. Todorov, and S.T. Yau. *Maximal unipotent monodromy for complete intersection CY manifolds.* math.AG/0008061.

[66] S.A. Merkulov. it Strongly homotopy algebras of a Kähler manifold. Internat. Math. Res. Notices, **3** (1999), 153–164.

[67] S. Novikov. *Multivalued functions and functional - an analogue of the Morse theory.* Sov. Math. Dokl., **24** (1981), 222–225.

[68] H. Ooguri and C. Vafa. *Summing up Dirichlet instantons.* Phys. Rev. Lett., **77(16)** (1996), 3296–3298.

[69] J. Roe. *Elliptic operators, topology and asymptotic methods.* Longman Scientific & Technical, Harlow, 1988.

[70] W-D. Ruan. *Lagrangian torus fibrations and mirror symmetry of Calabi-Yau manifolds.* preprint, math.DG/0104010.
[71] H. Shima. *Hesse Geometry.* Shookabou, Tokyo, 2001. in Japanese.
[72] A. Strominger, S.T. Yau, and E. Zaslow. *Mirror symmetry is T-duality.* Nuclear Phys. B, **479(1-2)** (1996), 243–259.
[73] G. Tian. *Gauge theory and calibrated geometry. I.* Ann. of Math., **151** (2000), 193–268.
[74] K. Uhlenbeck, and S.T. Yau. *On the existence of Hermitian-Yang-Mills connections in stable vector bundles.* Comm. Pure Appl. Math., **39** (1986), 257–293.
[75] A. Voros. *The return of the quartic oscillator: the complex WKB method.* Ann. Inst. H. Poincaré Sect. A (N.S.), **39(3)** (1983), 211–338.
[76] E. Witten. *Supersymmetry and Morse theory.* J. Differential Geom., **17(4)** (1983), 661–692 (1983).
[77] E. Witten. *Chern-Simons gauge theory as a string theory.* In "The Floer memorial volume", 637–678. Birkhäuser, Basel, 1995.
[78] T. Yamaguchi. *Collapsing and pinching under a lower curvature bound.* Ann. of Math., **133(2)** (1991), 317–357.
[79] I. Zharkov. *Torus fibrations of Calabi-Yau hypersurfaces in toric varieties.* Duke Math. J., **101(2)** (2000), 237–257.

DEPARTMENT OF MATHEMATICS KYOTO UNIVERSITY, KITASHIRAKAWA, SAKYO-KU, KYOTO JAPAN

COMBINATORIAL ASPECTS
OF DYNAMICS

Some Applications of Combinatorial Differential Topology

Robin Forman

Contents

1. Introduction
2. Discrete Vector Fields and Dynamical Systems
3. The Proof of the Morse Inequalities
4. A Refinement: Counting Closed Orbits According to their Homotopy Class
5. Discrete Morse Theory
6. A Simple Example
7. An Application to Computer Science
8. Topological Invariants via counting Gradient Paths
9. Supersymmetry and Discrete Morse Theory
10. Discrete Morse Theory and the Distance Function
11. Main Ingredient
References

1. Introduction

The title of this paper is perhaps a bit confusing, since it is not at all clear what is meant by the term "combinatorial differential topology". We use this expression to refer to the application of the standard concepts of differential topology, such as vector fields and their corresponding flows, to the study of combinatorial spaces such as simplicial complexes. This paper will be a survey of some of the author's work in this direction (see [15-22]). In order to aid us in our discussion, we will take as the unifying idea the notion of a combinatorial dynamical system.

The notion of a *dynamical system* has become one of the grand unifying ideas in all of science over the last century. The study of combinatorial dynamical systems is also rather classical. Indeed, the subject of *symbolic dynamics* is precisely that (see, e.g. [**47**], or [**62**] for a look at some of the numerous directions in which the theory has been applied). The main goal in this paper is to present a class of very topological combinatorial dynamical systems, the study of which allows one to recreate, in the context of combinatorial spaces, many phenomena which one might normally think of as being associated only to smooth manifolds.

We will be exploring a number of applications of these ideas to topology, geometry, mathematical physics and computer science. We will also be presenting some potential future applications which at this point are just fantasy. While much of this conference has focussed on combinatorial aspects of "quantum mathematics", most of this paper will be concerned with combinatorial aspects of classical mathematics. However, the examples discussed in sections 5 and 8 are closely related to issues discussed by others.

2. Discrete Vector Fields and Dynamical Systems

In this section we introduce what will be the central notions of this paper, the idea of a discrete vector field on a cell complex, and the dynamics of the corresponding flow. Most of the content of this section appeared first in [**18**].

Let K be a finite regular CW complex. (See Chapter III of [**42**] for definitions). The regularity of K is merely for technical convenience. If the reader is unfamiliar with such general cell complexes, then one may think of the special case of polyhedra or simplicial complexes.

DEFINITION 2.1. *A discrete vector field V is a set of pairs of cells of the form $\{\alpha \prec \beta\}$, where the notation $\alpha \prec \beta$, or $\beta \succ \sigma$, indicates that α is a codimension-one face of β, i.e. $\alpha \subset \overline{\beta}$ and $\dim(\alpha) + 1 = \dim(\beta)$, and such that no cell is in more than one pair.*

It is useful to picture a discrete vector field by drawing arrows on K (see Figure 2.2). If $\{\alpha \prec \beta\}$ is a pair in V, then we draw an arrow from α to β (i.e. an arrow whose tail is in α and whose head is in β).

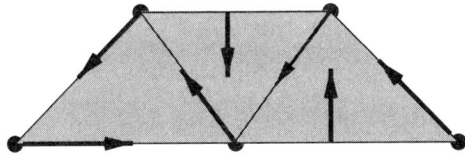

A discrete vector field.

FIGURE 2.2.

Since no cell is in more than one pair in V, every cell α of K satisfies exactly one of the following conditions:
(i) α is the head of exactly one arrow
(ii) α is the tail of exactly one arrow
(iii) α is neither the head nor the tail of any arrow.

A *zero* (or *rest point* or *critical point*) of V is a cell which satisfies condition (iii), i.e. a cell which is not in any pair of V.

We will show in the course of this paper that using this notion of a discrete vector field one can recover in a combinatorial setting many aspects of smooth vector fields and their dynamics. At the start, however, we should point out one essential difference between the smooth and combinatorial categories. There is an obvious involution on the space of smooth vector fields on a smooth manifold, sending any vector field V to $-V$. On the other hand, in the general combinatorial setting we have presented, if V is a discrete vector field on a cell complex K then there is no notion of "$-V$". One can recover this basic operation if K is a combinatorial manifold. In this case $-V$ does exist as a discrete vector field on the dual cell complex K^*. Namely, $-V$ is the collection of all pairs $\{\beta^* \prec \alpha^*\}$ such that $\{\alpha \prec \beta\}$ is a pair in V (where α^* denotes the cell in K^* which is dual to the cell α in K, and β^* is defined similarly.) This is closely related to the fact that the homology of a combinatorial manifold satisfies Poincaré duality, but there is no such symmetry in the homology of a general cell complex. We return to this issue again in section 5 and in part III of section 8.

Our goal is to investigate the topological properties of the dynamics of V. We can begin by observing the following simple combinatorial analogue of the Poincaré-Hopf Theorem.

THEOREM 2.3. *The Euler characteristic of K, $\chi(K)$, is equal to*
$$\sum_{\text{zeros } \alpha} (-1)^{\dim(\alpha)}.$$

This theorem follows from the simple observation that the cells which are not zeros are paired, with each cell paired with a cell whose dimension differs by one, and hence has opposite parity.

To proceed further, let us begin our discussion of the dynamics of V. We define a *V-path* to be a sequence γ of cells

(2.1) $$\gamma = \alpha_0^{(p)}, \beta_0^{(p+1)}, \alpha_1^{(p)}, \beta_1^{(p+1)}, \alpha_2^{(p)}, \ldots, \beta_{r-1}^{(p+1)}, \alpha_r^{(p)}$$

where the notation $\alpha^{(p)}$ indicates that the cell α is p-dimensional, and such that for each $i = 1, \ldots r$, $\{\alpha_i \prec \beta_i\}$ is in V, and $\beta_i \succ \alpha_{i+1} \neq \alpha_i$. We draw such a path in Figure 2.4. We say a V-path γ as in (2.1) is a (non-trivial) closed path if $r > 0$ and $\alpha_0 = \alpha_r$. We say that γ has length r and index p.

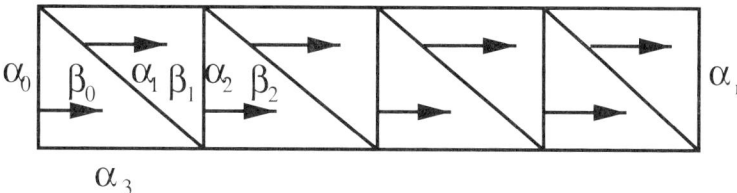

A V-path from α_0 to α_r.

FIGURE 2.4.

Note that a V-path is not determined by its starting cell. In Figure 2.4, the cells $\alpha_0, \beta_0, \alpha_3$ also form a V-path.

The main idea of this section is that the topology of K is carried by the closed orbits and rest points of V. Let us now make this precise. Define the *chain recurrent set* of V to be the set of cells which are either rest points of V or are contained in some non-trivial closed path of V. Here we see an important simplification that occurs when moving from the smooth to the combinatorial categories. In the smooth setting, the chain recurrent set also includes points which "almost" lie on closed orbits. In our discrete setting, there are no almost closed orbits.

We partition the chain recurrent set into *basic sets*, where α_1 and α_2 are in the same basic set if and only if there is a non-trivial closed V-path which contains both α_1 and α_2. For example, in Figure 2.5 we again show the discrete vector field of Figure 2.2, but this time we have labelled the cells in the chain recurrent set. For this discrete vector field the chain recurrent set $\mathcal{R}$ consists of two basic sets.

$$\mathcal{R} = \{v_1, e_1, v_2, e_2, v_3, e_3\} \cup \{f\}$$

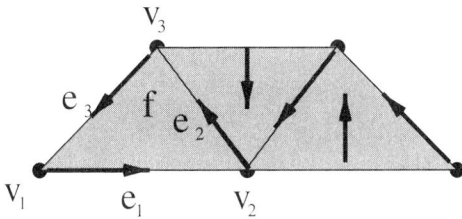

The chain recurrent set

FIGURE 2.5.

In general, as in this example, basic sets come in 2 types. If Λ is a basic set then either

(2.2) $$\Lambda = \{\alpha^{(p)}\} \text{ where } \alpha \text{ is a zero of } V$$

or

(2.3) $$\Lambda = \{\alpha_i^{(p)}, \beta_j^{(p+1)}\} \quad i = 1, 2 \ldots$$
$$j = 1, 2, \ldots$$

is the union of closed paths of index p. For any basic set Λ, let

$$\overline{\Lambda} = \cup_{\alpha \in \Lambda} \cup_{\beta \subseteq \overline{\alpha}} \beta$$

and

$$\dot{\Lambda} = \overline{\Lambda} - \Lambda.$$

We observe that if Λ consists of a single zero of V as in (2.2), then for any coefficient field $\mathbb{F}$,

$$H_i(\overline{\Lambda}, \dot{\Lambda}, \mathbb{F}) \cong \begin{cases} \mathbb{F} & \text{if } i = p \\ 0 & \text{if } i \neq p \end{cases}$$

While if Λ is a union of closed orbits as in (2.3), then our answer is not as precise, but we can say that

$$H_i(\overline{\Lambda}, \dot{\Lambda}, \mathbb{F}) \cong 0 \quad \text{if } i \neq p, p+1$$

We are now prepared to state the Morse inequalities for a discrete vector field. Define the Morse numbers m_i by

$$m_i = \sum_{\text{basic sets} \Lambda} \dim H_i(\overline{\Lambda}, \dot{\Lambda}, \mathbb{F})$$

where $\mathbb{F}$ is any fixed coefficient field. Let b_i denote the Betti numbers of K

$$b_i = \dim H_i(K, \mathbb{F}).$$

THEOREM 2.6. *The Strong Morse Inequalities.*

$$\begin{aligned} m_0 &\geq b_0 \\ m_1 - m_0 &\geq b_1 - b_0 \\ m_2 - m_1 + m_0 &\geq b_2 - b_1 + b_0 \\ &\cdots \\ m_n - m_{n-1} + \cdots \pm m_0 &= b_n - b_{n-1} + \cdots \pm b_0 = \chi(K) \end{aligned}$$

where $n = dim(K)$.

The Strong Morse Inequalities simply the following weaker, but simpler inequalities.

COROLLARY 2.7. *The Weak Morse Inequalities.*

$$m_i \geq b_i \quad \text{for each } i$$

This is precisely the sort of result we were looking for. The complex K can have no more topology than that carried by the closed orbits and rest points of V. In the next section we indicate how one proves these theorems.

3. The Proof of the Morse Inequalities

The main ingredient in the proof of the Morse inequalities is the idea of *Lyapunov function*. A Lyapunov function for V is a real-valued function on the set of cells of K with the property that it is monotone non-increasing along all V paths, and it is strictly decreasing off the chain recurrent set. More precisely,

DEFINITION 3.1. *A function*

$$f : \{cells \ of \ K\} \to \mathbb{R}$$

is a Lyapunov function for V if for any V path $\alpha_0, \beta_0, \alpha_1$, we have $f(\alpha_0) \geq f(\beta_0) \geq f(\alpha_1)$ and $f(\alpha_0) = f(\alpha_1)$ if and only if α_0 and α_1 are in the same basic set (which implies that β_0 is also in the same basic set).

In Figure 3.2 we illustrate a Lyapunov function for the discrete vector field shown in Figure 2.2.

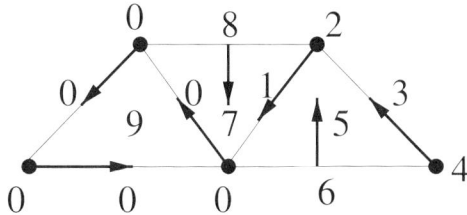

A Lyapunov Function for the discrete vector field shown in Figure 2.2

FIGURE 3.2.

In fact, we have the following theorem (see Theorem 2.4 of [**18**]).

THEOREM 3.3. *For every regular cell complex K and every discrete vector field V on K, there is a Lyapunov function for V.*

We can now indicate the proof of the Morse inequalities. The idea is to build K, beginning with the empty set, by adding the cells in the order prescribed by the Lyapunov function f. More precisely, for any real number c, define the level subcomplex $K(c)$ to be the union of all of the cells β of K satisfying $f(\beta) \leq c$, along with all of their faces. That is,

$$K(c) = \cup_{f(\beta) \leq c} \cup_{\alpha \leq \beta} \alpha.$$

In Figure 3.4 we illustrate all of the level subcomplexes of the example shown in Figure 3.2.

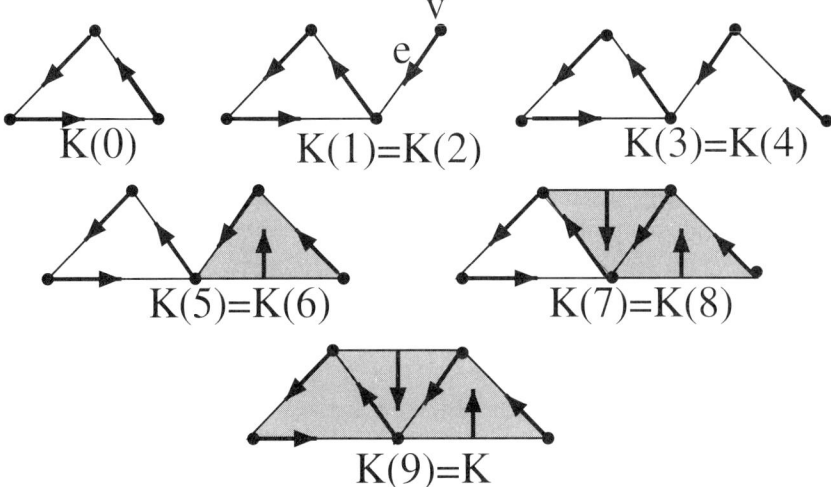

The level subcomplexes of the Lyapunov function shown in Figure 3.2

FIGURE 3.4.

Note that the discrete vector field V on K induces a discrete vector field on each of these level subcomplexes. Namely, for any pair of cells $\{\alpha \prec \beta\}$ in V, and any level subcomplex K', it is easy to see that α is in K' if and only if β is. We draw an arrow from α to β in any subcomplex K' in which α and β appear.

We can now prove the Morse inequalities by induction, showing that they are true for each simplicial complex appearing in this list with respect to the induced discrete vector fields (beginning the induction with $K(-1) = \emptyset$, for which the inequalities are trivially valid). One must show that if the Morse inequalities are true at one step then they are necessarily true at the next step. There are two cases, determined by whether or not we are adding any cells in the chain recurrent set (we are assuming, merely for simplicity, that the Lyapunov function is as 1-1 as possible, i.e. with the exception of being constant on each basic set, and this can always be arranged). Consider first the case when the added cells are not chain recurrent, see for example the transition from $K(0)$ to $K(1)$ in Figure 3.4. In this case the Morse numbers of the complex do not change, so we must see that the Betti numbers also stay the same. The edge $e = f^{-1}(1)$ is not in the chain recurrent set, and as a result, it has an endpoint v which is assigned a larger value. Thus, when we move from $K(0)$ to $K(1)$ and add the edge e, we must also add v. As a consequence, in the complex $K(1)$ the vertex v is a free face of e (i.e. a face which is not the face of any other simplex). We can deformation retract $K(1)$ onto $K(0)$

by pushing in the edge e starting at the vertex v. This special sort of deformation retract is called a "collapse". Not only can we say that the homotopy type of the level subcomplex does not change when a cell which is not in the chain recurrent set is added (so that, in particular, the Betti numbers do not change), but the new complex collapses onto the old complex. We recall that the equivalence relation generated by collapsing is called *simple-homotopy*.

Now consider the case in which we add a basic set, e.g. in the transition from $K(8)$ to $K(9)$. In this case one can check that the entire boundary of the basic set is in the original complex, and the new complex is built by simply gluing the basic set in along its boundary. The weak Morse inequalities follow from the fact that under this operation, the Betti numbers can grow by at most the homology of the basic set relative to its boundary. The Strong Morse Inequalities follow from an examination of the long exact sequence corresponding to the pair (in this case) $(K(9), K(8))$, and the excision property. The details can be found on page 21 of [**44**].

4. A Refinement: Counting Closed Orbits According to their Homotopy Class

In this section we discuss one way in which to make our counting of closed orbits more precise. Namely, we can count the closed orbits according to their free homotopy class. We will also indicate some possible connections to Seiberg-Witten theory, and symplectic geometry. However, these ideas are, at the moment, very much in the realm of fantasy. As the content of this section is somewhat independent of the rest of the paper, our presentation will be brief, and some terms will be left undefined. For details, see [**18**] and [**21**].

Let $\rho : \pi_1(K) \to O(n)$ be a representation. Consider the zeta function

$$\zeta_V(z,\rho) = \exp\left[\sum_{k=1}^{\infty} \frac{z^k}{k} \sum_{p=0}^{\dim K} (-1)^p \sum_{\gamma \in \mathcal{C}_k^p} m(\gamma) trace(\rho(\gamma))\right]$$

where $\mathcal{C}_k^p$ = closed V paths of length k and index p. The multiplicity $m(\gamma)$ of a closed V-path γ is $\pm$, depending on whether the orientation of a p-simplex in γ is preserved or reversed as one slides it along γ. See [**49**] and [**23**][**24**] for investigations of such zeta functions in the case of smooth dynamical systems.

THEOREM 4.1. *For any fixed K, V and ρ, the function $\zeta_V(z,\rho)$ has a positive radius of convergence and can be analytically continued to a meromorphic function on all of $\mathbb{C}$.*

Our main result is that a special value of this zeta function is related to the Reidemeister torsion $T(K,\rho)$ of K, a topological invariant of K. There is a slight complication, as one must also include a term built from the zeros of V.

THEOREM 4.2. *If $H_*(\overline{\Lambda}, \dot{\Lambda}, \rho) = 0$ for each non-trivial basic set Λ, then $\zeta_V(z,\rho)$ is analytic at $z = 1$. If, in addition, V has no zeros then*

$$T(K,\rho) = \zeta_V(1,\rho).$$

In the general case in which V may have zeros, then

$$T(K,\rho) = tor(\mathcal{M}^*(\rho))\zeta_V(1,\rho).$$

where $\mathcal{M}^*(\rho)$ indicates the Morse complex of K (a differential complex built from the zeros of V and the gradient paths that connect them - see section8), and $\text{tor}(\mathcal{M}^*(\rho))$ means the torsion of this complex.

The main ingredient in the proof of Theorem 4.1 will be presented in section 11. One proof of Theorem 4.2, based on the Witten deformation of the Laplace operator, will be discussed in section 9. Theorem 4.2 should be compared to the theorems of Fried [23],[24], Hutchings-Lee [32] [33], and Pazhitnov [50], where similar results are proved for the closed orbits of classes of smooth vector fields.

In the case of a smooth 3-manifold and a vector field dual, with respect to some auxiliary Riemannian metric, to a closed 1-form, the identity takes on particular significance, since in this case the result is related to the Seiberg-Witten invariant of the product of the 3-manifold and a circle.

The basic idea is the following. Taubes [60] has shown that in the case of a symplectic 4-manifold with a compatible almost complex structure, the Seiberg-Witten invariant is equal to an invariant constructed by counting pseudoholomorphic curves. Suppose we have a 3-manifold M with a harmonic 1-form α with no zeros. Then one can construct a symplectic form ω on $M \times S^1$ by setting

$$\omega = *_3\alpha + \alpha \wedge d\theta.$$

Endow M with a Riemannian metric, and $M \times S^1$ with the product metric. Let V be the vector field on M dual to α with respect to the Riemannian metric. If one has a closed orbit of V, then crossing that closed orbit with S^1 results in a pseudoholomophic curve on $M \times S^1$ (with respect to the almost complex structure induced by ω and Riemannian metric), and, conversely, any pseudoholomorphic curve on $M \times S^1$ whose homology class lies in $H_1(M) \times H_1(S^1)$ arises in this fashion. Thus, Taubes' theorem implies that in this case, i.e. with a vector field dual to a harmonic 1-form with no zeros, the Seiberg-Witten invariant of $M \times S^1$ is given by a count of closed orbits of V on M. In [32], Hutchings and Lee conjectured that such a relationship holds for all 1-forms, with the count modified by the torsion of the Morse complex. In fact, the conjecture follows from work of Turaev [61] (see [32][33] and [31] for precise statements of these ideas). A goal of much ongoing research in gauge theory is a combinatorial formula for Seiberg-Witten invariants. Perhaps our Theorem 4.2 can be useful for understanding the combinatorial nature of Seiberg-Witten at least in the case of a 3-manifold, or 4-manifold of the form $M^3 \times S^1$.

Taubes has demonstrated a very deep connection between Seiberg-Witten theory and the symplectic topology of a 4-manifold [60]. Perhaps one could begin an investigation of combinatorial Seiberg-Witten theory by developing a theory of combinatorial symplectic geometry (which would, of course, be extremely interesting in its own right). It is not clear to me what one might expect of such a theory.

Perhaps the hint of a starting point of such a theory can be found in [20], in which we present a combinatorial version of Novikov's generalization of Morse theory. Given a smooth manifold and a closed 1-form on the manifold, Novikov shows that one can relate the zeros of the 1-form to certain generalized Betti numbers. Standard Morse theory results from the case in which the 1-form is exact. In [20],we present a discrete version of Novikov's theory. In particular, we present a notion of a combinatorial differential form on a CW complex which is, in some sense, dual to the notion of discrete vector field which we have been discussing. Perhaps this notion of combinatorial differential form can serve as a starting point for the investigation of other facets of combinatorial differential topology and geometry.

Lastly, a question: Which smooth vector fields can be triangulated? That is, given a smooth manifold and a smooth vector field, when can the manifold be triangulated so that there exists a discrete vector field with the same dynamics. This can be done for gradient vector fields, and somewhat more generally (as will be shown in a later paper) but it is no doubt true for larger classes of vector fields. Dennis Sullivan has suggested that there might be a close relationship between a vector field being stable and being triangulable, and there seems to be much truth in this.

5. Discrete Morse Theory

In this section, we begin our examination of an important special case of our discussion. Namely, the case of discrete vector fields with no closed orbits. We refer to the theory in this case as *discrete Morse theory* ([**16**]).

DEFINITION 5.1. *We say that a discrete vector field V is a gradient vector field if the are no closed V-paths. In this case, we call a Lyapunov function for V a Morse function.*

Let us recall the main points in our proof of the Morse inequalities in section 3. We showed that when passing from one level subcomplex to the next, if what is added is not in the chain recurrent set then the larger subcomplex collapses onto the smaller. When we add a basic set it is glued in along its entire boundary. In the case of a gradient vector field, each basic set consists of a single cell, a zero of V. From this we can easily deduce the main theorem of discrete Morse theory.

THEOREM 5.2. *Let V be a gradient vector field on a finite regular CW complex K. Then K is (simple-)homotopy equivalent to a CW complex with exactly one cell of dimension p for each zero of V of dimension p.*

An important special case is when the gradient vector field has exactly one zero. We state this case separately for future reference.

THEOREM 5.3. *Let K be a regular CW complex with a gradient vector field with exactly one critical point. Then the critical point must be a vertex, and K collapses to that vertex.*

There have been other adaptations of Morse theory that are suitable for the study of combinatorial spaces. For example, a Morse theory of piecewise linear functions appears in [**39**] and the very powerful Stratified Morse Theory was developed by Goresky and MacPherson [**29**],[**30**]. These theories, especially the latter, have been successfully applied to prove some very striking results.

One strength of the theory we present here is, as we will indicate in this paper, we are able to recreate in the combinatorial setting just about every facet of smooth Morse theory. Moreover, as a result of its purely discrete nature, it can be applied to any simplicial complex, not just those with some special structure. There is a growing list of references in which Theorem 5.2 has been used to investigate the topology of some explicit, but complicated, simplicial complexes which arise in combinatorics. See, for example, the papers [**2**],[**3**], [**10**], [**13**], [**37**], [**38**], [**41**], [**54**].

These papers contain some beautiful mathematics in which the authors construct "by hand" gradient vector fields on the complexes under consideration.

So far, all of our results have been stated for general cell complexes. If the cell complexes under consideration have more structure, then we have additional tools at our disposal. For example, recall from section 1 that if K is a combinatorial manifold and V is a discrete vector field on K, then $-V$ is a discrete vector field on K^*. It is rather easy to see that $-V$ is a discrete gradient vector field if and only if V is. From this observation, along with the Morse complex presented in part I of section 8, one can give a proof of Poincaré duality for combinatorial manifolds. Moreover, for combinatorial manifolds one can sometimes strengthen the conclusion of Theorem 5.2 from a statement about homotopy equivalence to a statement about combinatorial equivalence. This idea rests on some beautiful work of J.H.C. Whitehead [63]. We present here a special case of the main theorem of [63].

THEOREM 5.4. *Let K be a combinatorial manifold with boundary which simplicially collapses to a vertex. Then K is a combinatorial ball.*

As one example of how this theorem map be applied, suppose that K is a cell complex with a gradient vector field with precisely two critical cells. Then Theorem 5.2 implies that K is homotopy equivalent to a sphere. Whitehead's theorem implies the following refinement.

THEOREM 5.5. *Let K be a combinatorial manifold with a gradient vector field with exactly two critical points. Then K is a combinatorial sphere.*

The proof of this Theorem is quite simple (given Whitehead's wonderful result as stated in Theorem 4.4). Suppose that K is an n-dimensional combinatorial manifold with a gradient vector field with exactly two critical points. Then it is easy to see that one of the critical cells must be an n-cell and the other a 0-cell (if $n = 0$ then they are both 0-cells). Let α denote the critical n-cell. Then $K - \alpha$ has only a single critical cell, so by Theorem 5.3 it collapses. By Whitehead's theorem, $K - \alpha$ must be a combinatorial ball, which implies that K is a combinatorial sphere. □

Given this sphere theorem, it is natural to wonder if this discrete Morse theory can be use to prove the Poincaré Conjecture. In fact, one can use the discrete Morse theory to give a proof of the higher dimensional Poincaré conjecture in the PL category (and its far-reaching generalization the PL s-cobordism theorem) exactly along the lines of the Morse theoretic proof of the h-cobordism theorem presented in [45] in the smooth category.

Very loosely speaking, to prove the Poincaré conjecture in dimensions higher than 4 (following Milnor's presentation in [45]) one begins with a combinatorial n-manifold M, $n \geq 5$, which is a homotopy n-sphere. Endow M with a gradient vector field V (in the combinatorial setting one may begin with the empty vector field consisting of no pairs at all). The goal is to show that it is possible to cancel out the critical points of V until one is left with a gradient vector field with exactly two critical simplices. It must then be a combinatorial sphere, by Theorem 5.5. The following is a combinatorial version of the First Cancellation Theorem of [45] (see Lemma 2.1).

THEOREM 5.6. *Let V be a gradient vector field. Suppose $\alpha^{(p)}$ and $\beta^{(p+1)}$ are zeros of V and there is exactly one V-path from the boundary of β to α. Then there* is a gradient vector field W such that

$$zeros(W) = zeros(V) - \{\alpha, \beta\}$$

The proof is again quite simple. Suppose that α and β are as in the top half of Figure 5.7, where we have drawn the unique gradient path from the boundary of β to α. Create a new vector field by reversing the arrows along this gradient path, while not changing any of the other arrows (see the bottom of Figure 5.7). It is easy to see that the hypothesis that the gradient path from β to α was unique implies that our construction has not resulted in the creation of any closed paths, and hence the new vector field is a gradient vector field. Moreover, it is also clear that neither α nor β is a critical cell of the new vector field, and we have not changed the criticality of any other cell. This completes the proof.

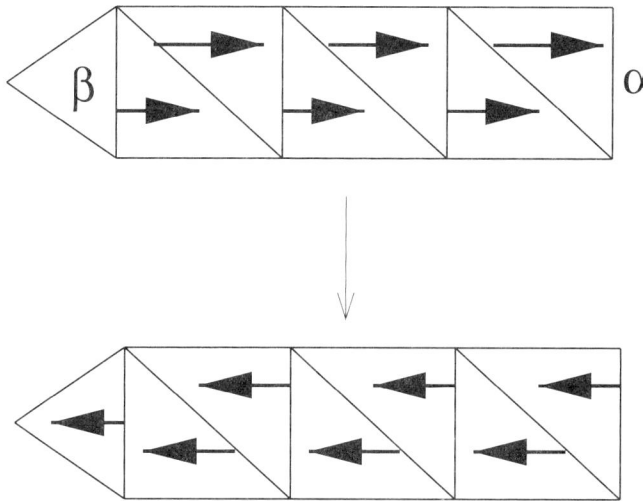

Cancelling the critical simplices α and β

FIGURE 5.7.

It is interesting, I think, to compare the proof of this combinatorial version to that of the classical smooth version, either as proved by Morse in [**46**] or Milnor [**45**]. The proof is essentially the same in both versions, in that one simply reverses the vector field along the distinguished path. However, when working in the smooth category one has more work to do, since the result must be a smooth vector field. Thus one must also turn all nearby vectors, making certain that the result is a gradient vector field, so that, for example, one must make certain that no closed orbits are introduced. All such difficulties are avoided in the combinatorial setting, because there are no nearby vectors.

This is a rather extreme case of the sort of simplification that one sees quite often when working in the combinatorial category. To be honest, while presenting some technical difficulties, the proof of the cancellation theorem is not one of the most challenging steps in the proof of the higher-dimensional Poincaré conjecture. As far as I know, the transition from the smooth category to the combinatorial category does not introduce any other significant simplifications in the proof of the higher dimensional Poincaré conjecture. In addition, the "difficulties" that occur in dimensions 3 and 4 in the smooth category definitely persist in the combinatorial category.

The cancelation theorem can be used to show that every connected combinatorial n-manifold (resp. n-manifold with nonempty boundary) has a gradient vector field with exactly one critical vertex, and one critical n-cell (resp. no critical n-cells). Combining this result with the Morse inequlaities yields the following general result, which we will apply in section 10.

THEOREM 5.8. *Let M be a connected combinatorial 3-manifold with boundary, satisfying $\chi(M) = 1$. Then M is collapsible if and ony if M has a discrete gradient vector field with no critical edges.*

This result can be used to simplify some existing proofs in the literature (e.g. [**11**]).

6. A Simple Example

In this section we demonstrate some of the ideas of the previous section with a simple example. Fix a positive integer n, and consider the following $(n-2)$-dimensional simplicial complex, which we denote M_n. Starting with the following expression
$$(x_0 x_1 x_2 \ldots x_n)$$
consider all ways of adding legal pairs of parentheses. An expression resulting from adding $p+1$ pairs of parentheses will be a p-simplex in our complex. The faces of this p-simplex are all expressions that result from removing corresponding pairs of parentheses.

For example, consider the case $n = 3$. The vertices of M_3 are the expressions
$$v_1 = ((x_0 x_1) x_2 x_3), v_2 = ((x_0 x_1 x_2) x_3), v_3 = (x_0 (x_1 x_2) x_3),$$
$$v_4 = (x_0 (x_1 x_2 x_3)), v_5 = (x_0 x_1 (x_2 x_3)),$$
and the edges are the expressions
$$e_1 = (((x_0 x_1) x_2) x_3), e_2 = ((x_0 (x_1 x_2)) x_3), e_3 = (x_0 ((x_1 x_2) x_3)),$$
$$e_4 = (x_0 (x_1 (x_2 x_3))), e_5 = ((x_0 x_1)(x_2 x_3)).$$
One can easily check the relations
$$e_1 = \{v_1, v_2\}, e_2 = \{v_2, v_3\} e_3 = \{v_3, v_4\}, e_4 = \{v_4, v_5\}, e_5 = \{v_5, v_1\},$$
so that M_3 is a circle triangulated with 5 edges and 5 vertices.

These complexes arise in a number of different settings. For example, they arise in the study of planar rooted trees. To illustrate by an example, the edge $(((x_0 x_1 x_2) x_3) x_4)$ of M_4 can naturally be associated with the planar rooted tree shown in Figure 6.1. From this point of view, the top dimensional simplices correspond to binary trees. (See [**6**] and the references therein for an extensive discussion of such issues.) Moreover, the complexes M_n arise in geometry, as they are closely related to the simplicial complex of subdivisions of an $(n+1)$-gon into subpolygons (see, e.g. [**40**]). In the study of homotopy associative algebras ([**58**][**59**]) one studies an algebra which is associative only up to homotopy. In that case, M_2, for example, arises from studying all ways of multiplying 3 elements, with $(x_0 x_1 x_2)$ representing a homotopy between $((x_0 x_1) x_2)$ and $(x_0 (x_1 x_2))$. Note that here we see a slight difference. From this point of view, one would like to think of $((x_0 x_1) x_2)$ and $(x_0 (x_1 x_2))$ as vertices, and $(x_0 x_1 x_2)$ as an edge between them. Thus, in this context, one is essentially working with the dual of the complex we have defined. We will say more about this a bit later (see the remarks following Theorem 5.3).

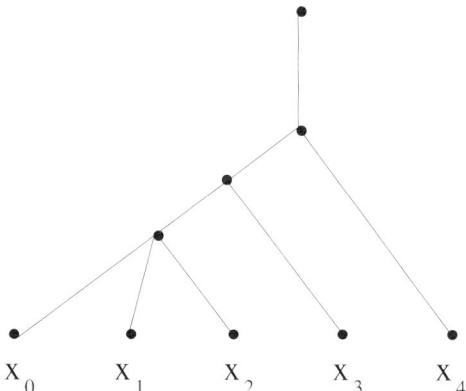

The planar rooted tree corresponding to $(((x_0x_1x_2)x_3)x_4)$

FIGURE 6.1.

The main goal of this section is to use discrete Morse theory to give a simple proof of the following result.

THEOREM 6.2. *The complex M_n is homotopy equivalent to an $(n-2)$-sphere.*

This result is well known, and it is only our proof that is new. We will prove this theorem by showing that one can easily construct a discrete gradient vector field on M_n which has precisely two critical simplices, namely one critical vertex and one critical $(n-2)$-simplex. The theorem then follows from Theorem 5.2. In fact, one can deduce more. We saw above that M_3 is not just a homotopy circle, but rather it is an actual combinatorial circle. One can easily see that the link of every vertex of M_n is isomorphic to a complex of the form $M_p * M_{n-p}$ (where $*$ denotes join). By induction, M_p and M_{n-p} are combinatorial spheres of dimension $p-2$ and $n-p-2$, respectively, so the link is a combinatorial sphere of dimension $n-3$ (see Proposition II.1 of [**28**]). Since the link of every vertex of M_n is a combinatorial $(n-3)$-sphere, it follows that M_n is a combinatorial $(n-2)$-manifold (see page 19 of [**28**]). Therefore we can apply Theorem 5.5 to learn the following stronger result.

THEOREM 6.3. *The complex M_n is a combinatorial $(n-2)$-sphere.*

Before beginning our proof, we return to our earlier comments about the complex arising in the study of homotopy associative algebras. As remarked above, in that case one considers what is essentially the dual of the complex M_n. However, there is a slight modification. Let M_n^* denote a combinatorial $(n-2)$-sphere endowed with the cell decomposition which is dual to that of M_n. In M_n the trivial expression $(x_0x_1\ldots x_n)$ corresponds to a simplex of dimension -1, i.e. the empty set. In the dual setting, $(x_0x_1,\ldots x_n)$ corresponds to a cell of dimension $n-1$, namely the interior of the cone on M_n^*. Adding in this cell to form the cone on M_n^* results in a complex, introduced in [**58**] (see also [**59**]) called the associahedron (or Stasheff polytope), and which is often denoted A_{n+1}. Thus we learn

COROLLARY 6.4. *The associahedron A_{n+1} is a combinatorial $(n-1)$-ball.*

A proof of this appears in [**58**], by very different methods, and numerous alternative proofs have also been presented. In fact, A_{n+1} is a polytope ([**40**]).

Let us now describe the construction of the desired gradient vector field V on M_n. Let s be a simplex of M_n. Suppose that there is not a pair of parenthesis around x_0 and x_1. If it is possible to legally add a pair of parenthesis around x_0 and x_1 do so and call the resulting simplex t. We then add the pair $\{s \prec t\}$ to V. For example, in M_4 the expression $((x_0 x_1 x_2)(x_3 x_4))$ is paired with $(((x_0 x_1) x_2)(x_3 x_4))$. After this step, the expressions which have not been paired with any other expression are those that have at least one parenthesis between x_0 and x_1, and it is simple to see that any such parenthesis must be a left parenthesis. There is one additional unpaired expression, namely the expression $s^* = ((x_0 x_1) x_2 x_3 \ldots x_n)$. According to our rule, this should be paired with the original expression $(x_0, x_1 \ldots x_n)$ with no added parentheses, but this is not permitted.

If s is any expression other than s^* that is currently unpaired, and a pair of parentheses can legally be added around the elements x_1 and x_2, do so and call the resulting simplex t. We then add the pair $\{s \prec t\}$ to V. After this step, the expressions which have not been paired with any other expression are s^* and those that have at least one left parenthesis between x_0 and x_1, and at least one left parenthesis between x_1 and x_2. Pair such an expression with the one resulting from adding a pair of parenthese around x_2 and x_3 if possible. Continue this process as long as possible. When it has terminated, the only expressions that have not been paired up with any other expression are s^* and the one that has a left parenthesis between every consecutive pair x_1 and x_{i+1} for $i = 0, 1, \ldots, n-1$, i.e. the expression $t^* = (x_0(x_1(x_2(\ldots(x_{n-2}(x_{n-1} x_n))))\ldots)$. Note that t^* is an $(n-2)$-simplex of the complex M_n.

This completes our construction of the vector field V. All that needs to be checked is there are no closed V-paths. Denote by V_k the discrete vector field that has been constructed after the kth step in the construction, i.e. after consideration of the pair x_{k-1}, x_k. Let $s_0^{(p)}, t_0^{(p+1)}, s_1(p)$ denote a V-path. This requires that s_0 and t_0 be paired in V. Suppose that s_0 and t_0 are paired in V_k. The reader can check that this implies that either s_1 is the head of an arrow in V_k (and hence the V-path cannot be continued) or s_1 is paired in V_{k-1}. Thus, by induction, there can be no closed V-paths.

7. An Application to Computer Science

We now present an application of discrete Morse theory to computer science. That such an application exists should not be surprising, since many algorithms can be naturally interpreted as a discrete dynamical system of the sort we have introduced. One idea of this work is that a typical impossibility result in computer science shows that a problem is impossible to solve by showing that any algorithm will fail to solve the problem in at least one case. In some settings a failure of the algorithm can be identified with a zero or closed orbit of a discrete vector field. Hence the ideas we have presented, in particular the Morse inequalities of Theorem 2.7, can be used to give a topological explanation for the necessary failure of any algorithm, as well as quantitative lower bounds on the number of failures. In this section we present a single example of this phenomenon. Hopefully we will be able to explore this topic further in a later paper. For a more complete treatment of the content of this section, see [**20**].

The problem we study is a topological version of a standard type of search problem. The generalized version that we will present first appeared in [**52**]. Let S be an n-dimensional simplex, with vertices $v_0, v_1, \ldots, v_n$, and K a subcomplex of S which is known to you. Let σ be a face of S which is not known to you. Your goal is to determine if σ is in K. In particular, you need not determine the face σ, just whether or not it is in K. You are permitted to ask questions of the form "Is v_i in σ?". You may use the answers to the questions you have already asked in determining which vertex to ask about next. Of course, you can determine if σ is in K by asking $n+1$ questions, since by asking about all $n+1$ vertices you can completely determine σ. You win this game if you answer the given question after asking fewer that $n+1$ questions.

Say that K is *nonevasive* if there is a winning strategy for this game, i.e there is a question algorithm that determines whether or not $\sigma \in K$ in fewer than $n+1$ questions, no matter what σ is. Say K is *evasive* otherwise.

Kahn, Saks and Sturtevant ([**36**]) proved the following relationship between the evasiveness of K and its algebraic topology.

THEOREM 7.1. *If $\tilde{H}_*(K) \neq 0$, where $\tilde{H}_*(K)$ denotes the reduced homology of K, then K is evasive.*

In fact, they proved something stronger, and we will come back to this point later. We illustrate the previous theorem with a simple example. Let S be the 2-simplex shown in Figure 7.2, spanned by the vertices $v_0, v_1,$ and v_2, with K the subcomplex consisting of the edge $[v_0, v_1]$ together with the vertex v_2.

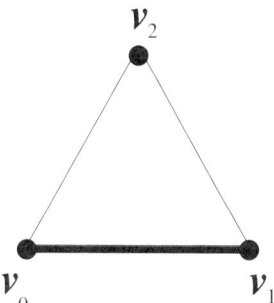

An example of an evasive subcomplex of the 2-dimensional simplex.

FIGURE 7.2.

A possible guessing algorithm is shown in Figure 7.3 (where we indicate which vertex to ask about at each stage, given the answers of the previous questions). Define an *evader* of a guessing algorithm to be a face σ of S with the property that when questions are asked in the order determined by the algorithm one must ask all three questions before it is known whether or not σ is in K. In particular, the evaders of the illustrated guessing algorithm are:

$$\sigma = [v_2], \quad [v_0, v_2]$$

Note that the subcomplex K has nonzero reduced homology, so the theorem of Kahn, Saks, and Sturtevant guarantees that every guessing algorithm has some evaders.

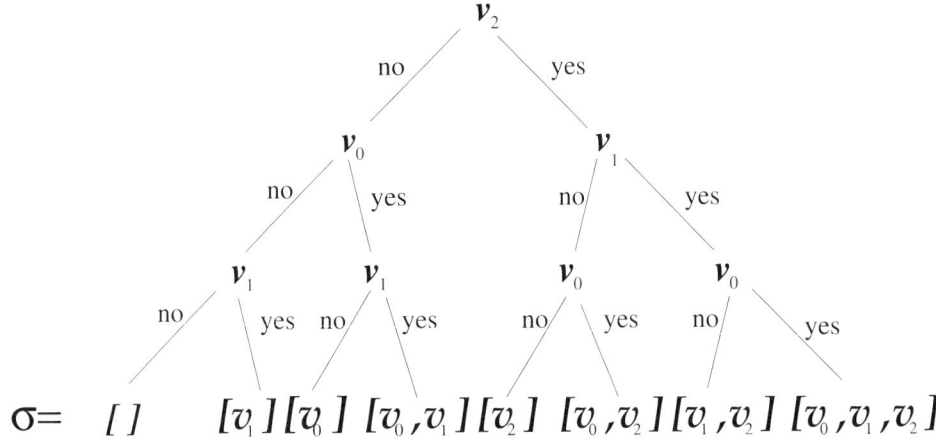

A guessing algorithm

FIGURE 7.3.

Discrete Morse theory comes to the fore when one observes that a guessing algorithm induces a discrete vector field on S. For example, the guessing algorithm shown in Figure 7.3 induces the vector field

$$V = \quad \{\emptyset < [v_1]\}\{[v_0] < [v_0, v_1]\}$$
$$\{[v_2] < [v_0, v_2]\}\{[v_1, v_2] < [v_0, v_1, v_2]\}$$

That is, V consists of those pairs of faces of S which are not distinguished by the guessing algorithm until the last question. There is slight subtlety here in that a guessing algorithm pairs a vertex with the empty simplex $\emptyset$, which is not permitted in our original definition a discrete vector field. Thus, to get a true discrete vector field, we must remove this pair from V. It is precisely this subtle point that results in the reduced homology of K being the relevant measure of topological complexity, rather than the nonreduced homology. However, for simplicity, from now on we will simply ignore this technical point.

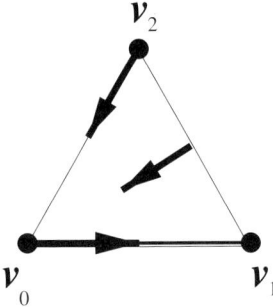

The vector field induced by the guessing algorithm shown in Figure 7.3.

FIGURE 7.4.

The reader can easily check that the discrete vector field illustrated in Figure 7.4 is a gradient vector field. This is true in general.

THEOREM 7.5. *The discrete vector field induced by any guessing algorithm is a gradient vector field.*

We will postpone the proof of this result until the end of this section.

Now restrict V to K (by taking only those pairs in V s.t. both simplices are in K). For example, in our example, this results in the vector field

$$V_K = \{\{[v_0] < [v_0, v_1]\}\}.$$

From the Theorem 7.5, V has no closed orbits. Any discrete vector field consisting of a subset of the pairs of V has fewer paths, and hence also has no close orbits. Therefore, V_K is a gradient vector field on K. Note that V pairs every face of S with another face, and hence there are no critical simplices (we are continuing to ignore for now the simplex which is paired with the emptyset). Thus, the critical simplices of V_K are precisely the simplices of K which are paired in V with a face of S which is not in K. These are precisely the simplices of K which are the evaders of the guessing algorithm.

The Morse inequalities imply that the number of evaders in K is at least $\dim \widetilde{H}_*(K)$. Evaders occur in pairs, with each pair having one face of K and one face not in K. This yields the following quantitative refinement of Theorem 7.1.

THEOREM 7.6. *For any guessing algorithm*

$$\# \text{ of evaders } \geq 2 \dim \widetilde{H}_*(K)$$

Suppose that K is nonevasive. Then there is some guessing algorithm which has no evaders. From our above discussion we seen that this implies that K has a gradient vector field with no critical simplices. Actually, this is not quite true. The gradient vector field must have a critical vertex - the vertex that is paired with the empty set - this is that minor technicality that we have been ignoring. Applying Lemma 5.3 yields the following strengthening of Theorem 7.1.

THEOREM 7.7. *If K is nonevasive, then K simplicially collapses to a point.*

This theorem appears in [**36**]. The interested reader can consult [**20**] for some additional refinements of this theorem.

We end this section with a proof of Theorem 7.5. Fix a subcomplex K of an n-simplex S, and a guessing algorithm. Associate to each p-simplex α of S the sequence of integers

$$n(\alpha) = n_0(\alpha) < n_1(\alpha) < \cdots < n_p(\alpha)$$

where the $n_i(\alpha)$'s are the numbers of the questions answered "yes" if $\sigma = \alpha$.

If V is the vector field induced by the guessing algorithm and

$$\alpha_0^{(p)}, \quad \beta_0^{(p+1)}, \quad \alpha_1^{(p)}$$

is a V-path, then $\{\alpha_0, \beta_0\}$ is in V, which means that α_0 and β_0 are not distinguished until the $(n+1)^{st}$ question. Thus,

$$n(\beta_0) = n_0(\alpha_0) < n_1(\alpha_0) < \cdots < n_p(\alpha_0) < n+1.$$

We now observe that the vertices of a_1 are a subset of the vertices of b_0. Suppose the vertex of β_0 which is not in α_1 is the vertex tested in question $n_i(\beta_0)$. Then we must have $i \neq n+1$ (since $\alpha_0 \neq \alpha_1$). This demonstrates that

$$n(\alpha_1) = n_0(\alpha_0) < n_1(\alpha_0) < \cdots < n_{i-1}(\alpha_0) < n_i(\alpha_1) < \ldots$$

for some $i < n+1$, and such that $n_i(\alpha_1) > n_i(\alpha_0)$. Thus $n(\alpha_1) > n(\alpha_0)$ in the lexicographic order, which is sufficient to prove that there are no closed orbits.

8. Topological Invariants via counting Gradient Paths

The First Cancellation Theorem (see Theorem 5.6) states that if these is a unique gradient path between two critical points, then the critical points are not essential and can be cancelled out. This hints at the fact that if we have extra information about the gradient paths travelling between critical points, then we may be able to strengthen our conclusions. In what follows we present three examples of this phenomenon.

I. The Morse Complex

In the first, and fundamental example, we will show how a complete knowledge of the gradient paths of a gradient vector field V on a simplicial complex K allows one to calculate the homology of K exactly.

Let $\mathcal{M}_p(K,\mathbb{Z}) \subseteq C_p(K,\mathbb{Z})$ denote the span of the critical p-cells (we call $\mathcal{M}_*(K,\mathbb{Z})$ the space of Morse chains). From Theorem 5.2 it follows that there is a chain differential $\tilde{\partial}$, so that the homology of the complex

$$(8.1) \qquad \mathcal{M}_* : 0 \longrightarrow \mathcal{M}^n \xrightarrow{\tilde{\partial}} \mathcal{M}_{n-1} \xrightarrow{\tilde{\partial}} \mathcal{M}_{n-2} \xrightarrow{\tilde{\partial}} \ldots \xrightarrow{\tilde{\partial}} \mathcal{M}_0 \longrightarrow 0$$

is isomorphic to the homology of K (where we have written $\mathcal{M}_p$ for $\mathcal{M}_p(K,\mathbb{Z})$).

Our main goal is to give an explicit description of this differential in terms of the gradient paths of V. Choose an orientation for each critical simplex. For any critical $(p+1)$-simplex β, set

$$(8.2) \qquad \tilde{\partial}\beta = \sum_{\text{critical } \alpha^{(p)}} c_{\alpha\beta} \alpha$$

where

$$(8.3) \qquad c_{\alpha\beta} = \sum_{\gamma \in \Gamma(\beta,\alpha)} m(\gamma).$$

The sum is over all $\Gamma(\beta,\alpha)$, the set of gradient paths from some maximal face of β to α. Given such a gradient path γ, an orientation on β induces an orientation on α. Rather than giving a precise definition here, we will simply note that the intuitive idea is that one slides the orientation along γ. For example, in Figure 8.2 we illustrate a single gradient path $\gamma \in \Gamma(\beta,\alpha)$. In this case, the indicated orientation on β induces the indicated orientation on α. We define the multiplicity of γ, $m(\gamma)$ to ± 1 depending on whether or not the chosen orientation on β indices the chosen orientation on α.

We can now present the main theorem.

THEOREM 8.1. *The operator $\tilde{\partial}$ defined in (8.2) and (8.3) satisfies $\tilde{\partial} \circ \tilde{\partial} = 0$. With this differential, the differential complex (8.1) has the same homology as the underlying simplicial complex K.*

We refer to this complex as the *Morse complex* (although a number of different names appear in the litereature). The main ingredient of the proof of Theorem 8.1 is presented in section 11. In the smooth setting this complex is studied in great detail in [**53**].

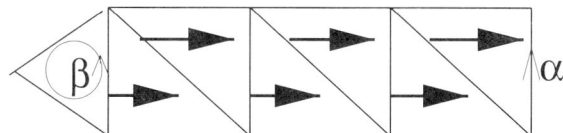

Counting such gradient paths leads to the differential of the Morse complex.

FIGURE 8.2.

Here we see another important simplification that arises in the transistion from the smooth to the combinatorial category. In the smooth setting, one must take care to ensure that gradient paths only travel from a critical point to critical points of smaller index. Moreover, the standard definitions of multiplicity in this setting requires that all gradient paths be nondegenerate in a certain sense. Gradient vector fields which satisfy these conditions are called *Morse-Smale gradient vector fields*. Such gradient vector fields are generic in the space of all gradient vector fields. In the combinatorial setting, however, such issues simply never arise.

Although we presented the Morse complex from the point of view of homology, we could equally well have presented a Morse cocomplex which calculates the cohomology of K. In this case, we let $\mathcal{M}_p(K, \mathbb{Z})$ denote the p-cochains in M which are supported on the critical p-simplices. Writing $\mathcal{M}_p$ for $\mathcal{M}_p(K, \mathbb{Z})$, the Morse cocomplex has the form

$$\mathcal{M}^* : 0 \longrightarrow \mathcal{M}^0 \xrightarrow{\tilde{\delta}} \mathcal{M}^1 \xrightarrow{\tilde{\delta}} \mathcal{M}^2 \xrightarrow{\tilde{\delta}} \ldots \xrightarrow{\tilde{\delta}} \mathcal{M}_n \longrightarrow 0$$

where $\tilde{\delta}$ is the adjoint of $\tilde{\partial}$.

II. The Cup Product

Now that we have a Morse representation of the cohomology of K, it is natural to ask how one can see the cup product, and other cohomological operations, from the point of view of Morse theory. In the smooth setting this question was answered by Betz-Cohen ([**5**]) and independently by Fukaya ([**25**],[**27**]), who show how one can express cohomology operations in terms of a counting of gradient paths. One can apply the same procedure here. In this section, we indicate only how one can construct the cup product as a map of the Morse cochains. Let V_1, V_2, and V_3 be any three gradient vector fields on a finite simplicial complex K. The question we answer is: Suppose we have a cohomology class c_1 represented in terms of the Morse cocomplex induced by V_1 (i.e. given as a closed cochain in the corresponding Morse cocomplex), and a cohomology class c_2 represented in terms of the Morse cocomplex induced by V_2. How can we express the cup product $c_1 \cup c_2$ in terms of the Morse cocomplex induced by V_3?

In this case, rather than just counting single gradient paths, we will have to count Morse embeddings of the configuration shown in Figure 8.3, i.e. triples of gradient paths. One a V_1-gradient path ending at a V_1 critical α_1, and another a V_2 gradient path ending at a V_2-critical simplex α_2, beginning at any simplices a_1 and a_2, resp., and the third a V_3 path from a V_3 critical simplex α_3 to any simplex a_3. We require that $a_1^*, a_2^*,$ and a_3^*, the cosimplices dual to $a_1, a_2,$ and a_3 resp., satisfy $a_1^* \cup a_2^* = \pm a_3^*$, where $\cup$ represents any map on cosimplices which induces the cup product on cohomology. We draw such a configuration in Figure 8.4. In this figure, the solid arrows represent the discrete gradient vector field V_3, and the dashed and dotted arrows represent V_2 and V_1, resp.

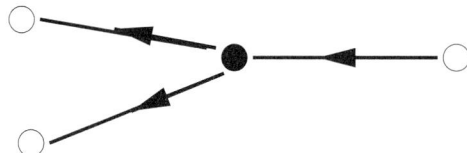

Counting Morse embeddings of this figure results in a map which induces the cup product.

FIGURE 8.3.

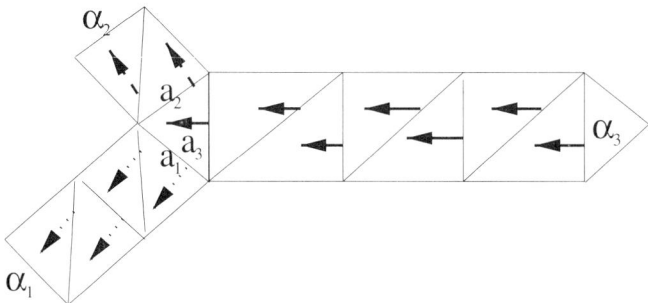

A more precise picture of the configuration shown in Figure 8.3.

FIGURE 8.4.

We now proceed more precisely. Define
$$\kappa : \mathcal{M}^*(V_1) \otimes \mathcal{M}^*(V_2) \longrightarrow \mathcal{M}^*(V_3).$$

By

(8.4) $$\kappa(\alpha_1, \alpha_2) = \sum_{\alpha_3} \kappa^{\alpha_3}_{\alpha_1, \alpha_2} \alpha_3$$

where α_i is critical for V_i and

(8.5) $$\kappa^{\alpha_3}_{\alpha_1, \alpha_2} = \sum_{\alpha_1 \cup \alpha_2 = \alpha_3} \sum_{\gamma_1, \gamma_2, \gamma_3} \pm 1$$

where

γ_1 is a V_1-gradient path' from α_1 to a_1

γ_2 is a V_2-gradient path from α_2 to a_2

and

γ_3 is a V_3-gradient path from a_3 to α_3

As usual, the $\pm$ is determined by a careful consideration of orientations.

In [**22**] we prove the following theorem.

THEOREM 8.5. *The map κ defined in (8.4) and (8.5) induces the cup product on cohomology.*

A corresponding result holds in the smooth case (as demonstrated by Betz-Cohen [**5**] and Fukaya [**25**] [**27**]) with an additional complication. Namely, in the smooth case these is again a genericity condition which requires that each of the gradient vector fields be Morse-Smale and, in addition, be in general position with respect to each other. There are no such conditions in the combinatorial setting. In particular, one may choose each of V_1, V_2 and V_3 to be the trivial gradient vector field, that is the vector field consisting of no pairs, so that the space of Morse cochains is equal to the space of all cochains. Then the map introduced above is just the original cup product on cochains.

III. Finite-Type Invariants

In this section we discuss some work in progress. As a consequence, we will not even state a precise theorem. (In fact, it would take quite a bit of work to state the precise theorem.)

In [**26**], Fukaya showed how to define a finite type invariant of a (smooth) rational homology S^3 by choosing a generic triple of Morse functions, (f_1, f_2, f_3), counting the Morse embeddings of the diagram shown in Figure 8.6, and adding correction terms corresponding to related diagrams. Much work goes into proving that the resulting invariant does not depend on the choice of the Morse functions.

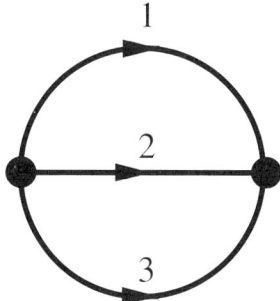

Counting Morse embeddings of this theta graph leads to a finite-type invariant.

FIGURE 8.6.

The corresponding construction using discrete Morse theory leads to a combinatorial invariant of combinatorial $\mathbb{Q}$-homology S^3's. At this point the reader may wonder why we must restrict attention to combinatorial manifolds. After all, up to this point everything was true for a general simplical complex, and even more generally. Let us take another look at Figure 8.6. Rather than viewing the diagram as three arcs beginning together and ending together, it is more accurate to view the figure as a set of loops. One loop is formed by going in the positive direction along the arc labelled 1, followed by going in the negative direction along the arc labelled 2. Another loop is formed by going in the positive direction along the arc labelled 2 and in the negative direction along the arc labelled 3, etc.

In the smooth category, there is no difficulty in flowing in the direction of a gradient vector field V or in the opposite direction, since flowing in the opposite direction is equivalent to flowing in the direction of the gradient vector field $-V$. In the combinatorial category, at least for general cell complexes, the negative of a vector field does not exist. For a combinatorial manifold, however, the negative of a discrete vector field does exist as a discete vector field on the combinatorial manifold with the dual cell decomposition. It is precisely for this reason that we restrict attention to combinatorial manifolds.

The upshot is that following Fukaya's work, one can show that the analogous construction in the combinatorial category yields a combinatorial invariant. At present, however, at certain points our proof rests on converting the combinatorial ingredients into continuous ingredients and applying Fukaya's proof. This is not completely satisfying. The main reason is that Fukaya's construction requires the three smooth gradient vector fields be in general position with respect to each other - in order to be able to easily count intersections of stable and unstable manifolds, and a similar condition remains in the combinatorial setting - in order to be able to define the intersections of the corresponding chains. However, just as no such condition was required to define a cup product map in Morse theory in Part II of this section, it seems clear that no such condition is really necessary for the construction of the combinatorial invariant. In order to provide a proof that holds for any triple of gradient vector fields, one should work entirely in the combinatorial category. Thus, what remains in this project is to use the ideas of Part II of this section to define all necessary intersections combinatorially, and then to give a direct combinatorial proof of the combinatorial invariance of the resulting invariant.

Finding a construction which does not require the three vector fields to be mutually transverse has more than just aesthetic value. In [1], Axelrod and Singer define a finite type invariant of a rational homology 3-sphere. The Axelrod-Singer invariant follows formally from Fukaya's construction if one chooses all three Morse functions to be constant functions, which of course are not Morse functions at all, let alone mutually transverse. Even though Fukaya's construction does not permit this choice of functions, it is natural to conjecture that his invariant is equal to that of Axelrod and Singer. (Fukaya presents some ideas - along the line of those presented in the next section of this paper- for how one might be able to go about proving the equality of Fukaya's invariant and that of Axelrod and Singer). One advantage of working in the combinatorial category in which all triples of gradient vector fields are permitted is that the combinatorial version of Fukaya's invariant and the combinatorial analogue of the invariant of Axelrod and Singer are all on equal footing, different special cases of the same construction, and thus is would follow immediately that they are equal. Of course the difficulty would then remain of showing that the combinatorial versions of these invariants are equal to the smooth versions.

9. Supersymmetry and Discrete Morse Theory

In the paper "Supersymmetry and Morse theory", Witten showed how Morse theory may be derived from supersymmetric considerations in quantum field theory. The same approach may be taken with discrete Morse theory, with, as usual, some simplifications. Perhaps most significantly, Witten derives the Morse complex from a mathematically ill-defined path integral. Following the same ideas in the combinatorial setting, we are lead instead to a well-defined discrete sum. Unfortunately, this last point is a bit too far afield to go into here (see section 8 of [17]). We will have to content ourselves here with a supersymmetric derivation of the Morse inequalities.

First, we must take another look at the idea of a discrete Morse function. Previously in these notes, such a function was defined to be a Lyapunov function associated to a gradient vector field. The following theorem gives a useful alternative definition.

THEOREM 9.1. *A function is a discrete Morse function if and only if for each simplex* α,
$$\#\{\beta \succ \alpha \mid f(\beta) < f(\alpha)\} \leq 1$$
and
$$\#\{\gamma \prec \alpha \mid f(\gamma) > f(\alpha)\} \leq 1.$$

In fact, in [**16**], Theorem 9.1 is used as the definition of a discrete Morse function, and the equivalence with the definition used here - a Lyapunov function corresponding to a gradient vector field with no closed orbits - is proved in Theorem 9.3 of [**16**]. It is quite useful, when doing the sort of analysis which follows, to restrict attention to a special class of discrete Morse functions. However, this restriction does not affect the generality of this approach, as we will now explain.

DEFINITION 9.2. *Say a discrete Morse function is flat if $\beta \succ \alpha$ implies that $f(\beta) \geq f(\alpha)$.*

DEFINITION 9.3. *Say that two discrete Morse functions f and g are equivalent if for every pair $\beta \succ \sigma$*

$$f(\beta) > f(\alpha) \iff g(\beta) > g(\alpha).$$

In particular, equivalent discrete Morse functions have the same critical simplices and gradient vector fields.

Figure 9.5(i) displays a discrete Morse function which is not flat. In Figure 9.5 (ii) we show a flat discrete Morse function which is equivalent to the function in 9.5(i). In fact, we have the following result (Theorem 1.4 of [**17**]).

THEOREM 9.4. *Every discrete Morse function is equivalent to a flat discrete Morse function.*

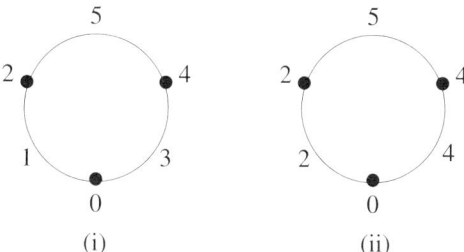

(i). A nonflat discrete Morse function. (ii). A flat discrete Morse fuction

FIGURE 9.5.

It follows from Theorem 9.4 that in order to prove the Morse inequalities for a general discrete Morse function, it is sufficient to restrict attention to flat Morse functions. With that in mind, let f denote a flat Morse function. Beginning with the real chain complex

$$0 \longrightarrow C_n(M, \mathbb{R}) \xrightarrow{\partial} C_{n-1}(M, \mathbb{R}) \xrightarrow{\partial} \cdots \xrightarrow{\partial} C_0(M, \mathbb{R}) \longrightarrow 0$$

we replace the boundary operator ∂ by

$$\partial_t = e^{tf} \partial e^{-tf}$$

where e^{tf} multiplies each simplex α by $e^{tf(\alpha)}$ and is extended linearly to all chains. The resulting differential complex has the same homology for all $t \in \mathbb{R}$. We will now investigate this homology using ideas from Hodge theory. Let ∂_t^* denote the adjoint of ∂_t with respect to the canonical inner product $<,>$ on chains, and $\Delta_p(t)$ the induced Laplacian on p-chains

$$\Delta_p(t) = \partial_t \partial_t^* + \partial_t^* \partial_t : C_p(M, \mathbb{R}) \longrightarrow C_p(M, \mathbb{R}).$$

It follows from basic linear algebra that for each $t \in \mathbb{R}$,
$$\operatorname{Ker} \Delta_p(t) \cong H_p(M, \mathbb{R}).$$

The next step is to derive a more explicit representation for $\Delta_p(t)$. Choose an orientation for each simplex. Then
$$\partial_t \sigma^{(p)} = \sum_{\gamma^{(p-1)} < \sigma} \pm e^{t(f(\gamma) - f(\sigma))} u$$

(where the $\pm$ sign is determined by whether the orientations on γ and σ agree or not),
$$\partial_t^* \sigma^{(p)} = \sum_{\beta^{(p+1)} > \sigma} \pm e^{t(f(\sigma) - f(\beta))} \beta,$$

and
$$\Delta_p(t) \sigma^{(p)} =$$
$$\sum_{\tilde{\sigma}^{(p)}} \left[\sum_{\gamma^{(p-1)} < \sigma\gamma < \tilde{\sigma}} \pm e^{t(2f(\gamma) - f(\sigma) - f(\tilde{\sigma}))} + \sum_{\beta^{(p+1)} > \sigma\beta > \tilde{\sigma}} \pm e^{t(f(\sigma) + f(\tilde{\sigma}) - 2f(\beta))} \right] \tilde{\sigma}.$$

It follows from our restrictions on f that if $\tilde{\sigma} \neq \sigma$, then the coefficient of $\tilde{\sigma}$ in the above expression is $O(e^{-tc})$ as $t \to \infty$, for some $c > 0$. Thus, up to exponentially small errors as $t \to \infty$, the Laplacian $\Delta_p(t)$ is diagonalized by the p-simplices. A closer look (see section 3 of [**17**]) reveals

(9.1) $$\left\langle \Delta_p(t) \sigma^{(p)}, \sigma^{(p)} \right\rangle = \begin{cases} O(e^{-tc}) & \text{if } \sigma \text{ is critical} \\ 1 + O(e^{-tc}) & \text{if } \sigma \text{ is not critical.} \end{cases}$$

In particular, we learn

THEOREM 9.6. *As $t \to \infty$,*

$$\#\{eigenvalues \text{ of } \Delta_p(t) \text{ which are } O(e^{-tc})\} = \#\{critical\ p - simplices\}.$$

All other eigenvalues of $\Delta_p(t)$ are $1 + O(e^{-tc})$.

That is, the dimension of the kernel of $\Delta_p(t)$ is equal to $b_p = \dim H_p(K, \mathbb{R})$ for each $t \in \mathbb{R}$, and the dimension of the kernel of $\Delta_p(\infty)$ is m_p, the number of critical p-simplices. Since the dimension of the kernel is an upper-semicontinuous function of t we learn the weak Morse inequalities

$$m_p \geq b_p.$$

In fact, Theorem 9.6 is sufficient to imply the strong Morse inequalities (over $\mathbb{R}$). See, for example, [**8**] (where the proof is discussed in the smooth category) or [**17**] for the completion of the argument.

We now take a closer look at the small eigenspaces. Let $W_p(t) \subseteq C_p(M, \mathbb{R})$ denote the span of the eigenvectors of Δ_t^p corresponding to the eigenvalues which are $O(e^{-tc})$. Equivalently, for t large enough one can define $W_p(t)$ to be the span of the eigenvectors of Δ_t^p corresponding to the eigenvalues which are less than $\frac{1}{2}$. Then

$$\partial_t(W_p(t)) \subseteq W_{p-1}(t)$$

so for all t we have a differential complex (the Witten complex) with the same homology as M

$$\mathcal{W}(t) : 0 \longrightarrow W_n(t) \xrightarrow{\partial_t} W_{n-1}(t) \xrightarrow{\partial_t} \cdots \xrightarrow{\partial_t} W_0(t) \longrightarrow 0.$$

It follows from (9.1) and the remarks directly preceeding (8.1) that

$$\lim_{t \to \infty} W_p(t) = \mathcal{M}_p,$$

where the limit is with respect to the natural topology on the Grassmannian of m_p-dimensional subspaces of $C_p(K, \mathbb{R})$, and where, as in the previous section, $\mathcal{M}_p$ denotes the Morse p-chains, i.e. the p-chains which are linear combinations of critical p-simplices. We can say more. In fact, after a simple change of coordinates the differential ∂_t on $\mathcal{W}(t)$ converges to the differential $\widetilde{\partial}$ on $\mathcal{M}$. To describe this change of coordinates, let π denote the orthogonal projection onto $\mathcal{M}_p$. For large t

$$\pi : W_p(t) \longrightarrow \mathcal{M}_p$$

is an isomorphism. Define a function θ on $W_p(t)$ by setting, for any critical simplex $\sigma^{(p)}$

$$\theta(\pi^{-1}(\sigma)) = e^{tf(\sigma)}$$

and extending linearly. Let $\widetilde{\partial}_t = \theta^{-1} \partial_t \theta$.

THEOREM 9.7. $\lim_{t \to \infty} (\mathcal{W}(t), \widetilde{\partial}_t) = (\mathcal{M}, \widetilde{\partial})$.

See Theorem 4.1 in [**17**]. In [**17**] we use this theorem to give a second proof, along the lines suggested by Witten in [**64**] that the discrete Morse complex $\mathcal{M}$ has the same homology as the underlying manifold.

We now return to the observation that all eigenvalues of $\Delta_p(t)$ which do not converge to 0, converge to 1. This may seem like just a curiosity, but it obviously has interesting implications when one is considering the determinant of the Laplace operator. Given a representation

$$\rho : \pi_1(M) \longrightarrow O(n, \mathbb{R})$$

we can twist the chain complex by ρ to define a complex

$$\mathcal{C}(\rho) : 0 \longrightarrow C_n(M, \rho) \xrightarrow{\partial} C_{n-1}(M, \rho) \xrightarrow{\partial} \cdots \longrightarrow C_0(M, \rho) \longrightarrow 0.$$

There is a canonical inner product on each $C_p(M, \rho)$ and hence we can define an adjoint operator ∂^* and a Laplacian

$$\Delta_p(\rho) = \partial \partial^* + \partial^* \partial : C_p(M, \rho) \longrightarrow C_p(M, \rho).$$

If each Laplacian $\Delta_p(\rho)$ is invertible, we can define the Reidemeister torsion of M with respect to the representation ρ by

$$T(M, \rho) = \prod_{p=0}^{n} (\text{Det } \Delta_p(\rho))^{(-1)^p \frac{p}{2}}$$

(this formula is due to Ray and Singer [**51**]). Now replace the differential ∂ by ∂_t as before, form the induced Laplacians, and consider

$$T(M, \rho, t) = \prod_{p=0}^{n} (\text{Det } \Delta_p(\rho, t))^{(-1)^p \frac{p}{2}}.$$

As with the untwisted Laplacian, all eigenvalues of $\Delta_p(\rho, t)$ which do not converge to 0, converge to 1 (this is essentially a local phenomenon which does not see the twisting). Thus, as $t \to \infty$, the right hand side collapses onto the complex $\mathcal{W}(\rho, t)$ of small eigenspaces. As $t \to \infty$, the complex of small eigenspaces converges, after a change of variables, to the Morse complex (suitably twisted by ρ). Keeping track of how $T(M, \rho, t)$ varies with t, and the effect of conjugating ∂_t by θ we prove

THEOREM 9.8. $T(M, \rho) = Torsion(\mathcal{M}(\rho), \widetilde{\partial})$

where $\mathcal{M}(\rho)$ denotes the twisted Morse complex, and the torsion of $\mathcal{M}(\rho)$ is

$$\prod_{p=0}^{n} (\operatorname{Det} \widetilde{\Delta}_p(\rho))^{(-1)^p \frac{p}{2}}$$

where

$$\widetilde{\Delta}_p(\rho) = \widetilde{\partial}\widetilde{\partial}^* + \widetilde{\partial}^*\widetilde{\partial} : \mathcal{M}_p(\rho) \longrightarrow \mathcal{M}_p(\rho).$$

Note that Theorem 9.8 is precisely Theorem 4.2, in the special case that V is a gradient vector field. The general case of Theorem 4.2 can be proved by an analogous technique. Given a general discrete vector field V, one can deform the Laplace operator via a flat Lyapunov function. In this case it is no longer true that the nonsmall eigenvalues converge to 1. However, they do converge, and taking a closer look at the limiting behavior of the Laplace operator as $t \to \infty$ enables one to equate the product of the nonsmall eigenvalues with the special value of the zeta function.

In section 4 we discussed the relevance of the smooth version of Theorem 4.2. The smooth version of the special case of Theorem 9.8 is also of great interest, as it plays a key role in recent proofs of the equality of the Reidemeister torsion and Ray and Singer's analytic torsion ([**7**], [**9**]).

10. Discrete Morse Theory and the Distance Function

In this section we present a recent application of discrete Morse theory to geometry. Just as in the smooth setting, Morse theory meets geometry when one tries to apply Morse theory to the study of the distance function on a metric space. Some of the fundamental theorems in Riemannian geometry which relate curvature to topology are proved in this fashion. In this section we present one such result in the combinatorial setting which is due to my graduate student Katherine Crowley [**12**]. Let K be a contractible triangulated 3-manifold with boundary. Our goal is to find some geometric conditions which imply that K collapses.

The main theorem is the following

THEOREM 10.1 (Crowley [**12**]). *Let K be a combinatorial 3-manifold with boundary. Endow K with the piecewise Euclidean metric resulting from declaring all edges to have length 1. If K is CAT(0) then K collapses.*

The rest of this section will be devoted to giving an outline of the proof, but we begin with some general remarks. The notion of CAT(0) is a global condition meaning (roughly) "simply-connected and with non-positive curvature". With this in mind, Crowley's theorem should be compared to Hadamard's theorem: A complete simply-connected nonpositively curved Riemannian manifold is contractible (in fact, is diffeomorphic to $\mathbb{R}^n$).

I find Crowley's result to be fascinating. Two different distance functions are being examined here. The hypothesis that K is Cat(0) is a restriction on the distance function induced by the piecewise Euclidean metric. The conclusion that

K collapses is more a statement about the combinatorial distance function on the vertex set, in which the distance between two vertices is defined to be the minimal number of edges needed to connect one vertex to the other. The proof of this theorem demonstrates some rather intriguing relationships between these two distance functions. The two distant functions seem to be most closely related when all of the edges have the same length, and that is the reason for hypothesis that all edges have length 1. However, this is not a rigig phenomenon. The theorem remains true, without any changes in the proof, if the ratio of the lengths of edges is sufficiently close to 1.

Recall that Hadamard's Theorem is proved by picking a point p in the manifold, and showing that the function "distance to p" has no critical points other than p. Crowley's theorem is proved in an analogous way. We pick a vertex v, and use the function "distance to v" to construct a gradient vector field on K with only a single critical simplex. By Theorem 5.8 it is sufficient to construct a gradient vector field with no critical edges.

The first step is to have all vertices flow to v along geodesics. That is, for every vertex $w \neq v$, choose a geodesic from w to v, that is, a minimal set of edges connecting w to v. Begin the construction of our gradient vector field by drawing an arrow from w into the first edge of the geodesic.

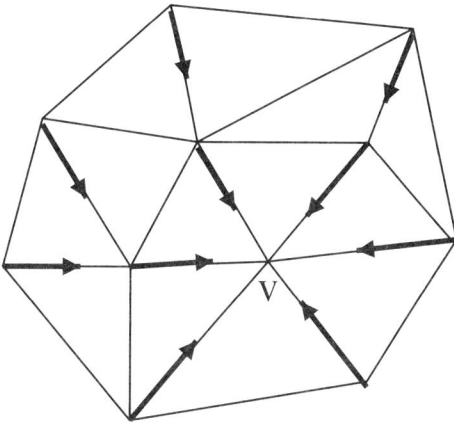

The first step in the construction of the gradient vector field.

FIGURE 10.2.

It is rather easy to check that this vector field has no closed orbits, and hence is a gradient vector field. The only critical vertex is v. However this gradient vector field will have lots of critical 1-simplices.

Let e be an edge that is is still critical (i.e. is not paired up with any vertex). We must now choose which 2-simplex to pair with e. The edge e, along with the geodesics from its endpoints to v, forms a loop (actually, the 2 geodesics may meet before reaching v)

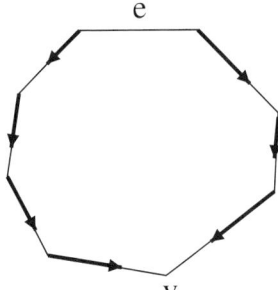

A look at a critical edge.

FIGURE 10.3.

Now we reach the main new idea of the proof. Fill in this loop with a minimal simplicial disc in K, i.e. a simplicial map from a triangulated disc into K which is minimal in the sense of having the fewest triangles.

Now draw an arrow from e pointing into this minimal disc. Doing this for every critical edge results in a gradient vector field with no critical 1-simplices.

It is quite non-trivial to show that the resulting vector field is a gradient vector field. In fact, it is not obvious that the resulting pairing is a discrete vector field at all. After all, what is to stop this procedure from pairing two edges with the same 2-simplex? It is at this point that we must use the hypotheses of the theorem. The key point is the following.

THEOREM 10.4 (Crowley). *A minimal simplicial disc spanning a loop in a 3-dim CAT(0) complex is CAT(0).*

In this context, a triangulated disc is CAT(0) if and only if every vertex in its interior has degree ≥ 6. This theorem is a combinatorial analogue of the fact that a minimal surface in R^3 always has nonpositive Gauss curvature. I believe that Crowley's work hints at the existence of a more complete theory of combinatorial differential geometry.

11. Main Ingredient

In this final section, we introduce what is the main ingredient in the proof of many of the results we have discussed. When given a smooth vector field on a smooth manifold, there are a couple different ways to extract information about the dynamics of induced flow. For example, one may fix a point in the manifold and let time vary. That is, fixing a point on the manifold, one may consider the flow line through that point. Another possibility is to fix a time and to consider all initial points. That is, given a real number t, to consider the diffeomorphism of the manifold resulting from letting every point flow along the vector fielld for time t. These two different points of view bring with them their own insights. We have already presented a combinatorial analog of the notion of a flow line, we have not yet seen a combinatorial analog of the induced diffeomorphism.

Before defining the flow, we must slightly change our notion of a discrete vector field. So far, we have treated a discrete vector field as a collection of pairs of simplices. Now we think of V as a map of oriented chains which has degree $+1$. Namely, if $\alpha^{(p)} < \beta^{(p+1)}$ are paired in V, then we define the map of oriented chains, which we also call V, by setting $V(\alpha) = \pm \beta$, with the sign chosen so that $<\alpha, \partial V(\beta)> = -1$. Given V, define the corresponding (discrete time t=1) flow

$$\Phi : C_p(K) \longrightarrow C_p(K); \qquad p = 0, 1, 2, \ldots$$

by the formula

(11.1) $$\Phi := 1 + \partial V + V \partial.$$

For example, in Figure 11.1 we illustrate the flow of an oriented edge.

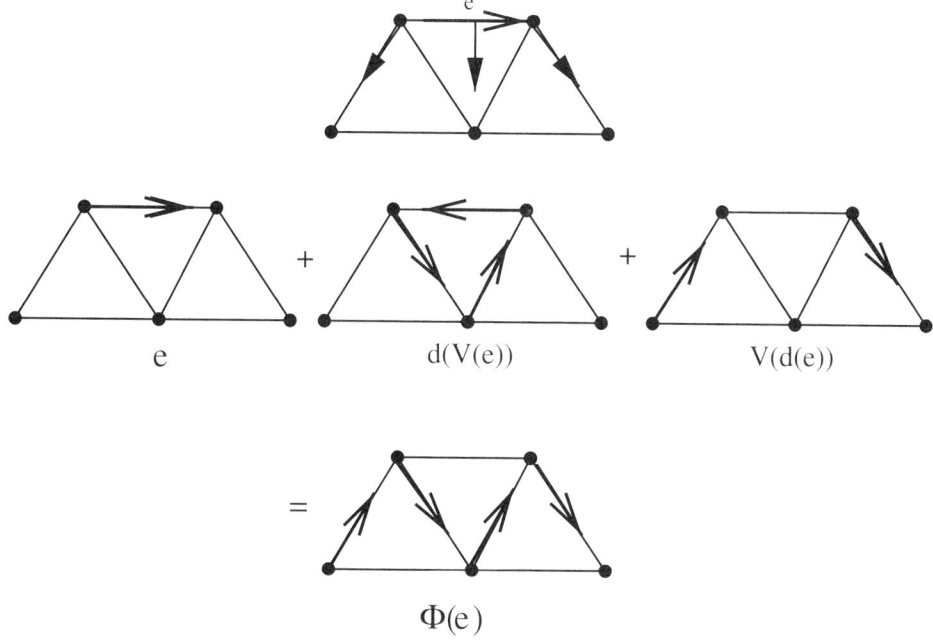

The flow of an oriented edge.

FIGURE 11.1.

We can now desrcibe briefly how this map enters into the proofs of results presented earlier. First consider Theorem 4.1 concerning the analytic continuation of the zeta function.

To give a hint of the proof it is perhaps best to consider the simple example of a shift of finite type. Let A be a finite set. Suppose that for each $a \in A$, we are given a subset $F(a) \subset A$, the set of letters which are permitted to follow a. Let Σ denote the set of all allowable double infinite sequences. That is, the set of all

$$x = \ldots, x_{-2}, x_{-1}, x_0, x_1, x_2, \ldots$$

where for each i, $x_i \in A$ and $x_{i+1} \in F(x_i)$. Say x is periodic with period k if $x_i = x_{i+k}$ for all I. Consider the corresponding zeta function

$$\zeta(z) = \exp[\sum_{k=0}^{\infty} \frac{z^k}{k} \mathcal{P}_k].$$

where $\mathcal{P}_k$ is the number of elements in Σ which are periodic with period k. It is one of the foundational results of symbolic dynamics that $\zeta(z)$ has an analytic continuation to all of $\mathbb{C}$. The proof is rather simple. Let $\mathcal{A}$ denote the square matrix whose rows and columns are identified with the elements of a, such that for i, j in A, the $(i, j)^{th}$ element of $\mathcal{A}$ is 1 if $j \in F(i)$, and 0 otherwise. It is then easy to check that $\mathcal{P}_k = trace(\mathcal{A}^k)$, so that

$$\zeta(z) = \det(I - z\mathcal{A})$$

from which the desired result follows.

Without going into details, suffice to say that the proof of Theorem 4.1 is along the same lines as that in the previous paragraph, with only minor complications. The key point is that the role of $\mathcal{A}$ played by the flow map Φ. (See section 6 of [**18**].)

It also remains to give some idea of the proofs of the results of section 7, in which we show how one can view the homology of a simplicial complex from the point of view of discrete Morse theory. The crucial point here is that it follows immediately from the formula (11.1) that the flow map commutes with the coboundary map, and induces the identity map on homology, with the discrete vector field V providing the chain homotopy. Similarly, one can consider the coflow

$$\Phi^* = 1 + \delta V^* + V^* \delta : C^*(K) \to C^*(K),$$

which commutes with the coboundary operator δ and induces the identity map on cohomology. In the case that V is the gradient vector field associated to a Morse function f, then Φ has many of the properties that one would expect from flowing along the negative of the gradient of a function. For example, if f is a discrete Morse function corresponding to the gradient vector field V and α is any simplex, then $\Phi(\alpha)$ is supported on simplices β with $f(\alpha) \leq f(\alpha)$. One major simplification that occurs in the combinatorial setting is that Φ stabilizes in finite time. That is, there is an N such that

$$\Phi^N = \Phi^{N+1} = \cdots = \Phi^\infty.$$

Theorem 8.1 is proved by showing that the map Φ^∞ maps the Morse complex into the standard cellular chain complex inducing the identity map on homology (See sections 6 and 7 of citeF2). Theorem 8.5 and the generalizations which appear in [**22**] are proved by examining this idea a bit closer.

References

[1] S. Axelrod and I. Singer, *Chern-Simons perturbation theory*, in Proceedings of the XXth International Conference of Differential Geometric Methods in Mathematical Physics (New York, 1991), Vol. 1, pp. 3-45, World Sci. Pub., River Edge, N.J., 1992.

[2] E. Babson, A. Björner, S. Linusson, J. Shareshian and V. Welker, *Complexes of not i-connected graphs*, Topology, **38** (1999), pp. 271-299.

[3] E. Babson, P. Hersh, *Discrete Morse functions from lexicographic orders*, preprint.

[4] E. Batzies and V. Welker *Discrete Morse theory for cellular resolutions*, to appear in J. Reine Angw. Math.

[5] M. Betz and R. Cohen, *Graph moduli space and cohomology operations*, Turk. J. of Math., **18** (1994), pp. 23-41.

[6] L. Billera, S. Holmes and K. Vogtmann *Geometry of the space of phylogenic trees*, Adv. in Appl. Math., **29** (2002), pp. 733-767.

[7] J.M. Bismut and W. Zhang *An Extension of a theorem by Cheeger and Müller*

[8] R. Bott *Morse theory indomitable*, Publ. Math. I.H.E.S. 68 (1988) pp. 99–117

[9] D. Burghelea, L. Friedlander and T. Kappeler *Asymptotic expansion of the Witten deformation of the analytic torsion*, J. Funct. Anal, **137** (1996). pp. 320-363.

[10] M. Chari, *On Discrete Morse functions and combinatorial decompositions*, Discrete Math, **217**, 2000, pp. 101-113.

[11] D. Chillingworth, *Collapsing three-dimensional convex polyhedra*, Proc. Cambridge Philos. Soc., **63** (1967), pp. 353-357.

[12] K. Crowley, *Discrete Morse Theory and the Geometry of Nonpositively Curved Simplicial Complexes*, thesis, Rice University, 2001.

[13] X Dong, *The Topology of Bounded Degree Graph Complexes and Finite Free Resolutions*, thesis, Univ. of Minn., 2001.

[14] A. Duval *A Combinatorial decomposition of simplicial complexes*, Israel J. of Math., to appear.

[15] R. Forman, *A Discrete Morse theory for cell complexes*, in Geometry, Topology & Physics for Raoul Bott, S.T. Yau (ed.), International Press, 1995.

[16] _____, *Morse theory for cell complexes*, Adv. in Math., **134** (1998), pp. 90-145.

[17] _____, *Witten-Morse theory for cell complexes*, Topology, **37** (1998), pp. 945-979.

[18] _____, *Combinatorial vector fields and dynamical systems*, Math. Zeit., **228** (1998), pp. 629-681.

[19] _____, *Combinatorial differential topology and geometry*, in New Perspectives in algebraic Combinatorics (Berkeley, CA. 1996-97), Math. Sci. Res. Inst. Publ. 38, Cambridge Univ. Press, Cambridge, 91999), pp. 177-206.

[20] _____, *Morse theory and evasiveness*, Combinatorica, **20** (2000), pp. 489-504.

[21] _____, *Combinatorial Novikov-Morse theory*, Int. J. of Math, **13** (2002), pp. 333-368.

[22] _____ *The cohomology ring and discrete Morse theory*, Trans. of the A.M.S., **354** (2002), pp. 5063-5085.

[23] D. Fried, *Homological identities for closed orbits*, Invent. Math., **71** (1983), pp. 419-442.

[24] _____, *Lefschetz formulas for flows*, Contemporary Mathematics, vol 58, part III, AMS, (1987), pp. 19-62.

[25] K. Fukaya, *Morse homotopy, A^∞-category, anf Floer homologies*, in Proc. Garc. Workshop on Geometry and Topology '93, (Seoul, 1993), Lecture Note ed. H. J. Kim, Lecture Note Ser. 18, Seoul National University.pp 1-102.

[26] _____, *Morse Homotopy and Chern-Simons perturbation theory*, Comm. Math. Phys., **181** (1996), pp. 37-90.

[27] _____, *Morse homotopy and its quantization*, in Geometric Topology (Athens, GA 1993),AMS/IP Stud. Adv. Math 21, Amer. Math. Soc., Prov. RI, 1997, pp. 409-440.

[28] L. Glaser *Geometrical Combinatorial Topology, Vol. 1*, Van Nostrand Reinhold Mathematical Studies # 27, 1970.

[29] M. Goresky and R. MacPherson *Stratified Morse theory*, in Singularities, Part I (Arcata, CA, 1981), Proc. Sympos. Pure Math., 40, Amer. Math. Soc., R.I., (1983) pp. 517-533.

[30] _____ *Stratified Morse Theory*, Ergebnisse der Mathematik und ihrer Grenzgebeite (3), 14, Springer Verlag

[31] M. Hutchings *Reidemeister torsion in generalized Morsetheory*, Forum Math., **14** (2002), pp. 209-244.

[32] M. Hutchings and Y. Lee. *Circle-valued Morse theory, F-torsion, and Seiberg-Witten invariants of 3-manifolds*, Topology, **38** (1999), pp. 861-888.

[33] _____ *Circle-Valued Morse theory and Reidemeister torsion*, Geom. Topol., **3** (1999), pp. 369-396 (electronic).

[34] J. Jonsson, *On the homology of some complexes of graphs*, preprint, 1998.

[35] _____, *The decision tree method*, preprint, 1999.

[36] J. Kahn, M. Saks and D. Sturtevant *A topological approach to evasiveness*, Combinatorica, **4** (1984), pp.297-306.

[37] D. Kozlov, *Collapsibility of $\Delta(\Pi_n)/S_n$ and some related CW complexes*, Proc. of the AMS, **128**, (2000), pp. 2253-2259.

[38] _____, *Topology of spaces of hyperbolic polynomials with multiple roots*, preprint, 2000.

[39] W. Kühnel *Triangulations of manifolds with few vertices* in Advances in Differential Geometry and Topology, World Sci. Publishing, N.J., (1990) pp. 59-114.

[40] C. Lee *The associahedron and triangulations of the n-gon*, Europ. J. of Comb., **10** (1989), pp. 551-560.

[41] S. Linusson and J. Shareshian, *Complexes of t-colorable graphs*, preprintitemLW A. Lundell and S. Weingram, *The Topology of CW Complexes*, Van Nostrand Reinhold Company, New York, 1969.

[42] A. Lundell and S. Weingram, *The Topology of CW Complexes*, Van Nostrand Reinhold Company, New York, 1969.

[43] V. Mathai and S.G. Yates, *Discrete Morse theory and extended L^2 homology*, preprint, 1999.

[44] J. Milnor *Morse Theory*, Annals of Mathematics Study No. 51, Princeton University Press, 1962.

[45] _____ *Lectures on the h-Cobordism Theorem* Princeton Mathematical Notes, Princeton University Press, 1965.

[46] M. Morse, *Bowls of a non-degenerate function on a compact differentiable manifold*, in Differential and Combinatorial Topology (A Symposium in Honor of M. Morse), Princeton University Press (1965), pp. 81-104.

[47] M. Morse and G. Hedlund, *Symbolic dynamics*, Amer. J. Math., **60**, (1938), pp. 815-866.

[48] S. Novikov, *Multivalued functions and functions: An analogue of the Morse theory*, Soviet Math. Dokl., **24** (1981), pp. 222-226.

[49] W. Parry and M. Pollicott *Zeta Functions and the Periodic Orbit Structure of Hyperbolic Dynamics*, Astérisque, **187-188**, Sociiété Mathématique de France, (1990).

[50] A. Pazhitnov, *Simple homotopy type of the Novikov complex and Lefschetz ζ-function of the gradient flow*, e-print dg-ga/9706014.

[51] D.B. Ray and I.M. Singer, *R-Torsion and the Laplacian on Riemannian manifolds*, Adv. in Math., **7** (1971), pp.145-210.

[52] R.L. Rivest and J. Vuillemin *On recognizing graph properties from adjacency matrices*, Theor. Comp. Sci, **3** (1976) pp. 371-384.

[53] M. Schwartz *Morse Homology* Progress in Mathematics, 111, Birkhäuser Verlag, Basel (1993).

[54] J. Shareshian *Discrete Morse theory for complexes of 2-connected graphs*, to appear in Topology.

[55] S. Smale, *On gradient dynamical systems*, Annals of Math. **74** (1961) pp. 199-206.more

[56] _____ *The generalized Poincaré conjecture in dimensions greater than four*, Annals of Math. **74** (1961) pp. 391-406.

[57] R. Stanley, *A combinatorial decomposition of acyclic simplicial complexes* Discrete Math. **118** (1993), pp. 175-182.

[58] J. Stasheff *Homotopy associativity of H-spaces*, Trans. of the A.M.S., **108** (1963), pp. 275-292.

[59] _____ The prehistory of operads. *in Operads: Proceedings of the Renaissance Conferences (Hartford, CT/Luminy, 1995)*, Cotemp. Math. 2002, A.M.S., pp. 9-14.

[60] C. Taubes, *Seiberg Witten and Gromov Invariants for Symplectic 4-manifolds*, First International Press Lecture Series, 2. International Press, Somerville, MA, 2000.

[61] V. Turaev, *A combinatorial formulation for the Seiberg-Witten invariants of 3-manifolds*, Math. Res. Lett., **5** (1998), pp. 583-598.
[62] P. Walters, ed., *Symbolic Dynamics and its Applications*, Contemporary Mathematics (AMS) vol. 135, Amer. Math. Soc., Providence, R.I., 1992.
[63] J.H.C. Whitehead, *Simplicial spaces, nuclei, and m-groups*, Proc. London Math. Soc. **45** (1939) pp. 243-327.
[64] E. Witten *Supersymmetry and Morse theory* J. Diff. Geom. **17** (1982) pp. 661–692.

DEPARTMENT OF MATHEMATICS, RICE UNIVERSITY, HOUSTON, TX 77251

Extensions, quotients and generalized pseudo-Anosov maps

André de Carvalho

ABSTRACT. We describe a circle of ideas relating the dynamics of 2-dimensional homeomorphisms to that of 1-dimensional endomorphisms. This is used to introduce a new class of maps generalizing that of Thurston's pseudo-Anosov homeomorphisms.

CONTENTS

1. Introduction
2. The natural extension
3. Another kind of quotient
4. Generalized train tracks
5. Generalized pseudo-Anosov maps
References

1. Introduction

In this paper we discuss a circle of ideas which is present in many different contexts in dynamical systems. It was first introduced by Williams in his study of expanding attractors, and has been used since by many authors. In its most basic form it can be stated as follows:

Collapsing segments of stable manifolds of a homeomorphism yields a lower dimensional endomorphism; the original homeomorphism may be recovered by taking the natural extension (inverse limit) of the quotient endomorphism.

The context we focus attention on is the interplay this creates between the dynamics of 1-dimensional endomorphisms — endomorphisms of trees and graphs — and that of 2-dimensional homeomorphisms — homeomorphisms of surfaces. In recent years, this interplay between graph endomorphisms and surface homeomorphisms has been used for different, although related, purposes. For example, it was used to give algorithmic proofs of Thurston's classification theorem for surface homeomorphisms up to isotopy. It also appeared in the construction of models for

---

1991 *Mathematics Subject Classification.* 37E30, 37E25, 37B40, 37B45, 37D45.
Key words and phrases. Natural extension, graph maps, train tracks, pseudo-Anosov maps.

families of surface homeomorphisms passing from trivial to chaotic dynamics as parameters are varied. A combination of these results led to a conjecture about the way in which *forcing* organizes the braid types of horseshoe periodic orbits.

Complexification should yield a closely related discussion — which will not be treated here — linking the dynamics of endomorphisms of branched (Riemann) surfaces to the dynamics of automorphisms of $\mathbb{C}^2$. The natural extension of an endomorphism of a branched surface is usually a homeomorphism of a surface lamination and such laminated spaces have already been found to be among the main objects in the study of complex Hénon maps. In both the real and complex cases, much is known about 1-dimensional endomorphisms, at least when the space is unbranched, whereas about homeomorphisms in dimension 2 much less is known.

This double interplay — between 1- and 2-dimensional dynamics and between the real and complex settings — seems to be an interesting approach to explore. In this paper we develop some of its aspects, mostly in the real setting.

We also introduce a different kind of quotient, taking a surface homeomorphism to another surface homeomorphism by collapsing "dynamically irrelevant" domains, i.e. those which do not carry topological entropy. This produces an interesting class of surface homeomorphisms which includes torus Anosovs and Thurston's pseudo-Anosov maps. These maps — called *generalized pseudo-Anosov maps* — preserve a pair of invariant measured foliations with (possibly infinitely many) singularities, giving the underlying surface a naturally defined complex structure. Regarded from the point of view of this complex structure, the invariant foliations become the horizontal and vertical trajectories of an integrable quadratic differential (with possibly infinitely many poles) and the generalized pseudo-Anosov map becomes a Teichmüller mapping.

The construction of generalized pseudo-Anosov maps is done by first introducing generalized train tracks: smooth branched 1-submanifolds, with possibly infinitely many branches, of the ambient surface. We describe how to find invariant generalized train tracks for certain surface homeomorphisms which, together with measure theoretic information obtained from transition matrices, are the main ingredients in the construction of generalized pseudo-Anosov maps. This is a variant of the same circle of ideas mentioned above.

The paper is organized as follows: in Section 2 the natural extension is defined, the class of *thick graph maps* is introduced and, in Lemma 1, the circle of ideas mentioned above is closed. In Section 3 the *zero-entropy* equivalence relation is introduced and a theorem (Theorem 2) is stated about the quotient maps so obtained — this is the first way in which generalized pseudo-Anosov maps appear. In Section 4 generalized train tracks are defined as the main tool to construct generalized pseudo-Anosov maps in the following Section 5, which is the longest of the paper. The complex structure is discussed in the last subsection of Section 5.

The writing style is Sullivanian as is fitting for a paper prepared for such an occasion. Many arguments are only sketched and some are omitted entirely.

Acknowledgments: I have worked on or around the main subjects in this paper for many years. Several people in the course of several conversations have influenced me. Dennis Sullivan, of course, was a major influence as my adviser. Others who have directly contributed to the ideas in this paper are Michael Handel, with whom I talked about many things, including generalized train tracks and generalized pseudo-Anosovs; Fred Gardiner, who taught me Teichmüller theory and how to show a puncture is a puncture; and my friend and coworker Toby Hall, whose work was the original stimulus for me to study this subject and who wrote a good part of Section 4 when we were engaged in a project which has now been put on hold[1]. I would also like to thank the referee for several suggestions for improvement and clarification.

2. The natural extension

In this paper a *dynamical system* will mean a continuous self-map of a topological space, at least. As we go along, we may require more of our maps, for example, that they be differentiable or be diffeomorphisms.

2.1. Definition and examples.

Let $f\colon X \to X$ be a continuous surjective map of a topological space X. If f is not invertible, there is a naturally associated invertible map: set

$$\hat{X} = \{(x_0, x_1, x_2, \ldots) \in \prod_0^\infty X; f(x_{i+1}) = x_i, \text{for } i = 0, 1, 2, \ldots\}$$

and define $\hat{f}\colon \hat{X} \to \hat{X}$ setting $\hat{f}(x_0, x_1, x_2, \ldots) = (f(x_0), x_0, x_1, \ldots)$. This map is called the *natural extension* of f. The space $\hat{X}$ is also known as the *inverse* or *projective limit space* and the map $\hat{f}$ as the *inverse* or *projective limit map* associated to $f\colon X \to X$.

Here are some prototypical examples of natural extensions which come up in a variety of contexts in the study of low-dimensional dynamical systems.

EXAMPLES 1. a) Let $X = \mathbb{S}^1$ be the unit circle in the complex plane and $f\colon \mathbb{S}^1 \to \mathbb{S}^1$ be the squaring map $f(z) = z^2$ in complex notation. Then $\widehat{\mathbb{S}^1}$ is the *dyadic solenoid*. It is a fiber bundle over $\mathbb{S}^1$ with fiber a dyadic Cantor set.

b) Let $\mathbb{C}^* = \mathbb{C} \setminus 0$ and $f\colon \mathbb{C}^* \to \mathbb{C}^*$ be again $f(z) = z^2$. Then $\widehat{\mathbb{C}^*}$ is the *complex dyadic solenoid* and is a fiber bundle over $\mathbb{C}^*$ with same fiber as above.

c) A variant of the previous example may be obtained by restricting f to $X = \mathbb{C}\setminus\mathbb{D}$, the exterior of the closed unit disk. Since the action of f on X has a fundamental domain (namely, any annulus of the form $A = \{z \in \mathbb{C}; 1 < R \leq |z| < R^2\}$), so does the action of $\hat{f}$ on $\hat{X}$ and we can take the quotient $\mathcal{S} = \hat{X}/\hat{f}$. This is *Sullivan's Riemann Surface Lamination*. $\mathcal{S}$ is a *lamination* as were the previous examples, but, unlike them, for which collapsing transversals produces a good old space, this time the quotient is the *branched surface* obtained as the quotient A/f of the annulus by identifying its boundary under the dynamics.

d) Let $I = [0,1]$ and f be the *tent map* $f(x) = 2x$ if $x \in [0, \frac{1}{2}]$ and $f(x) = 2 - 2x$ if $x \in [\frac{1}{2}, 1]$. The inverse limit space $\hat{I}$ is the *Knaster continuum*. Again, $\hat{I}$ is a laminated space and the quotient by collapsing transversals is I, but $\hat{I}$ is not a fiber bundle over I. We will talk more about examples like this below.

2.2. The natural extension in 2-dimensional dynamics.

In this section we introduce a class of maps which are suitable for several applications (see [**BH95, dCH01, FM93**]). They are called *thick graph maps* and, as the name suggests, they are essentially graph endomorphisms that have been thickened and made into

---

[1]This project has now been completed [**dCH**]. There it is possible to find details of many of the arguments which are only sketched in the present paper.

surface homeomorphisms. All of their interesting dynamics is contained in a subsurface called a *thick graph*, that is, a graph in which each point has been thickened up, either to a disk or an arc according as the point is a vertex or a regular point of the graph. The homeomorphisms are assumed to act on the thick graph in such a way that they induce endomorphisms of the underlying graph.

DEFINITIONS 1. A *thick graph* is a pair $(S, \mathbb{G})$, where S is a closed orientable surface endowed with a fixed metric compatible with its topology and $\mathbb{G}$ is a compact subsurface of S (with boundary) which is partitioned into compact *decomposition elements*, such that the following hold:
  i) Each decomposition element of $\mathbb{G}$ is either a *leaf* homeomorphic to $[0,1]$, or a *junction* homeomorphic to D^2 (the unit disk in $\mathbb{R}^2$).
  ii) The boundary in $\mathbb{G}$ of each junction is a finite number of disjoint arcs: if there are k such arcs, then the disk is called a k-*junction*.
  iii) The set of k-junctions with $k \neq 2$ is finite.
  iv) Each decomposition element which is not in the accumulation of a sequence of distinct 2-junctions is contained in a chart as depicted in Figure 1.
  v) Each component of $S \setminus \mathbb{G}$ is an open disk.

If $(S, \mathbb{G})$ is a thick graph, let $\sim$ be the equivalence relation on $\mathbb{G}$ given by $x \sim y$ if and only if x and y lie in the same decomposition element. Then $G = \mathbb{G}/\sim$ is a graph whose vertices (which may have valence 2) correspond to the junctions of $\mathbb{G}$: the canonical projection will be denoted by $\pi\colon \mathbb{G} \to G$. The vertex set of G will be denoted by V, and the union of the junctions of $\mathbb{G}$ will be denoted by $\mathbb{V}$: thus $\mathbb{V} = \pi^{-1}(V)$. The components of $\mathbb{G} \setminus \mathbb{V}$ are called *strips*: each strip is therefore homeomorphic to $(0,1) \times [0,1]$. The union of the closures of the strips will be denoted by $\mathbb{E}$ and the corresponding set in the quotient, the set of *edges* of G, will be denoted by E. Thus $\mathbb{E} \cap \mathbb{V}$ consists of a collection of closed arcs which are the boundary components (in $\mathbb{G}$) of both the junctions and the strips of $\mathbb{G}$.

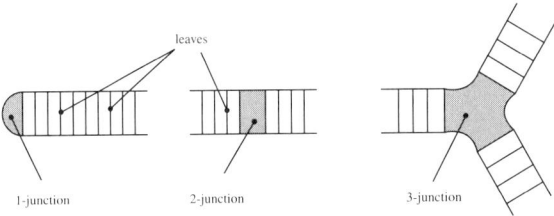

FIGURE 1. Charts in a thick graph.

REMARKS 1. a) We will often refer to the thick graph as $\mathbb{G}$ alone, although it should be kept in mind that there is a surface floating around.
  b) Notice that, at this point, we allow a thick graph to have infinitely many 2-junctions and, therefore, infinitely many strips (and, thus, the graph G to have infinitely many edges).

If $(S, \mathbb{G})$ is a thick graph and $F\colon (S, \mathbb{G}) \to (S, \mathbb{G})$ is a homeomorphism (i.e., a homeomorphism $F\colon S \to S$ with $F(\mathbb{G}) \subset \mathbb{G}$) under which the image of each decomposition element of $\mathbb{G}$ is contained in a decomposition element, then $F|_{\mathbb{G}}$ induces a graph endomorphism $f\colon G \to G$ such that $\pi \circ F|_{\mathbb{G}} = f \circ \pi$.

DEFINITION 2. A *thick graph map of* $(S, \mathbb{G})$ is an orientation-preserving homeomorphism $F \colon (S, \mathbb{G}) \to (S, \mathbb{G})$ such that:

i) $F(\mathbb{G}) \subset \text{Int}(\mathbb{G})$.
ii) If γ is a decomposition element of $\mathbb{G}$, then $F(\gamma)$ is contained in a decomposition element, and $\text{diam}(F^n(\gamma)) \to 0$ as $n \to \infty$.
iii) The induced graph endomorphism $f \colon G \to G$ is piecewise monotone (that is, there is a finite subset L of V such that $f^{-1}(x) \cap U$ is connected for each $x \in G$ and each component U of $G \setminus L$); and is strictly monotone away from the preimages of vertices (that is, every $x \in G \setminus f^{-1}(V)$ has a neighborhood on which f restricts to an embedding).
iv) For each component U of $S \setminus \mathbb{G}$ there is a (least) positive integer n_U for which either $F^{n_U}(U) \subset \mathbb{G}$ or $F^{n_U}(U) \cap U \neq \emptyset$, in which case U contains a period n_U point p_U of F, which is a source whose immediate basin contains U: that is, $F^{-kn_U}(x) \to p_U$ as $k \to \infty$ for all $x \in U$.

REMARKS 2. a) Item iv) in the definition says that the dynamics of a thick graph map in $S \setminus \mathbb{G}$ is easily understood and uninteresting.
b) If $F \colon (S, \mathbb{G}) \to (S, \mathbb{G})$ is a thick graph map, then so are all of its forward iterates F^n ($n \geq 1$), and the graph endomorphism induced by F^n is $f^n \colon G \to G$.
c) Let $f \colon G \to G$ be the quotient of a thick graph map. A point $x \in G$ is a *critical point* if f is not a local homeomorphism at x. Since thick graphs may have infinitely many 2-junctions, the forward orbit of critical points of f may be infinite.

EXAMPLE 2. The first example is Smale's horseshoe map which will be denoted by $F_1 \colon (\mathbb{S}^2, \mathbb{I}) \to (\mathbb{S}^2, \mathbb{I})$ here and in what follows. It is shown in Figure 2. The thick graph in this case is a thick tree — a thick interval, in fact — and is denoted by $\mathbb{I}$. The point at infinity in $\mathbb{S}^2$ is a repeller whose basin contains all points outside $\mathbb{I}$. The horseshoe has two saddle fixed points which are labeled x_0 and x_1 (shown as • and ∘, respectively, in Figure 2) and an attracting fixed point in the 1-junction on the left denoted by x (shown as ■). The quotient tree is the interval and the quotient map — the "flat top" tent map $f_1 \colon I \to I$ — is also shown in the figure: the image is shown slightly separated so it is possible to see what f_1 does to the interval. This way of representing graph maps will always be used in what follows.

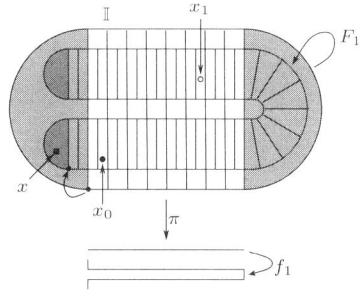

FIGURE 2. The horseshoe map.

EXAMPLE 3. In this example the map is again a sphere homeomorphism and the thick graph is again a thick tree as shown in Figure 3. It is denoted $F_2 \colon (\mathbb{S}^2, \mathbb{T}) \to (\mathbb{S}^2, \mathbb{T})$. The 1- and 2-junctions of $\mathbb{T}$ contain a periodic orbit of period 6 ($\circ$) and the 3-junctions contain a periodic orbit of period 2 ($\bullet$). The quotient tree and tree endomorphism $f_2 \colon T \to T$ are shown in Figure 4.

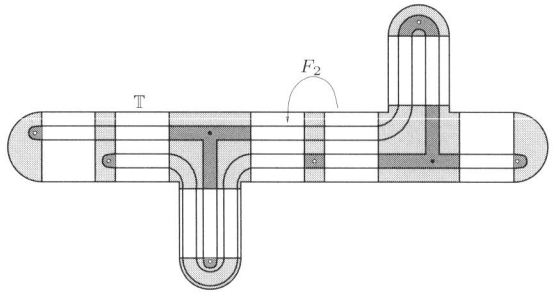

FIGURE 3. A thick tree map.

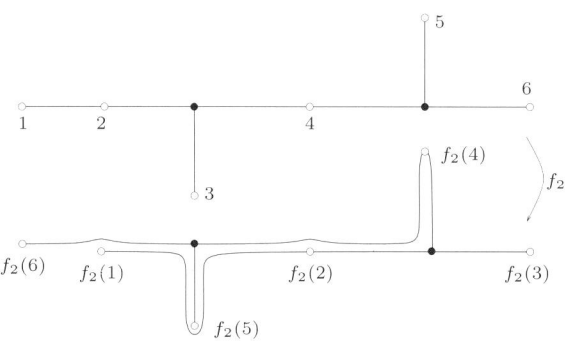

FIGURE 4. The quotient tree map of the thick tree map in Figure 3.

REMARK 3. Both maps F_1 and F_2 above can be made to be diffeomorphisms. This will be used below when we talk about stable and unstable manifolds of their periodic points.

Let $F \colon \mathbb{G} \to \mathbb{G}$ be a thick graph map and $f \colon G \to G$ its quotient. The infinite nested intersection $\Lambda = \bigcap_{n=0}^{\infty} F^n(\mathbb{G})$ is a compact subset of $\mathbb{G}$ which is invariant under F. We now relate the dynamics of F on Λ with the natural extension of f (see [**Bar86**]).

LEMMA 1. *The maps $F|_\Lambda \colon \Lambda \to \Lambda$ and $\hat{f} \colon \hat{G} \to \hat{G}$ are topologically conjugate, that is, there exists a homeomorphism $\hat{\pi} \colon \Lambda \to \hat{G}$ such that $\hat{\pi} \circ F = \hat{f} \circ \hat{\pi}$.*

PROOF. Let $\pi \colon \mathbb{G} \to G$ be the projection. The map $\hat{\pi}$ is defined by setting
$$\hat{\pi}(z) = (\pi(z), \pi(F^{-1}(z)), \pi(F^{-2}(z)), \ldots)$$

for each $z \in \Lambda$. It is straightforward to check that $\hat{\pi}$ is well defined, continuous and surjective. Injectivity follows from the assumption that $\operatorname{diam}(F^n(\gamma)) \to 0$ as $n \to \infty$. $\square$

3. Another kind of quotient

In this section we describe a way of modifying two-dimensional maps by a semi-conjugacy which collapses "irrelevant" dynamics, that is, parts of the space which do not carry entropy. The semi-conjugacy is defined quite generally and the space of maps it yields is interesting in its own right. A more thorough treatment of the equivalence relation can be found in [**dCP04**]. The space of quotient homeomorphisms will be treated in a future paper.

3.1. The zero-entropy equivalence relation.
We start by recalling Bowen's definition of topological entropy [**Bow71**]. If X is a metric space and $F\colon X \to X$ is a uniformly continuous map, we say that $x, y \in X$ are (n, ϵ)-*separated* if it is possible to distinguish between the orbits of x and y up to $n-1$ iterates with precision ϵ, that is, $x, y \in X$ are (n, ϵ)-separated if $d(F^j(x), F^j(y)) > \epsilon$ for some $0 \leq j < n$. The topological entropy of F is defined to be the limit as $\epsilon \to 0$ of the exponential growth rate of the number of (n, ϵ)-separated orbits as $n \to \infty$. If $K \subset X$ is a compact subset and we only count those orbits which start in K, we obtain the entropy of F in K, denoted $h_F(K)$. More precisely, if we denote by $s(n, \epsilon, K)$ the cardinality of a maximal (n, ϵ)-separated subset of K, then

$$h_F(K) = \lim_{\epsilon \to 0} \limsup_{n \to \infty} \frac{1}{n} \ln s(n, \epsilon, K)$$

and the entropy of F is defined by

$$h(F) = \sup\{h_F(K); K \subseteq X, K \text{ compact}\}.$$

DEFINITION 3. If $F\colon X \to X$ is a homeomorphism, we define two points x and y to be *zero-entropy equivalent* if there is a continuum (that is, a compact connected set) K which contains both points and for which

$$h_F(K) = 0 = h_{F^{-1}}(K).$$

REMARKS 4. a) That this indeed defines an equivalence relation follows from two facts: 1) $h_F(K \cup K') = \max\{h_F(K), h_F(K')\}$ and 2) the union of two continua containing a point in common is also a continuum.

b) Notice that if K is a proper subset of X, it is not necessarily the case that $h_F(K) = h_{F^{-1}}(K)$.

c) If F is not invertible we can consider the equivalence relation defined using only the first equality above. It would be interesting to understand this equivalence relation — and the ones mentioned below — for interval or, more generally, tree endomorphisms.

d) In general, we can consider the family of equivalence relations $\sim_\alpha$, indexed by a positive real α, declaring two points to be $\sim_\alpha$-equivalent if there is a continuum containing both and carrying entropy strictly smaller than α.

The zero-entropy equivalence relation is most interesting for two-dimensional systems.

EXAMPLE 4. Let us describe its equivalence classes for the horseshoe map F_1. Denote by $\mathcal{H}^u$ and $\mathcal{H}^s$ the closures of the unstable and stable manifolds of the fixed point 0 (or indeed of any other periodic point, since their closures coincide) and let $\mathcal{H} = \mathcal{H}^s \cup \mathcal{H}^u$. Equivalence classes are of four kinds:

a) Closures of connected components of $\mathbb{S}^2 \setminus \mathcal{H}$.
b) Closures of connected components of $\mathcal{H}^u \setminus \mathcal{H}^s$ (not already contained in sets in a)).
c) Closures of connected components of $\mathcal{H}^s \setminus \mathcal{H}^u$ (not already contained in sets in a)).
d) Single points which are in none of the sets in a), b) or c).

To see that these sets do not carry entropy, notice that all points in any connected component of $\mathbb{S}^2 \setminus \mathcal{H}$ (before taking the closure) converge to the attracting fixed point x. It is not hard to see that, after taking the closure nothing more exciting happens and this shows the sets in a) indeed carry no entropy. The same holds for sets of types b) and c). To see that any larger continuum must contain entropy, notice that if C is a connected set that contains two distinct sets among the ones described above, then it must intersect a Cantor set's worth of invariant manifolds, either stable or unstable (or both). It follows that one of its ω- or α-limit sets contains all the nonwandering set of the horseshoe and therefore one of $h_F(C)$ or $h_{F^{-1}}(C)$ equals $\ln 2$.

In [**dCP04**] it is shown that if F is a $C^{1+\epsilon}$ surface diffeomorphism, then the zero-entropy equivalence classes form an *upper semi-continuous monotone* decomposition of the surface. In particular, the quotient space is a *cactoidal* surface (roughly, a surface with nodes; see [**Moo62, RS38**] [2]). Since the equivalence is dynamically defined, F projects to a homeomorphism $F/\sim$ on the quotient space. Moreover, any nontrivial continuum in the quotient space carries entropy of either the quotient homeomorphism or its inverse. The quotient map by the zero-entropy equivalence relation should be thought of as a "tight" version of the original map in which all the wandering domains have been collapsed to points.

The quotient of the sphere by the zero-entropy horseshoe equivalence of Example 4 is shown in Figure 5. The quotient space is a sphere (we are collapsing everything outside the homoclinic tangle to a point), obtained by identifying the solid boundary in the figure along the dotted arcs from the mid-point at the top to the corner point on the lower left. The stable and unstable manifolds of the horseshoe project to two transverse foliations with singularities, represented by solid and dashed lines, respectively. In fact, these foliations carry transverse invariant measures whose product gives a measure on the sphere. The quotient map preserves both foliations, dividing one of the transverse measures by 2 and multiplying the other by 2, so that the product measure is invariant. This map is a *generalized pseudo-Anosov map*, which will be defined presently. We will also state a theorem that generalizes this construction to a class of thick graph maps.

3.2. The Markov story and generalized pseudo-Anosov maps.

We start by introducing the basic concepts in the Perron-Frobenius theory for non-negative matrices (see [**Bal00, Gan59**]). Let M be a square matrix with non-negative entries. M is said to be *reducible* if, by a permutation of the index set, it

---
[2] Compare with Voronov's article in this volume.

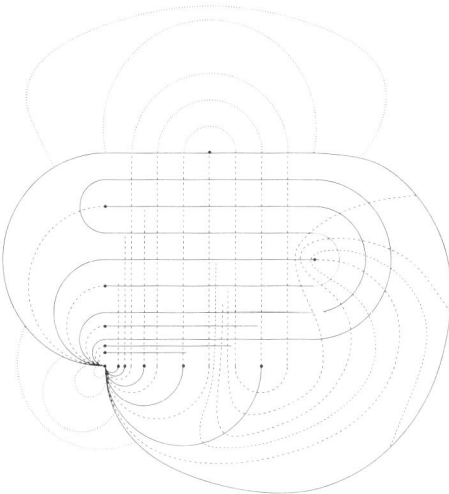

FIGURE 5. The quotient of the sphere under the zero-entropy equivalence relation for the horseshoe.

is possible to put it in triangular block form:

$$M = \begin{bmatrix} A & 0 \\ B & C \end{bmatrix}.$$

Otherwise, M is said to be *irreducible*. The matrix M is said to be *irreducible and aperiodic* if there exists a positive integer k such that M^k is positive, that is, all its entries are positive. A non-negative irreducible matrix M has a unique positive eigenvector (up to scaling) and the associated eigenvalue λ — called the *Perron-Frobenius eigenvalue* of M — equals the spectral radius of M. If $\lambda = 1$ then M is a cyclic permutation matrix. Otherwise, $\lambda > 1$ and, for every i,j, there exists a power k such that the ij-entry of M^k is arbitrarily large (in fact, the entries of M grow like $(\mathrm{const})\times\lambda^k$). If M is irreducible then λ is a simple root of the characteristic polynomial of M and if M is also aperiodic, then λ is the only eigenvalue on the circle $\{z \in \mathbb{C}\,;\,|z|=\lambda\}$.

DEFINITIONS 4. A thick graph map $F\colon \mathbb{G} \to \mathbb{G}$ with quotient $f\colon G \to G$ will be called *Markov* if it satisfies the following additional conditions:

i) $\mathbb{G}$ has a finite number of strips and junctions (i.e., the graph G has finitely many edges). For each strip s we fix a homeomorphism $h_s\colon \overline{s} \to [0,1]\times[0,1]$ from the closure of the strip to the closed unit square, so that the decomposition elements of s are of the form $h_s^{-1}(\{x\}\times[0,1])$, for $0 < x < 1$.
ii) F is linear with respect to the structure homeomorphisms h_s, that is, in each connected component of $s_i \cap F^{-1}(s_j)$, where s_i, s_j are strips, F contracts vertical coordinates uniformly by a factor $\mu_{ij} < 1$ and expands horizontal coordinates uniformly by a factor $\lambda_{ij} \geq 1$.
iii) If J, J' are junctions such that $F(J) \subset J'$ then $F(\partial_{\mathbb{G}} J) \subset \partial_{\mathbb{G}} J'$.
iv) If J is a periodic junction of least period n, then J has an attracting periodic point of least period n in its interior whose basin contains $\mathrm{Int}(J)$.

We can associate a *transition matrix* $M = [m_{ij}]$ to a Markov thick graph map: letting $E = \{e_1, e_2, \ldots, e_n\}$ be the edges of G, set

$$m_{ij} = \text{number of times } f(e_j) \text{ crosses } e_i.$$

REMARKS 5. a) Notice that Markov thick graph maps can be made differentiable and we will assume, whenever we talk about them, that they are diffeomorphisms of the surface S. In particular, we will talk freely about stable and unstable manifolds of their periodic points.

b) Let $F\colon (S, \mathbb{G}) \to (S, \mathbb{G})$ be a Markov thick graph map whose transition matrix is irreducible and aperiodic, let $\Lambda = \bigcap_{n=0}^{\infty} F^n(\mathbb{G})$ and $p \in \mathbb{G}$ be any saddle periodic point of F. Then it is easy to see that $\Lambda = \overline{W^u(p)}$.

DEFINITION 5. A surface homeomorphism $\Phi\colon S \to S$ is called a *generalized pseudo-Anosov map* if it satisfies the following conditions: there exist a pair $(\mathcal{F}^s, \mu^s)$, $(\mathcal{F}^u, \mu^u)$ of transverse measured foliations with singularities — either modeled on pronged singularities as in Figure 6 or accumulations of such, of which there are only finitely many — and a real number $\lambda > 1$ such that

$$\begin{aligned} \Phi(\mathcal{F}^s, \mu^s) &= (\mathcal{F}^s, \lambda \mu^s) \\ \Phi(\mathcal{F}^u, \mu^u) &= (\mathcal{F}^u, \tfrac{1}{\lambda} \mu^u). \end{aligned}$$

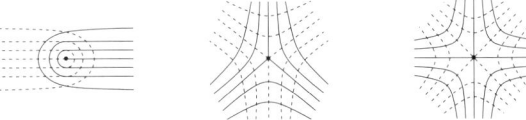

FIGURE 6. Pronged singularities of the invariant foliations.

Examples of generalized pseudo-Anosov maps include the torus Anosov maps and Thurston's pseudo-Anosov maps. We argue below that Markov thick graph maps also give rise to generalized pseudo-Anosovs. The definition above, however, probably includes many other maps and these Markov examples should form a dense set in the space of all generalized pseudo-Anosovs. A more thorough study of these issues, including a description of a uniform structure on the set of generalized pseudo-Anosovs, is currently under way and will hopefully appear in forthcoming papers.

THEOREM 2. *Let $F\colon (S, \mathbb{G}) \to (S, \mathbb{G})$ be a Markov thick graph map with $\mathbb{G}$ of the homotopy type of S minus a point. Assume the associated transition matrix is irreducible and aperiodic. Then the quotient of S by the zero-entropy equivalence relation is homeomorphic to S and F projects to a generalized pseudo-Anosov homeomorphism $\Phi\colon S \to S$.*

A more detailed account of the zero-entropy equivalence relation and its quotients is given in [**dCP04**]. Below we will present an alternative construction of the generalized pseudo-Anosov quotient maps.

4. Generalized train tracks

In this section we will only deal with *finite* thick graphs, that is, thick graphs with finitely many strips and junctions. To simplify the exposition and the statements, we also assume that if $(S, \mathbb{G})$ is a thick graph then $\mathbb{G}$ has the homotopy type of the punctured surface $S \setminus \{p\}$, where $p \in S \setminus \mathbb{G}$. We refer to p as the *point at infinity* and denote it by ∞. In general, $\mathbb{G}$ has the homotopy type of the several times punctured surface $S \setminus \{p_1, \ldots, p_k\}$. If this is the case, the discussion below has to be appropriately modified, but the ideas are essentially the same. The statements made here hold for the once punctured case.

4.1. Definitions.
We start by fixing some notation and presenting some definitions. Suppose that $f \colon X \to X$ is a homeomorphism. Then f is said to be *supported* on a subset U of X if f is the identity on $X \setminus U$. A second homeomorphism $g \colon X \to X$ is *isotopic to* f if there is a continuous map $\psi \colon X \times [0, 1] \to X$ such that each *slice map* $\psi_t \colon X \to X$ defined by $x \mapsto \psi(x, t)$ is a homeomorphism, and $\psi_0 = f$ and $\psi_1 = g$. The map ψ is called an *isotopy* from f to g. A *pseudo-isotopy* is a continuous map $\psi \colon X \times [0, 1] \to X$ such that, for $0 \le t < 1$, the slice maps ψ_t are homeomorphisms onto their images. The isotopy or pseudo-isotopy is said to be *supported* on a subset U of X, denoted $\operatorname{supp}(\psi) = U$, if the homeomorphisms ψ_t are all equal on $X \setminus U$, and is said to be *relative* to U if $\psi_t(U) \subset \psi_0(U)$ for all $t \in [0, 1]$. A map $g \colon X \to X$ is called a *near-homeomorphism* if it can be arbitrarily well approximated by homeomorphisms. Thus, if ψ is a pseudo-isotopy, the map $g = \psi_1$ is a near-homeomorphism.

Let $(S, \mathbb{G})$ be a finite thick graph, and $A \subset \mathbb{G}$ be a finite set, each of whose points lies in the interior of a junction and so that no junction contains more than one point of A. For each strip s of $\mathbb{G}$, let γ_s be an arc joining the two boundary components of s in $\mathbb{G}$ and intersecting each leaf of s exactly once. In terms of the structure homeomorphisms h_s introduced in Subsection 3.2, we can take $\gamma_s = h_s^{-1}([0, 1] \times \{1/2\})$. Let $R \subseteq \mathbb{G}$ be the union of the arcs γ_s. The endpoints of the arcs γ_s are called *switches* and we denote by L the set of switches.

DEFINITIONS 6. Let $\tau \subseteq \mathbb{G} \setminus A$ be a graph with vertex set L and countably many edges, each of which intersects $\partial \mathbb{V}$ only at L, such that

i) $\tau \cap \mathbb{E} = R$, and
ii) No two edges e_1, e_2 contained in a given junction J are *parallel*: that is, they do not bound a disk which contains no point of A or other edges.

The isotopy class of τ by isotopies supported on $\mathbb{V} \setminus A$ (the set of junctions of $\mathbb{G}$ minus the points in A) is called a *generalized train track*[3] for $(\mathbb{G}, A)$. We will always refer to τ itself as the generalized train track, but it should be kept in mind that we do not distinguish between τ and τ' if it is possible to deform one to the other without crossing over points of A.

The edges of τ which are contained in $\mathbb{E}$ (that is, the connected components of R) are called *real*, and those which are contained in $\mathbb{V}$ are called *infinitesimal*. Write I for the set of infinitesimal edges of τ.

A generalized train track τ is *finite* if it has only finitely many edges. An infinitesimal edge is called a *bubble* if its two endpoints coincide. An edge of τ is

---
[3] To be painfully precise, we should talk about the isotopy class of the inclusion map $\iota \colon \tau \hookrightarrow \mathbb{G}$, but we won't do it.

homotopically trivial if it is a bubble which bounds a disk containing no point of A, and is *homotopically non-trivial* otherwise.

Clearly a generalized train track τ for $(\mathbb{G}, A)$ is determined by its infinitesimal edges. It will sometimes be convenient to write $\tau(I)$ for the generalized train track whose set of infinitesimal edges is I, provided the thick graph and the set A are clear from the context.

A homotopy $\{\alpha_t\}$ of a path $\alpha \colon [0,1] \to X$ is said to be *relative to* $U \subset X$ if the points of $\alpha_t([0,1])$ that belong to U for $t = 0$ do not leave U for $0 \leq t \leq 1$. Let $[\alpha]$ be a homotopy class of paths in $S \setminus A$ relative to $\partial \mathbb{V}$, with endpoints in $\partial \mathbb{V}$. Then $[\alpha]$ is *carried* by a generalized train track τ if it can be realized by an edge-path in τ with alternating real and infinitesimal edges.

REMARK 6. Although for our purposes a train track is a combinatorial object, we think of it as a smooth branched 1-submanifold of S. From this standpoint, the homotopy class of a path is carried by a generalized train track if there is a smooth representative in the class which is contained in the train track.

Now let $F \colon (S, \mathbb{G}, A) \to (S, \mathbb{G}, A)$ be a thick graph map such that $F(A) = A$, where $A = A_F$ is the set of attracting periodic orbits of F. On each strip s of $\mathbb{G}$ define the pseudo-isotopy $\psi_s \colon \bar{s} \times [0,1] \to \bar{s}$ to be given in coordinates by $\psi_s(x, y, t) = (x, (1-t)y + t/2)$ so that $\psi_s(\cdot, 1)$ maps $\bar{s}$ onto γ_s. Extend these pseudo-isotopies to a pseudo-isotopy $\psi_0 \colon S \times [0,1] \to S$ in the following way: first extend the ψ_s to mutually disjoint disk neighborhoods $U_s \supset \bar{s}$, with $U_s \subset S \setminus A$, so that they are isotopies on $U_s \setminus \bar{s}$ and the identity on ∂U_s; then extend them to be the identity elsewhere.

Notice that, if τ is a generalized train track, then $\psi_0(F(\tau), 1)$ satisfies the definition of train track, except possibly b), that is, it may have parallel edges.

DEFINITIONS 7. Define $F_*(\tau)$ to be the generalized train track consisting of a maximal subset of the edges of $\psi_0(F(\tau), 1)$ which contains no pair of parallel edges.

The train track $\tau = \tau(I)$ is said to be *F-invariant* if $F_*(\tau)$ is isotopic to τ in $S \setminus A$ and I is minimal (under inclusion) with this property.

REMARK 7. Notice that $F_*(\tau)$ carries the homotopy class of $F(e)$ for each edge e of τ.

From the definition it follows that there exists a pseudo-isotopy ψ with the property that $\psi(F(\tau), 1) = \tau$.

DEFINITION 8. The *train track map* $\phi \colon \tau \to \tau$ associated to $F \colon \mathbb{G} \to \mathbb{G}$ is defined by $\phi(\cdot) = \psi(F(\cdot), 1)$.

4.2. Construction of invariant generalized train tracks.

Let $F \colon (S, \mathbb{G}, A) \to (S, \mathbb{G}, A)$ be as before. The following procedure constructs an F-invariant generalized train track τ.

Let $\tau_0 = R$, and for each $n \geq 0$ define $\tau_{n+1} = F_*(\tau_n)$. Since τ_0 is a subset of τ_1, each τ_n is naturally isotopic to a subset of τ_{n+1}: hence each τ_{n+1} can be adjusted, as it is constructed, by an isotopy relative to $A \cup \partial \mathbb{V}$ such that $\tau_n \subseteq \tau_{n+1}$. Define $\tau = \bigcup_{n \geq 0} \tau_n$. Then τ is a generalized train track by construction. It easy to see that it is F-invariant.

REMARK 8. It is clear that this construction provides the minimal F-invariant generalized train track τ. Also, given any generalized sub-track τ' of τ, it is clear that the same construction, starting with $\tau_0 = \tau'$, must generate $\tau = \bigcup_{n=0}^{\infty} F_*^n(\tau')$.

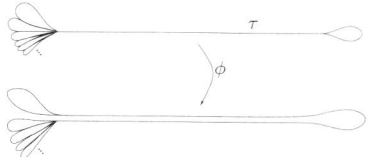

FIGURE 7. The invariant train track for the horseshoe.

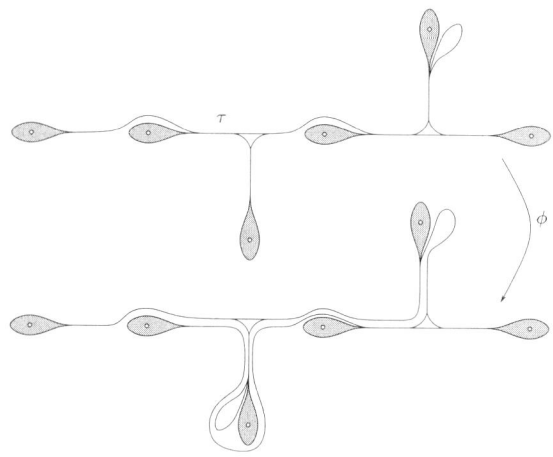

FIGURE 8. The invariant train track for the thick tree map F_2 of Example 3. Figure 9 shows a blow up of the shaded bubbles in this figure.

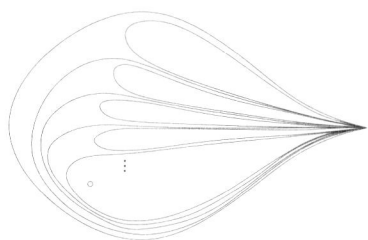

FIGURE 9. Detailed view of the bubbles in Figure 8. There and here ∘ represents a periodic point in the period 6 attracting orbit in Example 3.

EXAMPLE 5. The invariant train track for the horseshoe map F_1 is shown in Figure 7. The set A consists of the fixed point x contained in the left 1-junction of $\mathbb{I}$. No bubble encloses it and it is not shown in the figure.

EXAMPLE 6. The invariant train track for the thick tree map F_2 of Example 3 is shown in Figures 8 and 9. Here, A consists of the union of the period 6 and period 2 orbits contained in the junctions of $\mathbb{T}$.

The invariant train track τ and the train track map ϕ should be thought of as more careful 1-dimensional representations of the thick graph $\mathbb{G}$ and the thick graph map $F: \mathbb{G} \to \mathbb{G}$: whereas $f: G \to G$ does not pay attention to junctions — they are collapsed to points — the map $\phi: \tau \to \tau$ gives a careful account of the behavior of the images of strips under iterates of F inside the junctions.

We now gather some useful facts that are straightforward consequences of the constructions above.

REMARKS 9. a) ϕ is a near-homeomorphism, that is, it can be approximated arbitrarily well by homeomorphisms.
b) ϕ maps I into itself and each infinitesimal edge is mapped homeomorphically onto another infinitesimal edge.
c) At most finitely many infinitesimal edges are mapped onto a given infinitesimal edge under ϕ.
d) If e is a real edge, then $\phi(e)$ can only intersect finitely many infinitesimal edges (except possibly at endpoints).

DEFINITION 9. By an *infinitesimal polygon* we mean a component of the complement of τ bounded by finitely many infinitesimal edges. It is called an *n-gon* if it is bounded by n infinitesimal edges (see Figure 10).

FIGURE 10. Examples of n-gons for $n = 1, 3, 4$ and a 1- and a 3-gon together.

REMARKS 10. a) Bigons (2-gons) are not allowed unless they contain a point of A.
b) It is allowed that an n-gon have fewer than n vertices, as shown in the last diagram in Figure 10: two vertices in the 3-gon coincide.

The following lemma is an immediate consequence of the fact that ϕ is a near-homeomorphism on a surface.

LEMMA 3. *For each integer $n \geq 1$, ϕ induces a 1-1 map on the collection of n-gons. If an n-gon contains a periodic point in its interior, then it is periodic under ϕ (with the same period). Otherwise, it belongs to a semi-infinite orbit of n-gons $\{\phi^n(\Delta); n \geq 0\}$, where Δ is an n-gon which is not the image under ϕ of any other n-gon. Two such orbits either coincide or are disjoint. Moreover, in this case, for all but finitely many n, $\phi^n(\Delta)$ has at most two vertices.*

5. Generalized pseudo-Anosov maps

In this section we describe the construction of a generalized pseudo-Anosov map using an invariant train track, as defined in the previous section. The construction follows closely the one in [**BH95**] for finite invariant train tracks and reduces to that in case the train track is finite. We also describe an associated complex structure on the surface with respect to which the map becomes a Teichmüller mapping. We continue to assume that thick graphs have the homotopy type of a once punctured surface.

5.1. The construction. Let $F\colon \mathbb{G} \to \mathbb{G}$ be a thick graph map and $\phi\colon \tau \to \tau$ its associated invariant train track map as defined in Section 4. As before, R and I denote the real and infinitesimal edges of τ. We number the real edges $\{e_i; 1 \leq i \leq n\}$ and the infinitesimal edges $\{e_i; i > n\}$ and define a (possibly infinite) transition matrix $M = [m_{ij}]$ setting, as above,

$$m_{ij} = \text{number of times } \phi(e_j) \text{ crosses } e_i$$

Since $\phi(I) \subset I$, M has block form

$$M = \begin{bmatrix} N_{n \times n} & 0 \\ B & \Pi \end{bmatrix}$$

The matrix N records transitions between real edges and Π records transitions between infinitesimal edges, whereas B records transitions from real to infinitesimal edges. The (possibly infinite) square matrix Π has only 0's and 1's in its entries. Each of its columns has exactly one non-zero entry and each row has at most finitely many non-zero entries. The matrix B (which has n columns and possibly infinitely many rows) has at most finitely many non-zero entries in each column. These observations follow from those in Remarks 9.

EXAMPLE 7. For the horseshoe map, the transition matrix is infinite but is quite simple: $m_{11} = 2$, $m_{i\,i-1} = 1$ and $m_{ij} = 0$ otherwise.

Standing Assumption: It is assumed throughout this section that the matrix N is irreducible and aperiodic. This implies that its Perron-Frobenius eigenvalue $\lambda > 1$. It also follows that there is a positive integer k such that, for every real edge e, $\phi(e)$ contains all other real edges. In particular, τ is connected. This assumption is not necessary for all the results that follow, but it simplifies the discussion.

We think of M as an operator acting on the space l^1 of summable sequences of real numbers with norm $|y|_1 = \sum_{i \geq 1} |y_i|$. From the remarks above, it follows that M is a bounded operator with $\|M\|_1 \leq \max_j\{\sum_i |m_{ij}|\} < \infty$.

Let Y be an eigenvector of N associated to λ (and therefore unique, up to scale). It is possible to complete Y to an eigenvector $y = [Y\ Y']$ of M. Since the columns of Π have at most one non-zero entry which is 1, $\|\Pi\|_1 \leq 1$ and thus $\lambda I - \Pi$ is invertible. Setting $Y' = (\lambda I - \Pi)^{-1} BY$ we have

$$M \begin{bmatrix} Y \\ Y' \end{bmatrix} = \begin{bmatrix} N & 0 \\ B & \Pi \end{bmatrix} \begin{bmatrix} Y \\ Y' \end{bmatrix}$$
$$= \begin{bmatrix} NY \\ BY + \Pi Y' \end{bmatrix}$$
$$= \begin{bmatrix} \lambda Y \\ \lambda Y' \end{bmatrix}$$

In order to see that Y' is a positive vector, notice that

$$Y' = \frac{1}{\lambda}(I + \frac{1}{\lambda}\Pi + \frac{1}{\lambda^2}\Pi^2 + \ldots)BY$$
$$= \frac{1}{\lambda}(B + \frac{1}{\lambda}\Pi B + \frac{1}{\lambda^2}\Pi^2 B + \ldots)Y$$

The matrices $\Pi^k B$ that appear above represent transitions from a real edge to an infinitesimal edge under the $(k+1)$-st iterate of the map ϕ. In fact, they represent exactly those transitions which occurred from a real to an infinitesimal edge in the first iterate and which then remained among infinitesimal edges for the next k iterates (the other ways to get from a real to an infinitesimal edge under the $(k+1)$-st iterate of ϕ are represented by matrices of the form $\Pi^{k-j-1}BN^j$). But every infinitesimal edge is the image, under some iterate of ϕ, of an infinitesimal edge of τ_1 and these are the intersections of $\phi(\tau_0)$ with $\mathbb{V}$, that is, they are transitions from real to infinitesimal edges under the first iterate of ϕ. This means that, for every $i \geq 1$, there exists $k \geq 1$ such that the i-th row of $\Pi^k B$ is non-zero. Since Y is a positive vector, it follows that every entry of Y' is non-zero.

DEFINITION 10. A collection $\{y_i = y(e_i)\}_{i \geq 1}$ of non-negative real numbers, called *weights*, is said to satisfy the *switch conditions* if, for each switch q of τ, we have
$$y(e_{i_0}) = \sum y(e_i) + 2 \sum y(e_j)$$
where e_{i_0} is the real edge with endpoint at q and the first and second sums range over the set of infinitesimal edges having one or both endpoints at q respectively.

LEMMA 4. *Let M be the transition matrix associated to $\phi \colon \tau \to \tau$, λ its Perron-Frobenius eigenvalue and $y = [Y \; Y'] = [y_1 y_2 \ldots]$ an eigenvector associated to λ as constructed above. Then the set of weights $\{y_i\}_{i \geq 1}$ satisfy the switch conditions.*

Proof (cf. [**BH95**]). Fix a large positive integer k. The equality $M^k y = \lambda^k y$ written in coordinates states that for each $i \geq 1$
$$y_i = \frac{1}{\lambda^k} \sum_j y_j \cdot (\text{number of times } \phi^k(e_j) \text{ intersects } e_i)$$

If $\phi^k(e_j)$ crosses a switch, it must cross edges on both sides except at its endpoints. Thus, the contribution to both sides of the switch is the same up to a bounded amount. Letting $k \to \infty$ yields the result. □

Now, let X be an eigenvector of N^T associated to the Perron-Frobenius eigenvalue λ. If we try to complete X to an eigenvector $[X \; X']$ for the adjoint M^*, we are forced to set $X' = 0$. The reason is that X' is a solution to the equation $\Pi^* X' = \lambda X' \Leftrightarrow (\lambda I - \Pi^*) X' = 0$. Since $\|\Pi^*\|_\infty \leq 1$, the only solution is $X' = 0$. This is why infinitesimal edges are called such.

We now give a description of the construction of the generalized pseudo-Anosov homeomorphisms corresponding to $\phi \colon \tau \to \tau$. As was mentioned before, it follows closely that presented in [**BH95**].

Let $x = (x_1, x_2, \ldots, x_n, 0, 0, \ldots)$ and $y = (y_1, y_2, \ldots)$ be eigenvectors of M^* and M, respectively, associated to the Perron-Frobenius eigenvalue λ as just described. To each real edge e_i, $1 \leq i \leq n$ of τ we associate a Euclidean rectangle R_i of dimensions $x_i \times y_i$ endowed with foliations by horizontal and vertical line segments. Each foliation has a transverse measure induced by Lebesgue measure. The horizontal and vertical foliations will be called *unstable* and *stable* respectively and, under the map, unstable leaves will be stretched and stable leaves will be contracted by the factors λ and $1/\lambda$, respectively. Place (homeomorphic copies of) these rectangles on S along the real edges of τ. The infinitesimal edges of τ are used to define an equivalence relation on the vertical sides of the rectangles, as follows. Let e_j be an infinitesimal edge and y_j the corresponding entry of the eigenvector y. Identify segments of length y_i along the vertical sides of the rectangles which contain the endpoints of e_j (note that these rectangles could be the same). The facts that the train track is a subset of S and that the switch conditions are satisfied imply that there is exactly one way in which these identifications can be made without self-intersections. The quotient space $\mathcal{R}$ of the rectangles under these identifications is a compact topological surface (with boundary) of the homotopy type of $S \setminus \{p\}$. It is foliated by unstable "horizontal" lines, most of which are now infinite, that is, they are immersed copies of $\mathbb{R}$ (the exceptions being those that contain singularities of the foliations). Moreover, there is defined a map $\tilde{\Phi}\colon \mathcal{R} \to \mathcal{R}$ which stretches the unstable foliation by the factor λ, contracts the stable foliation (whose leaves are still finite segments) by the factor $1/\lambda$ and places $\tilde{\Phi}(\mathcal{R})$ inside $\mathcal{R}$ in the manner dictated by $\phi\colon \tau \to \tau$. Restricted to the interior of $\mathcal{R}$, $\tilde{\Phi}$ is a homeomorphism, but not along the boundary of $\mathcal{R}$. Notice that, by collapsing to points the stable segments, we obtain the graph G. Denoting the projection by $\tilde{\pi}\colon \mathcal{R} \to G$, $\tilde{\Phi}$ factors down to $f\colon G \to G$, that is, $f \circ \tilde{\pi} = \tilde{\pi} \circ \tilde{\Phi}$. The boundary $\partial \mathcal{R}$ has the structure of a smooth finite sided polygon, each side of which contains a periodic point. We identify segments of adjacent sides which are mapped to the same segment under some iterate of $\Phi\colon \mathcal{R} \to \mathcal{R}$. This usually leads to infinitely many identifications of smaller and smaller pieces of $\partial \mathcal{R}$ (for example, this is the case with the horseshoe, as will be seen below). The points of the periodic orbit on $\partial \mathcal{R}$ are all identified and become *the point at infinity*. Under these further identifications, the quotient of $\mathcal{R}$ is homeomorphic to S, the "vertical" segments become a foliation of S by stable leaves, most of which are immersed copies of $\mathbb{R}$, and the induced map, denoted by $\Phi\colon S \to S$, now becomes a homeomorphism: this is the generalized pseudo-Anosov homeomorphism. The stable and unstable foliations are denoted by $\mathcal{F}^s$ and $\mathcal{F}^u$ respectively. Both are preserved by Φ, the leaves of the stable foliation being contracted and those of the unstable foliation being stretched by the factors $1/\lambda$ and λ, respectively.

REMARKS 11. a) The foliations $\mathcal{F}^s$ and $\mathcal{F}^u$ are foliations with singularities: to each n-gon of τ there corresponds an n-pronged singularity of the foliations. There may be infinitely many such singularities, but they can accumulate on at most finitely many periodic orbits, which are in 1-1 correspondence with a subset of A. In the next section the orbits of 1-pronged singularities will be studied in greater detail.

b) The periodic orbits on $\partial \mathcal{R}$ described above have been found in different guises by other authors. See, for example, [**BL98, NP73**].

EXAMPLE 8. The invariant train track for the horseshoe has only one real edge and therefore the surface ($\mathbb{S}^2$) will be constructed by identifying the edges of one rectangle. The identifications are shown in Figure 11 by dotted lines. Notice that the equivalence class of the lower left corner contains infinitely many points. The points marked with • lie on one orbit of 1-pronged singularities which is forward and backward asymptotic to the lower left corner, which is also the point at infinity. The quotient sphere with the two invariant foliations is shown in Figure 5.

EXAMPLE 9. In Figure 12 are shown the identifications on seven rectangles dictated by the invariant train track for F_2. All the unstable identifications are indicated (dotted lines) whereas only the first set of stable identifications are shown (dashed-dotted lines). Further stable identifications are obtained from these by iterating the map backwards. The points of the period 2 orbit of infinitesimal triangles

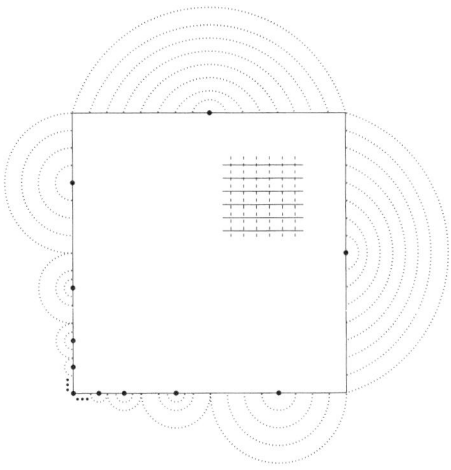

FIGURE 11. The identifications for the horseshoe.

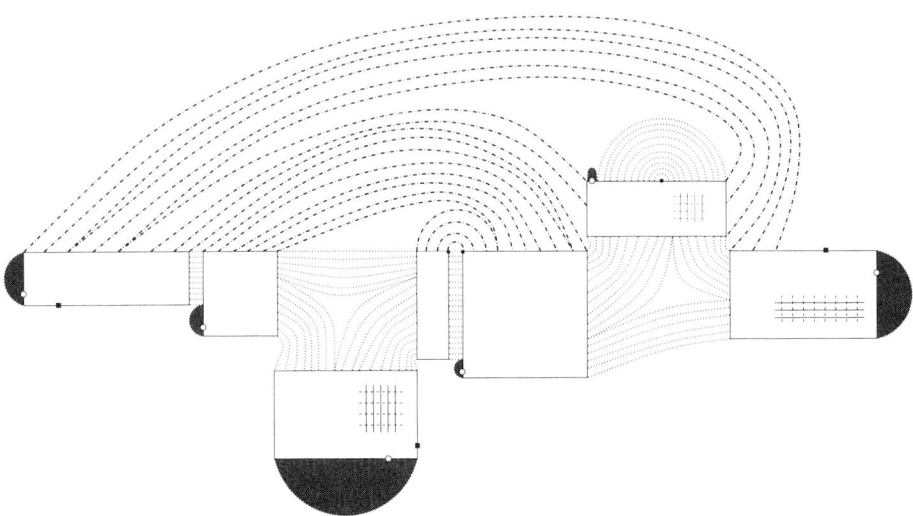

FIGURE 12. The identifications for F_2.

give rise to a period 2 orbit of 3-pronged singularities. There are infinitely many 1-pronged singularities (two of which come from the points marked with • in the figure) converging in the future to the period 6 orbit (indicated by ∘) corresponding to the period 6 attracting orbit of F_2 and converging in the past to the period 3 orbit at infinity (indicated by ■). There is also an orbit of 3-pronged singularities with the same forward and backward fates as the orbit of 1-pronged singularities. A detailed view of the shaded regions in Figure 12 is shown in Figure 13.

5.2. 1-pronged singularities. The generalized pseudo-Anosov maps constructed in the previous section preserve a pair of measured foliations with singularities. The 1-pronged singularities are the most important in many interrelated

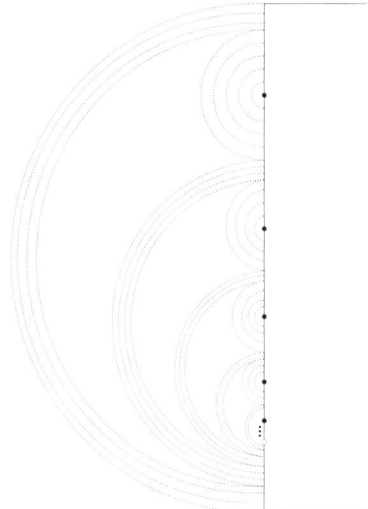

FIGURE 13. The identifications in the shaded regions in Figure 12.

ways. Dynamically, they play a role analogous to that played by the critical points for endomorphisms of the interval. They also "hold the map in place," so to say, in the sense that isotopies relative to the set of 1-pronged singularities cannot destroy any dynamics (this has been proved for the case of finitely many singularities [**Hal91, Han85, Thu88**]; some infinities have been dealt with in [**BH99**]; the general case seems to be technically delicate).

We now describe how to find the orbits of 1-pronged singularities of the invariant foliations $\mathcal{F}^{u,s}$. They come from infinitesimal 1-gons of τ so we need to be able to determine the orbits of these. It follows from Lemma 3 that if a 1-gon contains a periodic point, it is itself periodic under ϕ and therefore corresponds to a periodic 1-pronged singularity of the invariant foliations. If a 1-gon does not belong to a periodic orbit, then there exists a *first* 1-gon whose orbit contains the given 1-gon, that is, there exists a 1-gon Δ, which is not the image of any 1-gon under ϕ and whose orbit contains the given 1-gon. In this case, there exists either a real edge e such that $\phi(e) \supset \Delta$ or there is an infinitesimal edge e', which is not a 1-gon, that is, whose boundary points are distinct switches, such that $\phi(e') = \Delta$. In either case, there must exist a real edge e, an arc $\gamma \subset e$ and a smallest integer $k \geq 1$ such that $\phi^k(\gamma) = \Delta$. If there are several such arcs (possibly contained in several distinct real edges), we use the ambient surface or thick graph to choose γ to be *innermost* among of them. By this we mean the following. To the arc γ there corresponds a *thick arc* $\Gamma = [a,b] \times [0,1] \subset s$ where s is a strip of $\mathbb{G}$. Because $\phi^k(\gamma) = \Delta$, $F^n(\Gamma)$ is contained in the junction J that contains Δ and $F^n(\{a,b\} \times [0,1])$ are contained in the same component of ∂J, that is, $F^n(\Gamma)$ "makes a turn" inside J. We call γ innermost if $F^n(\Gamma)$ is an innermost turn among all thick arcs that map to J under F^n. Since infinitesimal edges are assigned 0 length in the construction of the invariant foliations, γ corresponds to a vertical segment in the rectangle R_e associated to the real edge e. Under the identifications required to make the leaves of the stable foliation infinite, R_e will be folded and one of the endpoints of this segment will become a 1-pronged singularity (the one that becomes innermost after folding). Since the "horizontal" sides of rectangles are contained in the unstable manifold of the point at infinity, it follows that this 1-pronged singularity is backward asymptotic to the point at infinity. To summarize, we have proved the

THEOREM 5. *Let $\mathcal{F}^{u,s}$ be the stable and unstable foliations of the generalized pseudo-Anosov map associated to a Markov thick graph map as constructed above. Then the 1-pronged singularities of $\mathcal{F}^{u,s}$ either belong to periodic orbits or are backward asymptotic to the point at infinity and forward asymptotic to one of finitely many periodic orbits.*

There is an easy way of tracking the orbits of 1-pronged singularities backwards using the graph map $f\colon G \to G$. Suppose $p \in G$ is a critical point of f (that is, a point at which f is not a local homeomorphism), so that $v = f(p)$ is a vertex. Assume, without loss of generality, that $f(p)$ is innermost in the junction $V = \pi^{-1}(v)$. Choose $p_{-1} \in f^{-1}(p)$ so that $f^2(p_{-1})$ is innermost in V, then choose $p_{-2} \in f^{-1}(p_{-1})$ so that $f^3(p_{-2})$ is innermost in V and proceed like this. If we reach a point p_{-n} which is not a vertex, then all subsequent ones are not vertices (since vertices map to vertices) and, in fact, at each subsequent step, there is exactly one innermost preimage (since the graph map is assumed to be onto). In this case, p_{-n} converges to the orbit at infinity. Otherwise, there are two possibilities: either we eventually return to p or we get to a point which has *no* innermost preimage. In the first case, we have found a periodic orbit of 1-pronged singularities and in the second, we were not following an orbit of 1-pronged singularities. Below we give examples explaining all three possibilities.

EXAMPLES 10. a) Consider the thick tree map F_2. In Figure 14, the quotient tree map is drawn. To avoid having too many symbols in the figure, the images of points in T under f are denoted by the same symbol on the image tree $f(T)$ drawn below. The only critical point is $p = 4$ and $v = f(4) = 5$. The preimage $f^{-1}(4)$ has two points: 2 and a point in edge e_5. It is the latter that becomes innermost under f^2, so $p_{-1} \in e_5$. The preimage $f^{-1}(p_{-1})$ again consists of two points, of which the one that becomes innermost under f^3 lies on edge e_2: this is p_{-2}. Again $f^{-1}(p_{-2})$ consists of two points and $p_{-3} \in e_1$. From here on things repeat, as shown in Figure 14: $p_{-4} \in e_7$, $p_{-5} \in e_3$, $p_{-6} \in e_1$, $p_{-7} \in e_7$, etc. The sequence p_n converges to the period 3 orbit at infinity.

b) In this example (see Figure 15, where only the induced tree endomorphism is drawn) $p = 3$ and $v = 5$. The preimage $f^{-1}(p)$ has two points, namely, 4 and a point on edge e_1. Under f^2 it is point 4 that becomes innermost so $p_{-1} = 4$. Again, $f^{-1}(4)$ contains two points, namely, 2 and a point on edge e_3, but it is 2 that becomes innermost under f^3 so that $p_{-2} = 2$. Continuing like this, we get $p_{-3} = 1$, $p_{-4} = 5$ and $p_{-5} = 3$. Thus, p gives rise to a periodic orbit of 1-pronged singularities. The associated generalized pseudo-Anosov is then a pseudo-Anosov map in the sense of Thurston.

c) The third possibility is shown in Figure 16. The top part of the figure contains parts of the graph containing the points p and $v = f(p)$ and two points in $f^{-1}(p)$; the bottom contains their images under f. Because no arc has image entering p through edge a and exiting through edge b (indicated by the dashed

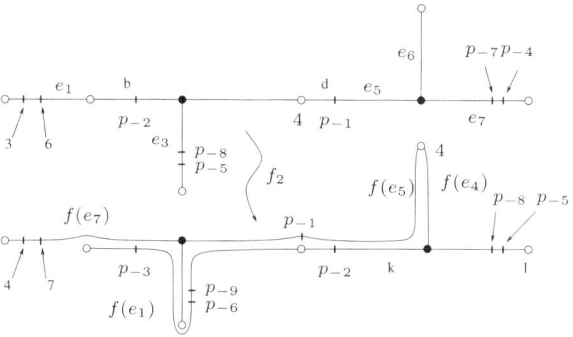

FIGURE 14. The backward orbit of 1-pronged singularities for F_2.

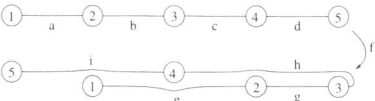

FIGURE 15. The quotient of a thick tree map whose associated generalized pseudo-Anosov has a periodic orbit of 1-pronged singularities.

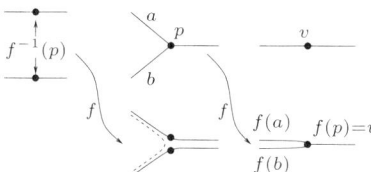

FIGURE 16. When a critical point of a tree map does not produce an orbit of 1-pronged singularities.

line in the figure), $f^{-1}(p)$ does not contain a point whose image under f^2 is innermost in v.

5.3. The complex structure. We now describe a natural complex structure associated to a generalized pseudo-Anosov map, with respect to which it becomes a Teichmüller mapping and the invariant foliations become the horizontal and vertical trajectories of an associated integrable quadratic differential.

As was seen above, S is obtained as the quotient of Euclidean rectangles under side identifications by Euclidean isometries. In the interior of the rectangles, the complex structure is the one determined by the Euclidean structure. At k-pronged singularities, use the maps $z \mapsto z^{2/k}$ to define coordinate charts. This produces a complex structure at all points of S, except at the accumulations of singularities. These are topologically isolated and it is necessary to decide whether the complex structure in the complement regards them as punctures or as holes. We now argue that they are in fact punctures so that the complex structure extends across them uniquely. To prove this we present a sequence of concentric annuli converging to an

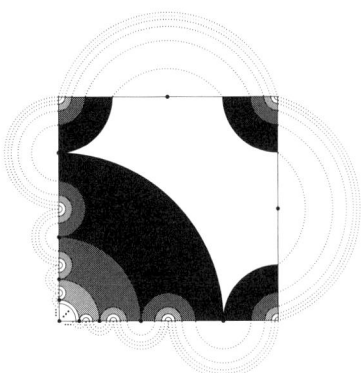

FIGURE 17. A sequence of concentric annular regions whose sum of moduli diverges.

accumulation of singularities and whose moduli add up to infinity. By length-area arguments (see [**Ahl73, LV73**]), the result follows[4].

The basic inequality used is

$$\mathrm{mod}(A) \geq \frac{\inf_{\gamma \in \Gamma} l(\gamma)^2}{\mathrm{Area}(A)}$$

where Γ is the set of all rectifiable curves joining the boundary components of the annular region A and $l(\gamma)$ is the length of the curve γ.

We first give the argument for the accumulation of singularities p in the generalized pseudo-Anosov associated to the horseshoe.

In Figure 17 are shown the first four concentric annular regions of a sequence $\{A_n\}_{n=1}^{\infty}$ converging to p. We argue that $\mathrm{mod}(A_n) > C/n$, for some positive constant C, implying thus that $\sum_{n=1}^{\infty} \mathrm{mod}(A_n)$ diverges. Annular region A_n is made up of three round quarter-annuli at the upper-left and right and lower-right corners, the region bounded by four circular arcs on the lower-left and $2n$ round half-annuli, n along each of the left and bottom sides of the square. All parts being (parts of) Euclidean annuli, they contribute to the extremal length of the family of all curves joining the boundary components of A_n at least like Euclidean annuli do. The infimum of the lengths of curves joining boundary components is $\propto 1/2^n$. The sum of the areas of the quarter annuli is $\propto 1/2^{2n}$, whereas the $2n$ half-annuli contribute $\propto n/2^{2n}$ to the total area. Therefore, we have

$$\begin{aligned} \mathrm{mod}(A_n) &\geq \frac{(\text{Distance between boundary components of } A_n)^2}{\mathrm{Area}(A_n)} \\ &\geq \frac{C_1}{C_2 + C_3 n} \end{aligned}$$

The general result is

THEOREM 6. *Let* $F\colon (S, \mathbb{G}) \to (S, \mathbb{G})$ *be a Markov thick graph map whose associated transition matrix is irreducible and aperiodic. Let* $\Phi\colon S \to S$ *be the associated generalized pseudo-Anosov and* $\{p_1, \ldots, p_k\}$ *the accumulation points of*

---

[4]I am grateful to Fred Gardiner for suggesting this argument (see also [**EG97**]).

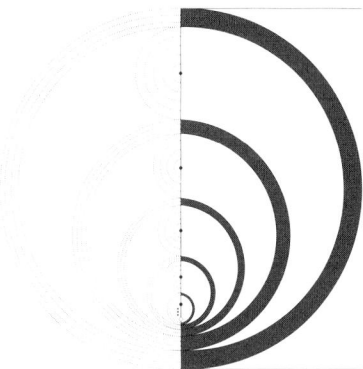

FIGURE 18. Another sequence of concentric annular regions whose sum of moduli diverges.

pronged singularities of the Φ-invariant foliations. Then the complex structure on $S \setminus \{p_1, \ldots, p_k\}$ induced by the Euclidean structure extends uniquely to a complex structure on the compact surface S.

Sketch of Proof. Let p be an accumulation of pronged singularities of the invariant foliations corresponding to an attracting periodic orbit of F, which we will also denote by p (the other possibility, namely, that p is the point at infinity, is treated similarly). Passing to an appropriate iterate, we may assume that p is an attracting fixed point of F. For simplicity, we assume it lies in a 1-junction V; the other cases are analogous.

There are two possibilities to consider: either the invariant train track τ has a bubble in V which encloses p or not. If there is a bubble β enclosing p, all its iterates also do. They then give rise to an infinite collection of disjoint concentric annular regions A_n, all enclosing p, with $\mod(A_n) \propto 1$ (see Figure 18). In case there is no bubble enclosing p in V, then p is the point at infinity and the situation is analogous to the horseshoe example explained above. □

References

[Ahl73] L. V. Ahlfors, *Conformal invariants: topics in geometric function theory*, McGraw-Hill Book Co., New York, 1973, McGraw-Hill Series in Higher Mathematics. MR 50 #10211

[Bal00] V. Baladi, *Positive transfer operators and decay of correlations*, World Scientific Publishing Co. Inc., River Edge, NJ, 2000. MR 1 793 194

[Bar86] M. Barge, *Horseshoe maps and inverse limits*, Pacific J. Math. **121** (1986), no. 1, 29–39. MR **87g:**58077

[BH95] M. Bestvina and M. Handel, *Train-tracks for surface homeomorphisms*, Topology **34** (1995), no. 1, 109–140. MR **96d:**57014

[BH99] P. Boyland and T. Hall, *Isotopy stable dynamics relative to compact invariant sets*, Proc. London Math. Soc. (3) **79** (1999), no. 3, 673–693. MR **2000k:**37026

[BL98] C. Bonatti and R. Langevin, *Difféomorphismes de Smale des surfaces*, Astérisque (1998), no. 250, viii+235, With the collaboration of E. Jeandenans. MR **99m:**58123

[Bow71] R. Bowen, *Entropy for group endomorphisms and homogeneous spaces*, Trans. Amer. Math. Soc. **153** (1971), 401–414. MR 43 #469

[dCH] A. de Carvalho and T. Hall, *Unimodal generalized pseudo-Anosov maps*, Submitted, arXiv:math.DS/0307211.

[dCH01] _____, *Pruning theory and Thurston's classification of surface homeomorphisms*, J. Eur. Math. Soc. (JEMS) **3** (2001), no. 4, 287–333. MR 1 866 160

[dCP04] A. de Carvalho and M. Paternain, *Monotone quotients of surface diffeomorphisms*, Math. Res. Lett. **10** (2004), 603–619.

[EG97] C. J. Earle and F. P. Gardiner, *Teichmüller disks and Veech's $\mathcal{F}$-structures*, Extremal Riemann surfaces (San Francisco, CA, 1995), Amer. Math. Soc., Providence, RI, 1997, pp. 165–189. MR **97k:**32031

[FM93] J. Franks and M. Misiurewicz, *Cycles for disk homeomorphisms and thick trees*, Nielsen theory and dynamical systems (South Hadley, MA, 1992), Amer. Math. Soc., Providence, RI, 1993, pp. 69–139. MR **95e:**58133

[Gan59] F. R. Gantmacher, *The theory of matrices. Vols. 1, 2*, Chelsea Publishing Co., New York, 1959. MR 21 #6372c

[Hal91] T. Hall, *Unremovable periodic orbits of homeomorphisms*, Math. Proc. Cambridge Philos. Soc. **110** (1991), no. 3, 523–531. MR **92i:**58146

[Han85] M. Handel, *Global shadowing of pseudo-Anosov homeomorphisms*, Ergodic Theory Dynam. Systems **5** (1985), no. 3, 373–377. MR **87e:**58172

[LV73] O. Lehto and K. I. Virtanen, *Quasiconformal mappings in the plane*, second ed., Springer-Verlag, New York, 1973, Translated from the German by K. W. Lucas, Die Grundlehren der mathematischen Wissenschaften, Band 126. MR 49 #9202

[Moo62] R. L. Moore, *Foundations of point set theory*, American Mathematical Society, Providence, R.I., 1962. MR 27 #709

[NP73] S. Newhouse and J. Palis, *Hyperbolic nonwandering sets on two-dimensional manifolds*, Dynamical systems (Proc. Sympos., Univ. Bahia, Salvador, 1971), Academic Press, New York, 1973, pp. 293–301. MR 49 #4044

[RS38] J. Roberts and N. Steenrod, *Monotone transformations of two-dimensional manifolds*, Ann. of Math. **39** (1938), 851–862.

[Thu88] W. P. Thurston, *On the geometry and dynamics of diffeomorphisms of surfaces*, Bull. Amer. Math. Soc. (N.S.) **19** (1988), no. 2, 417–431. MR **89k:**57023

INSTITUTE FOR MATHEMATICAL SCIENCES, STATE UNIVERSITY OF NEW YORK AT STONY BROOK, NY 11794-3660, USA

E-mail address: `andre@math.sunysb.edu`

Current address: Departamento de Matemática Aplicada, IME - USP, Rua do Matão 1010, Cidade Universitária, 05508-090 São Paulo, SP, Brazil

E-mail address: `andre@ime.usp.br`

Unimodal maps and hierarchical models

Michael Yampolsky

CONTENTS

1. Renormalization in Statistical Mechanics
2. Unimodal maps

References

Foreword. This note came from an attempt by the author to understand the basis for the analogy between renormalization and universality in statistical mechanics, and in one-dimensional dynamics, which ever since the original works of Feigenbaum [**Feig**], and Collet and Tresser [**CT**] has motivated the development of the latter. We present the evidence that the two fields are related directly. We begin by discussing the best understood statistical-mechanical model of phase transitions - the Dyson's Hierarchical model studied by Bleher and Sinai [**BS1, BS2**] and others. We then proceed to discuss various constructions of Hamiltonians corresponding to renormalizable unimodal maps, beginning with the works of [**VSK**], and [**Sul1**], and show how some recent works in the field may be used to construct a thermodynamical analogue of the Feigenbaum-Coullet-Tresser renormalization.

1. Renormalization in Statistical Mechanics

1.1. A brief review of phase transitions. The purpose of this section is to broadly describe the phenomenology of phase transitions in statistical mechanics, and in this way to set the stage for further discussion. We will not attempt to give an introduction to the subject of statistical mechanics here, and will instead refer the reader to one of the numerous existing textbooks. We recall that the central object of study of the theory is the definition and properties of Gibbs probability distributions. For an infinite set $\Lambda \subset \mathbb{R}^d$ (most often Λ is the lattice $\mathbb{Z}^d$, but not in the examples considered in this note), which is invariant under the action of the symmetry group G, the set of spins on Λ is the collection Φ of functions mapping Λ to some set A (which for our purposes can be assumed to be finite). A Hamiltonian

---

1991 *Mathematics Subject Classification.* 37E20, 82B28.
Key words and phrases. universality, renormalization, unimodal map, hierarchical model.
The author gratefully acknowledges the support of NSERC.

$\mathcal{H}$ is a G-equivariant functional $\Phi \to \mathbb{R}$, which should be interpreted as the energy of the spin configuration ϕ. In the case when Λ is finite, the Gibbs measure $\mu_{\mathcal{H},\beta}$ defined for every $\beta > 0$ is the probability distribution on Φ given by

$$\mu_{\mathcal{H},\beta}(\phi) = \frac{\exp(-\beta\mathcal{H}(\phi))}{\sum_{\phi \in \Phi} \exp(-\beta\mathcal{H}(\phi))} \tag{1.1}$$

When Λ is infinite, one chooses a sequence of finite subsets $\Lambda^n \to \Lambda$ and considers the set of measures obtained as weak limits of the expressions of the form (1.1) with ϕ restricted to Λ^n. These measures are called the Gibbs measures. They form a convex set, whose extreme points are called the extreme, or pure Gibbs measures. Other quantities describing the system can be introduced using a similar thermodynamical limit procedure. For instance, the function

$$\mathbf{f}(\beta) = -\beta^{-1} \lim_{n\to\infty} |\Lambda^n|^{-1} \log \sum_{\phi|_{\Lambda^n}} \exp(-\beta\mathcal{H}(\phi))$$

is the free energy per unit volume which relates the canonical and microcanonical ensembles.

One possible way to define a phase transition is as a value of $\beta = \beta_{\mathrm{cr}}$ for which the structure of the set of pure Gibbs distributions changes. There are two kinds of phase transitions: in the first kind there is a discontinuous change in the thermodynamical parameters defining the system; in the second the change is continuous, but not smooth. It is the latter that will concern us. The simplest example, the kind of which occurs in ferromagnetics, would be a system in which for $\beta < \beta_{\mathrm{cr}}$ there is a single pure Gibbs distribution, and for $\beta > \beta_{\mathrm{cr}}$ there are two. The phenomenological picture of a phase transition is characterized by the appearance of critical exponents. Let us describe a typical such picture, following the exposition in Sinai's book [**Sin**]. For the Gibbs distribution μ_{cr} corresponding to $\beta = \beta_{\mathrm{cr}}$ the correlation length becomes infinite, and we have

$$E_{\mu_{\mathrm{cr}}}\phi(x) \cdot \phi(y) \sim \frac{\mathrm{const}}{\|x-y\|^\xi}$$

for the two-point correlation function. This implies, in particular, that for a finite volume $V \subset \Lambda$

$$E_{\mu_{\mathrm{cr}}} \left(\sum_{x \in V} \phi(x) \right)^2 \sim \mathrm{const} \cdot |V|^\alpha, \text{ for } \alpha = \alpha(\xi) > 1.$$

On the other hand, for $\beta < \beta_{\mathrm{cr}}$ the corresponding right-hand side is $\sigma(\beta)|V|$. One then expects that

$$\sigma(\beta) \sim \mathrm{const} \cdot (\beta_{\mathrm{cr}} - \beta)^{-\gamma} \xrightarrow[\beta \to \beta_{\mathrm{cr}}]{} \infty,$$

which gives another example of a critical exponent. Also, for $\beta > \beta_{\mathrm{cr}}$ the two Gibbs distributions produce two distinct average values for the spin $\phi(x)$. Denoting them $a_1(\beta)$ and $a_2(\beta)$ we expect that

$$|a_1(\beta) - a_2(\beta)| \sim \mathrm{const} \cdot |\beta - \beta_{\mathrm{cr}}|^\omega.$$

The list of critical exponents can be continued, there are also various algebraic connections between them, which we will not dwell upon. The main goal of understanding a phase transition lies in estimating the values of the critical exponents.

To this end a renormalization transformation $\mathcal{R}$ is introduced. Without attempting to give a general definition (indeed, finding an appropriate definition is usually the crux of the matter!) let us summarize its properties. The operator $\mathcal{R}$ acts on Hamiltonians $\mathcal{H}$ by averaging out some of the degrees of freedom in the system. It is expected to preserve the sets of Gibbs measures, and the correlation length ξ. On an appropriate Banach manifold of Hamiltonians it is expected to have a fixed point $\mathcal{H}_*$ which is hyperbolic. In the simplest scenario, it has a single unstable eigenvalue $\lambda \in \mathbb{C} \setminus \mathbb{D}$, whose value determines all of the critical exponents.

To conclude this section let us discuss hierarchical models, which are the main subject of this note. They are somewhat different from the usual lattice models. We will not attempt to give a definitive description of such models here, but just summarize the common features of the models that we will consider. A model is hierarchical if the symmetry group G acts on $\Lambda \subset \mathbb{R}^n$ by piecewise-linear transformations which properly map Λ into itself. We thus obtain a hierarchy of images $\Lambda_{g_1} \subset \Lambda_{g_1 g_2} \subset \cdots$. It is worth noting that in this context it is reasonable to expect $\mathcal{R}$ to be defined in such a way as to propagate a Hamiltonian from Λ^n to a larger set $\cup_{g \in G} g^{-1}(\Lambda^n)$.

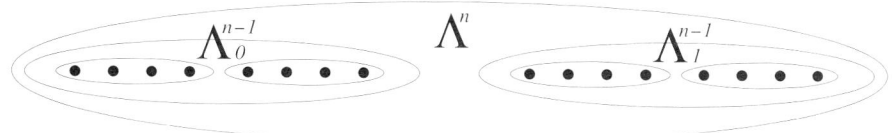

FIGURE 1

1.2. Dyson's Hierarchical models. The principal thermodynamical example which will concern us is the Dyson's Hierarchical model. These are well-studied objects, and their renormalization theory is presented in the books of Sinai [**Sin**], and Collet and Eckmann [**CE**]. Sinai's book, in particular, gives a clear and concise account of the theory, to which we do not have anything to add, therefore in this section we will simply recap its main points for future reference. The main results in the study of Dyson's Hierarchical models were obtained by Bleher and Sinai [**BS1, BS2**], these papers will serve as our general references.

Let us denote Λ^n the set of binary sequences of length n

$$\Lambda^n = \{(x_1, x_2, \ldots, x_n) \mid x_i \in \{0,1\}\}.$$

This set is naturally partitioned in two, $\Lambda_0^{n-1} = (0, x_2, \ldots, x_n)$ and $\Lambda_1^{n-1} = (1, x_2, \ldots, x_n)$; and more generally, for every $k \leq n$ and every sequence $(y_1, \ldots, y_k) \in \{0,1\}^k$ we can define

$$\Lambda_{y_1, \ldots, y_k}^{n-k} = \{(y_1, \ldots, y_k, x_{k+1}, \ldots, x_n) \mid x_i \in \{0,1\}\}$$

to partition Λ^n into 2^{n-k} subsets of equal size. The space of spins on Λ^n is $\Phi_n = \{\phi : \Lambda^n \to \{-1,1\}\}$. A Dyson's Hierarchical model is a Hamiltonian $\mathcal{H}^n : \Phi_n \to \mathbb{R}_{\geq 0}$. To define it we need the following set of data: $n_0 \leq n$; a "start-up" Hamiltonian $\mathcal{H}^{n_0} : \Phi_{n_0} \to \mathbb{R}_{\geq 0}$, and a number $c \in (1, 2)$. The functional $\mathcal{H}^n$ is then defined inductively by the following formula:

$$(1.2) \qquad \mathcal{H}^n(\phi|_{\Lambda^n}) = \mathcal{H}^{n-1}(\phi|_{\Lambda_0^{n-1}}) + \mathcal{H}^{n-1}(\phi|_{\Lambda_1^{n-1}}) + \frac{c^n}{2^{2n}} \left[\Sigma_{\zeta \in \Lambda^n} \phi(\zeta)\right]^2.$$

We shall denote the triple $(n_0, \mathcal{H}^{n_0}, c) = \Theta$, and sometimes write $\mathcal{H}^n_\Theta$ to denote the Dyson's Hamiltonian given by this data. To draw an analogy with the unimodal maps further in the paper, we will interpret the formula (1.2) as the definition of a renormalization transformation

$$\mathcal{R}_{\text{Dyson}} : \mathcal{H}^{n-1}_\Theta \to \mathcal{H}^n_\Theta.$$

We caution the reader, that this is not the usual block-spin renormalization transofrmation as applied to the Dyson's Hierarchical model. The discussion of the latter may be found in [**Ble**].

Given a Hamiltonian $\mathcal{H}^n$ and a value $\beta > 0$ we obtain a probability measure $\mu_{\mathcal{H}^n, \beta}$ on Φ_n given by

$$\mu_{\mathcal{H}^n, \beta}(\phi) = \frac{\exp[-\beta \mathcal{H}^n(\phi)]}{\Sigma_{\phi \in \Phi_n} \exp[-\beta \mathcal{H}^n(\phi)]}.$$

For an even integer $t \in [-2^n, 2^n]$ let us define the function $f_n(t; \beta)$ as the probablity that the sum $\Sigma_{y \in \Lambda^n} \phi(y)$ takes the value t. Some calculations lead to the following recurrent relations:

$$(1.3) \qquad f_n(t; \beta) = P_n(\beta) \exp(\beta c^n 2^{-2n} t^2) \sum_{t_1 = -2^{n-1}}^{2^{n-1}} f_{n-1}(t_1; \beta) f_{n-1}(t - t_1; \beta),$$

where $P_n(\beta)$ is a normalizing factor. Further considerations depend on the following heuristic understanding of the behaviour of the model near the critical β_{cr}. For $\beta < \beta_{cr}$ one expects that the typical values of the sum $\Sigma_{y \in \Lambda^n} \phi(y)$ are of order $t \sim 2^{n/2}$. For such t the multiple $\exp(\beta c^n 2^{-2n} t^2) \to 1$ as $n \to \infty$, and the equation (1.3) for large n behaves like the convolution.

On the other hand, for $\beta > \beta_{cr}$, the typical values of t become of order n, and the multiple $\exp(\beta c^n 2^{-2n} t^2)$ becomes dominating. We then assume that at β_{cr} the typical values of t are such that $c^n 2^{-2n} t^2 \sim 1$. If that is so, it is natural to introduce a new variable z by $t = z 2^n c^{-n/2}$ and expect that $z \sim 1$ for $\beta = \beta_{cr}$. Setting $g_n(z; \beta) = f_n(z 2^n c^{-n/2}; \beta) 2^{n+1} c^{-n/2}$, we have

$$g_n(z; \beta) = L_n(\beta) e^{\beta z^2} \sum_{(z_1 + z_2)/2 = z/\sqrt{c}} g_{n-1}(z_1; \beta) g_{n-1}(z_2; \beta) c^{n/2} 2^{-n-1}.$$

Our assumptions lead us to expect that $g_n(z; \beta)$ has a limit $g(z; \beta)$ as $n \to \infty$. Letting $n \to \infty$ in the above equation and dropping the normalizing factor we obtain the following integral equation for $g(z; \beta)$:

$$(1.4) \qquad g(z; \beta) = e^{\beta z^2} \int_{-\infty}^{\infty} g\left(\frac{z}{\sqrt{c}} + u; \beta\right) g\left(\frac{z}{\sqrt{c}} - u; \beta\right) du$$

Note that if $g(z; \beta_0)$ is a solution of (1.4) for some $\beta_0 > 0$, then

$$g(z; \beta) = g\left(\sqrt{\frac{\beta}{\beta_0}} z; \beta_0\right) \sqrt{\frac{\beta}{\beta_0}}.$$

Hence we may fix β_0 as convenient, and solve the integral equation to find a function of z.

Solving the integral equation. It is natural at this point to interpret the right side of the equation (1.4) as an integral operator $\mathcal{R}_{\text{int}} g(z; \beta)$ and also call it a renormalization transformation. Bleher and Sinai observe, that for every $c \in (1,2)$ the operator $\mathcal{R}_{\text{int}}$ has a Gaussian fixed point

$$g^{(0)}(z;\beta) = \sqrt{\frac{a_0}{\pi}} \exp[-a_0(\beta)z^2] \text{ where } a_0(\beta) = \beta c/(2-c).$$

They then show that in an appropriate Hilbert space of functions, for $c \in (\sqrt{2}, 2)$ this fixed point is hyperbolic with a single unstable eigenvalue $\lambda = 2/c > 1$. For $c < \sqrt{2}$ other unstable eigenvalues appear for the linearization of $\mathcal{R}_{\text{int}}$ at $g^{(0)}$. However, for ϵ small enough they are able to show for $c = \sqrt{2} - \epsilon$ the existence of a different fixed point $g_\epsilon(z;\beta)$ with a single unstable eigenvalue. The numerical evidence suggests that such a solution exists for every $c \in (1,2)$, and there exists a computer-assisted proof of Koch and Wittwer [**KW**] for $c = 2^{1/3}$.

For a given $n_0 \in \mathbb{N}$ the space of real-valued functionals on Φ_n is a real vector space; the space of start-up Hamiltonians $\mathcal{H}^{n_0}$ with the property that $\mathcal{H}^{n_0}(\phi) = \mathcal{H}^{n_0}(-\phi)$ is a closed subset of this space.

DEFINITION 1.1. We will say that a fixed point $g(z;\beta)$ is thermodynamically stable if there exists n_0 and an open set U of start-up Hamiltonians $\mathcal{H}^{n_0}$ such that the following holds.

- For each $\mathcal{H}^{n_0} \in U$ there exists $\beta_{\text{cr}}(\mathcal{H}^{n_0})$ such that the probability distribution $g_n(z;\beta_{\text{cr}})$ weakly converges to $g(z;\beta_{\text{cr}})$ as $n \to \infty$.
- For $\beta < \beta_{\text{cr}}$ there exists a function $h(\beta)$ such that as $n \to \infty$

$$f_n(t;\beta) \sim \frac{1}{\sqrt{2\pi h(\beta)2^n}} \exp\left\{-\frac{t^2}{2h(\beta)2^n}\right\}$$

  for t such that $|t2^{-n/2}| \leq A$ for every A. There exists γ depending only on $g(z;\beta)$ such that $h(\beta) \sim \text{const} \cdot (\beta_{\text{cr}} - \beta)^{-\gamma}$ for $\beta \nearrow \beta_{\text{cr}}$.
- For $\beta > \beta_{\text{cr}}$ there exists a function $m(\beta)$ such that the average spint $2^{-n}\sum_{x \in \Lambda^n} \phi(x)$ converges to $\pm m(\beta)$ as $n \to \infty$ in probability generated by the Hamiltonian $\beta\mathcal{H}^n$. Moreover, there exists ω depending only on $g(z;\beta)$ such that $m(\beta) \sim \text{const} \cdot |\beta - \beta_{\text{cr}}|^\omega$.

One expects that for $c \in (\sqrt{2}, 2)$ the Gaussian fixed point $g^{(0)}$ is thermodynamically stable. This is indeed so, as was shown by Bleher and Sinai:

THEOREM 1.1. *For every $\beta_{cr}^{(0)} > 0$, $\epsilon > 0$, $c \in (\sqrt{2}, 2)$ there exists $n_0 \in \mathbb{N}$ and an open set U in the space of start-up Hamiltonians $\mathcal{H}^{n_0}$ such that for every $\mathcal{H}^{n_0} \in U$ there exists $\beta_{cr}(\mathcal{H}^{n_0})$ such that $|\beta_{cr} - \beta_{cr}^{(0)}|$; and $g_n(z;\beta_{cr})$ converges weakly to the Gaussian distribution. The Gaussian distribution is thermodynamically stable with respect to the start-up Hamiltonians in the set U.*

2. Unimodal maps

2.1. A review of the Feigenbaum Renormalization Theory after Sullivan, McMullen, and Lyubich.
We briefly recall here the main results and conjectures of the celebrated Feigenbaum renormalization theory of unimodal maps. We will only discuss the case of period-doublings, for a more detailed exposition including a review of the general case we refer the reader to Lyubich's paper [**Lyu1**].

Let us fix $\gamma > 1$. Consider an even unimodal map $f : [-1,1] \to [-1,1]$, $f(-1) = f(1) = -1$. We will require that f be C^2 smooth, except at the critical point. The latter will be further assumed to be of order γ, that is $f(x) = \phi(|x|^\gamma)$ near 0, where $\gamma > 1$, and ϕ is a local diffeomorphism.

DEFINITION 2.1. The map f is Feigenbaum-Collet-Tresser (FCT) renormalizable if the following holds: the critical value $f(0)$ lies above the diagonal $x = y$, which means, in particular, that f has a fixed point $p_f = 1/\alpha_f \in (0, 1)$; and the iterate f^2 maps the interval $[-p_f, p_f]$ into itself.

If f is FCT renormalizable, then the rescaled second iterate

$$-\alpha_f f \circ f(-x/\alpha_f) : [-1, 1] \to [-1, 1]$$

is again a unimodal map, we shall call it the FCT renormalization of f, and denote it $\mathcal{R}_{FCT} f$.

We shall say that a map f is of Feigenbaum combinatorial type, or has Feigenbaum combinatorics, if the above procedure may be repeated indefinitely.

Feigenbaum-Collet-Tresser renormalization hyperbolicity conjecture. *For every $\gamma > 1$ there exists a Banach manifold $\mathcal{B}_\gamma$ of unimodal maps such that $\mathcal{R}_{FCT}$ is a smooth mapping of an open subset of $\mathcal{B}_\gamma$ into $\mathcal{B}_\gamma$. The mapping $\mathcal{R}_{FCT}$ has a fixed point $f_\gamma \in \mathcal{B}_\gamma$ with the following hyperbolicity property: $D_{f_\gamma} \mathcal{R}_{FCT}$ is a compact operator with a single eigenvalue $\delta_\gamma > 0$ outside the closed unit disk, and the rest of the spectrum inside $\mathbb{D}$. The stable manifold of f_γ consists of unimodal maps with Feigenbaum combinatorics, and for every f with Feigenbaum combinatorics and the critical point of order γ,*

$$\mathcal{R}_{FCT}^n f \to f_\gamma$$

in the uniform topology.

The eigenvalue δ_γ is called the Feigenbaum exponent, and the limit

$$\alpha = \lim_{n \to \infty} \alpha_{\mathcal{R}_{FCT}^n f}$$

is called the Feigenbaum scaling factor.

The real renormalization theory. The main achievement of this theory is Lanford's computer-assisted proof [**Lan**] of the existence of a fixed point of $\mathcal{R}_{FCT}$ for $\gamma = 2$ with the right hyperbolicity properties. Without the assistance of a computer, it has been shown that the renormalizations $\{\mathcal{R}_{FCT}^n f\}$ belong to a compact set in the C^1 topology. In particular, there exists a bound $C = C(\gamma) > 1$ such that $1/C < 1/\alpha_g < C$ for $g = \mathcal{R}_{FCT}^n$ for all n sufficiently large. As a consequence, the ω-limit set of the critical point of f is a dyadic Cantor set with bounded geometry.

The complex renormalization theory. This theory has been developed by Sullivan [**Sul2, MvS**], and later by McMullen [**McM**], and Lyubich [**Lyu1, Lyu2**]. It applies when γ is an even integer, we will only discuss the case $\gamma = 2$. Denote $\mathcal{A}_D$ the Banach manifold of analytic maps f defined in an $\mathbb{R}$-symmetric topological disk $D \subset \mathbb{C}$ with the sup norm, and such that 0 is a simple critical point of f.

Renormalization hyperbolicity theorem. *In the space of C^2-unimodal maps with $\gamma = 2$ there exists a unique fixed point f_* of $\mathcal{R}_{FCT}$. For every C^2-unimodal f with $\gamma = 2$ and Feigenbaum combinatorics, $\mathcal{R}_{FCT}^n f \to f_*$ in the C^1 topology. The mapping f_* is an analytic mapping of a domain $D \subset \mathbb{C}$ which is quadratic-like (that is $f_* : D \to f_*(D)$ is a double-covering, branched at 0, and $D \Subset f_*(D)$). There is an open subset $\mathcal{QL} \subset \mathcal{A}_D$ which consists of quadratic-like maps, such that $\mathcal{R}_{FCT}$*

extends to an analytic operator of an open neighborhood of f_* in $\mathcal{QL}$ to $\mathcal{QL}$, such that $D_{f_*}\mathcal{R}_{FCT}$ has the right hyperbolicity properties.

2.2. Vul-Sinai-Khanin Hamiltonians. The paper of Vul, Sinai, and Khanin [**VSK**] offered a direct link between the Feigenbaum renormalization and the world of statistical mechanics. We will take it as the departure point for our investigation, and will briefly recap the main points of the construction below. We will begin by fixing an even unimodal mapping $f : [-1,1] \to [-1,1]$ with Feigenbaum combinatorics, with a critical exponent $\gamma > 1$ and will assume that the Feigenbaum-Collet-Tresser Hyperbolicity Conjecture holds for γ. That is, our statements will be founded for $\gamma = 2\ell$, $\ell \in \mathbb{N}$, and should be treated as conjectures for other values of γ. We note that the authors of [**VSK**] actually take f to be the Feigenbaum renormalization fixed point with $\gamma = 2$, however, such a restriction is not necessary for their construction.

Let us begin by introducing some supporting notation. We will denote $\Delta_0^{(1)} = [-p_f, p_f]$, and $f(\Delta_0^{(1)}) = \Delta_1^{(1)}$, so that $f(\Delta_0^{(1)}) = [p_f, f^2(0)] \subset \Delta_0^{(1)}$. Similarly, we will let $\Delta_0^{(n)} \ni 0$ be the symmetric renormalization interval of level n, that is, the maximal symmetric interval around 0 on which the iterate f^{2^n} is unimodal. The cycle of intervals of level n will be denoted $\Delta_k^{(n)} = f^k(\Delta_0^{(n)})$. The invariant Cantor set of f is then the intersection

$$\mathbf{C}_f = \omega_f(0) = \bigcap_{n \geq 1} \bigcup_{k=0}^{2^n - 1} \Delta_k^{(n)}.$$

For an integer number $k \geq 0$ let us also denote

$$\text{Bin}(k) = \underline{\epsilon_0 \epsilon_1 \epsilon_2 \cdots \epsilon_{n-1}}, \; \epsilon_i \in \{0, 1\}$$

its binary extension, $k = \epsilon_0 + \epsilon_1 \cdot 2 + \cdots + \epsilon_{n-1} \cdot 2^{n-1}$. We obviously have:

PROPOSITION 2.1. *Suppose $\Delta_k^{(n)} \supset \Delta_m^{(n+1)}$. If $\text{Bin}(k) = \underline{\epsilon_0 \epsilon_1 \epsilon_2 \cdots \epsilon_{n-1}}$, then $\text{Bin}(m) = \underline{\epsilon_0 \epsilon_1 \epsilon_2 \cdots \epsilon_n}$.*

In view of the previous proposition, we have a well-defined one-to-one map $\Psi : x \mapsto \underline{\epsilon_0 \epsilon_1 \cdots \epsilon_n \cdots}$ from $\mathbf{C}_f$ to the space of one-sided binary sequences $\Sigma_{\{0,1\}}$.

PROPOSITION 2.2. *The map Ψ conjugates the action of f to the adding one transformation*

$$\text{Add}_1 : \underline{\epsilon_0 \epsilon_1 \cdots \epsilon_n \cdots} \mapsto \underline{\epsilon_0 \epsilon_1 \cdots \epsilon_n \cdots} + 100000 \cdots.$$

Let us define the lattices $\Lambda = \mathbb{N}$ and $\Lambda^n = \{1, \ldots, n\}$, and the spaces of spins $\Phi = \{\phi : \Lambda \to \{0,1\}\}$, $\Phi_n = \{\phi : \Lambda^n \to \{0,1\}\}$. The Vul-Sinai-Khanin Hamiltonian which we define below is a functional $\mathcal{H}_{VSK} : \Phi \to \mathbb{R}_{\geq 0}$. Let $\Delta_{p \cdot 2^{n-k}+1}^{(n)} \subset \Delta_{p' \cdot 2^{n-k-1}+1}^{(n-1)}$ for $0 \leq p < 2^k$. The map Ψ carries them respectively into

$$\overbrace{1000 \cdots 0}^{n-k} \epsilon^k \epsilon^{k-1} \cdots \epsilon^1 \quad \text{and} \quad \overbrace{1000 \cdots 0}^{n-k} \epsilon^k \epsilon^{k-1} \cdots \epsilon^2$$

Let

$$\log\left(\frac{|\Delta_{p \cdot 2^{n-k}+1}^{(n)}|}{|\Delta_{p' \cdot 2^{n-k-1}+1}^{(n-1)}|}\right) \equiv U_k^n(\epsilon^1, \epsilon^2, \ldots, \epsilon^k).$$

PROPOSITION 2.3. *For a fixed sequence $\epsilon^1, \epsilon^2, \ldots, \epsilon^k$ the function $U_k^n(\epsilon^1, \epsilon^2, \ldots, \epsilon^k)$ has a limit as $n \to \infty$ which we denote $U_k(\epsilon^1, \epsilon^2, \ldots, \epsilon^k)$. In fact, there exists $\sigma > 1$ such that for a fixed k*

$$(2.1) \qquad |U_k^n(\epsilon^1, \epsilon^2, \ldots, \epsilon^k) - U_k(\epsilon^1, \epsilon^2, \ldots, \epsilon^k)| \leq \text{const} \cdot \sigma^{-n}$$

Moreover, the functions $U_k(\epsilon^1, \epsilon^2, \ldots, \epsilon^k)$ converge to a limit as $k \to \infty$, which will be denoted $U(\epsilon^1, \epsilon^2, \ldots, \epsilon^k, \ldots)$, and

$$(2.2) \qquad |U(\epsilon^1, \epsilon^2, \ldots, \epsilon^k, \ldots) - U_k(\epsilon^1, \epsilon^2, \ldots, \epsilon^k)| \leq \text{const} \cdot \sigma^{-k}.$$

We have a natural identification of Φ with $\Sigma_{0,1}$. In view of this, the authors of [**VSK**] proposed to view the function $U(\epsilon^1, \epsilon^2, \ldots, \epsilon^k, \ldots)$ as the interaction potential of the first coordinate with the remaining coordinates. They have thus obtained a sequence of Hamiltonians $\mathcal{H}_{VSK}^n : \Phi_n \to \mathbb{R}_{\geq 0}$ given by

$$\mathcal{H}_{VSK}^n(\epsilon_1, \ldots, \epsilon_n) = -\Sigma_{s=1}^n U(\epsilon_s, \epsilon_{s-1}, \ldots, \epsilon_2, 1, 0, \ldots, 0, \ldots).$$

In view of (2.2) the Hamiltonian $\mathcal{H}_{VSK} : \Phi \to \mathbb{R}_{\geq 0}$ may be naturally defined as the limit of the above sequence of finite Hamiltonians. Observe that the Feigenbaum-Collet-Tresser renormalization hyperbolicity implies that

PROPOSITION 2.4. *The values of the Hamiltonians $\mathcal{H}_{VSK}^n$ and $\mathcal{H}_{VSK}$ do not depend on the choice of a particular mapping f, but only on its universality class, that is, the value of the critical exponent $\gamma > 0$.*

This property of $\mathcal{H}_{VSK}$ makes it a convenient tool of the study of the geometric properties of $\mathbf{C}_f$, as will be seen below. However, the independence from the choice of f makes $\mathcal{H}_{VSK}$ a poor candidate for defining a statistical-mechanical analogue of the Feigenbaum renormalization. We will dwell on the later point further, for the moment, let us proceed with describing the results of [**VSK**].

The following estimate of [**VSK**] is fundamental for their analysis: there exists a constant $C > 1$ such that

$$(2.3) \qquad C^{-1} \leq \frac{|\Delta_k^{(n)}|}{\exp\{-\mathcal{H}_{VSK}^n(\epsilon_0, \ldots, \epsilon_{n-1})\}} \leq C,$$

where $\text{Bin}(k) = \overline{\epsilon_0 \ldots \epsilon_{n-1}}$, and $\epsilon_0 = 1$. The free energy of $\mathcal{H}_{VSK}$ is, by definition, the following limit

$$\mathbf{f}(\beta) = \lim_{n \to \infty} \frac{1}{n} \log \left[\Sigma_{\epsilon_2, \ldots, \epsilon_n} \exp\{\beta \Sigma_{s=1}^n U(\epsilon_s, \epsilon_{s-1}, \ldots, \epsilon_2, 1, 0, 0, \ldots, 0, \ldots)\} \right].$$

This limit is seen to exist, and to produce a smooth and monotonously decreasing function with $\mathbf{f}(0) = \log 2$, $\lim_{\beta \to \infty} \mathbf{f}(\beta) = -\infty$. The unique value of $\beta_0 > 0$ for which $\mathbf{f}(\beta_0) = 0$ corresponds to a phase transition. For this value of β the sum in the square brackets is bounded by above and below by two positive constants, which in conjunction with (2.3) can be shown to imply that

$$\text{Hdim}(\mathbf{C}_f) = \beta_0.$$

Sullivan's observation on $C^{1+\alpha}$-self-similar Cantor sets. The existence of the limiting scaling ratios $U(\epsilon^1, \epsilon^2, \ldots, \epsilon^n, \ldots)$ may be seen at this point as a property intrinsic to the Feigenbaum functional equation. However, their nature is more general as seen from the work of Sullivan [**Sul1**]. Firstly, let g be a fixed point of the Feigenbaum renormalization transformation with the critical exponent γ, set

$$(2.4) \qquad \sigma(x) = \begin{cases} -\alpha x, & x \in \Delta_0^{(1)} \\ -\alpha g(x), & x \in \Delta_1^{(1)} \end{cases}$$

where α is the Feigenbaum scaling factor, as before. The function $\sigma(x)$ will play a significant role in what follows, at the moment, however, let us simply note that it is smooth, has no critical points, and leaves $\mathbf{C}_g$ invariant. The reader is invited to verify the last property, which is a ready consequence of the Feigenbaum functional equation. More specifically, the function $\sigma(x)$ shifts the binary sequences in $\Sigma_{0,1}$, parametrizing the points in $\mathbf{C}_f$ to the left, thus transforming $(\epsilon^1, \epsilon^2, \ldots)$ into $(\epsilon^2, \epsilon^3, \ldots)$. Sullivan then asks a question, when would the shift operation on a Cantor set be realized as a smooth mapping. Before formulating the answer, let us give several definitions. Let us consider a general dyadic Cantor set $\mathbf{C}$ which is the intersection of nested intervals $\cap_{n \geq 1} \cup_{k=0}^{2^n-1} \Delta_k^{(n)}$ indexed in such a way that Proposition 2.1 holds. Recall that the Cantor set $\mathbf{C}$ has bounded geometry if for every finite binary sequence $\omega_n = \{\epsilon^i\}_{i=1}^n$ the triple of scaling ratios

$$\mathfrak{g}_n(\omega_n) = \left\{ \frac{|\Delta_{\epsilon^n \epsilon^{n-1} \ldots \epsilon^1}|}{|\Delta_{\epsilon^n \epsilon^{n-1} \ldots \epsilon^1 1}|}, \frac{|\Delta_{\epsilon^n \epsilon^{n-1} \ldots \epsilon^1}|}{|\Delta_{\epsilon^n \epsilon^{n-1} \ldots \epsilon^1 0}|}, \frac{|\Delta_{\epsilon^n \epsilon^{n-1} \ldots \epsilon^1}|}{|G_{\epsilon^n \epsilon^{n-1} \ldots \epsilon^1}|} \right\}$$

is bounded from above and below by positive constants independent of ω_n and n, where $G_{\epsilon^n \epsilon^{n-1} \ldots \epsilon^1}$ is the middle gap between the two sub-intervals of $\Delta_{\epsilon^n \epsilon^{n-1} \ldots \epsilon^1}$. For $\beta \in (0,1]$ let $C^{1+\beta}$ denote the class of smooth local diffeomorphisms with β-Hölder continuous derivatives. Change of coordinates by such diffeomorphisms have the following effect on the scaling ratios:

PROPOSITION 2.5. *Suppose $\mathbf{C}$ has bounded geometry. Then for every $\beta > 0$, the values of the scaling ratios $\mathfrak{g}_n(\omega_n)$ become independent of $C^{1+\beta}$-coordinate changes exponentially fast in n.*

Let us say that the set of scaling functions $\{\mathfrak{g}_n(\epsilon^1, \epsilon^2, \epsilon^3, \ldots, \epsilon^n)\}$ for the Cantor set $\mathbf{C}$ is Hölder continuous if the value of each scaling function is determined exponentially fast in k by the first k arguments. In particular, in this case, we have limiting scaling functions $\mathfrak{g}(\epsilon^1, \epsilon^2, \ldots, \epsilon^k, \ldots)$ associated to every sequence in $\Sigma_{0,1}$. Moreover, if $\mathbf{C}$ has bounded geometry, Proposition 2.5 implies that these scaling functions are independent of $C^{1+\beta}$ coordinate changes, and thus only depend on the choice of the $C^{1+\beta}$ differentialble structure on the Cantor set.

THEOREM 2.6. *Suppose a bounded geometry Cantor set $\mathbf{C}$ is such that its scaling functions are Hölder continuous. Then the action of the binary shift on the Cantor set is realized as a $C^{1+\beta}$-expanding map in some smooth metric.*

Conversely, if the shift on the Cantor set is $C^{1+\beta}$ smooth in a differentiable structure on the interval in which the geometry of the Cantor set is bounded, then its scaling functions are Hölder continuous, and, in particular, the limiting scaling functions $\{\mathfrak{g}(\epsilon^1, \epsilon^2, \epsilon^3, \ldots)\}$ exist for every infinite sequence $(\epsilon^1, \epsilon^2, \ldots) \in \Sigma_{0,1}$.

Since the functions $U(\epsilon^1, \epsilon^2, \ldots, \epsilon^n, \ldots)$ introduced in [**VSK**] are nothing but logs of particular limiting scaling functions, the Proposition 2.3 is a corollary of the above theorem.

"Naïve" Hamiltonians. The above considerations, in particular, the estimate (2.3) suggest a different approach in associating Hamiltonians to a unimodal map f with the Feigenbaum combinatorics. We may define $\mathcal{H}_N^n(f) : \Phi_n \to \mathbb{R}_{\geq 0}$ simply by
$$\mathcal{H}_N^n(f)(\epsilon_0, \ldots, \epsilon_{n-1}) = -\log|\Delta_k^{(n)}|,$$
where $\text{Bin}(k) = \epsilon_0 \epsilon_1 \cdots \epsilon_{n-1}$. Such "naïve" approach has the advantage that the Hamiltonians $\mathcal{H}_N^n(f)$ depend on the mapping f, and thus are eligible candidates for defining renormalization. We shall see in the next section, that more complex definitions are required, however, this is a step in the correct direction. For the moment, let us make note, that the estimate (2.3) may be shown to imply that denoting β_0^n the critical values of β for $\mathcal{H}_N^n(f)$, we have $\beta_0^n \to \beta_0 = \text{Hdim}(\mathbf{C}_f)$. However, the Hamiltonians $\mathcal{H}_N^n(f)$ themselves do not converge to any particular limit.

2.3. Martens' result in thermodynamical terms. As we shall see in this section, a thermodynamical definition for renormalization of unimodal maps has already appeared in the literature. We will present an interpretation of a paper of Martens [**Mar**], which ties the Feigenbaum-Coullet-Tresser Renormalization operator with a hierarchical renormalization transformation acting on Hamiltonians related to the ones considered above. Before formulating the main result of Martens' work, let us make several definitions, which generalize the Feigenbaum-Coullet-Tresser renormalization to unimodal maps with periodic intervals of periods higher than 2. Namely, a unimodal map $f : [-1, 1] \to [-1, 1]$ is renormalizable if there exists a symmetric interval L_0 around 0 and $m > 1$ such that the intervals $L_i = f^i(L_0)$ have pairwise disjoint interiors for $i < m$, $f^m(L_0) \subset L_0$, and $f^m(\partial L_0) \subset \partial L_0$. The smallest such m is the renormalization period of f. In this case, selecting the maximal such interval $L_0 \equiv \Delta_0$, and denoting $\phi : \Delta_0 \to [-1, 1]$ the linear rescaling, which maps the boundary point $f^m(\partial \Delta_0)$ to -1, we set
$$\mathcal{R}f = \phi \circ f^m|_{\Delta_0} \circ \phi^{-1}.$$
This unimodal map is the renormalization of f. Note, that when $m = 2$, we have $\mathcal{R} = \mathcal{R}_{FCT}$. Let us denote $\tau(f) \in S_m$ the permutation of $\{1, \ldots, m\}$ which corresponds to i the position of the interval $f^{i-1}(\Delta_0) \equiv \Delta_i$ among the intervals of the cycle with respect to the linear order. Let us call admissible any permutation which is induced in this way by a renormalizable unimodal map. As before, a map f is infinitely renormalizable, if $\mathcal{R}$ can be iterated indefinitely on f. The combinatorial type of an infinitely renormalizable map is then the infinite sequence $\aleph(f) = (\tau(f), \tau(\mathcal{R}f), \ldots, \tau(\mathcal{R}^n f), \ldots)$. We will now state the main theorem of [**Mar**]:

THEOREM 2.7 (**Periodic points of renormalization**). *For every $\gamma > 1$, every $p \geq 2$ and every periodic sequence $\aleph = (\tau_1, \ldots, \tau_p, \tau_1, \ldots, \tau_p, \ldots)$ where τ_i is an admissible permutation there exists an even unimodal map f which is infinitely renormalizable with combinatorial type $\aleph(f) = \aleph$, and whose critical point 0 has the order γ, such that*
$$\mathcal{R}^p f = f.$$

While uniqueness of f is not shown, an upshot of the argument is that f belongs to a very restrictive class of maps (the so-called Epstein class). H. Epstein has obtained the same results by a different method in [**Ep**]. Of main interest to us will be not the theorem itself, but the rather remarkable method which Martens used to prove it. We will therefore confine ourselves exclusively to the case of the Feigenbaum combinatorics.

Let us begin by fixing $\gamma > 1$, and denoting $q_t(x) : [-1,1] \to [-1,1]$ for each $t \in [0,1]$ the unimodal mapping

$$q_t(x) = -2t|x|^\gamma + 2t - 1. \tag{2.5}$$

For a closed oriented interval $I = [a,b] \subset \mathbb{R}$ let us denote $\iota_I : \mathbb{R} \to \mathbb{R}$ the affine orientation preserving mapping which transforms $[-1,1]$ into I. An elementary property of the folding mappings q_t which is useful to note is the following:

PROPOSITION 2.8. *Let $J \equiv [-p,p]$ be a proper subinterval of $[-1,1]$, and $I = [q_t(p), l]$ such that $l \geq q_t(0) = 2t - 1$. Then there exists $s = s(l,p)$ such that*

$$(\iota_I)^{-1} \circ q_t|_J \circ \iota_J = q_s.$$

Let $f = \phi \circ q_t$, where $\phi : [-1,1] \to [-1,1]$ is a diffeomorphism, be an even infinitely renormalizable unimodal mapping of the Feigenbaum combinatorial type, whose critical point has the order γ. As before, let us denote $\{\Delta_k^{(n)}\}_{k=0}^{2^n-1}$ the cycle of renormalization intervals of level n. The standard considerations of one-dimensional dynamics (see e.g. [**MvS**]) imply that the restrictions of f to these intervals are approximated geometrically fast in n by restrictions of q_t in the smooth metric. This will motivate to some extent the following discussion.

DEFINITION 2.2. For every $n \in \mathbb{N}$ a Martens' Hamiltonian[1] of level n is a mapping $\mathcal{H}_{\text{Mar}}^n : \Phi_n \to [-1,1]^2$ which corresponds to every finite binary sequence $\{\epsilon_0, \epsilon_1, \ldots, \epsilon_{n-1}\}$ a closed interval

$$\mathcal{H}_{\text{Mar}}^n(\epsilon_0, \epsilon_1, \ldots, \epsilon_{n-1}) = J_{\epsilon_0 \epsilon_1 \cdots \epsilon_{n-1}} \subset [-1,1] \setminus \{0\}.$$

Similarly, a Martens' Hamiltonian $\mathcal{H}_{\text{Mar}} : \Phi \to [-1,1]^2$ corresponds an interval to each infinite binary sequence. An extended Martens' Hamiltonian $\tilde{\mathcal{H}}_{\text{Mar}}^n$ is a pair $(\mathcal{H}_{\text{Mar}}^n, t)$ with $t \in [0,1]$.

Let us denote $\mathbf{Mar}^n$ the set $\left([-1,0)^2 \cup (0,1]^2\right)^n$ viewed as the collection of Martens' Hamiltonians of level n. The extended Hamiltonians then correspond to points in $\mathbf{Mar}^n \times [0,1]$. For the moment, we shall view them simply as topological spaces, whose topology in a natural way comes from the topology of $\mathbb{R}$.

DEFINITION 2.3. For an infinitely renormalizable mapping $f = \phi \circ q_t$ of Feigenbaum combinatorial type we obtain a sequence of Martens' Hamiltonians $\mathcal{H}_{\text{Mar}}^n(f)$ in the following fashion. For $i \neq 0$ we set $J_i \equiv \Delta_i^{(n)}$, and for $i = 0$ we set $J_0 = q_t(\Delta_0^{(n)})$. To define an extended Hamiltonian $\tilde{\mathcal{H}}_{\text{Mar}}^n(f)$, we set $\tilde{\mathcal{H}}_{\text{Mar}}^n(f) = (\mathcal{H}_{\text{Mar}}^n(f), s)$ where

$$\left(\iota_{q_t(\Delta_0^{(n)})}\right)^{-1} \circ q_t|_{\Delta_0^{(n)}} \circ \iota_{\Delta_0^{(n)}} = q_s.$$

Before proceeding any further, let us note the following:

REMARK 2.1. We can obtain the naïve Hamiltonian $\mathcal{H}_N^n(f)$ from the pair $\mathcal{H}_{\text{Mar}}^n(f), t$ by the forgetful transformation φ which replaces each of the intervals J_k, $k \neq 0$ with its length, and replaces J_0 with $|q_t^{-1}(J_0)|$.

---

[1] We deviate from the convention somewhat by calling $\mathcal{H}_{\text{Mar}}$ a Hamiltonian, even though its values are pairs of real numbers. This minor abuse of notation should not lead to any confusion.

More importantly, there is also a way to recover a unimodal map $f_{\tilde{\mathcal{H}}^n_{\mathrm{Mar}}}$ from an extended Martens' Hamiltonian. To that end, let us first define a composition operation $\mathrm{Comp}^n : \mathcal{H}^n_{\mathrm{Mar}} \mapsto \phi \in \mathrm{Diff}([-1,1])$ as follows.

DEFINITION 2.4. The composition
$$\mathrm{Comp}^n(\mathcal{H}^n_{\mathrm{Mar}}) = \prod_{k=1}^{2^n-1} \left((\iota_{q_t(J_k)})^{-1} \circ q_t|_{J_k} \circ \iota_{J_k} \right)$$

It is not difficult to see that the result of this construction is independent of the value of t used.

Now, for an extended Martens Hamiltonian $\tilde{\mathcal{H}}^n_{\mathrm{Mar}} = (\mathcal{H}^n_{\mathrm{Mar}}, s)$ we define
$$f_{\tilde{\mathcal{H}}^n_{\mathrm{Mar}}} = \mathrm{Comp}^n(\mathcal{H}^n_{\mathrm{Mar}}) \circ q_s : [-1,1] \to [-1,1].$$

This definition is natural in the following sense:

PROPOSITION 2.9. *Let $g = q_{t*}$ be the map in the folding family (2.5) with the Feigenbaum combinatorics. For every $n \in \mathbb{N}$ we have*
$$f_{\tilde{\mathcal{H}}^n_{\mathrm{Mar}}(g)} = \mathcal{R}^n_{FCT}(g).$$

Of course, the composition operation behaves well with respect to the usual norm on $\mathbf{Mar}^n$. However, we will be interested also in making sense out of composing infinite Hamiltonians. To that end, a different notion of distance needs to be introduced. Let us recall that the nonlinearity of a C^2-diffeomorphism $\phi : [-1,1] \mapsto [-1,1]$ is
$$\eta(\phi)(x) = \frac{D^2\phi(x)}{D\phi(x)} = D|\ln D\phi(x)|.$$

This map is a bijection between $\mathrm{Diff}^2_+([-1,1])$ and the space of continuous functions $C([-1,1])$, the inverse being given by
$$\phi(\eta)(x) = 2\frac{\int_{-1}^x e^{\int_{-1}^s \eta(y)dy} ds}{\int_{-1}^1 e^{\int_{-1}^s \eta(y)dy} ds} - 1.$$

We will denote $\|\cdot\|$ the norm on Diff^2_+ which is the pull-back of the sup-norm on $C([-1,1])$ under this bijection. Suppose $\mathcal{H}^n_{\mathrm{Mar}} : \{\epsilon_0,\ldots,\epsilon_{n-1}\} \to J_{\epsilon_0\cdots\epsilon_{n-1}}$ and $\widehat{\mathcal{H}^n_{\mathrm{Mar}}} : \{\epsilon_0,\ldots,\epsilon_{n-1}\} \to \hat{J}_{\epsilon_0\cdots\epsilon_{n-1}}$ are two Martens' Hamiltonians. We will define a pseudo-distance between them as
$$\mathrm{dist}(\mathcal{H}^n_{\mathrm{Mar}}, \widehat{\mathcal{H}^n_{\mathrm{Mar}}}) = \sum_{k=1}^{2^n-1} \|(\iota_{q_t(J_k)})^{-1} \circ q_t|_{J_k} \circ \iota_{J_k} - (\iota_{q_t(\hat{J}_k)})^{-1} \circ q_t|_{\hat{J}_k} \circ \iota_{\hat{J}_k}\|.$$

Distinct Hamiltonians may become indistinguishable with respect to dist. In order to understand the situation better, let us note that
$$\eta(q_t)(x) = (\gamma - 1)\frac{1}{x}.$$

Hence, the nonlinearity of each of the terms $(\iota_{q_t(J_k)})^{-1} \circ q_t|_{J_k} \circ \iota_{J_k}$ for a Martens' Hamiltonian is an expression of the form $d/(ax+b)$. We may thus denote $\mathbf{Mar}^n_*$ the space of n triples $a,b,d \geq 0$ such that $b^2 - a^2 > 0$, and naturally identify
$$\Psi : \mathbf{Mar}^n/\mathrm{dist} \to \mathbf{Mar}^n_*.$$

Denote **Mar**$_1$ the subset of **Mar** on which the l_1 semi-norm induced by $\|.\|$ is finite. Martens shows that the composition operator Comp may be defined on **Mar**$_1$ as the limit of finite compositions. Further:

PROPOSITION 2.10. *The composition operation* Comp *is continuous with respect to the above distance. Moreover, it is Lipschitz on every bounded set.*

REMARK 2.2. It is appropriate to note here that there are other choices of distance on **Mar**^{n} which allow the composition to be extended to the case of infinitely many maps. One example would be to use the weigthed l^1-norm

$$\|\mathcal{H}_{\mathrm{Mar}}^n\|_\rho = \sum_k \sum_{x \in \partial \Delta_k} \rho^n |x|, \text{ for } \rho > 1.$$

Uniform boundedness in this norm would imply that the intervals shrink in size geometrically fast – while such an approach would be essentially equivalent to what we did above, it may still be technically useful.

Renormalization. Let us say that an extended Martens' Hamiltonian $\tilde{\mathcal{H}}_{\mathrm{Mar}}^n = (\mathcal{H}_{\mathrm{Mar}}, t)$ is renormalizable if the unimodal map $f_{\tilde{\mathcal{H}}_{\mathrm{Mar}}^n}$ is renormalizable with period 2 in the usual sense. In this case, $f_{\tilde{\mathcal{H}}_{\mathrm{Mar}}}$ has a pair of periodic intervals $\Delta_0 = [-p, p] \ni 0$ and Δ_1 whose boundaries touch at a repelling fixed point $p > 0$.

DEFINITION 2.5. The renormalization of a renormalizable extended Martens' Hamiltonian $\tilde{\mathcal{H}}_{\mathrm{Mar}}^n = (\mathcal{H}_{\mathrm{Mar}}^n, t)$ is a pair $\mathcal{R}\tilde{\mathcal{H}}_{\mathrm{Mar}}^n = (\widehat{\mathcal{H}_{\mathrm{Mar}}^{n+1}}, \hat{t})$ defined as follows.

- The intervals $\hat{J}_{\underline{\epsilon_0 \cdots \epsilon_n}} = \widehat{\mathcal{H}_{\mathrm{Mar}}^{n+1}}(\epsilon_0, \cdots, \epsilon_n)$ are given by

$$\hat{J}_{\underline{\epsilon_0 \cdots \epsilon_n}} = \left(\prod_{k > \underline{\epsilon_0 \cdots \epsilon_{n-1}}} (\iota_{q_t(J_k)})^{-1} \circ q_t|_{J_k} \circ \iota_{J_k} \right)^{-1} \Delta_{\epsilon_n}.$$

- The value of $\hat{t}$ is the parameter of the folding map $q_{\hat{t}}$ (2.5) such that

$$\mathcal{R}_{FCT} f_{\tilde{\mathcal{H}}_{\mathrm{Mar}}^n} = \mathrm{Comp}^{n+1}(\widehat{\mathcal{H}_{\mathrm{Mar}}^{n+1}}) \circ q_{\hat{t}}.$$

This definition may also be extended to the infinite Hamiltonians in **Mar**$_1$, although we will not make use of this fact.

PROPOSITION 2.11. *By virtue of the construction,* $\mathcal{R}_{FCT} f_{\tilde{\mathcal{H}}_{\mathrm{Mar}}^n} = f_{\mathcal{R}\tilde{\mathcal{H}}_{\mathrm{Mar}}^n}$.

Proof of the existence of a fixed point. The reader should see the virtue of using Martens' Hamiltonians to represent unimodal maps at this stage. While general unimodal mappings form an unwieldy infinite-dimensional real Banach space, those represented by Martens' Hamiltonians of level n are an open subset of $\mathbb{R}^k$. The only obstacle to defining finite-dimensional approximation to the renormalization is that a Hamiltonian of level n is transformed by $\mathcal{R}$ into a Hamiltonian of level $n + 1$. A way to circumvent it is the following. Let us define a Martens' tree of depth k as a sequence $\mathcal{T}^k = \{\mathcal{H}_{\mathrm{Mar}}^n\}_{n=1}^k$, $k \in \mathbb{N} \cup \{\infty\}$ and an extended tree as a pair $(\mathcal{T}^k, t)$. Renormalization naturally acts on extended trees of infinite depth, and transforms trees of depth k into trees of depth $k + 1$. A truncation of level $l < k$ is the tree $\mathrm{trunc}^l(\mathcal{T}^k) = \{\mathcal{H}_{\mathrm{Mar}}^n\}_{n=1}^l$. Let us consider the set of trees bounded with respect to the l_1 semi-norm, induced by $\|.\|$. Modulo the corresponding pseudo-distance, this space of trees may be identified with an open set in a real Banach space $\mathcal{B}$.

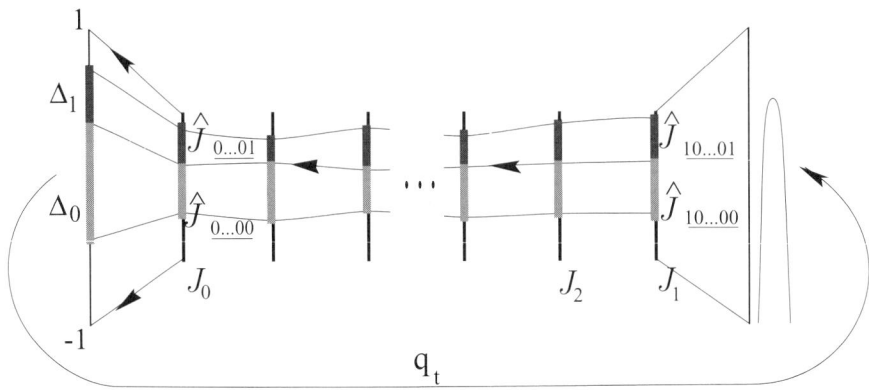

Figure 2

REMARK 2.3. Let us first note that an infinitely-renormalizable C^2-smooth map f generates an infinite tree $\mathcal{T}(f) = \{\mathcal{H}_{\text{Mar}}^n(f)\}_{n=1}^\infty$ which has a finite l_1 semi-norm.

We also remark that using the norm $||\cdot||_\rho$ from Remark 2.2 to define the l^1 norm on trees has the following geometric advantage. If ρ is sufficiently large, and $\mathcal{T}(f)$ is the infinite tree corresponding to an infinitely renormalizable map, then all trees in its sufficiently small neighborhood may be interpreted as dyadic Cantor sets with bounded geometry.

Martens proved the following theorem:

THEOREM 2.12. *For every $k > 1$ there exists an extended tree $(\mathcal{T}^k, t_k)$ such that*
$$\text{trunc}^k(\mathcal{R}(\mathcal{T}^k, t_k)) = (\mathcal{T}^k, t_k).$$
Moreover, there is an infinite depth tree $\mathcal{T}^\infty = \{\mathcal{H}_{Mar}^n\}$, and a subsequence n_k such that
$$t_{n_k} \to t, \text{ and } \text{dist}(\mathcal{T}^{n_k}, \text{trunc}^{n_k}(\mathcal{T}^\infty)) \to 0.$$
Hence, the extended infinite tree $(\mathcal{T}^\infty, t)$ is a fixed point of $\mathcal{R}$.

As a consequence, the unimodal map $f = \lim f_{(\mathcal{H}_{\text{Mar}}^{n_k}, t)}$ is a fixed point of $\mathcal{R}_{FCT}$. At the heart of the argument is the following topological statement [2]:

LEMMA 2.13 (**"Bottom goes down, top goes up"**). *Suppose D_n is a closed n-dimensional ball, and let $F : D_n \times [0,1] \to D_n \times \mathbb{R}$ be a continuous map. If*
$$F(D_n \times \{0\}) \subset D_n \times (-\infty, 0) \text{ and } F(D_n \times \{1\}) \subset D_n \times (1, \infty),$$
then F has a fixed point in $D_n \times [0,1]$.

It is not difficult to understand where the bottom goes down, top goes up condition comes into play. If we denote $t_-(\mathcal{T}^k)$, $t_+(\mathcal{T}^k)$ the minimal and the maximal values of t for which $(\mathcal{T}^k, t)$ is renormalizable, then $(\mathcal{T}^k, t)$ is renormalizable for all $t \in$

---
[2] Martens attributes the idea to Dennis Sullivan

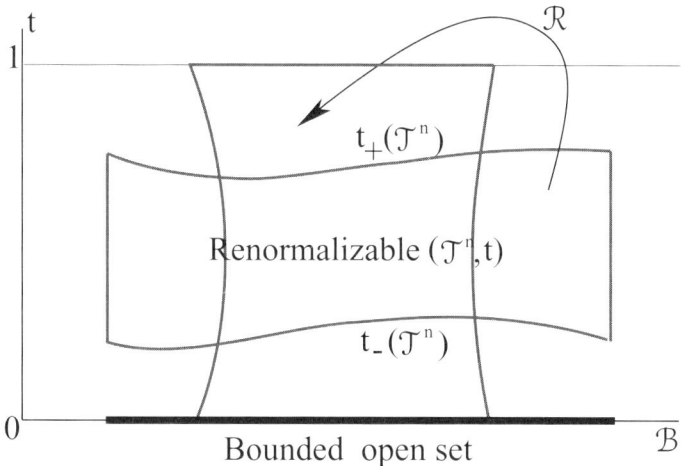

FIGURE 3

$[t_-(\mathcal{T}^k), t_+(\mathcal{T}^k)]$, and the considerations of maximality imply that this interval of values is transformed by $\mathcal{R}$ into $[-1,1]$. To show that there is an open and bounded set of trees which is mapped inside itself, Martens applies a version of real *a priori* bounds to get an *a priori* bound on the nonlinearity of renormalized maps. The same bounds also imply the existence of a convergent subsequence of $\mathcal{T}^k$ with the desired property. We will postpone our comments on the proof and on the method till the final chapter. At this point, let us note that we now have a new object to study - a termodynamical renormalization operator which is functorially related to the Feigenbaum-Coullet-Tresser renormalization. As we shall see in the next section, there has been a previous instance when such a renormalization appeared in the literature.

2.4. Geometric interpretation of renormalization: a Perron–Frobenius type operator. We will discuss in this section the results of Jiang, Morita, and Sullivan from [**JMS**]. The central objective of [**JMS**] was to attempt to justify the existence of an expanding eigenvalue for the Feigenbaum-Coullet-Tresser renormalization operator at a fixed point.

Let us fix $\gamma = 2\ell$, $\ell \in \mathbb{N}$, and let $g(x)$ denote the period doubling fixed point with critical point of order γ. Being an analytic mapping, $g(x)$ extends to an open topological disk U of $[-1,1]$. Let $\sigma(x)$ be the expanding mapping (2.4). Denote $J_0 = [g^2(0), -g^2(0)]$ and $J_1 = g(J_1) = [g^3(0), g(0)]$. Using the expansiveness of σ, we may select a real-symmetric topological disk $\Omega \subset g(U)$, $\Omega \supset [-1,1]$, small enough so that $J_0 \cup J_1 \Subset \Omega$, and there exist disjoint open neighborhoods $\Omega_i \supset J_i$ such that $\Omega_i \Subset \Omega$, and $\sigma : \Omega_i \mapsto \Omega$ is a conformal mapping. Let us further denote by V the connected component of $g^{-1}(\Omega)$ which contains the origin. Denote $\mathcal{A}_V$ the Banach space of analytic functions on V continuous up to the boundary equipped with the uniform norm. Let $\mathcal{B}_V \subset \mathcal{A}_V$ be the Banach manifold of unimodal mappings f with a single critical point at the origin of order γ which

have analytic extensions to V continuous up to the boundary, and such that
$$f^3(0) > -f^2(0) > f^4(0) > 0 > f^2(0).$$
Let $\mathcal{V}_\Omega^\omega$ denote the space of real-symmetric analytic vector fields on Ω with a continuous extension to the boundary, again with a uniform norm. The following proposition is straightforward:

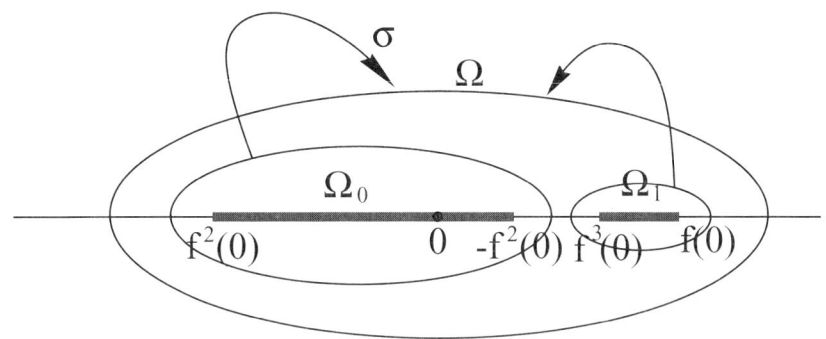

FIGURE 4

PROPOSITION 2.14. *The mapping g_* from $\mathcal{V}_\Omega^\omega$ into $T_g\mathcal{B}_V$ defined by*
$$g_*(v)(x) = v(g(x)) \text{ for } x \in \Omega \text{ and } v \in \mathcal{V}_\Omega^\omega$$
is an isomorphism.

DEFINITION 2.6. The Perron-Frobenius type operator $\mathcal{L}_\sigma : \mathcal{V}_\Omega^\omega \to \mathcal{V}_\Omega^\omega$ is defined as
$$(\mathcal{L}_\sigma v)(z) = \sum_{w \in \sigma^{-1}(z)} \sigma'(w)v(w).$$

The next statement relates the Perron-Frobenius type operator $\mathcal{L}_\sigma$ with the Feigenbaum-Coullet-Tresser renormalization:

PROPOSITION 2.15. *The operators $\mathcal{L}_\sigma$ and $D_g\mathcal{R}_{FCT}$ (the latter viewed as an operator on $T_g\mathcal{B}_V$) have the same eigenvalues except for the value 1. More precisely,*
$$\mathcal{L}_\sigma = (g_*)^{-1} \circ D_g\mathcal{R}_{FCT} \circ g_* + e_1,$$
where e_1 is a projection onto the eigenspace of eigenvalue one generated by the vector $(g_*)^{-1}(g'(x)x - g(x))$.

As a remark, the authors of [**JMS**] note that for every $m \in \mathbb{N}$ the vector
$$v_{2m-1} = (g_*)^{-1}[g'(x)x^{2m-1} - (g(x))^{2m-1}] \in \mathcal{V}_\Omega^\omega$$
is an eigenvector of $\mathcal{L}_\sigma$ with eigenvalue $\lambda_{2m-1} = \alpha^{-(2m-2)}$.

The statement of the last proposition should not come as a surprise after the results of the previous sections. Indeed, according to the Remark 2.3, Martens' trees in a small neighborhood of the tree of g in an appropriate norm may be viewed as Cantor sets with bounded geometry. Since σ acts as a shift on the Cantor set $\mathbf{C}_g$, the operator $\mathcal{L}_\sigma$ is naturally interpreted as a linearization of Martens' renormalization at the fixed point in a suitable smooth structure. We thus have a geometric interpretation for the statistical-mechanical renormalization of the previous section given by the embedding of the invariant Cantor set in the plane.

Since the operator $\mathcal{L}_\sigma$ is not positive, the Ruelle-Perron-Frobenius theory does not apply to show the existence of the single expanding eigenvalue. The main result of [**JMS**] is the following:

THEOREM 2.16. *Suppose $g : [-1,1] \to [-1,1]$ is a concave function. Then $\mathcal{L}_\sigma$ has an eigenvalue $\lambda > 1$.*

We remark that the assumption of concavity is supported by the numerical experiments for $\gamma = 2$, and is false if γ is sufficiently large.

Remark on critical circle maps. We would like to remark here that a similar construction may be carried out for the commuting pair (η, ξ) which is the golden-mean fixed point for the renormalization of critical circle maps (see [**Ya**] for a description of the renormalization theory of critical circle maps). Namely, we may define an expanding map

$$\sigma(x) = \begin{cases} -\alpha x, & x \in [\xi(0), \xi \circ \eta(0)] \\ -\alpha \xi(x), & x \in [\xi \circ \eta(0), 1 = \eta(0)] \end{cases}$$

where the scaling factor $\alpha = -1/\xi(0)$. In this case, the operator $\mathcal{L}_\sigma$ is negative,

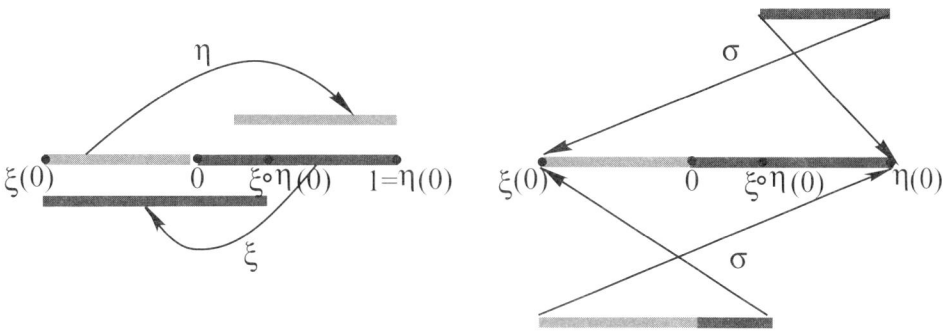

FIGURE 5

and the Ruelle-Perron-Frobenius theory applies to establish the existence of a single expanding eigenvalue. This aspect of the renormalization theory of critical circle maps, however, is quite trivial. On the other hand, constructing a Banach manifold in which the renormalization is an analytic operator is highly nontrivial. One may hope therefore that there is a possibility for an alternative approach to the hyperbolicity problem in the renormalization theory of critical circle maps – the study of the spectrum of a linear operator $\mathcal{L}_\sigma$ in a real Banch space.

2.5. Concluding remarks.

Comparison of the two cases. Let us summarize the obvious similarities between the two examples considered in this note: the Dyson's Hierarchical Hamiltonians, and the Hamiltonians for the unimodal maps. In both cases one may begin by considering an action of a renormalization transformation acting on Hamiltonians: in the first instance a candidate for such renormalization is $\mathcal{R}_{\text{Dyson}}$, in the second, the Martens' renormalization $\mathcal{R}$. For a fixed point of such a renormalization, the self-similarity produces functional equations for the thermodynamical parameters of the system (in the first case, $g(z;\beta)$, in the second, the unimodal map f). For the Dyson's models it is the equation (1.4), for the unimodal maps, the FCT fucntional equation. The equation is then interpreted as a renormalization operator acting on an appropriate function space. We may call these two approaches to renormalization a microscopic and a macroscopic ones. In both instances the macroscopic renormalization is studied by analytic methods: the properties of the Gauss integral operator in the first, the Sullivan-McMullen-Lyubich theory in the second. For the unimodal maps there is now Martens' proof of the existence of a fixed point for the microscopic renormalization, as well as the Jiang-Morita-Sullivan approach to proving its hyperbolicity. To our knowledge, such microscopic approach is missing in the Dyson's case.

Dyson's Hierarchical models	**Unimodal maps**
Dyson's Hamiltonian $\mathcal{H}_\Theta$	Martens' Hamiltonian $\tilde{\mathcal{H}}_{\text{Mar}}$
Microscopic renormalization $\mathcal{R}_{\text{Dyson}}$ (?)	Microscopic renormalization $\mathcal{R}$
?	Martens' proof of the existence of fixed point; JMS analysis of hyperbolicity
Distribution $g(z;\beta)$	Unimodal map f
$\mathcal{R}_{\text{int}} g(z;\beta) = \int g(\frac{z}{\sqrt{c}} + u;\beta) g(\frac{z}{\sqrt{c}} - u;\beta) du$	$\mathcal{R}_{FCT}(f) = -\alpha_f f \circ f(-x/\alpha_f)$
Bleher-Sinai analysis of Gauss integral operator	Sullivan-McMullen-Lyubich theory

Some open problems. Several questions on hierarchical models were posed in Sinai's book [**Sin**]. The main open problem is:

- Extend the existence of the hyperbolic fixed point of $\mathcal{R}_{int}$ with a single unstable eigenvalue to the whole interval $c \in (1, \sqrt{2})$.

In view of our analysis we will add:

- Find a critical fixed point of the microscopic renormalization. In a forthcoming article with M. Benedicks we will show how such a problem may be attacked with Martens' approach to renormalization.

In the case of the dynamical renormalization it is natural to pose the following problems:

- Develop a complete hyperbolic picture for the Martens' renormalization. A possible approach to this is given by the study of the Perron-Frobenius type operator of [**JMS**].
- Use the same approach, as indicated in the previous section, to construct an alternative renormalization theory for critical circle maps.

Finally, we would like to speculate that the microscopic approach to renormalization could prove useful in other contexts, both statistical-mechanical, and dynamical (such as the conservative mappings of $\mathbb{T}^2$, for instance).

References

[Ble] P. Bleher. Construction of non-Gaussian self-similar random fields with hierarchical structure. *Comm. Math. Phys.* **84**(1982), 557-578.

[BS1] P. M. Bleher, Ya. G. Sinai. Investigation of the critical points in models of the type of Dyson's hierarchical models. *Comm. Math. Phys.* **33**(1973), 23-42

[BS2] P. M. Bleher, Ya. G. Sinai. Critical indices for Dyson's asymptotically hierarchical models. *Comm. Math. Phys.*, **45**(1975), 247-278.

[CE] P. Collett, J.-P. Eckmann. A renormalization group analysis of the hierarchical model in statistical mechanics. Lecture Notes in Physics, Vol. 74. Springer-Verlag, Berlin-New York, 1978.

[CT] P. Coullet, C. Tresser. Itération d'endomorphismes et groupe de renormalisation. *J. Phys. Colloque C* 539, C5-C25, 1978.

[Ep] H. Epstein. Existence and properties of p-tupling fixed points. *Comm. Math. Phys.* **215**(2000), no. 2, 443–476.

[Feig] M. Feigenbaum. Quantitative metric properties for a class of non-linear transformations. *J. Stat. Phys.*, **19**(1978), 25-52.

[KW] H. Koch, P. Wittwer. A non-Gaussian renormalization group fixed point for hierarchical scalar lattice field theories. Comm. Math. Phys. **106**(1986), no. 3, 495–532.

[Lan] O. E. Lanford III. A computer assisted proof of the Feigenbaum conjectures. *Bull. Amer. Math. Soc.* **6**(1982), 427-434.

[Lyu1] M. Lyubich. Feigenbaum-Coullet-Tresser Universality and Milnor's Hairiness Conjecture. *Annals of Math.*, **149**(1999), 319 - 420.

[Lyu2] M. Lyubich. Almost every real quadratic map is either regular or stochastic. *Ann. of Math.* (2) **156**(2002), no. 1, 1-78.

[Mar] M. Martens. The periodic points of renormalization. *Ann. of Math.* (2) **147**(1998), no. 3, 543-584.

[McM] C. McMullen. Renormalization and 3-Manifolds Which Fiber over the Circle. Princeton University Press, 1996.

[MvS] W. de Melo, S. van Strien. One-dimensional dynamics. Vol. 25, Springer Verlag 1993

[JMS] Y. P. Jiang, T. Morita, D. Sullivan. Expanding direction of the period doubling operator. *Comm. Math. Phys.* **144**(1992), no. 3, 509-520.

[Sin] Ya. G. Sinai. Theory of phase transitions: Rigorous results. Pergamon Press, 1982.

[Sul1] D. Sullivan. Differentiable structures on fractal-like sets, determined by intrinsic scaling functions on dual Cantor sets. The mathematical heritage of Hermann Weyl (Durham, NC, 1987), 15–23, *Proc. Sympos. Pure Math.*, 48, Amer. Math. Soc., Providence, RI, 1988.

[Sul2] D. Sullivan. Bounds, quadratic differentials, and renormalization conjectures. American Mathematical Society centennial publications, Vol. II (Providence, RI, 1988), 417–466, Amer. Math. Soc., Providence, RI, 1992.

[VSK] Vul, E. B.; Sinaĭ, Ya. G.; Khanin, K. M. Feigenbaum universality and thermodynamic formalism. (Russian) *Uspekhi Mat. Nauk* **39**(1984), no. 3(237), 3-37.

[Ya] M. Yampolsky. Hyperbolicity of renormalization of critical circle maps. *Publ. Math. IHÉS*, **96**(2002), 1-41.

MATHEMATICS DEPARTMENT UNIVERSITY OF TORONTO 100 ST GEORGE STREET TORONTO, ONTARIO, M5S 3G3 CANADA

PHYSICS

Quantum geometry in action: big bang and black holes

Abhay Ashtekar

ABSTRACT. Over the last three years, a number of fundamental issues in quantum gravity were addressed in the framework of quantum geometry, discussed extensively by John Baez at this conference. In particular, these include: A statistical mechanical derivation of the horizon entropy, encompassing astrophysically interesting black holes cosmological horizons, and a natural resolution of the big-bang singularity. The goal of this article is to communicate these advances in general terms.

Contents

1. Introduction
2. A bird's eye view of loop quantum gravity
2.1. Viewpoint
2.2. Quantum geometry
3. Applications of quantum geometry
3.1. Big bang
3.2. Black holes
4. Conclusion
References

1. Introduction

Major paradigm shifts in theoretical physics have required mathematical arenas that were, at the time, new to physics. Newton's mechanics and theory of gravitation could not have been introduced without calculus; Maxwell's electrodynamics required partial differential equations; Einstein had to learn differential geometry to develop general relativity; and quantum mechanics needed the theory of Hilbert spaces and operator algebras. It is widely believed that quantum gravity will lead to the next profound paradigm shift in physics. What would be the required mathematical arenas? The answers to this question vary. For example, Roger Penrose's twistor theory posits that space-time would be a secondary, derived concept, arising

---

1991 *Mathematics Subject Classification.* 83C45, 51P05, 81T99, 28Exx.
Key words and phrases. Geometry and physics, quantum gravity, cosmology, black holes.

from a 4-dimensional *complex* space; the fundamental theory would be based on complex manifolds, sheaf-cohomology and algebraic geometry [1]. Alain Connes' approach aims at describing fundamental physics through non-commutative geometry [1]. In the loop quantum gravity approach I will discuss here, the basic tool is Riemannian quantum geometry. Just as differential geometry provides the mathematical language to formulate classical gravitational theories, such as general relativity, a specific quantum Riemannian geometry provides the required setting for quantum gravitational theories. Since this subject was covered in detail by John Baez, I will only provide a semi-qualitative introduction and focus, rather, on applications of this quantum geometry. (For details, see the review articles in [1].) However, let me make a general remark in this Introduction. As Dennis Sullivan emphasized during discussions at the conference, from the perspective of graph theory, freedom in the construction of a (background independent) quantum theory of geometry is very limited. Thus, the mathematical structures, definitions and constructions we use are not only natural from this perspective, but essentially unique.

Let us now turn to physics. What are some of the central *physical* and conceptual questions of quantum gravity? I would like to outline a few of these to give a flavor of the subject. Since this article is addressed to non-specialists, I will select questions that arise from what we already know about Nature, and what we expect based on physical theories which are *firmly* based on observations, avoiding issues —such as higher dimensions and supersymmetry– that are internal to specific quantum gravity programs and which have yet to receive observational support. Further discussion of the background material can be found in Section 3.

• *Big-Bang and other singularities:* It is widely believed that the prediction of a singularity, such as the big-bang of classical general relativity, is primarily a signal that the physical theory has been pushed beyond the domain of its validity. A key question to any quantum gravity theory, then, is: What replaces the big-bang? Is there a mathematically consistent description of the evolution of the quantum state of the universe which is singularity free? General relativity predicts that the space-time curvature must grow as we approach the big-bang but we expect the quantum effects, ignored by general relativity, to intervene, making quantum gravity indispensable before infinite curvatures are reached. If so, what is the upper bound on curvature? How close to the big-bang can we "trust" classical general relativity? What can we say about the "initial conditions", i.e., the quantum state of geometry and matter that correctly describes the big-bang? If they have to be imposed externally, is there a *physical* guiding principle?

• *Black holes:* In the early seventies, using imaginative thought experiments, Jacob Bekenstein [11] argued that black holes must carry entropy proportional to their area. About the same time, Jim Bardeen, Brandon Carter and Stephen Hawking [11] showed that black holes in equilibrium obey two basic laws, which have the same form as the laws of thermodynamics, provided one replaces the area of the black hole horizon a_{hor} by a multiple of the entropy S in thermodynamics and black hole surface gravity κ by a corresponding multiple of the temperature T.[1]

---

[1]One can think of the horizon as the "surface" of the black hole. In classical general relativity, one can not send causal signals from the region within the horizon to the region outside. Surface gravity κ is, roughly, the black hole analog of the acceleration g due to gravity on the surface of the earth.

However, at first this similarity was thought to be only a formal analogy because the BCH analysis was based on *classical* general relativity and simple dimensional considerations show that the proportionality factors must involve Planck's constant $\hbar$. Two years later, using quantum field theory on a black hole background space-time, Hawking [11] showed that black holes in fact radiate quantum mechanically as though they are black bodies at temperature $T = \hbar\kappa/2\pi$. Using the analogy with the first law, one can then conclude that the black hole entropy should be given by $S_{\rm BH} = a_{\rm hor}/4G\hbar$, where G is Newton's gravitational constant. This conclusion is striking and deep because it brings together the three pillars of fundamental physics — general relativity, quantum theory and statistical mechanics. However, the argument itself is a rather hodge-podge mixture of classical and semi-classical ideas, reminiscent of the Bohr theory of atom. A natural question then is: what is the analog of the more fundamental, Pauli-Schrödinger theory of the Hydrogen atom? More precisely, what is the statistical mechanical origin of black hole entropy? What is the nature of a quantum black hole and what is the interplay between the quantum degrees of freedom responsible for entropy and the exterior curved geometry? Can one derive the Hawking effect from first principles of quantum gravity?

• *Planck scale physics and the low energy world:* Perhaps the central lesson of general relativity is that *gravity is geometry.* There is no longer a background metric, no inert stage on which dynamics unfolds. Geometry itself is dynamical. Therefore, one expects that a fully satisfactory quantum gravity theory would also be free of a background space-time geometry. However, of necessity, a background independent description must use physical concepts and mathematical tools that are quite different from those of the familiar, low energy physics formulated in flat space-time. A major challenge then is to show that this low energy description does arise from the pristine, Planckian world in an appropriate sense.[2] In this "top-down" approach one starts with physics at the fundamental Planck scale where everything including geometry is quantum mechanical and wishes to arrive at the familiar world rooted in the space-time continuum. Does the fundamental theory constructed at the Planck scale admit a "sufficient number" of semi-classical states? Do these semi-classical sectors provide enough of a background geometry to anchor low energy physics? can one recover the familiar description? Furthermore, can one pin point why the standard "bottom-up" perturbative approach fails? That is, what is the essential feature which makes the fundamental description mathematically coherent, but is absent in the standard perturbative quantum gravity?

Of course, this is by no means a complete list of challenges. There are many others. Since there is no background space-time metric, what does "time evolution" mean? Without a fixed space-time at one's disposal, what is one to make of quantum measurement theory and the associated questions of interpretation of quantum mechanics? What role does space-time topology play and can it change? .... Recent advances within loop quantum gravity have led to illuminating answers to many of these questions and opened up avenues to address others. As in my talk, in this report, I will focus on issues related to black holes and big-bang.

---

[2]The characteristic length scale in this regime is the Planck length $\ell_{\rm Pl} \sim 10^{-33}$cm while the smallest distance we can probe with our highest energy accelerators today is $\sim 10^{-17}$cm.

2. A bird's eye view of loop quantum gravity

In this section, I will briefly summarize the salient features and current status of quantum geometry. The emphasis is on structural and conceptual issues; details can be found in references [1-9].[3]

2.1. Viewpoint. In this approach, one takes the central lesson of general relativity seriously: gravity *is* geometry whence, in a fundamental theory, there should be no background metric. In quantum gravity, geometry and matter should *both* be "born quantum mechanically". Thus, in contrast to approaches developed by particle physicists, one does not begin with quantum matter on a background geometry and use perturbation theory to incorporate quantum effects of gravity. There *is* a manifold but no metric, or indeed any other physical fields, in the background.[4]

In classical gravity, Riemannian geometry provides the appropriate mathematical language to formulate the physical, kinematic notions as well as the final dynamical equations. This role is now taken by *quantum* Riemannian geometry, discussed below. In the classical domain, general relativity stands out as the best available theory of gravity, some of whose predictions have been tested to an amazing accuracy, surpassing even the legendary tests of quantum electrodynamics. However, if one applies to general relativity the standard perturbative techniques of quantum field theory, one obtains a "on-renormalizable" theory, i.e., a theory with uncontrollable infinities. Therefore, it is natural to ask: *Does quantum general relativity, coupled to suitable matter* (or supergravity, its supersymmetric generalizations) *exist as a consistent theory non-perturbatively ?* There is no a priori implication that such a theory would be the final, complete description of Nature. Nonetheless, this is a fascinating open question at the level of mathematical physics.

In the particle physics circles, the answer is often assumed to be in the negative, not because there is concrete evidence against non-perturbative quantum gravity, but because of an analogy to the theory of weak interactions, where non-renormalizability of the initial "Fermi theory" forced one to replace it by the renormalizable Glashow-Weinberg-Salam theory. However this analogy overlooks the crucial fact that, in the case of general relativity, there is a qualitatively new element. Perturbative treatments pre-suppose that the space-time can be assumed to be a continuum *at all scales* of interest to physics under consideration. Since this is a safe assumption for weak interactions, non-renormalizability was a genuine problem. However, in the gravitational case, the scale of interest is given by the Planck length ℓ_{Pl} and there is no physical basis to pre-suppose that the continuum picture should be valid down to that scale. The failure of the standard perturbative treatments may simply be due to this grossly incorrect assumption, and a non-perturbative treatment which correctly incorporates the physical micro-structure of geometry may well be free of these inconsistencies.

As indicated above, even if quantum general relativity did exist as a mathematically consistent theory, there is no *a priori* reason to assume that it would be

---

[3] For a very recent and mathematically self-contained review of the subject, see the last reference in [**1**].

[4] In 2+1 dimensions, although one begins in a completely analogous fashion, in the final picture one can get rid of the background manifold as well. Thus, the fundamental theory can be formulated *combinatorially* [**3, 1**]. To achieve this goal in 3+1 dimensions, one needs a much better understanding of the theory of links in 3 dimensions.

the "final" theory of all known physics. In classical general relativity, while requirements of background independence and general covariance do restrict the form of interactions between gravity and matter fields and among matter fields themselves, they do not *determine* these interactions. Quantum general relativity would have the same limitation. Put differently, such a theory would not be a satisfactory candidate for unification of all known forces. However, just as general relativity has had powerful implications in spite of this limitation in the classical domain, quantum general relativity should have qualitatively new predictions, pushing further the existing frontiers of physics. Indeed, unification does not appear to be an essential criterion for usefulness of a theory even in other interactions. Quantum chrmodynamics (QCD) for example, is a powerful theory of strong interactions even though it does not unify them with electro-weak ones. Furthermore, the fact that we do not yet have a viable candidate for the grand unified theory does not make QCD any less useful.

2.2. Quantum geometry. Although there is no natural unification of dynamics of all interactions in loop quantum gravity, it does provide a kinematic unification. More precisely, in this approach one begins by formulating general relativity in the mathematical language of connections, the basic variables of gauge theories of electro-weak and strong interactions. Thus, now the configuration variables are not metrics (as in the older 'geometrodynamics' program of Paul Dirac, Peter Bergmann and John Wheeler), but certain *spin connections*; the emphasis is shifted from distances to holonomies [2, 1]. Consequently, the basic kinematic structures are the same as those used in gauge theories. A key difference, however, is that while a background space-time metric is available and crucially used in gauge theories, now there are no background fields whatsoever. This absence is forced on us by the requirement of diffeomorphism invariance.

This is a key difference and it causes a host of conceptual as well as technical difficulties in the passage to quantum theory. For, most of the techniques used in the familiar, Minkowskian quantum theories are deeply rooted in the availability of a flat background metric. It is this structure that enables one to single out the vacuum state, perform Fourier transforms to decompose fields canonically in to creation and annihilation operators, define masses and spins of particles and carry out regularizations of products of operators. Already when one passes to quantum field theory in curved space-times, extra work is needed to construct mathematical structures that are adequate for physics. In our case, the situation is much more drastic: there is no background metric whatsoever. Therefore new physical ideas and mathematical tools are now necessary. Fortunately, they were constructed by a number of researchers in the mid-nineties and have given rise to a detailed quantum theory of geometry [4, 5, 6, 7, 8].

Because the situation is conceptually so novel and because there are no direct experiments to guide us, reliable results require mathematical precision to ensure that there are no hidden infinities. Achieving this precision has been a high priority in the program. Thus, while one is inevitably motivated by heuristic, physical ideas and formal manipulations, the final results are mathematically rigorous. In particular, due care is taken in constructing function spaces, defining measures and functional integrals, regularizing products of field operator, and calculating eigenvectors and eigenvalues of geometric operators. The final results are all free of divergences, well-defined, and respect the background independence and diffeomorphism invariance.

Let me now turn to specifics. It is perhaps simplest to begin with a Hamiltonian or symplectic description of general relativity. Let us fix an orientable 3-manifold Σ, which will represent "space". As in gauge theories of strong and electro-weak forces, the configuration space **C** consists of connections on an appropriate G-bundle. The structure group G for general relativity is SU(2). Since all SU(2) bundles on Σ are trivial, one fixes a trivialization and represents each connection through a 1-form A on Σ, which take values in the Lie-algebra of SU(2). Holonomies of A dictate how SU(2)-spinors are parallel transported along curves in Σ. Physically, these spinors can be thought of as 'left handed' fermions.

The phase space $\mathbf{\Gamma} = T^*\mathbf{C}$ is the cotangent bundle over **C**. The momenta, i.e., the cotangent vectors to **C**, are represented by the "electric field" 2-forms E, which take values in the dual of the Lie-algebra of SU(2). In the present gravitational context, the momenta acquire a geometrical significance: their Hodge-duals $^\star E$ can be naturally interpreted as orthonormal frame fields (with density weight 1) and determine the dynamical, Riemannian geometry of Σ. In the older "geometrodynamics" program of Dirac, Bergmann and Wheeler, the Riemannian metrics on Σ played the role of configuration variable and the momentum were represented by the extrinsic curvature (i.e., the second fundamental form) of Σ (see e.g. Wheeler's writings in [1]). Consequently, general relativity was treated very differently from theories of other basic forces of Nature. In quantum geometry, general relativity is cast as a dynamical theory of connections and thus brought closer to theories of other fundamental interactions. The Riemannian structures on Σ are now built from *momentum* variables. Here, I will focus on the gravitational field by itself. Matter couplings to gravity have also been studied extensively [2, 1].

Before summarizing the mathematical structure of quantum geometry, let me provide the underlying heuristic physical pictures. In quantum theory, the fundamental excitations of geometry are most conveniently expressed in terms of holonomies [3, 4]. They are thus *one-dimensional*, polymer-like and, in analogy with gauge theories, can be thought of as "flux lines of the electric field". More precisely, they turn out to be flux lines of areas: an elementary flux line deposits a quantum of area on any 2-surface S it intersects. Thus, if quantum geometry were to be excited along just a few flux lines, most surfaces would have zero area and the quantum state would not at all resemble a classical geometry. Semi-classical geometries can result only if a huge number of these elementary excitations are superposed in suitably dense configurations [13, 14]. The state of quantum geometry around you, for example, must have so many elementary excitations that $\sim 10^{68}$ of them intersect the sheet of paper you are reading, to endow it an area of $\sim 100 \text{cm}^2$. Even in such states, the geometry is still distributional, concentrated on the underlying elementary flux lines; but if suitably coarse-grained, it can be approximated by a smooth metric. Thus, the continuum picture is only an approximation that arises from coarse graining of semi-classical states.

The mathematical structure is the following. (For details, see [4], particularly the most recent review therein.) First, one introduces a $C^\star$ algebra of functions of the SU(2) connections A, constructed from matrix elements of holonomies along piecewise analytic paths ("edges") on Σ. The Gel'fand spectrum of this unital $C^\star$-algebra is a compact Hausdorff space $\mathcal{A}$. The space of smooth classical connections admits a natural embedding in $\mathcal{A}$ which is dense in the Gel'fand topology. Elements of $\mathcal{A}$ are therefore called *generalized connections*. It is possible to give an explicit description of $\mathcal{A}$ which is useful in practice. It turns out that $\mathcal{A}$ admits a natural, regular, Borel measure μ_o which is induced on it by the Haar measure on SU(2). The group of diffeomorphisms on Σ has a natural action on $\mathcal{A}$ and μ_o is invariant with respect to this action. The quantum states of geometry are elements of the Hilbert space $\mathcal{H} = L^2(\mathcal{A}, d\mu_o)$. This space is very large. However, it can be conveniently decomposed in to a family of orthonormal, *finite* dimensional sub-spaces

$$(2.1) \qquad \mathcal{H} = \oplus_{\gamma,\vec{j}} \mathcal{H}_{\gamma,\vec{j}},$$

labelled by finite graphs γ each edge of which itself is labelled by a non-trivial irreducible representation of SU(2) (or, a half-integer, or a spin j) [5]. $\mathcal{H}_{\gamma,\vec{j}}$ can be regarded as the Hilbert space of a "spin-system". These spaces are extremely simple to work with; this is why very explicit calculations are feasible. Elements of $\mathcal{H}_{\gamma,\vec{j}}$ are referred to as *spin-network states* [5].

As one would expect from the structure of the classical theory, the basic quantum operators are the holonomies $\hat{h}_p$ along paths p in Σ and the triads $\widehat{{}^\star E}$ [6]. Both sets of operators are densely defined and self-adjoint on $\mathcal{H}$. Furthermore, a striking result is that *all eigenvalues of the triad operators are discrete*. This key property is, in essence, the origin of the fundamental discreteness of quantum geometry. For, just as the classical Riemannian geometry of Σ is determined by the triads ${}^\star E$, all Riemannian geometry operators —such as the area operator $\hat{A}_S$ associated with a 2-surface S or the volume operator $\hat{V}_R$ associated with a region R— are constructed from $\widehat{{}^\star E}$. However, since even the classical quantities A_S and V_R are non-polynomial functionals of the triads, the construction of the corresponding $\hat{A}_S$ and $\hat{V}_R$ is quite subtle and requires a great deal of care. But their final expressions are rather simple [6].

In this regularization, the underlying background independence turns out to be a blessing. For, diffeomorphism invariance constrains the possible forms of the final expressions *severely* and the detailed calculations then serve essentially to fix numerical coefficients and other details. Let us illustrate this point with the example of the area operators $\hat{A}_S$. Since they are associated with 2-surfaces S while the states have 1-dimensional support, the diffeomorphism covariance requires that the action of $\hat{A}_S$ on a state $\Psi_{\gamma,\vec{j}}$ must be concentrated at the intersections of S with γ. The detailed expression bears out this fact: the action of $\hat{A}_S$ on $\Psi_{\gamma,\vec{j}}$ is dictated simply by the spin labels j_I attached to those edges of γ which intersect S. For all surfaces S and 3-dimensional regions R in Σ, $\hat{A}_S$ and $\hat{V}_R$ are densely defined, self-adjoint operators. *All their eigenvalues are discrete* [6]. Naively, one might expect that the eigenvalues would be uniformly spaced, given by, e.g., integral multiples of the Planck area or volume. This turns out *not* to be the case; the distribution of eigenvalues is quite subtle. In particular, the eigenvalues crowd rapidly as areas and volumes increase. In the case of area operators, the complete spectrum is known in a closed form, and the first several hundred eigenvalues have been explicitly computed numerically. For a large eigenvalue a_n, the separation $\Delta a_n = a_{n+1} - a_n$ between consecutive eigenvalues decreases exponentially: $\Delta a_n \leq \ell_{\text{Pl}}^2 \exp -(\sqrt{a_n}/\ell_{\text{Pl}})$! Because of such strong crowding, the continuum approximation becomes excellent quite rapidly just a few orders of magnitude above the Planck scale. At the Planck scale, however, there is a precise and very specific replacement. This is the arena of quantum geometry. The premise is that the standard perturbation theory fails because it ignores this fundamental discreteness (see Section 2.1).

There is however a key mathematical subtlety [**2, 7**]. This non-perturbative quantization has a one parameter family of ambiguities labeled by $\gamma > 0$. This γ is called the *Barbero-Immirzi parameter* (and is rather similar to the well-known θ-parameter of QCD). In the classical theory, γ is irrelevant but in quantum theory different values of γ correspond to unitary inequivalent representations of the algebra of geometric operators. The overall mathematical structure of all these sectors is very similar; the only difference is that the eigenvalues of all geometric operators scale with γ. For example, the simplest eigenvalues of the area operator $\hat{A}_S$ in the γ quantum sector is given by

$$(2.2) \qquad a_{\{j\}} = 8\pi\gamma\ell_{\text{Pl}}^2 \sum_I \sqrt{j_I(j_I+1)}$$

where $I = 1, \ldots N$ for some integer N and each j_I is a half-integer. Since the representations are untruly inequivalent, as usual, one must rely on Nature to resolve this ambiguity: Just as Nature must select a specific value of θ on QCD, it must select a specific value of γ in loop quantum gravity. With one judicious experiment —e.g., measurement of the lowest eigenvalue of the area operator $\hat{A}_S$ for a 2-surface S of any given topology— we could determine the value of γ and fix the theory. Unfortunately, such experiments are hard to perform! However, we will see in Section 3.2 that the Bekenstein-Hawking formula of black hole entropy provides an indirect measurement of this lowest eigenvalue of area for the 2-sphere topology and can therefore be used to fix the value of γ.

3. Applications of quantum geometry

In this section, I will summarize two recent developments that answer several of the questions raised under first two bullets in the Introduction.[5]

3.1. Big bang. Let us first recall how the big-bang singularity arises in classical general relativity. Observations have shown that, in the rest frame defined by the cosmic microwave background, the universe is *spatially* homogeneous and isotropic on a cosmological scale. The presence of this rest frame leads one to assume that the 4-manifold representing space-time is foliated by 3-dimensional spatial manifolds and homogeneity and isotropy is incorporated in the assumption that each 3-dimensional leaf is equipped with a metric of constant curvature. Thus there are three possibilities: the leaves of the preferred foliation are either metric 3-spheres, or flat, or metric 3-hyperboloids, depending on whether the scalar curvature (which is constant on each leaf) is positive, zero or negative. Suggestions have been made that the physical space may actually be a quotient of one of these spaces under the action of a discrete sub-group of the corresponding isometry groups. While this possibility can be not be ruled out definitively at this stage, observational evidence does not favor it.

---

[5]An up to date and more detailed review of this material can be found in the last paper in [**1**]. For a mathematically oriented treatment of spatially homogeneous and isotropic quantum cosmology, see the last paper in [**10**]. For the most recent developments on black hole entropy, see the last paper in [**11**].

To be specific, let us consider the 3-sphere case. Then, although the scalar curvature is constant on any one spatial slice, it changes in time, giving rise to an overall expansion or contraction. The radius a of the 3-sphere encodes the full information of the 3-metric at that instant of time and is called the *scale factor*. One then suitably models matter-sources —galaxies and the observed radiation fields— and seeks solutions of Einstein's equation with these symmetries. The equation implies that the universe must have "originated from a big-bang": if we evolve the solution backward in time, the scale factor a must eventually go to zero and the curvature must diverge as $1/a^2$. At this "initial instant", Einstein's equation breaks down; classical physics stops. As discussed in the Introduction, the general belief is that this singular behavior is an artifact of our insistence of applying general relativity beyond the domain of its validity. Quantum effects are thought to intervene and dominate the "real physics" in the high curvature regions. The question then is: what replaces the big-bang in this new, more accurate theory?

This question has been discussed for over thirty years in a framework called "quantum cosmology". Traditionally, one has proceeded by first imposing spatial symmetries —such as homogeneity and isotropy— to freeze out all but a finite number of degrees of freedom *already at the classical level* and then quantizing the reduced system. In the simplest case, the basic variables of the reduced classical system are the scale factor a and matter fields ϕ. One then asks: in the theory so quantized, do the singularities of the classical general relativity disappear? Unfortunately, without an additional input, they do not: typically, to resolve the singularity one either had to introduce matter with unphysical properties or introduce boundary conditions by invoking new principles.

In a series of seminal papers [**10**], Martin Bojowald has shown that the situation in loop quantum cosmology is quite different: the underlying quantum geometry makes a *qualitative* difference very near the big-bang and naturally resolves the singularity. In the standard procedure summarized above, the reduction is carried out at the classical level and this removes all traces of the fundamental discreteness. Therefore, the key idea in Bojowald's analysis is to retain the essential features of quantum geometry by first quantizing the kinematics of the *full theory* as in Section 2.2 and then restricting oneself to *quantum* states which are spatially homogeneous and isotropic. As a result, the scale factor operator $\hat{a}$ has *discrete eigenvalues*. The continuum limit is reached rapidly. For example, the gap between an eigenvalue of $\hat{a}$ of order 1cm and the next one is at most of order 10^{-63}cm! Nonetheless, near $a \sim \ell_{\rm Pl}$ there are surprises. Predictions of loop quantum cosmology are very different from those of traditional quantum cosmology.

The first surprise occurs already at the kinematic level. Recall that, in the classical theory curvature is essentially given by $1/a^2$, and blows up at the big-bang. What is the situation in quantum theory? Denote the Hilbert space of spatially homogeneous, isotropic kinematic quantum states by $\mathcal{H}_{\rm HI}$. A self-adjoint operator $\widehat{\rm curv}$ corresponding to curvature can be constructed on $\mathcal{H}_{\rm HI}$, and turns out that its spectrum has an *upper bound*. This is very surprising because $\mathcal{H}_{\rm HI}$ admits an eigenstate of the scale factor operator $\hat{a}$ with a discrete, zero eigenvalue. At first, it may appear that this could happen only by an artificial trick in the construction of $\widehat{\rm curv}$ and that this quantization can not possibly be right because it seems to represent a huge departure from the classical relation $({\rm curv})\, a^2 = 1$. However, these

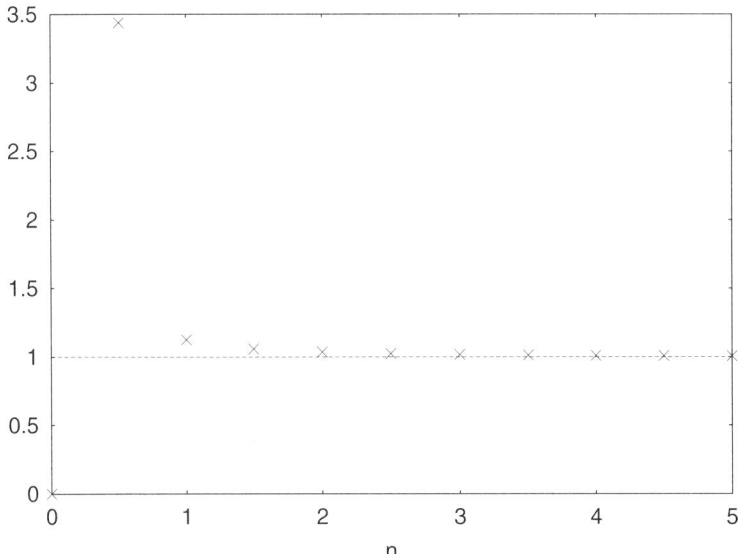

FIGURE 1. The product $a_n \cdot b_n$ as a function of n. The corresponding classical product $a \cdot \sqrt{\text{curv}}$ equals 1.

concerns turn out to be misplaced. The procedure for constructing $\widehat{\text{curv}}$ is natural and, furthermore, descends from full quantum theory.

Let us examine the properties of $\widehat{\text{curv}}$. Its upper bound u_{curv} is finite but absolutely huge:

$$(3.1) \qquad u_{\text{curv}} \sim \frac{256}{81} \frac{1}{\ell_{\text{Pl}}^2} \equiv \frac{256}{81} \frac{1}{G\hbar}$$

or, about 10^{77} times the curvature at the horizon of a solar mass black hole. The functional form of the upper bound is also illuminating. Recall first the Pauli-Schrödinger treatment of the hydrogen atom in non-relativistic quantum mechanics. Because the Coulomb potential between the proton (nucleus of the atom) and the electron diverges as $-1/r$, in the classical theory the energy is unbounded from below. However, thank to the Planck's constant $\hbar$, in the quantum theory, we obtain a finite value, $E_0 = -(me^4/\hbar^2)$. Similarly, u_{curv} is finite because $\hbar$ is non-zero and tends to the classical answer as $\hbar$ tends to zero.

At curvatures as large as u_{curv}, it is natural to expect large departures from classical relations such as $(\text{curv}) a^2 = 1$. But is this relation recovered in the semi-classical regime? The answer is in the affirmative. In fact it is somewhat surprising how quickly this happens. As one would expect, one can simultaneously diagonalize $\hat{a}$ and $\widehat{\text{curv}}$. If we denote their eigenvalues by a_n and b_n respectively, then $a_n \cdot b_n - 1$ is of the order 10^{-4} at $n = 100$ and decreases rapidly as n increases. These properties show that, in spite of the initial surprise, the quantization procedure is viable. Furthermore, one can apply it also to more familiar systems such as a particle moving on a circle and obtain results which at first seem surprising but are in complete agreement with the standard quantum theory of these systems.

Since the curvature is bounded above in the entire Hilbert space, one might hope that the quantum evolution may be well-defined right through the big-bang singularity. Is this in fact the case? Because of our assumption of spatial homogeneity and isotropy, in classical general relativity, the evolution is governed by a single, ordinary differential equation. In quantum theory, the corresponding equation governs the evolution of quantum states. In the geometrodynamic approach, this is the so-called Wheeler-De Witt equation [1] for which the evolution fails to be regular at the singularity unless one resorts to matter fields which are physically unacceptable or introduces a new input "by hand." The second surprise of the quantum geometry based approach is that, although for large a the quantum evolution is close to that of the Wheeler-De Witt equation, there are dramatic differences near the big-bang which makes the evolution well defined even *at* the big-bang, *without any additional input*. To solve the quantum Einstein equation, Bojowald again follows, step by step, the procedure introduced (by Thomas Thiemann) in the full theory.

Let us expand the full quantum state as $|\Psi> = \sum_n \psi_n(\phi) | n>$ where $| n>$ are the eigenstates of the scale factor operator and ϕ denotes matter fields. Then, the quantum Einstein equation takes the form:

(3.2) $\quad c_n \psi_{n+8}(\phi) + d_n \psi_{n+4}(\phi) + e_n \psi_n(\phi) + f_n \psi_{n-4}(\phi) + g_n \psi_{n-8}(\phi) = \gamma \ell_{\rm Pl}^2 \, \hat{H}_\phi \psi_n(\phi)$

where $c_n, \ldots g_n$ are fixed numerical coefficients, γ the Barbero-Immirzi parameter (introduced at the end of section 2) and $\hat{H}_\phi$ is the matter Hamiltonian. (Again, using the Thiemann regularization [9] one can show that the matter Hamiltonian is a well-defined operator.)

As one would expect from the phase space-formulation of classical general relativity, primarily, equation (3.2) serves to constrain the coefficients $\psi_n(\phi)$ of physically permissible quantum states. However, *if* we choose to interpret the scale factor (more precisely, the square of the scale factor times the determinant of the triad) as a time variable, equation (3.2) can be interpreted as an "evolution equation" which evolves the state through discrete time steps. In a (large) neighborhood of the big-bang singularity, this notion of time is viable. For the choice of factor ordering used in the Thiemann regularization, one can evolve in the past through $n = 0$, i.e. right through the classical singularity. Thus, the infinities predicted by the classical theory at the big-bang are indeed artifacts of assuming that the classical, continuum space-time approximation is valid right up to the big-bang. In the quantum theory, the state can be evolved through the big-bang without any difficulty. However, the classical space-time description fails near the big-bang; quantum evolution is well-defined but the classical space-time "dissolves".

The "evolution" equation (3.2) has other interesting features. To begin with, the space of solutions is 16 dimensional. Can we single out a preferred solution by imposing a *physical* condition? One possibility is to impose a *pre-classicality* condition, i.e., to require that the quantum state not oscillate rapidly from one step to the next at *late* times when we know our universe behaves classically. Although this is an extra input, it is not a theoretical prejudice about what should happen at (or near) the big-bang but an observationally motivated condition that is clearly satisfied by our universe. The coefficients $c_n, \ldots g_n$ of (3.2) are such that this condition singles out a solution uniquely. One can ask what this state does at negative times, i.e., before the big-bang. (Time becomes negative because triads flip orientation on the "other side".) Preliminary indications are that the state does not become pre-classical there. If this is borne out by detailed calculations, then the "big-bang" separates two regimes: on "our" side, classical geometry is both meaningful and useful at late times while on the "other" side, it is not.[6] Another interesting feature is that the standard Wheeler-De Witt equation [1] is recovered if we take the limit $\gamma \to 0$ and $n \to \infty$ such that the eigenvalues of $\hat{a}$ take on continuous values. This is completely parallel to the limit we often take to coarse grain the quantum description of a rigidly spinning rotor to "wash out" discreteness in angular momentum eigenvalues and arrive at the classically allowed continuous angular momenta. From this perspective, then, one is led to say that the most striking of the consequences of loop quantum gravity are not seen in standard quantum cosmology because it "washes out" the fundamental discreteness of quantum geometry.

Finally, the detailed calculations have revealed another surprising feature. The fact that the quantum effects become prominent near the big bang, completely invalidating the classical predictions, is pleasing but not unexpected. However, prior to these calculations, it was not clear how soon after the big-bang one can start trusting semi-classical notions and calculations. It would not have been surprising if we had to wait till the radius of the universe became, say, a few million times the Planck length. These calculations strongly suggest that few hundred Planck lengths should suffice. This is fortunate because it is now feasible to develop quantum numerical relativity; with computational resources commonly available, grids with $(10^6)^3$ points are hopelessly large but one with $(100)^3$ points are readily available.

3.2. Black holes. Loop quantum cosmology illuminates dynamical ramifications of quantum geometry but within the context of mini-superspaces where all but a finite number of degrees of freedom are frozen. In this sub-section, I will discuss a complementary application where one considers the full theory but probes consequences of quantum geometry which are not sensitive to full quantum dynamics —the application of the framework to the problem of black hole entropy. This discussion is based on joint work with Baez, Corichi and Krasnov [11] which itself was motivated by earlier work of Krasnov, Rovelli and others.

Let us begin with classical general relativity. Consider first the simplest solution of Einstein's equation: a manifold which is topologically $\mathbb{R}^4$, equipped with a flat metric (of signature $-+++$). It has the property that an observer near infinity can receive a causal signal from *any* point in the interior, sometime along its infinite world-line. Thus, no part of space-time is permanently hidden from infinity. However, Einstein's equation admits solutions which do not share this property. In such solutions, the portion of space-time which is hidden from all asymptotic observers is called a *black hole region*. Physically, since gravity is attractive, the space-time metric around dense, compact astrophysical objects is such that the light cones in their vicinity are "bent toward the object". Since causal signals propagate with a speed less than or equal to that of light, it is "harder for the information to leak out" from their vicinity. A black hole region results if light cones are bent so much that they are "tilted completely inward"; i.e., no causal signal can leave the region of space-time in question. A space-like surface representing an "instant of time"

---

[6]There is thus a qualitative similarity to the phenomenon of phase transitions in magnets. "Our side" of the big-bang is analogous to the ferro-magnetic phase (the role of the "magnetization" mean field —the vector pointing from the south to the north pole of a ferro-magnet— being played by the classical geometry) and the "other side" is analogous to the para-magnetic phase (where "magnetization" is no longer a useful concept.

intersects the boundary of the black-hole region in a 2-sphere. We will refer to this 2-sphere boundary as the *black hole horizon*. (This terminology is not the standard one but is more convenient for our purposes here.)

As explained in the Introduction, since mid-seventies, a key question in the subject has been: What is the statistical mechanical origin of the black hole entropy $S_{\rm BH} = a_{\rm hor}/4\ell_{\rm Pl}^2$? What are the microscopic degrees of freedom that account for this entropy? This relation implies that a solar mass black hole must have $\sim (\exp 10^{77})$ quantum states, a number that is *huge* even by the standards of statistical mechanics. Where do all these states reside? To answer these questions, in the early nineties John Wheeler suggested the following heuristic picture, which he christened "It from Bit". Divide the black hole horizon in to elementary cells, each with one Planck unit, $\ell_{\rm Pl}^2$, of area and assign to each cell two microstates, or one "bit". Then the total number of states $\mathcal{N}$ is given by $\mathcal{N} = 2^n$ where $n = (a_{\rm hor}/\ell_{\rm Pl}^2)$ is the number of elementary cells, whence entropy is given by $S = \ln \mathcal{N} \sim a_{\rm hor}$. Thus, apart from a numerical coefficient, the entropy ("It") is accounted for by assigning two states ("Bit") to each elementary cell. This qualitative picture is simple and attractive. Therefore it is natural to ask if it can be made precise. Can these heuristic ideas be supported by a systematic analysis from first principles? What is the rationale behind dividing the black hole horizon in to elementary cells of unit Planck area? Why are there exactly two quantum states associated with each cell? It turned out that quantum geometry could supply answers to these questions through a detailed analysis.[7]

A systematic approach requires that we first specify the class of black holes of interest. Since the entropy formula is expected to hold unambiguously for black holes in equilibrium, most analysis were confined to space-times with "eternal" black holes admitting a global time-translation isometry, rather than the astrophysical ones which result from a gravitational collapse. From a physical viewpoint however, this assumption seems overly restrictive. After all, in statistical mechanical calculations of entropy of ordinary systems, one only has to assume that the given system is in equilibrium, not the whole world. Therefore, it should suffice to assume that the black hole itself is in equilibrium; the exterior geometry should not be forced to be time-independent. Finally, it has been known since the mid-seventies that the thermodynamical considerations apply not only to black holes but also to so-called "cosmological horizons". A natural question is: Can these diverse situations be treated in a single stroke? Within the quantum geometry approach, the answer is in the affirmative. The idea that the black hole (or the cosmological horizon) is itself in equilibrium is captured by certain boundary conditions which ensure that the horizon itself is *isolated*, allowing time-dependent space-time geometry and matter fields in the exterior region. Entropy associated with an isolated horizon refers to the family of observers in the exterior region for whom the isolated horizon is a physical boundary that separates the region which is accessible to them from the one which is not. (This point is especially important for cosmological horizons where, without reference to observers, one can not even define these horizons.) States which contribute to this entropy are the ones

---

[7]I should add, however, that this account does not follow chronology. Black hole entropy was computed in quantum geometry quite independently and the realization that the "It from Bit" picture works so well was somewhat of a surprise.

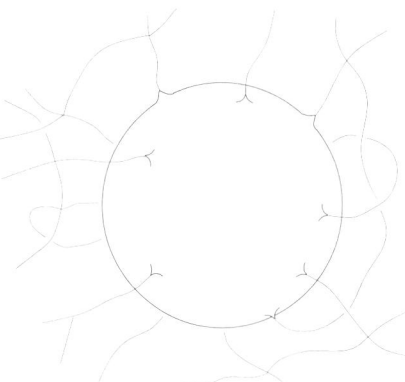

FIGURE 2. Quantum Horizon. Polymer excitations in the bulk puncture the horizon, endowing it with quantized area. Intrinsically, the horizon is flat except at punctures where it acquires a quantized deficit angle. These angles add up to endow the horizon with a 2-sphere topology.

which can interact with the states in the exterior; in this sense, they "reside" on the horizon.

In the detailed analysis, one considers space-times admitting an isolated horizon as inner boundary and carries out a systematic quantization. The quantum geometry framework can be naturally extended to this case. The isolated horizon boundary conditions imply that the intrinsic geometry of the quantum horizon is described by the so called U(1) Chern-Simons theory on the horizon. This is a well-developed, topological quantum field theory (also known as the "$a \wedge da$ theory of A. Schwarz). A deeply satisfying feature of the analysis is that there is a seamless matching of three otherwise independent structures: the isolated horizon boundary conditions which come from classical general relativity; the quantum geometry in the bulk; and the Chern-Simons theory on the horizon. In particular, one can calculate eigenvalues of certain physically interesting operators using purely bulk quantum geometry without any knowledge of the Chern-Simons theory, or using the Chern-Simons theory without any knowledge of the bulk quantum geometry. The two theories have never heard of each other. Yet, thanks to the isolated horizon boundary conditions, the two infinite sets of numbers match exactly, providing a coherent description of the quantum horizon. (For details, see the paper by Ashtekar, Baez and Krasnov in [**11**].)

In this description, the polymer excitations of the bulk geometry —i.e., the spin-work states of section 2.2— each labelled by a spin j_I, pierce the horizon, endowing it an elementary area a_{j_I} given by (2.2). Each intersection is called a *puncture*. The sum $\sum_I a_{j_I}$ adds up to the total horizon area a_{hor}. The intrinsic geometry of the horizon is flat except at these punctures. However, the Chern-Simons connection has a non-trivial holonomy around each puncture. Geometrically, this means that there is a deficit angle at each puncture. These angles add up to 4π thereby endowing the horizon with a 2-sphere topology. For a solar mass black hole, a typical horizon state would have 10^{77} punctures, each contributing a tiny deficit angle. So, although the quantum geometry *is* distributional, it can be well approximated by a smooth metric. The intrinsic quantum geometry is described in detail by the elements of the Hilbert space of the U(1)-Chern-Simons theory on the punctured sphere. This is called the *surface Hilbert space*.

The counting of states can be carried out as follows. First one constructs a micro-canonical ensemble by restricting oneself only to those states for which the total area, angular momentum, and charges lie in small intervals around fixed values $a_{\text{hor}}, J_{\text{hor}}, Q^i_{\text{hor}}$. (As is usual in statistical mechanics, the leading contribution to the entropy is independent of the precise choice of these small intervals.) For each set of punctures, one can compute the dimension of the surface Hilbert space, consisting of Chern-Simons states compatible with that set. One allows all possible sets of punctures (by varying both the spin labels and the number of punctures), subject to the constraint that the total area a_{hor} be fixed, and adds up the dimensions of the corresponding surface Hilbert spaces to obtain the number $\mathcal{N}$ of permissible surface states. One finds that the horizon entropy S_{hor} is given by

$$(3.3) \qquad S_{\text{hor}} := \ln \mathcal{N} = \frac{\gamma_o}{\gamma} \frac{a_{\text{hor}}}{\ell_{\text{Pl}}^2} + O(\frac{\ell_{\text{Pl}}^2}{a_{\text{hor}}}), \quad \text{where} \quad \gamma_o = \frac{\ln 2}{\sqrt{3}\pi}$$

Thus, for large black holes, entropy is indeed proportional to the horizon area. This is a non-trivial result; for example, early calculations often led to proportionality to the square-root of the area. However, even for large black holes, one obtains agreement with the Hawking-Bekenstein formula *only* in the sector of quantum geometry in which the Barbero-Immirzi parameter γ takes the value $\gamma = \gamma_o$. Thus, while all γ-sectors are equivalent classically, the standard quantum field theory in curved space-times is recovered in the semi-classical theory only in the γ_o-sector of quantum geometry. It is quite remarkable that thermodynamic considerations involving *large* black holes can be used to fix the quantization ambiguity which dictates such Planck scale properties as eigenvalues of geometric operators. Note however that the value of γ can be fixed by demanding agreement with the semi-classical result just in one case — e.g., a spherical horizon with zero charge, or a cosmological horizon in the de Sitter space-time. Once the value of γ is fixed, the theory is completely fixed and we can ask: Does this theory yield the Hawking-Bekenstein value of entropy of *all* isolated horizons, irrespective of the values of charges, angular momentum, and cosmological constant or the amount of distortion in the horizon geometry. The answer is in the affirmative. Thus, the agreement with quantum field theory in curved space-times holds in *all* these diverse cases.

Why does γ_o not depend on other horizon parameters such as the charges Q^i_{hor}? This important property can be traced back to a key consequence of the isolated horizon boundary conditions: detailed calculations show that only the gravitational part of the symplectic structure has a surface term at the horizon; the matter symplectic structures have only volume terms. (Furthermore, the gravitational surface term is insensitive to the value of the cosmological constant.) Consequently, in the geometric quantization procedure used in this analysis, there are no independent surface quantum states associated with matter. This provides a natural explanation of the fact that the Hawking-Bekenstein entropy depends only on the horizon geometry and is independent of electro-magnetic (or other) charges.

Finally, let us return to Wheeler's "It from Bit". One can ask: what are the states that dominate the counting? Perhaps not surprisingly, they turn out to be the ones which assign to each puncture the smallest quantum of area (i.e., spin value $j = \frac{1}{2}$), thereby maximizing the number of punctures. In these states, each puncture defines one of Wheeler's "elementary cell" and his two states correspond to the $j_z = \pm 1/2$ states, i.e. to whether the deficit angle is positive or negative. However, in the complete theory, all values of j (and hence of j_z) must be allowed to obtain a complete description of the geometry of the quantum horizon. If one is only interested in counting states for large black holes, however, the leading contribution comes from the $j = 1/2$ states.

To summarize, quantum geometry naturally provides the micro-states responsible for the huge entropy associated with horizons. In this analysis, all black holes —including the ones of direct astrophysical interest— and cosmological horizons are treated in an unified fashion. The sub-leading term has also been calculated and shown to be proportional to $\ln a_{\text{hor}}$ [11].

4. Conclusion

In this brief report, I have summarized two of the recent advances which have answered some of the long standing questions of quantum gravity raised in the Introduction. There have been two other notable advances: i) the development of "spin-foam models" [12] discussed briefly by Baez at the conference which provide a new, non-perturbative path integral approach to quantum gravity and have led to a variety of interesting and intriguing mathematical results on state sum models, extending in certain ways the very interesting work on Turaev and Viro in 3-dimensions; and, ii) the introduction of new measures on the space of connections relating the quantum geometry framework to the standard Fock description of photons and gravitons, which is paving the way to relate the Planck scale calculations of quantum geometry to the more familiar world of "low energy physics" [14]. The vitality of the program is reflected in the fact that many of the key ideas in all these developments came from young researchers —from Bojowald in quantum cosmology [10], Krasnov in the understanding of quantum black holes [11], Perez in spin foams [12] and Varadarajan in the relation to low energy physics [14].

Throughout the development of loop quantum gravity, unforeseen simplifications have arisen regularly, leading to surprising solutions to seemingly impossible difficulties. Progress could occur because some of the obstinate problems which had slowed developments in background independent approaches, sometimes for decades, evaporated when "right" perspectives were found. I will conclude with a few examples.

• Up until the early nineties, it was widely believed that spaces of connections do not admit non-trivial diffeomorphism invariant measures. This would have made it impossible to develop our background independent approach. Quite surprisingly, such a measure could be found by looking at connections in a slightly more general perspective. It is simple, natural, and has just the right structure to support quantum geometry. This geometry, in turn, supplied some missing links, e.g., by providing just the right expressions that Ponzano-Regge had to postulate without justification in their celebrated, early work on 3-dimensional gravity.

• Fundamental discreteness first appeared in a startling fashion in the construction of the so-called "weaves", quantum states which approximate given classical 3-geometries. In this construction, quantum states based on finite graphs were introduced as a starting point, with the goal of taking a "continuum limit" as in lattice gauge theories. It came as a major surprise that, if one wants to recover a given classical geometry *on large scales*, one can not take this limit, i.e., one can not refine the underlying graph arbitrarily; there is an in-built discreteness.

• Traces of holonomies of a suitably defined connection around a smooth loop define a natural set of functions of connections. At a heuristic level, it was found that they automatically solve the most difficult part of the quantum Einstein's equation. No one expected to find such simple and natural solutions even heuristically. This calculation suggested that the action of this part of the quantum Einstein equation is concentrated at vertices of graphs, which in turn led to strategies for its regularization.

• As I indicated in some detail, unforeseen insights arose in the well-studied subject of quantum cosmology essentially by taking an adequate account of the quantum nature of geometry, i.e., by respecting the fundamental discreteness of the eigenvalues of the scale factor operator. Similarly, in the case of black holes, three quite distinct structures —the isolated horizon boundary conditions, the bulk quantum geometry and the surface Chern-Simons theory— blended together unexpectedly to provide a coherent theory of quantum horizons.

Repeated occurrence of such "unreasonable" simplifications suggest that the ideas underlying loop quantum gravity may have captured an essential germ of truth.

Acknowledgment I would like to thank John Baez, Martin Bojowald, Rodolfo Gambini, Jerzy Lewandowski, Alejandro Perez, Jorge Pullin, Carlo Rovelli and Thomas Thiemann for numerous discussions. This work was supported in part by the NSF grant PHY-0090091 and the Eberly research funds of Penn State.

References

[1] *Books and recent reviews:*
J. A. Wheeler, *Geometrodynamics*, (Academic Press, New York, 1962)
J. A. Wheeler, Superspace and the nature of quantum geometrodynamics, in *Battelle rencontres: 1967 lectures in mathemartivcs and physics*, (Benjamin, New York, 1968)
R. Penrose and W. Rindler, *Spinors and space-times, Vol 2*, (cambridge University Press, Cambridge 1988)
A. Ashtekar, *Lectures on non-perturbative canonical gravity*, (World Scientific, Singapore, 1991)
A. Connes, *Non-commutative geometry*, (Academic Press, New York, 1994)
R. Gambini and J. Pullin, *Loops, knots, gauge theories and quantum gravity*, (CUP, Cambridge, 1996)
T. Thiemann, Lectures on loop quantum gravity, to appear in *Aspects of quantum gravity: From theory to experimental search* Lecture Notes in Physics, *Preprint* `gr-qc/0210094`
A. Ashtekar and J. Lewandowski, Background independent quantum gravity: A status report *pre-print* `gr-qc/0404018`

[2] *Reformulation of general relativity:*
A. Ashtekar, New variables for classical and quantum gravity, Phys. Rev. Lett. **57**, 2244-2247, (1986)
New Hamiltonian formulation of general relativity, Phys. Rev. **D36**, 1587-1602 (1987)
A. Ashtekar, J. D. Romano and R. S. Tate, New variables for gravity: Inclusion of matter, Phys. rev. **D40**, 2572-2587 (1989)
F. Barbero, Real Ashtekar variables for Lorentzian signature space-times, Phys. Rev. **D51**, 5507-5510 (1996)

[3] *Connections and loops:*
C. Rovelli and L. Smolin, Loop representation for quantum general relativity, Nucl. Phys. **B331** 80-152 (1990)
A. Ashtekar, V. Husain, C. Rovelli, J. Samuel and L. Smolin, 2+1 quantum gravity as a toy model for the 3+1 theory, Class. Quant. Grav. **6**, L185-L193 (1989)

A. Ashtekar, C. Rovelli and L. Smolin, Gravitons and loops, Phys. Rev. **D44**, 1740-1755 (1991)

[4] *Quantum geometry: basics*
A. Ashtekar and C. J. Isham, Representation of the holonomy algebras of gravity and non-Abelian gauge theories, Class. Quant. Grav. **9**, 1433-1467 (1992)
A. Ashtekar and J. Lewandowski, Representation theory of analytic holonomy algebras, in *Knots and Quantum Gravity*, edited by J. C. Baez, Oxford U. Press, Oxford, (1994)
J. C. Baez, Generalized measures in gauge theory, Lett. Math. Phys. **31**, 213-223 (1994)
D. Marolf and J. Mourão, On the support of the Ashtekar-Lewandowski measure, Commun. Math. Phys. **170**, 583-606 (1995)
A. Ashtekar and J. Lewandowski, Projective techniques and functional integration, Jour. Math. Phys. **36**, 2170-2191 (1995)
J. C. Baez and S. Sawin, Functional integration on spaces of connections, Jour. Funct. Analysis **150**, 1-27 (1997)

[5] *Spin networks:*
R. Penrose, Angular momentum: an approach to combinatorial space-time, in *Quantum Theory and Beyond*, edited by Ted Bastin, (Cambridge University Press, 1971)
C. Rovelli and L. Smolin, Spin networks and quantum gravity, Phys. Rev. **D52**, 5743-5759 (1995)
J. C. Baez, Spin networks in non-perturbative quantum gravity, in *The Interface of Knots and Physics*, edited by L. Kauffman, American Mathematical Society, Providence, 1996, pp. 167-203
J. C. Baez, Spin networks in gauge theory, Adv. Math. **117**, 253-272 (1996)
J. C. Baez and S. Sawin, Diffeomorphism-invariant spin network states, Jour. Funct. Analysis **158**, 253-266 (1998)

[6] *Geometric operators and their properties*
C. Rovelli and L. Smolin, Discreteness of area and volume in quantum gravity, Nucl. Phys. **B442**, 593-622 (1995); Erratum: Nucl. Phys. **B456**, 753 (1995)
R. Loll, The volume operator in discretized quantum gravity, Phys. Rev. Lett. **75** 3048-3051 (1995)
A. Ashtekar and J. Lewandowski, Differential geometry on the space of connections using projective techniques, Jour. Geo. & Phys. **17**, 191-230 (1995)
A. Ashtekar and J. Lewandowski, Quantum theory of geometry I: Area operators, Class. Quant. Grav. **14**, A55-A81 (1997)
Quantum theory of geometry II: Volume Operators, Adv. Theo. Math. Phys. **1**, 388-429 (1997)
R. Loll, Further results on geometric operators in quantum gravity, Class. Quant. Grav. **14** 1725-1741 (1997)
T. Thiemann, A length operator for canonical quantum gravity, Jour. Math. Phys. **39**, 3372-3392 (1998)

[7] *Barbero-Immirzi ambiguity*
G. Immirzi, Quantum gravity and Regge calculus, Nucl. Phys. Proc. Suppll. **57**, 65-72 (1997)
C. Rovelli and T. Thiemann, The Immirizi parameter in quantum general relativity, Phys. Rev. **D57** 1009-1014 (1998)
R. Gambini O. Obregon and J. Pullin, Yang-Mills analogs of the Immirzi ambiguity, Phys. Rev. **D59** 047505 (1999)

[8] *Quantum Einstein's equation I*
D. Marolf, Refined algebraic quantization: Systems with a single constraint, `gr-qc/9508015`
A. Ashtekar, J. Lewandowski, D. Marolf, J. Mourão, and T. Thiemann, Quantization of diffeomorphism invariant theories of connections with local degrees of freedom, Jour. Math. Phys. **36**, 6456-6493 (1995)
J. Lewandowski and T. Thiemann, Diffeomorphism invariant quantum field theories of connections in terms of webs, Class. Quant. Grav. **16**, 2299-2322 (1999)

[9] *Quantum Einstein's equation II*
C. Rovelli and L. Smolin, The physical Hamiltonian in nonperturbative quantum gravity,

Phys. Rev. Lett. **72**, 446-449 (1994)

T. Thiemann, Anomaly-free formulation of non-perturbative, four-dimensional Lorentzian quantum gravity, Phys. Lett. **B380** 257-264 (1996)

Quantum Spin Dynamics (QSD), Class. Quant. Grav. **15** 839-873 (1998)

QSD III : Quantum Constraint Algebra and Physical Scalar Product in Quantum General Relativity, Class. Quant. Grav. **15** 1207-1247 (1998)

QSD V : Quantum Gravity as the Natural Regulator of Matter Quantum Field Theories, Class. Quant. Grav. **15** 1281-1314 (1998)

R. Gambini, J. Lewandowski, D. Marolf and J. Pullin, On the consistency of the constraint algebra in spin network quantum gravity, Int. J. Mod. Phys. **D7**, 97-109 (1998)

J. Lewandowski and D. Marolf, Loop constraints: A habitat and their algebra, Int. J. Mod. Phys. **D7**, 299-330 (1998)

C. Di Bartolo, R. Gambini, J. Griego, J. Pullin, Consistent canonical quantization of general relativity in the space of Vassiliev knot invariants, Phys. Rev. Lett. **84**, 2314-2317 (2000)

[10] *Big-bang:*
M. Bojowald, Absence of singularity in loop quantum cosmology, Phys. Rev. Lett. **86**, 5227-5230 (2001)

Dynamical initial conditions in quantum cosmology, *Phys. Rev. Lett.* **87**, 121301 (2001)

Inverse scale factor in isotropic quantum geometry, Phys. Rev. **D64** 084018 (2001)

Loop quantum cosmology III: Wheeler-DeWitt operators, Class. Quant. Grav. **18** 1055–1070 (2001)

Loop Quantum Cosmology IV: Discrete Time Evolution, Class. Quant. Grav. **18** 1071–1088 (2001)

Quantum Geometry and Symmetry, (Shaker-Verlag, Aachen, 2001)

A. Ashtekar, M. Bojowald and J. Lewandowski, Mathematical structure of loop quantum cosmology, Adv. Theor. Math. Phys. **7** 233–268 (2003)

[11] *Black holes:*
J. D. Bekenstein, Black holes and entropy Phys. Rev. **D7** 2333–2346 (1973)

J. D. Bekenstein, Generalized second law of thermodynamics in black hole physics *Phys. Rev.* **D9** 3292–3300

J. W. Bardeen, B. Carter and S. W. Hawking, The four laws of black hole mechanics *Commun. Math. Phys.* **31** 161-170 (1973)

S. W. Hawking, Particle creation by black holes *Commun. Math. Phys.* **43** 199–220 (1975)

A. Ashtekar, J. Baez, A. Corichi, K. Krasnov, Quantum geometry and black hole entropy, Phys. Rev. Lett. **80** 904-907 (1998)

A. Ashtekar, A. Corichi and K. Krasnov, Isolated horizons: the classical phase space, Adv. Theor. Math. Phys. **3**, 418-471 (1999)

A. Ashtekar, J. Baez, K. Krasnov, Quantum geometry of isolated horizons and black hole entropy, Adv. Theo. Math. Phys., **4**, 1-95 (2000)

R. K. Kaul and P. Majumdar, Logarithmic corrections to the Bekenstein-Hawking entropy, Phys. Rev. Lett. **84**, 5255-5257 (2000)

A. Ashtekar, A. Corichi, Non-minimal couplings, quantum gravity and black hole entropy, Class. Quantum Grav. **20**, 4473-4484 (2003)

[12] *Spin foams and finiteness:*
A. Perez, Finiteness of a spinfoam model for Euclidean quantum general relativity, Nucl. Phys. **B599**, 427 (2001)

L. Crane, A. Perez, and C. Rovelli, Perturbative finiteness in spin-foam quantum gravity, Phys. Rev. Lett. **87** 181301, (2001)

A. Perez and C. Rovelli, Spin foam model for Lorentzian general relativity, Phys. Rev. **D63**, 041501 (2001)

J. W. Barrett, L. Crane, Relativistic spin networks and quantum gravity, J. Math. Phys. **39**, 3296 (1998)

J. Baez, Spin foam models, Class. Quant. Grav. **15**, 1827-1858 (1998)

J. W. Barrett, L. Crane, A Lorentzian signature model for quantum general relativity, Class. Quant. Grav. **17** 3101-3118 (2000)

O. Lauscher and M. Reuter, Is quantum Einstein gravity non-perturbatively renormalizable?, `hep-th/0110021`

[13] *Semi-classical states:*
A. Ashtekar, C. Rovelli and L. Smolin, Weaving a classical geometry with quantum threads, Phy. Rev. Lett. **69**, 237-240 (1992)
M. Arnsdorf, S. Gupta, Loop quantum gravity on non-compact spaces, Nucl. Phys. **B577** 529-546 (2000)
T. Thiemann, Gauge field theory coherent states (GCS) : I. general properties, Class. Quant. Grav. **18** 2025-2064 (2001)
H. Sahlmann, T. Thiemann and O. Winkler, Coherent states for canonical quantum general relativity and the infinite tensor product extension, `gr-qc/0102038`

[14] *Fock states in the polymer picture:*
M. Varadarajan, Fock representations from U(1) holonomy algebras, Phys. Rev. **D61**, 104001 (2000)
M. Varadarajan, Photons from quantized electric flux representations, `gr-qc/0104051`
J. M. Velhinho, Invariance properties of induced Fock measures for $U(1)$ holonomies, `math-ph/0107002`
A. Ashtekar and J. Lewandowski, Relation between plymer and Fock excitations, Class. Quant. Grav. **18**, L117-L127 (2001)
L. Bombelli, Statistical geometry of random weave states, `gr-qc/0101080`.

CENTER FOR GRAVITATIONAL PHYSICS AND GEOMETRY, PHYSICS DEPARTMENT, PENN STATE, UNIVERSITY PARK, PA 16802, USA

Supersymmetry, supergravity, superspace and BRST symmetry in a simple model

Peter van Nieuwenhuizen

Contents

1. Introduction
2. Rigid $N = 1$ supersymmetry in x-space
3. $N = 1$ supergravity in x-space
4. Rigid and local $N = 1$ superspace
5. Extended rigid supersymmetries
6. BRST quantization in a Hamiltonian approach

References

1. Introduction

In these lectures we shall introduce rigid supersymmetry, supergravity (which is the gauge theory of supersymmetry) and superspace, and apply the results to BRST quantization. We assume that the reader has never studied these topics. For readers who want to read more about these "super" subjects, we give a few references at the end of this contribution, but the whole point of these lectures is that one does not need additional references for a self-contained introduction. The reader should just sit down with paper and pencil. We could have decided to begin with the usual models in $3 + 1$ dimensional Minkowski space with coordinates x, y, z and t; this is the standard approach, but we shall instead consider a much simpler model, with only one coordinate t. We interpret t as the time coordinate. For physicists, the $3 + 1$ dimensional models are the ones of interest because they are supposed to describe the real world. For mathematicians, however, the simpler model may be of more interest because the basic principles appear without the dressing of physical complications. Let us begin with three definitions which should acquire meaning as we go on.

2000 *Mathematics Subject Classification.* 81T60.

Key words and phrases. Supersymmetry, superspace, action, Hamiltonian, spinors, conserved charges, constraints, gauge fixing, Faddeev-Popov ghosts, supergravity, BRST quantization.

Supersymmetry is a symmetry of the action (to be explained) with a rigid (constant) anticommuting parameter[1] (usually denoted by ϵ) between bosonic (commuting) and fermionic (anticommuting) fields (again to be explained). It requires that for every bosonic particle in Nature there exists a corresponding fermionic particle, and for every fermionic particle there should exist a corresponding bosonic particle. So supersymmetry predicts that there are twice as many particles as one might have thought. One may call these new particles supersymmetric particles. These supersymmetric particles will be looked for at CERN (the European high-energy laboratory) in the coming 8 years. So far not a single supersymmetric particle has been discovered: supersymmetry is a theoretical possibility, but whether Nature is aware of this possibility remains to be seen.[2]

Supergravity is the gauge theory of supersymmetry. Its action is invariant under transformation rules which depend on a local (space- and time-dependent) anticommuting parameter $\epsilon(x, y, z, t)$, and there is a gauge field for supersymmetry which is called the gravitino field. It describes a new hypothetical particle, the gravitino. The gravitino is the fermionic partner of the graviton. The graviton is the quantum of the gravitational field (also called the metric). The astonishing discovery of 1976 was that a gauge theory of supersymmetry requires gravity: Einstein's 1916 theory of gravity (called general relativity) is a product of local supersymmetry. Phrased differently: local supersymmetry is the "square root of general relativity", see (3.18). (Likewise, supersymmetry is the square root of translation symmetry, see (2.26)).

Superspace. In Nature fields can be divided into bosonic (commuting) fields and fermionic (anti-commuting) fields. This is one of the fundamental discoveries of the quantum theory of the 1920's. The anticommuting fields are described by spinors and the bosonic fields by tensors according to the spin-statistics theorem of the 1930's. (Spinors and tensors refers to their transformation properties under Lorentz transformations). One can also introduce in addition to the usual coordinates x^μ anticommuting counterparts θ^α. The space with coordinates x and θ is called superspace. In the case of a four-dimensional Minkowski space (our world) there are four coordinates (x, y, z and t) and also four θ's, namely θ^1, θ^2, θ^3 and θ^4, but in other dimensions the number of x's and θ's are not the same.[3] These θ's are Grassmann variables [1], for example $\theta^1 \theta^2 = -\theta^2 \theta^1$ and $\theta^1 \theta^1 = 0$. In our case we shall have one x^μ (namely t) and one θ^α (which we denote by θ). In superspace one can introduce superfields: fields which depend both on x and θ. Because $\theta^2 = 0$, the superfields we consider can be expanded as $\phi(t, \theta) = \varphi(t) + \theta \psi(t)$. This concludes the three definitions.

Supersymmetric quantum field theories have remarkable properties. Leaving aside the physical motivations for studying these theories, they also form useful toy models. We present here an introduction to supersymmetric field theories with rigid

---

[1]Technically: a Grassmann variable. "Constant" means "independent of the spacetime coordinates x, y, z and t".

[2]This is not the first time a doubling of the number of particles has been predicted. In 1931 Dirac predicted that for every fermionic particle a fermionic antiparticle should exist, and these antiparticles were discovered in 1932. We consider in these lectures only real fields, and the particles corresponding to real fields are their own antiparticles. Hence in these lectures the notion of antiparticles plays no role.

[3]The x^μ transform as vectors under the Lorentz group while the θ^α transform as spin 1/2 spinors.

and local supersymmetry, both in x-space and in superspace, in the simplest possible model[4]. To avoid the complications due to "Fierz rearrangements" (recoupling of four fermionic fields A, B, C, D from the structure $(AB)(CD)$ to $(AD)(CB)$)[5] we consider one-component (anticommuting) spinors. Then $(AB)(CD)$ is simply equal to $-(AD)(CB)$. The simplest case in which spinors have only one component is a one-dimensional spacetime, i.e., quantum mechanics. The corresponding superspace has one commuting coordinate t and one anticommuting coordinate θ. Both are real.

We repeat and summarize: one can distinguish between rigidly supersymmetric field theories, which have a constant symmetry parameter, and locally supersymmetric field theories whose symmetry parameter is an arbitrary space-time dependent parameter. For a local symmetry one needs a gauge field. For supersymmetry the gauge field has been called the gravitino. (The local symmetry on which Einstein's theory of gravitation is based is diffeomorphism invariance. The gauge field is the metric field $g_{\mu\nu}(x)$). Gauge theories of supersymmetry (thus theories with a local supersymmetry containing the gravitino) need curved spacetime. In other words, gravity is needed to construct gauge theories of supersymmetry, and for that reason local supersymmetry is usually called supergravity. In curved space the quanta of the metric $g_{\mu\nu}$ are massless particles called gravitons. They are the bosonic partners of the gravitinos. Neither gravitons nor gravitinos have ever been directly detected. Classical gravitational radiation may be detected in the years ahead, but the gravitons (the quantized particles of which the gravitational field is composed) are much harder to detect individually. The detection of a single gravitino would have far-reaching consequences.

In the last chapter we quantize the supergravity action which we obtained in chapter 3. There are several methods of quantization, all in principle equivalent, but we shall only discuss the BRST method. It yields the "quantum action", which is the action to be used in path integrals. This method has a beautiful and profound mathematical structure, and that is one of the reasons we chose to include it.

The author wrote in 1976 with D.Z. Freedman and S. Ferrara the first paper on supergravity, soon followed by a paper by S. Deser and B. Zumino. However, we will not discuss past work and give references; rather, the present account may serve as a simplified introduction to that work. For readers who want to read further, we include a few references at the end.

2. Rigid $N = 1$ supersymmetry in x-space

The model we consider contains in x-space (or rather t-space) two point particles which correspond to a real bosonic field $\varphi(t)$ and a real fermionic field $\lambda(t)$. We view them as fields whose space-dependence (the dependence on x, y, z) is suppressed. The function $\varphi(t)$ is a smooth function of t, so its derivative is well-defined, but for every value of t the expression $\lambda(t)$ is an independent Grassmann

---

[4]Actually, an even simpler model than the one we present in these lectures exists. It contains constant fields, so fields which do not even depend on t and θ. These so-called matrix models are important in string theory, but they do not have enough structure for our purposes, so we do not discuss them.

[5]Here (AB) means contraction of spinor fields A and B.

number [1].[6] So $\lambda(t_1)\lambda(t_2) = -\lambda(t_2)\lambda(t_1)$. We assume that the concept of a derivative of $\lambda(t)$ with respect to t can be defined, and that we may partially integrate. As action for these "fields" we take $S = \int L dt$ with

(2.1) $$L \text{ (rigid)} = \frac{1}{2}\dot\varphi^2 + \frac{i}{2}\lambda\dot\lambda.$$

The $\dot\lambda = \frac{d}{dt}\lambda$ are independent Grassmann variables, so $\dot\lambda(t)\lambda(t) = -\lambda(t)\dot\lambda(t)$ and $\dot\lambda(t_1)\dot\lambda(t_2) = -\dot\lambda(t_2)\dot\lambda(t_1)$. In particular they anticommute with themselves and with each other at equal t:

(2.2) $$\{\lambda(t), \lambda(t)\} = 0, \quad \{\lambda(t), \dot\lambda(t)\} = 0, \quad \{\dot\lambda(t), \dot\lambda(t)\} = 0.$$

The symbol $\{A, B\}$ is by definition $AB + BA$. Later we shall define Poisson brackets and Dirac brackets, which we denote by $\{A, B\}_P$ and $\{A, B\}_D$ to avoid confusion. At the quantum level the Poisson and Dirac brackets are replaced by commutators for commuting fields and anticommutators for anticommuting fields, which we denote by $[A, B]$ and $\{A, B\}$, respectively, and which are defined by $[A, B] = AB - BA$ and $\{A, B\} = AB + BA$.

We introduce a concept of hermitian conjugation under which $\varphi(t)$ and $\lambda(t)$ are real: $\varphi(t)^\dagger = \varphi(t)$ and $\lambda(t)^\dagger = \lambda(t)$. Also $\dot\lambda(t)$ is real. Furthermore $(AB)^\dagger = B^\dagger A^\dagger$ for any A and B. We define the action by $S = \int L(t) dt$. The action should be hermitian according to physical principles (namely unitarity[7]). We need then a factor i in the second term in (2.1) in order that $(\frac{i}{2}\lambda\dot\lambda)^\dagger = -\frac{i}{2}\dot\lambda\lambda$ be equal to $\frac{i}{2}\lambda\dot\lambda$. In the action we need $\lambda(t)$ at different t. We repeat that for different t the $\lambda(t)$ are independent Grassmann variables. Thus we need an infinite basis for all Grassmann variables.[8]

Physical intermezzo which can be skipped by mathematicians: The term $\frac{1}{2}\dot\varphi^2$ is a truncation of the Klein-Gordon action to an xyz independent field, and the term $\frac{i}{2}\lambda\dot\lambda$ is the truncation of the Dirac action for a real[9] spinor to one of its components. In higher dimensions the Dirac action in curved space reads (as discussed in detail in 1929 by H. Weyl)

(2.3) $$\mathcal{L} \text{ (Dirac)} = -(\det e_\mu{}^m) \bar\lambda \gamma^m e_m{}^\mu D_\mu \lambda,$$

where $D_\mu \lambda = \partial_\mu \lambda + \frac{1}{4}\omega_\mu{}^{mn}\gamma_{mn}\lambda$ with $\gamma_{mn} \equiv \frac{1}{2}[\gamma_m, \gamma_n]$ the Lorentz generators (constant matrices) and $\omega_\mu{}^{mn}$ the spin connection (a complicated function of the vielbein fields $e_\mu{}^m$). The matrices γ^m (with $m = 0, 1, 2, \ldots, d-1$) in d spacetime dimensions satisfy Clifford algebra relations, $\{\gamma^m, \gamma^n\} = 2\eta^{mn}$. The "vielbein" fields $e_\mu{}^m$ are the square root of the metric $g_{\mu\nu}$ in the sense that $e_\mu{}^m e_\nu{}^n \eta_{mn} = g_{\mu\nu}$, where η_{mn} is the Lorentz metric (a diagonal matrix with constant entries $(-1, +1, +1, \cdots, +1)$). Furthermore, $e_m{}^\mu$ is the matrix inverse of $e_\mu{}^m$, and $\bar\lambda$ is defined by $\lambda^\dagger i\gamma^0$. However, in one dimension there are no Lorentz transformations,

---

[6]For a more recent mathematical treatment, see "Five Lectures on Supersymmetry" by D. Freed (AMS) and articles by Deligne & Morgan and by Deligne & Freed in "Quantum Fields and Strings: A course for mathematicians" (AMS)

[7]"Unitarity" means "conservation of probability": the total sum of the probabilities that a given system can decay into any other system should be one.

[8]In some mathematical studies one takes a finite-dimensional basis for the Grassmann variables. This is mathematically consistent, but physically unacceptable: it violates unitarity.

[9]Real spinors are called Majorana spinors, and complex spinors are called Dirac spinors. Already at this point one can anticipate that λ must be real because we took φ to be real, and we shall soon prove that there exists a symmetry between λ and φ.

hence in our toy model $D_\mu \lambda$ is equal to $\dot{\lambda}$. Furthermore, $\det e_\mu{}^m = e_\mu{}^m$ in one dimension, and this cancels the factor $e_m{}^\mu$. Thus even in curved space, the Dirac action in our toy model reduces to $\frac{i}{2}\lambda\dot{\lambda}$ (for real λ; the factor $\frac{1}{2}$ is used for real fields, just as for $\frac{1}{2}\dot{\varphi}^2$ in (2.1)). As a consequence, the gravitational stress tensor, which is by definition proportional to $\frac{\delta}{\delta e_\mu^m(x)}S$, vanishes in this model for $\lambda(t)$. Also the canonical Hamiltonian [10] $H = \dot{q}p - L$ vanishes for L given in (2.1) and $q = \lambda(t)$. This will play a role in the discussions below. The sign of the term $\frac{1}{2}\dot{\varphi}^2$ is positive because it represents the kinetic energy, but the sign of the fermion term could have been chosen to be negative instead of positive. (Requiring λ to be real, we cannot redefine $\lambda \to i\lambda$ in order to change the sign of the second term and still keep real λ.) The $+$ sign in (2.1) will lead to the susy anticommutator $\{Q, Q\} = 2H$ instead of $\{Q, Q\} = -2H$ with a hermitian Q. End of physical intermezzo.

The supersymmetry transformations should transform bosons into fermions, and vice-versa, so φ into λ, and λ into φ. Since φ is commuting and λ anticommuting, the parameter must be anticommuting. We take it to be a Grassmann number ϵ, although other choices are also possible.[11] One might then be tempted to write down $\delta\varphi = i\epsilon\lambda$ and $\delta\lambda = \varphi\epsilon$ (where the factor i is needed in order that $\delta\varphi$ be real, taking ϵ to be real) but this is incorrect as one might discover by trying to prove that the action is invariant under these transformation rules. There is a more fundamental reason why in particular the rule $\delta\lambda = \varphi\epsilon$ is incorrect, and that has to do with the dimensions of the fields and ϵ as we now explain.

The dimension of an action $S \equiv \int L dt$ is zero (for $\hbar = 1$).[12] Hence $L = \frac{1}{2}\dot{\varphi}^2$ should have dimension $+1$, taking the dimension of t to be -1 as usual for a coordinate. It follows that the dimension of φ is $-1/2$ and that of λ is 0

(2.4) $$[\varphi] = -1/2; \quad [\lambda] = 0.$$

From $\delta\varphi = i\epsilon\lambda$ we then conclude that ϵ has dimension $-1/2$

(2.5) $$[\epsilon] = -1/2.$$

Thus $\delta\varphi = i\epsilon\lambda$ is dimensionally correct: $[\delta\varphi] = -1/2$ and $[\epsilon\lambda] = -1/2$. Consider now the law for $\delta\lambda$. The proposal $\delta\lambda = \varphi\epsilon$ has a gap of one unit of dimension: $[\delta\lambda] = 0$ but $[\varphi\epsilon] = -1/2 - 1/2 = -1$. To fill this gap we can only use a derivative (we are dealing with massless fields so we have no mass available). Thus $\delta\lambda \sim \dot{\varphi}\epsilon$. We claim that the correct factor is minus unity, thus

(2.6) $$\delta\varphi = i\epsilon\lambda; \quad \delta\lambda = -\dot{\varphi}\epsilon.$$

By correct we mean that (2.6) leaves the action invariant as we now show. It is easy to show that S (rigid) is invariant under these transformation rules if ϵ is constant (rigid supersymmetry). Let us for future purposes already consider a local

---

[10]The momenta are defined by left-differentiation: $p = \frac{\partial}{\partial \dot{q}}S$. Hence $p(\phi) = \dot{\phi}$ and $\pi(\lambda) = -\frac{i}{2}\lambda$, where π denotes the conjugate momentum of λ.

[11]The author has proposed long ago with J. Schwarz to consider θ's which satisfy a Clifford algebra, $\{\theta^\alpha, \theta^\beta\} = \gamma_\mu^{\alpha\beta} x^\mu$.

[12]More precisely, the dimension of H and L is an energy, and t has of course the dimension of time. In quantum mechanics the Planck constant $\hbar \equiv h/2\pi$ has the dimension of an energy $\times$ time (discovered by Planck in 1900). Since S has the dimension of an energy $\times$ time, one can define dimensionless exponents of the action by $\exp \frac{i}{\hbar}S$. Such exponents appear in path integrals. Physicists often choose a system of units such that $\hbar = 1$.

ϵ (meaning $\epsilon(t)$) and also keep boundary terms due to partial integration. One finds then if one successively varies the fields in S according to (2.6)

$$\delta S = \int \left(\dot{\varphi} \delta \dot{\varphi} + \frac{i}{2} \delta \lambda \dot{\lambda} + \frac{i}{2} \lambda \delta \dot{\lambda} \right) dt$$

$$= \int \left[\dot{\varphi} \frac{d}{dt}(i\epsilon\lambda) - \frac{i}{2}(\dot{\varphi}\epsilon)\dot{\lambda} - \frac{i}{2} \lambda \frac{d}{dt}(\dot{\varphi}\epsilon) \right] dt$$

$$= \int \left[\dot{\varphi} i \dot{\epsilon} \lambda + \dot{\varphi} i \epsilon \dot{\lambda} - \frac{i}{2} \dot{\varphi} \epsilon \dot{\lambda} - \frac{i}{2} \frac{d}{dt}(\lambda \dot{\varphi} \epsilon) + \frac{i}{2} \dot{\lambda} \dot{\varphi} \epsilon \right] dt$$

(2.7)
$$= \int \left[\dot{\epsilon}(i\dot{\varphi}\lambda) - \frac{i}{2} \frac{d}{dt}(\lambda \dot{\varphi} \epsilon) \right] dt.$$

We performed a partial integration in the third line and used $\dot{\lambda}\epsilon = -\epsilon\dot{\lambda}$ in the fourth line. We now assume that "fields" (and their derivatives) tend to zero at $t = \pm\infty$. (If there would also be a space dimension σ, we could consider a finite domain $0 \leq \sigma \leq \pi$, and then we should specify boundary conditions at $\sigma = 0, \pi$. This happens in "open string theory".) It is clear that neglecting boundary terms at $t = \pm\infty$, and taking ϵ constant ($\dot{\epsilon} = 0$), the action is invariant. (A weaker condition which achieves the same result is to require that the fields at $t \to +\infty$ are equal to the fields at $t \to -\infty$). This assumption that fields vanish at $t = \pm\infty$ is not at all easy to justify, but we just accept it.

The algebra of rigid supersymmetry transformations reveals that supersymmetry is a square root of translations, in the sense that two susy tranformations (more precisely, a commutator) produce a translation. On φ this is clear

(2.8)
$$[\delta(\epsilon_2), \delta(\epsilon_1)]\varphi = \delta(\epsilon_2)i\epsilon_1\lambda - \delta(\epsilon_1)i\epsilon_2\lambda$$
$$= i\epsilon_1(-\dot{\varphi}\epsilon_2) - i\epsilon_2(-\dot{\varphi}\epsilon_1) = (2i\epsilon_2\epsilon_1)\dot{\varphi}.$$

We recall that the symbol $[A, B]$ is defined by $AB - BA$, so $[\delta(\epsilon_2), \delta(\epsilon_1)]$ is a commutator of two supersymmetry transformations. We used in the last step that $\epsilon_1\epsilon_2 = -\epsilon_2\epsilon_1$. The result is a translation ($\dot{\varphi}$) over a distance $\xi = 2i\epsilon_2\epsilon_1$. The same result is obtained for λ

(2.9)
$$[\delta(\epsilon_2), \delta(\epsilon_1)]\lambda = -\delta(\epsilon_2)\dot{\varphi}\epsilon_1 + \delta(\epsilon_1)\dot{\varphi}\epsilon_2$$
$$= -\frac{d}{dt}(i\epsilon_2\lambda)\epsilon_1 + \frac{d}{dt}(i\epsilon_1\lambda)\epsilon_2 = (2i\epsilon_2\epsilon_1)\dot{\lambda}.$$

We used that ϵ_2 is constant, so $\frac{d}{dt}\epsilon_2 = 0$, and $\lambda\epsilon_1 = -\epsilon_1\lambda$.

In higher dimensional theories this commutator on a fermion yields in addition to a translation also a term proportional to the field equation of the fermion, and to eliminate this extra term with the field equation, one introduces auxiliary fields. (Auxiliary fields are fields which appear in the action without derivatives; they are usually bosonic fields which enter as $a(t)^2$). Here, however, the translation $\dot{\lambda}$ and the field equation of λ are both equal to $\dot{\lambda}$. So the result could still have been a sum of the same translation as on φ, and a field equation, because both are proportional to $\dot{\lambda}$. This is not the case: the coefficient of $\dot{\lambda}$ is the same as the coefficient of $\dot{\varphi}$. There is a simple counting argument that explains this and that shows that no auxiliary fields are needed in this model. Off-shell (by which physicists mean: when the field equations are not satisfied) the translation operator is invertible (the kernel of $\frac{\partial}{\partial t}$ with the boundary conditions mentioned before is empty), hence the commutator $[\delta(\epsilon_2), \delta(\epsilon_1)]$ cannot vanish on field components. It follows that under rigid supersymmetry if "the algebra closes" (meaning if $[\delta(\epsilon_2), \delta(\epsilon_1)]$ is uniformly equal to only a translation but no further field equations), each bosonic field component must be mapped into a fermionic one, and vice-versa. Then the number of bosonic field components must be equal to the number of fermionic field equations. In our toy model there is one bosonic field component (φ) and one fermionic field component (λ). Thus there are no auxiliary fields needed in this model.[13]

We can construct charges Q and H which produce susy and time-translation transformations. This requires equal-time Poisson brackets for φ, and Dirac brackets for λ, which become at the quantum level commutators and anticommutators. For φ these results are standard: the conjugate momentum p of φ is defined by $p = \frac{\partial}{\partial \dot{\varphi}} S$ and this yields $p = \dot{\varphi}$. The quantum commutator is given by

$$(2.10) \qquad p = \dot{\varphi}; \quad [p(t), \varphi(t)] = \frac{\hbar}{i}.$$

For λ the conjugate momentum is (we use left-derivatives) $\pi = \frac{\partial}{\partial \lambda} S = -\frac{i}{2}\lambda$. The relation $\pi = -\frac{i}{2}\lambda$ is a constraint between the coordinates and the conjugate momenta, called by Dirac a primary constraint

$$(2.11) \qquad \Phi = \pi + \frac{i}{2}\lambda = 0.$$

The naive Hamiltonian is $H_L = \dot{Q}P - L = \dot{q}p - \frac{1}{2}\dot{q}^2 + \dot{\lambda}\pi - \frac{i}{2}\lambda\dot{\lambda} = \frac{1}{2}p^2$. Here Q and P denote the total set of fields and their canonically conjugates. (We must put $\dot{\lambda}$ in front of π, if we define π by left-differentiation, $\pi = \frac{\partial}{\partial \dot{\lambda}} L$, because only then H_L is independent of $\dot{q}$ and $\dot{\lambda}$. Namely δH_L contains no terms with $\delta \dot{Q}$ but only with δQ and δP.) According to Dirac, one must then consider the naive Hamiltonian plus all possible primary constraints

$$(2.12) \qquad H = \frac{1}{2}p^2 + \alpha \left(\pi + \frac{i}{2}\lambda \right),$$

where $\alpha(t)$ is an arbitrary anticommuting parameter. Requiring that the constraint $\pi + \frac{i}{2}\lambda = 0$ be maintained in time requires $[H, \pi + \frac{i}{2}\lambda] = 0$ modulo the constraints, which is indicated by the symbol $\approx$

$$(2.13) \qquad \left[H, \pi + \frac{i}{2}\lambda \right]_P \approx 0.$$

The subscript P indicates that we use here Poisson brackets. We define the Poisson bracket by

$$(2.14) \qquad \{f(p,q), g(p,q)\}_P = -\partial f/\partial p \frac{\partial}{\partial q} g + (-)^{\sigma} \partial g / \partial p \frac{\partial}{\partial q} f,$$

where $\sigma = +1$ except when both f and g are anticommuting, in which case $\sigma = -1$. The basic relations are $\{p, q\}_P = -1$ and $\{\pi, \lambda\}_P = -1$. Of course $\frac{1}{2}p^2$ commutes

---

[13] In the 4-dimensional Wess-Zumino model there are 2 propagating real scalars (A and B) and a real 4-component spinor. Hence there one needs two real bosonic auxiliary fields (F and G). In the 2-dimensional heterotic string the right-handed spinors λ_R do not transform under rigid supersymmetry, $Q\lambda_R = 0$. It follows that on the right-hand side of the susy commutator evaluated on λ_R the field equation ($\dot{\lambda}$) exactly cancels the translation $P\lambda = \dot{\lambda}$.

with $\pi + \frac{i}{2}\lambda$, but

(2.15) $\quad \{\Phi, \Phi\}_P = \left\{\pi + \frac{i}{2}\lambda, \pi + \frac{i}{2}\lambda\right\}_P = -\frac{i}{2} - \frac{i}{2} = -i.$

Note that the Poisson bracket $\{p, q\}_P$ is -1 for bosons and fermions alike.[14] It follows that $[H, \pi + \frac{i}{2}\lambda] = -i\alpha$, and hence $\alpha = 0$. Thus, with $\alpha = 0$, there are no further (secondary) constraints, and we have

(2.16) $\quad H = \frac{1}{2}p^2.$

Whenever a set of constraints ϕ^α satisfies $\{\phi^\alpha, \phi^\beta\} = M^{\alpha\beta}$ with $\mathrm{sdet} M^{\alpha\beta} \neq 0$, we call these constraints second class constraints.[15] It follows that $\phi = \pi + \frac{i}{2}\lambda$ is a second class constraint. The Dirac bracket is defined by

(2.17) $\quad \{A, B\}_D = \{A, B\}_P - \{A, \Phi\}_P \{\Phi, \Phi\}_P^{-1} \{\Phi, B\}_P,$

where $\{A, B\}_P$ denotes the Poisson bracket. Its definition is chosen such that $\{A, \Phi\}_D = 0$ for any A and any second-class constraint Φ. Since in our toy model $\{\Phi, \Phi\} = -i$, we find

(2.18) $\quad \{A, B\}_D = \{A, B\}_P - i\{A, \pi + \frac{i}{2}\lambda\}_P \{\pi + \frac{i}{2}\lambda, B\}_P.$

We can now compute the basic equal-time Dirac brackets

$$\{\lambda(t), \lambda(t)\}_D = 0 - i(-1)(-1) = -i,$$
$$\{\pi(t), \lambda(t)\}_D = -1 - i\left(\frac{-i}{2}\right)(-1) = -\frac{1}{2},$$
(2.19) $\quad \{\pi(t), \pi(t)\}_D = 0 - i\left(\frac{-i}{2}\right)\left(\frac{-i}{2}\right) = \frac{i}{4}.$

Recalling that $\pi = -\frac{i}{2}\lambda$, we see that these relations are consistent: we may replace π by $-\frac{i}{2}\lambda$ on the left-hand side. At the quantum level, as first proposed by Dirac, we

---

[14]In quantum mechanics the sign of the quantum commutator $[p, q] = -i\hbar$ or the quantum anticommutator $\{\pi, \lambda\} = -i\hbar$ is not a matter of convention but follows from the compatibility of the field equations with the Heisenberg equations. For example, for a Dirac spinor ψ with mass m one has $L = i\psi^\dagger \dot\psi + m\psi^\dagger \psi$ and $\pi = -i\psi^\dagger$. For ψ the field equation is $i\dot\psi + m\psi = 0$, and the Heisenberg equation is $\dot\psi = \frac{i}{\hbar}[H, \psi]$ with $H = -m\psi^\dagger \psi$. Compatibility of the field equation with the Heisenberg equation requires $\{\psi, \pi\} = \{\psi, -i\psi^\dagger\} = -i\hbar$ which agrees with $\{\pi, \psi\} = \frac{\hbar}{i}$ in quantum brackets.

[15]The expression $\mathrm{sdet} M$ denotes the superdeterminant of a supermatrix $M = \begin{pmatrix} A & B \\ C & D \end{pmatrix}$, where A and D contain commuting entries and B and C anticommuting entries. Any matrix can be written as the product of diagonal matrices $\begin{pmatrix} A & 0 \\ 0 & B \end{pmatrix}$ and triangular matrices $\begin{pmatrix} I & C \\ 0 & I \end{pmatrix}$ and $\begin{pmatrix} I & 0 \\ D & I \end{pmatrix}$. Namely, $\begin{pmatrix} A & B \\ C & D \end{pmatrix} = \begin{pmatrix} I & BD^{-1} \\ 0 & I \end{pmatrix} \begin{pmatrix} A - BD^{-1}C & 0 \\ 0 & D \end{pmatrix} \begin{pmatrix} I & 0 \\ D^{-1}C & I \end{pmatrix}$. The superdeterminant of the product of supermatrices is the product of the superdeterminants of these supermatrices, and $\mathrm{sdet} \begin{pmatrix} A & 0 \\ 0 & D \end{pmatrix} = \det A / \det D$ while $\mathrm{sdet} \begin{pmatrix} I & B \\ 0 & I \end{pmatrix} = 1$. Hence $\mathrm{sdet} M = \det(A - BD^{-1}C)/\det D$.

add a factor $i\hbar$ to the Poisson brackets to obtain the quantum (anti) commutators. Hence

$$\{\lambda(t), \lambda(t)\} = \hbar,$$
(2.20)
$$[p(t), \varphi(t)] = \frac{\hbar}{i}.$$

We now construct the susy charge Q as a Noether charge. A Noether charge is the space integral of the time component of the Noether current, but since there is no space in our toy model, the Noether current is equal to the Noether charge. We want to obtain an expression for the Noether charge in terms of p's and q's, and therefore we rewrite (2.1) in Hamiltonian form, namely as $L = \dot{q}p - H$ where H depends only on p, π, φ, λ but not on their time derivatives. Since, as we shall discuss, the terms proportional to a derivative of the symmetry parameter yield the Noether current, the latter will only be a function of p's and q's but not of derivatives of p's and q's.

The action in Hamiltonian form reads

(2.21)
$$L = \dot{\varphi}p + \dot{\lambda}\pi - \frac{1}{2}p^2.$$

where we took the Dirac Hamiltonian $H = \frac{1}{2}p^2$ as discussed above. This action is invariant under

$$\delta\varphi = \frac{i}{2}\epsilon\lambda - \epsilon\pi, \quad \delta\lambda = -p\epsilon,$$
(2.22)
$$\delta p = 0, \quad \delta\pi = \frac{i}{2}p\epsilon.$$

These rules follow by requiring invariance of the action, but one can also derive them by adding equation of motion symmetries to the original rules. For example, $p = \dot{\varphi}$ leads to $\delta p = \frac{i}{2}\epsilon\dot{\lambda} - \epsilon\dot{\pi}$, and to remove $\dot{\lambda}$ one may add $\delta(extra)p = \frac{i}{2}\epsilon\frac{\partial}{\partial\pi}S$ and $\delta(extra)\pi = -\frac{i}{2}\frac{\delta S}{\delta p}\epsilon$. These extra transformation rules form a separate symmetry of any action, so we may add them to the original rules. Then $\delta p = -\epsilon\dot{\pi}$, and also $\delta\pi = \delta(-\frac{i}{2}\lambda) = \frac{i}{2}\dot{\varphi}\epsilon$ is modified into $\delta\pi = \frac{i}{2}\dot{\varphi}\epsilon - \frac{i}{2}(\dot{\varphi} - p)\epsilon = \frac{i}{2}p\epsilon$. To also remove the term $-\epsilon\dot{\pi}$ in δp, one adds another equation of motion symmetry: $\delta(extra)p = \frac{\delta}{\delta\lambda}S\epsilon$ and $\delta(extra)\lambda = \frac{\partial}{\partial p}S$. The final result is (2.22).

The standard way of obtaining the Noether charge follows from letting the rigid parameter become local and collecting terms proportional to $\dot{\epsilon}$. The terms with $\dot{\epsilon}$ in δL for local $\epsilon(t)$ are contained in

$$\delta L = p\frac{d}{dt}\left(\frac{i\epsilon}{2}\lambda - \epsilon\pi\right) - \pi\frac{d}{dt}(-p\epsilon) + \dot{\lambda}\left(\frac{i}{2}p\epsilon\right)$$
(2.23)
$$= \frac{i}{2}p\dot{\epsilon}\lambda - p\dot{\epsilon}\pi + \frac{d}{dt}(\pi p\epsilon).$$

So, defining Q as the coefficient of $i\dot{\epsilon}$, we find

(2.24)
$$Q = p\left(\frac{1}{2}\lambda + i\pi\right).$$

This reproduces (2.22); for example, $\frac{1}{\hbar}[\varphi, \epsilon Q] = \delta\varphi$ follows from (2.20). Using the constraint $\pi = -\frac{i}{2}\lambda$ we see that Q becomes equal to $Q = p\lambda$, and (2.22) reduces to (2.6). However, if one uses (anti) commutators one needs a Hamiltonian treatment, and then one needs (2.24).

The charge Q which appears in brackets such as $[\varphi, \epsilon Q]$ is clearly an operator, so Q in (2.24) is an operator expressed in terms of Heisenberg fields. The latter satisfy their own equations of motion. On the other hand, in the action the fields are off-shell. So, in principle one might need extra terms proportional to the equations of motion to obtain the correct off-shell transformations. In this case we do not need such terms. In section 6 we shall discuss the Hamiltonian approach with off-shell fields. This is a very general approach which yields the action in Hamiltonian form and the quantum BRST charge, starting only from the set of first class constraints.

The other way of obtaining Q (more precisely, $i\epsilon Q$) is to write it as a sum of terms of the form $\delta\varphi p$ for all fields, plus $-K$ where $\delta L = \frac{d}{dt}K$. From (2.23) we read off that $K = \pi p \epsilon$. Hence

$$i\epsilon Q = \delta\varphi p + \delta\lambda\pi - K = \left(\frac{i}{2}\epsilon\lambda - \epsilon\pi\right)p - p\epsilon\pi - \pi p\epsilon$$

(2.25)
$$= i\epsilon\left(\frac{1}{2}\lambda + i\pi\right)p,$$

which is indeed the same result as in (2.24).

The supersymmetry algebra (rather a superalgebra with commutators and anticommutators) is now easy to evaluate. Using the quantum brackets of (2.20) finds

$$\{Q,Q\} = \left\{\left(\frac{1}{2}\lambda + i\pi\right)p, \left(\frac{1}{2}\lambda + i\pi\right)p\right\} = \{\lambda p, \lambda p\} = \hbar p^2 = 2\hbar H,$$

(2.26) $\quad [H,Q] = 0 \quad$ (via Jacobi, or directly).

Thus the generators Q and H form a closed superalgebra, and supersymmetry is the square root of (the generator of time-) translations.

Dirac was the first to take the square root of the Laplace operator $\Box$, and this led to the famous Dirac equation of 1927. This equation led to the prediction that for every fermionic particle there is a fermionic antiparticle. These antiparticles have been found in the laboratories. Likewise, the square root $Q \sim \sqrt{H}$ predicts that for every particle there should be a superpartner. Not a single superpartner has been found so far, but that may change.

3. $N = 1$ supergravity in x-space

Having discussed the rigid supersymmetry (= susy) of the action S (rigid) $= \int L dt$ with $L = \frac{1}{2}\dot\varphi^2 + \frac{i}{2}\lambda\dot\lambda$, we turn to local susy. We let ϵ become time-dependent and find then (see (2.7)).

(3.1) $$\delta S \text{ (rigid)} = \int_{-\infty}^{\infty} \dot\epsilon(i\dot\varphi\lambda)dt.$$

The boundary terms at $t = \pm\infty$ vanish if we require that $\epsilon(t)$ vanishes at $t = \pm\infty$. To cancel this variation we introduce the gauge field for local susy, the gravitino $\psi(t)$. The transformation rule of a gauge field begins always with a derivative of the local parameter. We then couple ψ to the Noether current of rigid supersymmetry, using $\delta\psi = \dot{\epsilon} + \cdots$ to fix the overall constant of this new term

$$
(3.2) \qquad S\,(\text{Noether}) = \int_{-\infty}^{\infty} (-i\psi\dot{\varphi}\lambda)dt;\ \delta\psi = \dot{\epsilon} + \cdots
$$

If we vary ψ in (3.2), then the variation δS (Noether) cancels δS (rigid), but the fields $\dot{\varphi}$ and λ in S (Noether) must also be varied. This yields two further variations

$$
(3.3) \qquad \delta S\,(\text{Noether}) = \int_{-\infty}^{\infty} \left[-i\psi \left\{ \frac{d}{dt}(i\epsilon\lambda) \right\} \lambda + i\psi\dot{\varphi}\epsilon\dot{\varphi} \right] dt.
$$

In the first variation the $\frac{d}{dt}$ must hit the field λ because otherwise one would be left with $\lambda\lambda$ which vanishes. Hence the remaining variations to be canceled are

$$
(3.4) \qquad \delta S\,(\text{Noether}) = \int_{-\infty}^{\infty} i\psi(\dot{\varphi}\dot{\varphi} + i\lambda\dot{\lambda})\epsilon dt.
$$

The last term can be canceled by adding a new term in $\delta\lambda$ (because this variation is porportional to the field equation of λ). However, the first term can only be canceled by introducing a new field h (the graviton) and coupling it to $\dot{\varphi}\dot{\varphi}$. Thus the coupling of rigidly supersymmetric matter to the supergravity gauge fields requires for consistency (invariance of the whole action under local susy) also the coupling to gravity. *Local susy is a theory of gravity, and this explains the name supergravity.*

There appears, however, an ambiguity at this point: we can also couple this new field h to $-i\lambda\dot{\lambda}$, and the most general case is a linear combinations of both possibilities. Hence we add

$$
S\,(\text{stress}) = -\int_{-\infty}^{\infty} h[\dot{\varphi}\dot{\varphi} - i\lambda\dot{\lambda}x]dt,
$$

$$
(3.5) \qquad \delta h = -i\epsilon\psi, \quad \delta\lambda = -\dot{\varphi}\epsilon + i(1+x)\psi\lambda\epsilon,
$$

where x is a free constant real parameter. We must now evaluate the old variations in the new action, the new variations in the old action, and the new variations in the new action. The aim is to use these variations to cancel (3.4).

The new variation $\delta h = -i\epsilon\psi$ in the new action $-\int h\dot{\varphi}\dot{\varphi}dt$ cancels the first term in (3.4). The new variation $\delta\lambda = i(1+x)\psi\lambda\epsilon$ in the old action $\frac{i}{2}\lambda\dot{\lambda}$ in (2.1) and the new variation $\delta h = -i\epsilon\psi$ in the new action S (stress) cancel the variation $-\psi\lambda\dot{\lambda}\epsilon$ of δS (Noether)

$$
i\lambda\frac{d}{dt}[i(1+x)\psi\lambda\epsilon] - i\epsilon\psi(i\lambda\dot{\lambda}x) - \psi\lambda\dot{\lambda}\epsilon
$$
$$
(3.6) \qquad = (1+x)\lambda\dot{\lambda}\psi\epsilon - \psi\epsilon\lambda\dot{\lambda}x - \psi\epsilon\lambda\dot{\lambda} = 0.
$$

(We partially integrated the first term, and the last term comes from (3.4)). We find thus at this moment a free parameter x in the action and transformation rules; this frequently happens in the construction of supergravity models, and usually these parameters get fixed at a later stage, or they can be removed by field redefinitions. We demonstrate this later explicitly in our toy model.

We are again in the same situation as before: we canceled δS (Noether) by introducing a new term in the action, namely S (stress). We already took into account the variation of the gauge field h in this new term, but we must still vary the matter fields in S (stress). We first set $x = 0$ and later consider the case $x \neq 0$. If $x = 0$, we need only vary the $\dot\varphi$ in S (stress) and this yields

$$\delta L \text{ (stress)} = -2h\dot\varphi \frac{d}{dt}(i\epsilon\lambda) = -2h\dot\varphi i\dot\epsilon\lambda - 2h\dot\varphi i\epsilon\dot\lambda. \tag{3.7}$$

The first term is proportional to the Noether current $\dot\varphi\lambda$ in (3.2) and can thus be canceled by a new term in the gravitino law

$$\delta \text{ (new) } \psi = -2h\dot\epsilon. \tag{3.8}$$

Substituting this vatiation into (3.2), the first term in (3.7) is cancelled. The second term in (3.7) is proportional to the free field equation of λ and can be canceled by adding a new term to the transformation law of $\lambda, \delta\lambda = 2h\dot\varphi\epsilon$, because then $\delta(\frac{i}{2}\lambda\dot\lambda) = -i\dot\lambda(2h\dot\varphi\epsilon)$ cancels the second term in (3.7).

The new transformation law $\delta\lambda = 2h\dot\varphi\epsilon$ produces a new variation in the Noether action (3.2)

$$\delta L \text{ (Noether due to } \delta\lambda = 2h\dot\varphi\epsilon) = -i\psi\dot\varphi(2h\dot\varphi\epsilon). \tag{3.9}$$

Since this term is proportional to $\dot\varphi\dot\varphi$ it can be canceled by a final extra term in δh, namely $\delta h = 2ih\epsilon\psi$. Then the new variation of h used in (3.5) cancels (3.9).

Because each time when we replace an $\dot\epsilon$ in a variation by ψ we loose a time derivative, this process of adding further terms to the action and transformation laws is guaranteed to stop. Of course it is not guaranteed that an invariant action exists. Examples are known where this process does not yield an invariant action, for example adding a cosmological constant to supergravity in $10 + 1$ dimensions.

We have now canceled all variations for the case $x = 0$, hence we have constructed a locally susy action. The final results read

$$L = \frac{1}{2}\dot\varphi^2 + \frac{i}{2}\lambda\dot\lambda - i\psi\dot\varphi\lambda - h\dot\varphi^2,$$
$$\delta\varphi = i\epsilon\lambda, \quad \delta\lambda = -\dot\varphi\epsilon + i\psi\lambda\epsilon + 2h\dot\varphi\epsilon,$$
$$\delta\psi = \dot\epsilon - 2h\dot\epsilon, \quad \delta h = -i\epsilon\psi + 2ih\epsilon\psi. \tag{3.10}$$

Before going on, we make three comments.

1) There is no gauge action for gravity or local supersymmetry in one dimension as one might expect, since the scalar curvature R and its linearization, the Fierz-Pauli action,[16] vanishes in one dimension. (Also the gravitino gauge action vanishes in one and $1 + 1$ dimensions. The gravitational action is a total derivative in $1 + 1$ dimensions, where it yields the Euler invariant). A gauge action for supergravity in one dimension would have to start with $L = \frac{1}{2}h\dot h + \frac{i}{2}\psi\dot\psi$ and it is indeed invariant under the rigid symmetries $\delta h = i\epsilon\psi, \delta\psi = -h\epsilon$, see (2.1) and (2.6). However, for local $\epsilon(t)$ the rules were already fixed by the matter coupling, see (3.10), and these rules do not leave this action invariant.

---

[16]The linearized form of the Einstein-Hilbert action eR is called the Fierz-Pauli action and is given in n-dimensions by

$$L = -\frac{1}{2}\varphi_{\mu\nu,\lambda}^2 + \varphi_\mu^2 - \varphi_\mu\varphi_{,\mu} + \frac{1}{2}\varphi_{,\mu}^2,$$

where $\varphi_\mu = \partial^\nu\varphi_{\mu\nu}$, $\varphi = \eta^{\mu\nu}\varphi_{\mu\nu}$ and the metric $g_{\mu\nu} \equiv \eta_{\mu\nu} + \kappa h_{\mu\nu}$ is related to $\varphi_{\mu\nu}$ by $(\sqrt{-g}g^{\mu\nu})_{\text{lin}} - \eta^{\mu\nu} = -\kappa(h^{\mu\nu} - \frac{1}{2}\eta^{\mu\nu}h) = \varphi^{\mu\nu}$. In one dimension the first term in L cancels the last term, and the second term cancels the third term.

2) The term $\int g\varphi^n dt$ with g a coupling constant cannot be made supersymmetric. Yukawa couplings do not exist in this model because $\lambda\varphi\lambda$ vanishes. However, one can make λ a complex Dirac spinor and then supersymmetric interactions exist. One can also supersymmetrize a term $f(\varphi)\dot\varphi\dot\varphi$; the action becomes $f(\varphi)(\dot\varphi\dot\varphi + i\lambda\dot\lambda)$ and is called a susy nonlinear σ model because $f(\varphi)$ can be nonlinear, for example $\exp\varphi$.

3) One can also couple the first-order action in (2.21) to supergravity. Denoting the graviton and gravitino fields by H and Ψ, the result is

$$L = \dot\varphi p + \dot\lambda\pi - \frac{1}{2}p^2 - i\Psi\left(\frac{1}{2}p\lambda + ip\pi\right) - Hp^2,$$

$$\delta p = 0, \quad \delta\pi = \frac{i}{2}p\epsilon, \quad \delta\lambda = -p\epsilon,$$

(3.11) $$\delta\varphi = \frac{i}{2}\epsilon\lambda - \epsilon\pi, \quad \delta\Psi = \dot\epsilon, \quad \delta H = -i\epsilon\Psi.$$

Note that $\delta(\pi + \frac{i}{2}\lambda) = 0$ agrees with the constraint $\pi + \frac{i}{2}\lambda = 0$. So we may replace π by $-\frac{i}{2}\lambda$ in the action and transformation rules. Furthermore we can eliminate p by integrating in the path integral over a Gaussian with p^2 [**2**]. The result of these manipulations is the following action

(3.12) $$L = \frac{1}{2}\frac{1}{1+2H}\dot\varphi^2 - \frac{i}{1+2H}\dot\varphi\Psi\lambda + \frac{i}{2}\lambda\dot\lambda.$$

Comparison with the second-order action in (3.10) (the action without conjugate momenta) we can read off how H is related to h, and Ψ to ψ.

(3.13) $H = \dfrac{h}{1-2h}, \quad \Psi = \dfrac{1}{1-2h}\psi,$ or $h = \dfrac{H}{1+2H}, \quad \psi = \dfrac{1}{1+2H}\Psi.$

The Jacobian for the change of variables from (H, Ψ) to (h, ψ) is $\frac{1}{1-2h} = 1 + 2H$. (One needs a super Jacobian, in particular $\partial\delta\psi/\partial\psi$ is equal to $1 - 2h$, and not simply equal to $(1-2h)^{-1}$). The transformation rules in (3.11) go over into (3.10) if one uses these redefinitions.

For physicists: The Jacobian $J = 1/1 - 2h$ can be exponentiated using a new kind of ghosts, introduced by Bastianelli and the author and playing the same role as the Faddeev-Popov ghosts. In order that the theories with h and ψ and H and Ψ are equivalent at the quantum level (by which we mean that they should give the same Feynman graphs) one needs those new ghosts. The propagators of h and ψ come from the gauge fixing terms.

We now return to the model in (3.10) and evaluate the local susy algebra. On φ one finds

$$[\delta(\epsilon_2), \delta(\epsilon_1)]\varphi = i\epsilon_1(-\dot\varphi\epsilon_2 + i\psi\lambda\epsilon_2 + 2h\dot\varphi\epsilon_2) - i\epsilon_2(-\dot\varphi\epsilon_1 + i\psi\lambda\epsilon_1 + 2h\dot\varphi\epsilon_1)$$

(3.14) $$= [2i(1-2h)\epsilon_2\epsilon_1]\dot\varphi + i[-2i\epsilon_2\epsilon_1\psi]\lambda.$$

The right-hand side contains a general coordinate transformation $\delta\varphi = \hat\xi\dot\varphi$ with $\hat\xi = 2i(1-2h)\epsilon_2\epsilon_1$; this is clearly the gravitational extension of the nongravitational rigid translation with parameter $\xi = 2i\epsilon_2\epsilon_1$ which we found in the rigid susy commutator. The second term is a local susy transformation $i\hat\epsilon\lambda$ of φ with parameter $\hat\epsilon = -2i\epsilon_2\epsilon_1\psi$. Note that the composite parameters $\hat\xi$ and $\hat\epsilon$ are field-dependent. The structure constants are no longer constants! This has led to a new development in group theory.

On λ one finds after somewhat lengthy algebra

$$[\delta(\epsilon_2),\delta(\epsilon_1)]\lambda = \hat{\xi}\dot{\lambda} - \dot{\varphi}\hat{\epsilon} + 2h\dot{\varphi}\hat{\epsilon}. \tag{3.15}$$

and the terms with $\hat{\epsilon}$ agree with (3.10) (because $i\psi\lambda\hat{\epsilon} = 0$ due to $\psi\psi = 0$). The term with $\hat{\xi}$ constitutes a general coordinate transformation on λ. Clearly, the same algebra is found on λ as on ψ!

On ψ one finds

$$\begin{aligned}[\delta(\epsilon_2),\delta(\epsilon_1)]\psi &= -2(-i\epsilon_2\psi + 2ih\epsilon_2\psi)\dot{\epsilon}_1 + 2(-i\epsilon_1\psi + 2ih\epsilon_1\psi)\dot{\epsilon}_2 \\ &= -2i\left[\frac{d}{dt}(\epsilon_2\epsilon_1)\right]\psi + 4ih\left[\frac{d}{dt}(\epsilon_2\epsilon_1)\right]\psi \\ &= \frac{d}{dt}\hat{\epsilon} + 2i\epsilon_2\epsilon_1\dot{\psi} - 2h\frac{d}{dt}\hat{\epsilon} - 4ih\epsilon_2\epsilon_1\dot{\psi} \\ &= \frac{d}{dt}\hat{\epsilon} - 2h\frac{d}{dt}\hat{\epsilon} + \hat{\xi}\dot{\psi}. \end{aligned} \tag{3.16}$$

So also on ψ the same local algebra is realized.

Finally also on h the same algebra is realized[17]

$$[\delta(\epsilon_2),\delta(\epsilon_1)]h = -i\epsilon_1(\dot{\epsilon}_2 - 2h\dot{\epsilon}_2) - 2i[\delta(\epsilon_2)h\psi]\epsilon_1 - (1 \leftrightarrow 2)$$
$$= -i\epsilon_1(\dot{\epsilon}_2 - 2h\dot{\epsilon}_2) - 2i(-i\epsilon_2\psi + 2ih\epsilon_2\psi)\psi\epsilon_1 - 2ih(\dot{\epsilon}_2 - 2h\dot{\epsilon}_2)\epsilon_1 - (1 \leftrightarrow 2)$$
$$= -i\hat{\epsilon}\psi + 2ih\hat{\epsilon}\psi + \hat{\xi}\dot{h} - \dot{\hat{\xi}}h + \frac{1}{2}\dot{\hat{\xi}}. \tag{3.17}$$

The terms with $\hat{\epsilon}$ clearly agree with (3.10). The terms with $\hat{\xi}$ in (3.17) yield a general coordinate transformation of h as we shall discuss below. Hence, the local susy algebra closes on all fields uniformly. We can write this as

$$[\delta_s(\epsilon_2),\delta_s(\epsilon_1)] = \delta_s(-2i\epsilon_2\epsilon_1\psi) + \delta_g((1-2h)2i\epsilon_2\epsilon_1). \tag{3.18}$$

The commutator of δ_s with δ_g, and with itself, close (they are proportional to δ_s and δ_g).

We can understand the closure of the local supersymmetry algebra by using the same argument as used for the rigid supersymmetry algebra. There is one bosonic gauge field component (h) and one fermionic gauge field component (ψ), hence no auxiliary fields in the gauge sector are needed.

We now consider the case $x \neq 0$. Here a simple argument suffices. Rescaling

$$\lambda = (1+2hx)^{1/2}\tilde{\lambda}, \tag{3.19}$$

we obtain as action from (3.10)

$$\mathcal{L} = \frac{1}{2}\dot{\varphi}^2 + \frac{i}{2}(1+2hx)\tilde{\lambda}\frac{d}{dt}\tilde{\lambda} - i\tilde{\psi}\dot{\varphi}\tilde{\lambda} - h\dot{\varphi}^2, \tag{3.20}$$

where we also rescaled ψ according to

$$\psi(1+2hx)^{1/2} = \tilde{\psi}. \tag{3.21}$$

We have produced the action with $x \neq 0$ from the action with $x = 0$ by a simple rescaling of the fields λ and ψ. It follows that this action is also locally susy. The precise susy transformation rules follow from this rescaling

$$\delta\lambda = (1+2hx)^{1/2}\delta\tilde{\lambda} + (1+2hx)^{-1/2}x\delta h\tilde{\lambda}$$
$$= -\dot{\varphi}\epsilon + i\tilde{\psi}\tilde{\lambda}\epsilon + 2h\dot{\varphi}\epsilon. \tag{3.22}$$

---

[17] Here and below "1 ↔ 2" means "switch indices 1 and 2 in the preceding expression."

By dividing (3.22) by $(1+2hx)^{1/2}$ and using $\delta h = -i\epsilon\psi + 2ih\epsilon\psi$ we find $\delta\tilde\lambda$. It reads

(3.23) $\qquad \delta\tilde\lambda = -\dot\varphi\tilde\epsilon + i\tilde\psi\tilde\lambda\tilde\epsilon + 2h\dot\varphi\tilde\epsilon - x(1+2hx)^{-1}(-i\tilde\epsilon\tilde\psi + 2ih\tilde\epsilon\tilde\psi)\tilde\lambda,$

where we also rescaled ϵ according to $(1+2hx)^{-1/2}\epsilon = \tilde\epsilon$. (We used $\epsilon\psi = \tilde\epsilon\tilde\psi$ in δh). By rearranging these terms as

$$\delta\tilde\lambda = -(1-2h)\dot\varphi\tilde\epsilon + i\left(1 + \frac{x-2hx}{1+2hx}\right)\tilde\psi\tilde\lambda\tilde\epsilon$$

(3.24) $\qquad\qquad = -(1-2h)\dot\varphi\tilde\epsilon + i\left(\frac{1+x}{1+2hx}\right)\tilde\psi\tilde\lambda\tilde\epsilon,$

we find a polynomial result for $x=-1$

(3.25) $\qquad \delta\tilde\lambda = -(1-2h)\dot\varphi\tilde\epsilon, \quad \delta\varphi = i\tilde\epsilon(1-2h)\tilde\lambda.$

For δh and $\delta\tilde\psi$ we find in a similar manner

$$\delta h = -i\tilde\epsilon\tilde\psi + 2ih\tilde\epsilon\tilde\psi, \delta\tilde\psi = \sqrt{(1-2h)}\delta\psi = \sqrt{(1-2h)}(1-2h)\dot\epsilon$$

(3.26) $\qquad\qquad = (1-2h)[(1-2h)\dot{\tilde\epsilon} - \dot h\tilde\epsilon].$

We used that $\delta h\psi$ vanishes. Thus for $x = -1$ all laws become again polynomial.

We can, in fact, apply the Noether method also directly to gravity. This will explain the $\hat\xi$ terms in (3.17). Starting with

(3.27) $\qquad\qquad L\text{ (rigid)} = \frac{1}{2}\dot\varphi^2 + \frac{i}{2}\lambda\dot\lambda,$

we find from the translation symmetry rules $\delta\varphi = \xi\dot\varphi$ and $\delta\lambda = \xi\dot\lambda$ for local ξ

(3.28) $\qquad \delta S\text{ (rigid)} = \int\left[\dot\varphi\frac{d}{dt}(\xi\dot\varphi) + \frac{i}{2}\lambda\frac{d}{dt}(\xi\dot\lambda) + \frac{i}{2}\xi\dot\lambda\dot\lambda\right]dt.$

The third term vanishes as $\dot\lambda\dot\lambda = 0$, and the second term vanishes after partial integration, whereas the first term yields after partial integration of one-half of this term (this $\frac{1}{2} - \frac{1}{2}$ trick cancels the $\ddot\varphi$ terms)

(3.29) $\qquad \delta S\text{ (rigid)} = \int\left[\frac{1}{2}\dot\xi\dot\varphi\dot\varphi + \frac{d}{dt}\left(\frac{1}{2}\xi\dot\varphi\dot\varphi + \frac{i}{2}\xi\lambda\dot\lambda\right)\right]dt.$

We see that $\dot\varphi\dot\varphi$ is the Noether current for translations for the (φ, λ) system. (Also the other way to construct the Noether current gives the same result).

Introducing the gauge field h for gravity and defining $\delta h = \frac{1}{2}\dot\xi + \cdots$ we obtain

(3.30) $\qquad\qquad L\text{ (Noether)} = -h\dot\varphi\dot\varphi, \quad \delta h = \frac{1}{2}\dot\xi.$

Note that there is no term of the form $h\lambda\dot\lambda$ in L (Noether). (However the field redefinition (3.19) produces a $h\lambda\dot\lambda$ term in the action in (3.20)). As Noether current for λ one might have expected $\pi\dot\lambda$ where π is the conjugate momentum of λ, but $\dot\lambda$ vanishes on-shell, and anyhow time-derivatives of fields should not appear in the Hamiltonian formalism.

Variation of $\dot\varphi$ in L (Noether) yields

(3.31) $\qquad\qquad \delta S\text{ (Noether)} = \int -2h\dot\varphi\frac{d}{dt}(\xi\dot\varphi)dt,$

which can be canceled, using the $\frac{1}{2} - \frac{1}{2}$ trick, by $\delta h = \xi \dot h - h \dot \xi$. We could also have proceeded in another way: if we would partially integrate we would obtain $2 \dot h \dot \varphi \xi \dot \varphi + 2h \ddot \varphi \xi \dot \varphi$, and the first variation could be removed by $\delta h = 2\xi \dot h$ while the second variation could be eliminated by a suitable $\delta \varphi = 2\xi h \dot \varphi$ in $\frac{1}{2} \dot \varphi^2$. However, the variation $\delta \varphi = 2\xi h \dot \varphi$ produces in S (Noether) with the $\frac{1}{2} - \frac{1}{2}$ trick a new variation $-2\dot \xi h \dot \varphi h \dot \varphi$, which can be canceled in two ways, etc. All these ambiguities (and more) are equivalent to field redefinitions.

We have thus shown that

$$(3.32) \qquad L = \frac{1}{2} \dot \varphi^2 + \frac{i}{2} \lambda \dot \lambda - h \dot \varphi^2$$

is invariant under

$$(3.33) \qquad \delta \varphi = \xi \dot \varphi, \delta \lambda = \xi \dot \lambda, \delta h = \frac{1}{2} \dot \xi + \xi \dot h - \dot \xi h.$$

In particular the result for δh shows that the $\hat \xi$ dependent terms in (3.17) are indeed a general coordinate transformation.

In higher dimension the coupling of a scalar field to gravity is given by

$$L = -\frac{1}{2} (-\det g)^{\frac{1}{2}} g^{\mu \nu} \partial_\mu \varphi \partial_\nu \varphi,$$

so in one dimension e_m^μ corresponds to $1 - 2h$. Then (3.33) agrees with the usual transformation rules of general relativity.

Adding the coupling to the gravitino $-i\psi \dot \varphi \lambda$, we obtain invariance under ξ transformations, provided we choose $\delta \psi$ appropriately,

$$(3.34) \qquad \delta(-i\psi \dot \varphi \lambda) = -i \delta \psi \dot \varphi \lambda - i\psi \left[\frac{d}{dt}(\xi \dot \varphi)\right] \lambda - i\psi \dot \varphi \xi \dot \lambda.$$

Partially integrating the last two terms, we obtain

$$(3.35) \qquad -i\delta \psi \dot \varphi \lambda + i \dot \psi \xi \dot \varphi \lambda,$$

and these two terms can be canceled by choosing $\delta \psi = \xi \dot \psi$. In general, Lagrangian densities in general relativity transform as coordinate densities, which means that

$$(3.36) \qquad \delta(-i\psi \dot \varphi \lambda) = \frac{d}{dt}(-i\xi \psi \dot \varphi \lambda).$$

This is indeed achieved if ψ transforms like

$$(3.37) \qquad \delta \psi = \xi \dot \psi.$$

Thus the model in (3.10) has now also been shown to be invariant under general coordinate transformation. The result in (3.37) can be explained by noting that the field ψ in the Noether coupling in (3.2) corresponds to $(\det g)^{\frac{1}{2}} g^{\mu \nu} \psi_\nu \sim e_m{}^\mu \psi_\mu$ in one dimension. This field $e_m^\mu \psi_\mu$ is a coordinate scalar, in agreement with (3.37).

Let us again rescale $\lambda = \sqrt{1 + 2hx} \tilde \lambda$ to obtain an action with a term $h \tilde \lambda \dot{\tilde \lambda}$. We obtain from (3.32)

$$L = \frac{1}{2} \dot \varphi^2 + \frac{i}{2} \tilde \lambda \dot{\tilde \lambda} - h\left(\dot \varphi^2 - ix\tilde \lambda \frac{d}{dt}\tilde \lambda\right),$$

$$(3.38) \qquad \delta \tilde \lambda = \frac{\delta \lambda}{\sqrt{1 + 2hx}} - \frac{\delta hx\lambda}{(1+2hx)^{3/2}} = \xi \frac{d}{dt} \tilde \lambda - \frac{1}{2} x \dot \xi \left(\frac{1 - 2h}{1 + 2hx}\right) \tilde \lambda.$$

For $x = -1$ we find again a polynomial result $\delta\tilde{\lambda} = \xi\dot{\tilde{\lambda}} + \frac{1}{2}\dot{\xi}\tilde{\lambda}$. The second term indicates that $\tilde{\lambda}$ is a half-density, in agreement with our earlier observation that $1 - 2h = e_m{}^\mu = \det e_m{}^\mu$ is a density, and $\tilde{\lambda} = \sqrt{1-2h}\lambda$ is a half-density.

We already saw that it is natural to rescale ψ as in (3.21). After the rescaling $\psi\sqrt{1+2hx} = \tilde{\psi}$ according to which $-i\psi\dot{\varphi}\lambda = -i\tilde{\psi}\dot{\varphi}\tilde{\lambda}$, the field $\tilde{\psi}$ transforms for $x = -1$ as

(3.39) $$\delta\tilde{\psi} = \xi\frac{d}{dt}\tilde{\psi} - \frac{1}{2}\dot{\xi}\tilde{\psi}.$$

Collecting all results for $x = -1$, and dropping the tildas, we have found the following action and transformation rules for general coordinate transformations

$$L = \frac{1}{2}\dot{\varphi}^2 + \frac{i}{2}\lambda\dot{\lambda} - h(\dot{\varphi}^2 + i\lambda\dot{\lambda}) - i\psi\dot{\varphi}\lambda,$$

(3.40) $\quad \delta\varphi = \xi\dot{\varphi}, \quad \delta\lambda = \xi\dot{\lambda} + \frac{1}{2}\dot{\xi}\lambda, \quad \delta h = \frac{1}{2}\dot{\xi} + \xi\dot{h} - \dot{\xi}h, \quad \delta\psi = \xi\dot{\psi} - \frac{1}{2}\dot{\xi}\psi.$

The susy transformation rules for this model were given in (3.25) and (3.26),

$$\delta\varphi = i\epsilon(1-2h)\lambda, \quad \delta\lambda = -(1-2h)\dot{\varphi}\epsilon,$$
(3.41)
$$\delta\psi = (1-2h)[(1-2h)\dot{\epsilon} - \dot{h}\epsilon], \quad \delta h = -(1-2h)i\epsilon\psi.$$

Hence we have found two polynomial formulations of this supergravity model, one without a coupling $h\lambda\dot{\lambda}$ in (3.10), and one with it in (3.40). If one varies the susy Noether current $\dot{\varphi}\lambda$ in flat space under rigid susy, one finds

(3.42) $$\delta(\epsilon)\dot{\varphi}\lambda = \frac{d}{dt}(i\epsilon\lambda)\lambda - \dot{\varphi}\dot{\varphi}\epsilon = -(\dot{\varphi}\dot{\varphi} + i\lambda\dot{\lambda})\epsilon.$$

This is the current which couples to h in the model with a $h\lambda\dot{\lambda}$ coupling. The action of this model can suggestively be written as

(3.43) $$L = \frac{1}{2}(1-2h)(\dot{\varphi}^2 + i\lambda\dot{\lambda}) - i\psi\dot{\varphi}\lambda.$$

It is this formulation of the model which is easiest to write in superspace.

4. Rigid and local $N = 1$ superspace

We now turn to the superspace description. The coordinates of the superspace for our toy model are t and θ; both are real, and θ is a Grassmann variable. A superfield Φ depends on t and θ, but a Taylor expansion of Φ in terms of θ contains only two terms because $\theta\theta = 0$. We begin with

(4.1) $$\Phi = \varphi + i\theta\lambda,$$

where φ and λ are arbitrary functions of t. Since φ is real, also Φ is real, and this requires the factor i. Susy transformations in flat space are generated by the hermitian[18] generator Q

(4.2) $$Q = \frac{\partial}{\partial\theta} + i\theta\frac{\partial}{\partial t},$$

---

[18]To show that Q is hermitian, note that from the anticommutator $\{\frac{\partial}{\partial\theta},\theta\} = 1$ if follows that $\partial/\partial\theta$ is hermitian. Likewise, it follows from the commutator $[\frac{\partial}{\partial t},t] = 1$ that $\partial/\partial t$ is antihermitian.

because

$$\epsilon Q\Phi = \epsilon\left(\frac{\partial}{\partial\theta} + i\theta\frac{\partial}{\partial t}\right)(\varphi + i\theta\lambda) = i\epsilon\lambda + i\theta(-\epsilon\dot\varphi)$$
(4.3)
$$= \delta\varphi + i\theta\delta\lambda.$$

The results for $\delta\varphi$ and $\delta\lambda$ agree with (2.6).

The susy covariant derivative, by definition, anticommutes with Q, $\{Q, D\} = 0$, and this determines $D = \frac{\partial}{\partial\theta} - i\theta\frac{\partial}{\partial t}$. The action can be written as

$$S = \frac{i}{2}\int dt d\theta (\partial_t\Phi)(D\Phi)$$
(4.4)
$$= \frac{i}{2}\int dt d\theta (\dot\varphi + i\theta\dot\lambda)(i\lambda - i\theta\dot\varphi) = \frac{1}{2}\int[\dot\varphi\dot\varphi + i\lambda\dot\lambda]dt.$$

We used $\int d\theta\theta = 1$ and $\int d\theta c = 0$ if c is independent of θ. This definition of integration is called the Berezin integral [1] and it follows from translational invariance in θ. Namely requiring that $\int d\theta f(\theta) = \int d\theta f(\theta + c)$, and using that $f(\theta) = a + b\theta$ because $\theta\theta = 0$, one finds $\int d\theta c = 0$. We normalize θ such that $\int d\theta\theta = 1$.

The susy invariance of the action follows from $\delta\Phi = [\epsilon Q, \Phi]$, $\partial_t\epsilon Q = \epsilon Q\partial_t$ and $D\epsilon Q = \epsilon Q D$. One gets then for the variation of the action an expression of the form $\delta S = \frac{i}{2}\int dt d\theta Q[\ldots]$. The $\frac{\partial}{\partial\theta}$ in Q yields zero because $\partial_\theta[\ldots]$ contains no θ and $\int d\theta c = 0$. The $i\theta\frac{\partial}{\partial t}$ in Q yields zero because of the total t-derivative. This shows that any action built from superfields and derivatives $\frac{\partial}{\partial t}$ and D is always supersymmetric.

After performing the θ integration, we have obtained the correct t-space action. We can also show before the θ integration that $\partial_t\phi D\phi$ is the correct Lagrangian density by checking that it has the correct dimension: the dimension of ϕ is that of φ, so $[\phi] = -1/2$, and $[dt] = -1$ and $d\theta = 1/2$. The action should be dimensionless, so we need derivatives acting on ϕ with dimension $+3/2$. The only derivatives we have available are $\frac{\partial}{\partial t}$ and D, so the Lagrangian density is unique. If one considers generalized unitary group elements $e^{\theta Q + itH}$, one finds by left-multiplication the vielbein fields

$$e^{\epsilon Q + iaH}e^{\theta Q + itH} = e^{\epsilon Q + iaH + \theta Q + itH + \frac{1}{2}[\epsilon Q, \theta Q]}$$
(4.5)
$$\equiv \exp[\{z^\Lambda + dz^M E_M{}^\Lambda(Z)\}]T_\Lambda.$$

We used the Baker-Campbell-Hansdorff formua, and $z^\Lambda = (t, \theta), T_\Lambda = (H, Q)$ and $dz^M = (a, \epsilon)$. Using $[\epsilon Q, \theta Q] = -\epsilon\{Q, Q\}\theta = -2\epsilon H\theta$ we find in the exponent $(\theta + \epsilon)Q + i(t + a + i\epsilon\theta)H$. Thus $\delta(\epsilon)\theta = \epsilon$ and $\delta(\epsilon)t = i\epsilon\theta$. This is a nonlinear representation of the susy algebra

(4.6) $$\{Q, Q\} = 2H, \ [Q, H] = 0, \text{ and } [H, H] = 0.$$

in terms of coordinates. The field representation in (2.6) is induced by this coordinate representation, namely $\Phi'(t', \theta') = \Phi(t, \theta)$.[19]

One can repeat the same procedure as in (4.5) but now using right multiplication. Right multiplication yields D, and now we understand why $\epsilon_1 D$ and $\epsilon_2 Q$ commute: because left and right multiplications commute.

---

[19]Actually, one finds in this way minus the result of (2.6). To get (2.6), we would have replaced ϵ (and a) in (4.5) by $-\epsilon$ (and $-a$). In general, coordinates transform contragradiently (opposite) to fields.

In 4 dimensions it is better to define x-space components by D derivatives of Φ at $\theta = 0$. Here this makes no difference since there are only two components in Φ

$$\Phi\big|_{\theta=0} = \varphi(t), \quad D\Phi\big|_{\theta=0} = i\lambda(t). \tag{4.7}$$

Since $\int dt \int d\theta$ can be replaced by $\int dt D$ (recall that $\int d\theta = \frac{\partial}{\partial \theta}$ and $\int dt i\theta \frac{\partial}{\partial t} L = 0$ because it is a total t-derivative) we get

$$\begin{aligned} S &= \frac{i}{2} \int dt [(\partial_t D\Phi) D\Phi + \partial_t \Phi DD\Phi] \\ &= \frac{i}{2} \int dt [(i\dot{\lambda})(i\lambda) - \dot{\varphi} i\dot{\varphi}]. \end{aligned} \tag{4.8}$$

The extended susy algebra of Q, D and $H = i\frac{\partial}{\partial t}$ reads

$$\begin{aligned} \{Q,Q\} &= 2H, \quad \{D,D\} = -2H, \quad \{Q,D\} = 0, \\ [H,Q] &= [H,D] = 0. \end{aligned} \tag{4.9}$$

To formulate also the supergravity model in superspace, the x-space action of (3.43)

$$L = \frac{1}{2}(1 - 2h)(\dot{\varphi}^2 + i\lambda\dot{\lambda}) - i\psi\dot{\varphi}\lambda, \tag{4.10}$$

suggests to introduce as superfield for the gauge fields of supergravity

$$H = (1 - 2h) + 2i\theta\psi, \tag{4.11}$$

and to write

$$L = \frac{i}{2} \int d\theta H (\partial_t \Phi) D\Phi. \tag{4.12}$$

This is the action in the so-called prepotential approach. This action is not manifestly covariant w.r.t. general coordinate transformations in superspace, but we shall soon give another formulation which is manifestly covariant and equivalent, giving the same t-space action. Let us first check that the expression in (4.12) yields indeed the t-space action

$$\begin{aligned} L &= \frac{i}{2} \int d\theta [(1 - 2h) + 2i\theta\psi](\dot{\varphi} + i\theta\dot{\lambda})(i\lambda - i\theta\dot{\varphi}) \\ &= \frac{1}{2}(1 - 2h)(\dot{\varphi}\dot{\varphi} + i\lambda\dot{\lambda}) - i\psi\dot{\varphi}\lambda. \end{aligned} \tag{4.13}$$

This agrees with (3.43).

We now turn to the covariant approach. One introduces "flat covariant derivatives"

$$\begin{aligned} \mathcal{D}_{\bar{t}} &= E_{\bar{t}}{}^{\theta} D_\theta + E_{\bar{t}}{}^{t} \partial_t, \\ \mathcal{D}_{\bar{\theta}} &= E_{\bar{\theta}}{}^{\theta} D_\theta + E_{\bar{\theta}}{}^{t} \partial_t. \end{aligned} \tag{4.14}$$

The superfields $E_{\bar{t}}{}^{\theta}$ etc are the inverse super vielbeins. The bars on $\bar{\theta}$ and $\bar{t}$ denote that these are flat indices, while θ and t are curved indices. Because in rigid susy the natural derivatives are ∂_t and D_θ rather than ∂_t and ∂_θ, one introduces vielbeins and parameters on this basis. We introduce the notation $D_\Lambda = \{D_\theta, \partial_t\}$

and $\mathcal{D}_M = \{\mathcal{D}_{\bar\theta}, \mathcal{D}_{\bar t}\}$ where $\Lambda = \{\theta, t\}$ and $M = \{\bar\theta, \bar t\}$. Then the super-vielbeins on the basis $\partial_\Lambda = (\partial_\theta, \partial_t)$ are related to the vielbeins on the basis $\mathcal{D}_\Lambda$ as follows

$$\tilde E_M{}^\Lambda \partial_\Lambda = E_M{}^\Lambda D_\Lambda, \quad \tilde E_M{}^\theta = E_M{}^\theta, \quad \text{but} \quad \tilde E_M{}^t = E_M{}^t - E_M{}^\theta i\theta,$$
(4.15)
$$\tilde\Xi^\Lambda \partial_\Lambda = \Xi^\Lambda D_\Lambda, \quad \tilde\Xi^\theta = \Xi^\theta, \quad \text{but} \quad \tilde\Xi^t = \Xi^t - \Xi^\theta i\theta.$$

A supercoordinate transformation of Φ becomes $\delta\Phi = \tilde\Xi^\Lambda \partial_\Lambda \Phi = \Xi^\Lambda D_\Lambda \Phi$. The (inverse) super vielbein $\tilde E_M{}^\Lambda$ transforms as in ordinary general relativity.

$$\delta\tilde E_M{}^\Lambda = \tilde\Xi^\Pi(\partial_\Pi \tilde E_M{}^\Lambda) - \tilde E_M{}^\Pi(\partial_\Pi \tilde\Xi^\Lambda)$$
(4.16)
$$= \Xi^\Pi(D_\Pi \tilde E_M{}^\Lambda) - E_M{}^\Pi(D_\Pi \tilde\Xi^\Lambda).$$

Contracting with ∂_Λ on both sides of the equation and then using (4.15) one obtains

(4.17) $$\delta E_M{}^\Sigma = \Xi^\Pi D_\Pi E_M{}^\Sigma + E_M{}^\Pi D_\Pi \Xi^\Sigma + E_M{}^\Pi \Xi^\Lambda T^{(0)\Sigma}_{\Lambda\Pi},$$

where $[D_\Lambda, D_\Pi] = T^{(0)\Sigma}_{\Lambda\Pi} D_\Sigma$. In our case only the $\theta\theta$ component of the supertorsion is nonvanishing, $T^{(0)\,t}_{\theta\theta} = -2i$ (see (4.9)), and we get

$$\delta E_M{}^t = \Xi^\Pi D_\Pi E_M{}^t - E_M{}^\Pi(D_\Pi \Xi^t + 2i\Xi^\theta \delta_\Pi{}^\theta),$$
(4.18)
$$\delta E_M{}^\theta = \Xi^\Pi D_\Pi E_M{}^\theta - E_M{}^\Pi D_\Pi \Xi^\theta.$$

In x-space we only have the field content of H. To reduce the extra superfields we shall now impose a constraint on the super-vielbein and choose a gauge. Afterward we shall come back to the general super coordinate transformations in the presence of this constraint and gauge.

To avoid cumbersome notation we write $\mathcal{D}$ for $\mathcal{D}_{\bar\theta}$ and parametrize $\mathcal{D}$ as follows

(4.19) $$\mathcal{D} = ED + Xi\partial_t, \quad \text{with} \quad D = D_\theta = \frac{\partial}{\partial\theta} - i\theta\frac{\partial}{\partial t}.$$

Thus $E_{\bar\theta}{}^\theta$ is denoted by E, and $E_{\bar\theta}{}^t$ by iX. The fields in $\mathcal{D}_{\bar t}$ are determined in terms of the fields in $\mathcal{D}$ by imposing the following *constraint*.

(4.20) $$\tfrac{1}{2}\{\mathcal{D}, \mathcal{D}\} = -i\mathcal{D}_{\bar t}.$$

This is the curved space extension of $\{D, D\} = -2i\partial_t$. It is a so-called conventional constraint; in 4 dimensions it expresses the bosonic Lorentz connection in terms of the fermionic Lorentz connections, but here no connection is present, and the constraint expresses the super vielbeins with flat index $\bar t$ in terms of those with flat index $\bar\theta$.

The anticommutator in (4.20) yields

$$\tfrac{1}{2}\{ED + Xi\partial_t, ED + Xi\partial_t\}$$
$$= [-E^2 + E(DX) + iX\dot X]i\partial_t + [E(DE) + iX\dot E]D$$
(4.21)
$$\equiv -i[E_{\bar t}{}^\theta D + E_{\bar t}{}^t \partial_t].$$

Hence

$$E_{\bar t}{}^t = E^2 - E(DX) - iX\dot X,$$
(4.22)
$$E_{\bar t}{}^\theta = iE(DE) - X\dot E.$$

As a matrix, the inverse super vielbein is thus given by

$$(4.23) \quad E_M{}^\Lambda = \begin{pmatrix} E^2 - E(DX) - iX\dot{X} & iE(DE) - X\dot{E} \\ iX & E \end{pmatrix}.$$

The superdeterminant of $\begin{pmatrix} A & B \\ C & D \end{pmatrix}$ being given by[20]

$$(4.24) \quad s\det\begin{pmatrix} A & B \\ C & D \end{pmatrix} = \frac{1}{D}(A - BD^{-1}C),$$

and we find

$$s\det E_M{}^\Lambda = \frac{1}{E}\left\{E^2 - E(DX) - iX\dot{X} + [E(DE) + iX\dot{E}]\frac{1}{E}X\right\}$$
$$(4.25) \quad\quad = E - (DX) - i\frac{X\dot{X}}{E} + \frac{DE}{E}X,$$

where we used that $XX = 0$.

We now choose *the gauge which sets* $iX = E_{\bar{\theta}}{}^t$ *to zero*. This will fix some of the gauge symmetries in superspace as we discuss later. In the covariant formalism, an invariant action (an action invariant under super general coordinate transformation, there are no Lorentz transformations) is given by ($s\det E_\Lambda{}^M$) times covariant derivatives on tensors (scalars in our case) with flat indices. We find then for the covariant action in superspace, which generalizes (4.12),

$$S = \frac{i}{2}\int dt d\theta (s\det E_\Lambda{}^M)(\mathcal{D}_{\bar{t}}\Phi)(\mathcal{D}_{\bar{\theta}}\Phi)$$
$$(4.26) \quad\quad = \int dt d\theta \left(\frac{i}{2E}\right)(E^2\partial_t\Phi)(ED\Phi).$$

(A term $iE(DE)(D\Phi)$ in $\mathcal{D}_{\bar{t}}\Phi$ cancels out). Identifying $E^2 = H$ we find the same action as in (4.12).

Let us now discuss the local symmetries in superspace. On Φ a general supercoordinate transformation would read

$$(4.27) \quad \delta\Phi = \tilde{\Xi}^t\partial_t\Phi + \tilde{\Xi}^\theta\partial_\theta\Phi.$$

To incorporate rigid susy as the flat superspace limit $E = 1, X = 0$, we expand instead in terms of ∂_t and D_θ. Then

$$(4.28) \quad \delta\Phi = \Xi^t\partial_t\Phi + \Xi^\theta D_\theta\Phi = \Xi\Phi \text{ with } \Xi = \Xi^t\partial_t + \Xi^\theta D_\theta.$$

We used again (4.15). But Ξ^θ is not a free super parameter as we now show. We can also write $\Xi\Phi$ as the operator equation $[\Xi, \Phi] = 0$.

The transformation rules of the super vielbein follow from

$$(4.29) \quad \delta\mathcal{D}_M = [\Xi, \mathcal{D}_M].$$

In particular for $\mathcal{D}_{\bar{\theta}} = \mathcal{D} = ED + Xi\partial_t$ we find

$$(4.30) \quad \delta[ED + Xi\partial_t] = [\Xi^t\partial_t + \Xi^\theta D, ED + Xi\partial_t].$$

---

[20]Use $\begin{pmatrix} A & B \\ C & D \end{pmatrix} = \begin{pmatrix} A - BD^{-1}C & BD^{-1} \\ 0 & 1 \end{pmatrix}\begin{pmatrix} 1 & 0 \\ C & D \end{pmatrix}$ and take the product of the two superdeterminants.

If the gauge $X = 0$ has been reached, the terms with ∂_t on the right-hand side should vanish because they are absent on the left-hand side. This yields the following relation between Ξ^θ and Ξ^t

$$-\Xi^\theta E 2i\partial_t - E(D\Xi^t)\partial_t = 0 \Rightarrow \Xi^\theta = \frac{i}{2}(D\Xi^t). \tag{4.31}$$

Expanding

$$\Xi^t = \xi(t) - 2i\theta\epsilon(t), \tag{4.32}$$

we find

$$\Xi^\theta = \frac{i}{2}(-2i\epsilon(t) - i\theta\dot{\xi}) = \epsilon(t) + \frac{1}{2}\theta\dot{\xi}. \tag{4.33}$$

The transformation of Φ becomes then

$$\delta\Phi = (\xi - 2i\theta\epsilon)\dot{\Phi} + \left(\epsilon + \frac{1}{2}\theta\dot{\xi}\right)D\Phi, \tag{4.34}$$

which reads in components

$$\begin{aligned}\delta\varphi &= \xi\dot{\varphi} + i\epsilon\lambda, \\ \delta\lambda &= \xi\dot{\lambda} + \frac{1}{2}\dot{\xi}\lambda - 2\epsilon\dot{\varphi} + \epsilon\dot{\varphi}.\end{aligned} \tag{4.35}$$

These are the correct x-space results of (3.38) and (3.41) after defining $\epsilon = \epsilon(x - \text{space})(1 - 2h)$.

To obtain the transformation rules of the supergravity gauge fields in E, we return to

$$\delta\mathcal{D}_M = [\Xi, \mathcal{D}_M] \text{ with } \Xi = \Xi^t\partial_t + \Xi^\theta D. \tag{4.36}$$

From this relation δE follows if one takes $\mathcal{D}_{\bar\theta} = \mathcal{D} = ED$ and collects all terms with D,

$$\begin{aligned}(\delta E)D &= (\Xi E)D - E(D\Xi^\theta)D, \\ \delta E &= \Xi^t \dot{E} + \Xi^\theta(DE) - E(D\Xi^\theta), \\ \delta H &= 2E\delta E = \Xi^t \dot{H} + \Xi^\theta DH - 2H(D\Xi^\theta).\end{aligned} \tag{4.37}$$

Substituting the component expressions of (4.11) leads to

$$\begin{aligned}-2\delta h + 2i\theta\delta\psi &= (\xi - 2i\theta\epsilon)(-2\dot{h} + 2i\theta\dot{\psi}) + \left(\epsilon + \frac{1}{2}\theta\dot{\xi}\right)(2i\psi + 2i\theta\dot{h}) \\ &\quad -2[(1-2h) + 2i\theta\psi]\left(\frac{1}{2}\dot{\xi} - i\theta\dot{\epsilon}\right).\end{aligned} \tag{4.38}$$

This leads to

$$\begin{aligned}\delta(-2h) &= -2\xi\dot{h} + 2i\epsilon\psi - (1-2h)\dot{\xi}, \\ \delta(2\psi) &= 4\epsilon\dot{h} + 2\xi\dot{\psi} + \dot{\xi}\psi - 2\dot{h}\epsilon \\ &\quad -2\dot{\xi}\psi + 2(1-2h)\dot{\epsilon}.\end{aligned} \tag{4.39}$$

These results agree with (3.41) using again the identification $\epsilon = \epsilon(x - \text{space})(1 - 2h)$.

Instead of choosing a gauge we can also find a redefinition of the fields which leads to the same result. We could demonstrate this in the most general case, but to simplify the analysis we still impose the constraint $\frac{1}{2}\{\mathcal{D},\mathcal{D}\} = -i\mathcal{D}_t$. Hence $E_{\hat t}{}^t$ and $E_{\hat t}{}^\theta$ are given in terms of E and X. The action in the gauge $X = 0$ is given by

$$L = \int d\theta \left(\frac{i}{2}\right) E^2 (\partial_t \Phi)(D\Phi). \tag{4.40}$$

whereas the action without this gauge choice reads

$$L = \frac{i}{2}\int d\theta \left(E - DX - iX\dot X/E + \frac{DE}{E}X\right)^{-1}$$
$$[(E^2 - E(DX) - iX\dot X)\dot\Phi + (iEDE - X\dot E)D\Phi][ED\Phi + iX\dot\Phi]. \tag{4.41}$$

The question we ask is this: which field redefinition (redefinition of E, X, Φ) leads from (4.41) to (4.40)? The solution of this question is most easily found by first making a general symmetry transformation on the fields in (4.41) and then requiring that the new field X' vanishes. This will express the symmetry parameter in terms of E and X, and provide an explicit expression for the new field E' in terms of E such that the action takes the form in (4.40).

The symmetries of the theory are in this case general supercoordinate transformations. In classical general relativity a contravariant supervector F^Λ transforms, by definition, as follows: $(F^\Lambda)'(Z') = F^\Pi(Z)\frac{\partial}{\partial Z^\Pi}(Z')^\Lambda$. Since field redefinitions do not change Z into Z', we need a way to write general supercoordinate transformations as relations between $(F^\Lambda)'(Z)$ and $F^\Lambda(Z)$ *at the same point Z*. Such a formalism exists and we now explain it and then will use it. One may call it the "operator approach to diffeomorphisms".

For a scalar superfield (a scalar with respect to general supercoordinate transformations) we define, as it is customary for internal symmetries in Yang-Mills theory,

$$\phi'(Z) = e^{\hat X(Z)D}\phi(Z). \tag{4.42}$$

Here $Z = \{\theta, t\}$ and $\hat X(Z)$ is an arbitrary superfield parameter which we will later write as a complicated expression in terms of E and X. The most general transformation would be

$$\phi'(Z) = e^{\hat X D + \hat Y \partial_t}\phi(Z). \tag{4.43}$$

but we will reach our aim with $\hat Y = 0$. There will then still remain super-diffeomorphisms with one superfield parameter which keep $X' = 0$. They correspond to ϵ and ξ transformations in x-space.

Usually one writes a general supercoordinate transformation as

$$\phi'(Z') = \phi(Z), \quad Z' = Z'(Z). \tag{4.44}$$

The relation between (4.44) and (4.42) becomes clear if one writes the latter as

$$e^{-\hat X(Z)D}\phi'(Z) = \phi'(Z'). \tag{4.45}$$

A particular case is $e^{-\hat{X}(Z)D}Z = Z'$. This yields the relation between $Z'(Z)$ and $\hat{X}(Z)$. To work this out we expand

$$\begin{aligned}
e^{\hat{X}(Z)D} &= 1 + \hat{X}D + \frac{1}{2}\hat{X}D\hat{X}D + \frac{1}{3!}\hat{X}D\hat{X}D\hat{X}D \\
&= 1 + \hat{X}D + \frac{1}{2}\hat{X}(D\hat{X})D + \frac{1}{3!}\hat{X}(D\hat{X})(D\hat{X})D + \cdots \\
&= 1 + \hat{X}\left(\frac{e^{(D\hat{X})} - 1}{(D\hat{X})}\right)D \equiv 1 + f\hat{X}D,
\end{aligned} \tag{4.46}$$

where $f = f((D\hat{X})) = (e^{(D\hat{X})} - 1)/(D\hat{X})$. We used that $\hat{X}\hat{X} = 0$, because $\hat{X}$ is anticommuting. It follows that

$$Z' = e^{-\hat{X}D}Z = (1 - \bar{f}\hat{X}D)Z = Z - \bar{f}\hat{X}(DZ), \tag{4.47}$$

where $\bar{f} \equiv f(-(D\hat{X})) = 1 - \frac{1}{2}(D\hat{X}) + \cdots$. More explicitly,

$$\theta' = \theta - \bar{f}\hat{X}(D\theta) = \theta - \bar{f}\hat{X}, \quad t' = t - \bar{f}\hat{X}(Dt) = t + \bar{f}\hat{X}(i\theta). \tag{4.48}$$

Then we find indeed (4.45)

$$\begin{aligned}
e^{-\hat{X}D}\phi'(Z) &= (1 - \bar{f}\hat{X}D)\phi' = \phi' - \bar{f}\hat{X}\left(\frac{\partial}{\partial\theta} - i\theta\frac{\partial}{\partial t}\right)\phi' \\
&= \phi'(\theta - \bar{f}\hat{X}, t + \bar{f}\hat{X}i\theta) = \phi'(Z').
\end{aligned} \tag{4.49}$$

Consider now a transformation with $\hat{X}$ which is such that the new field X' vanishes. In terms of

$$\mathcal{D} = ED + iX\partial_t, \tag{4.50}$$

this means that

$$\mathcal{D}' = e^{\hat{X}D}\mathcal{D}e^{-\hat{X}D} = E'(Z)D. \tag{4.51}$$

Thus in $\mathcal{D}'$ all terms with free derivatives $i\partial_t$ and $i\partial_t D$ must cancel, and the coefficient of D is by definition $E'(Z)$. Requiring that the coefficients of $i\partial_t$ and $i\partial_t D$ vanish fixes $\hat{X}$ as a function of X and E.

The details are as follows. We begin with

$$\begin{aligned}
e^{\hat{X}D}\mathcal{D}e^{-\hat{X}D} &= (1 + f\hat{X}D)(ED + Xi\partial_t)(1 - \bar{f}\hat{X}D) \\
&= E'(Z)D + X'(Z)i\partial_t.
\end{aligned} \tag{4.52}$$

Somewhat laborious algebra using $DD = -i\partial_t$ yields the following result for this expression,

$$\begin{aligned}
&[\{E + f\hat{X}(DE)\}D + \{X - f\hat{X}E + f\hat{X}(DX)\}i\partial_t \\
&\quad - f\hat{X}Xi\partial_t D](1 - \bar{f}\hat{X}D) \\
&= [E + f\hat{X}(DE - ED(\bar{f}\hat{X})) - f\hat{X}(DE)\bar{f}(D\hat{X}) \\
&\quad - Xi\partial_t(\bar{f}\hat{X}) - f\hat{X}E\bar{f}(i\partial_t\hat{X}) - f\hat{X}(DX)\bar{f}(i\partial_t\hat{X})]D \\
&\quad + [X - f\hat{X}E + f\hat{X}(DX) - E\bar{f}\hat{X}]i\partial_t \\
&\quad + [-f\hat{X}X - X\bar{f}\hat{X} + f\hat{X}X\bar{f}(D\hat{X})]Di\partial_t.
\end{aligned} \tag{4.53}$$

The terms with $Di\partial_t$ should vanish identically, and one can check that this is the case. The vanishing of the terms with $i\partial_t$ shows that X is proportional to $\hat{X}$, and then the terms with $Di\partial_t$ also vanish. Hence

$$X = f\hat{X}E - f\hat{X}(DX) + E\bar{f}\hat{X},$$
$$(1 + f\hat{X}D)X = (f + \bar{f})E\hat{X},$$
$$e^{\hat{X}D}X = (f + \bar{f})E\hat{X},$$
(4.54) $$X = e^{-\hat{X}D}[(f + \bar{f})\hat{X}E].$$

This relation expresses X in terms of $\hat{X}$ and E, but we need $\hat{X}$ in terms of X and E. We can invert by expanding the right-hand side and then solve for $\hat{X}$ iteratively

$$\begin{aligned}X &= -(f + \bar{f})\hat{X}E + \hat{X}D[(f + \bar{f})\hat{X}E] + \cdots \\ &= -\left(2 + \frac{1}{3}(D\hat{X})^2 + \frac{1}{60}(D\hat{X})^4 + \cdots\right)\hat{X}E \\ &\quad + \hat{X}D\left[2\hat{X}E + \frac{1}{3}(D\hat{X})^2\hat{X}E + \frac{1}{60}(D\hat{X})^4\hat{X}E + \cdots\right] \\ &= -2\hat{X}E + 2\hat{X}(D\hat{X})E.\end{aligned}$$
(4.55)

Inversion of $\hat{X} - \hat{X}(X\hat{X}) = -\frac{1}{2}X/E \equiv -\tilde{X}$ yields

$$\hat{X} = -\tilde{X} + \tilde{X}(D\tilde{X}) - 2\tilde{X}(D\tilde{X})^2$$
(4.56) $$5\tilde{X}(D\tilde{X})^3 + 14\tilde{X}(D\tilde{X})^4 + \cdots \quad \text{with} \quad \tilde{X} \equiv \frac{1}{2}\frac{X}{E}.$$

This expression for $\hat{X}$ can then be used to find the new field E'

$$\begin{aligned}E' &= E + f\hat{X}(DE) - ED(\bar{f}\hat{X}) + \cdots \\ &= E - \frac{X}{E}(DE) + ED\left(\frac{X}{E}\right).\end{aligned}$$
(4.57)

If one then substitutes this E' and the Φ' in (4.42) into (4.40), the result is equal to (4.41).

It may be helpful for becoming more familiar with the operator approach to diffeomorphisms to check that one gets the same results as from the usual approach. In the usual approach one begins with supervielbeins (supervectors in supercoordinate space) $\tilde{E}_M{}^\Lambda$ and general coordinate transformations $Z^\Lambda \to Z^\Lambda + \Xi^\Lambda$ where we take Ξ^Λ as infinitesimal. Then

$$\begin{aligned}\delta\tilde{E}_M{}^\Lambda &= \tilde{\Xi}^\Pi \partial_\Pi \tilde{E}_M{}^\Lambda - \tilde{E}_M{}^\Pi \partial_\Pi \tilde{\Xi}^\Lambda \\ &= \Xi^\Pi D_\Pi \tilde{E}_M{}^\Lambda - E_M{}^\Pi D_\Pi \tilde{\Xi}^\Lambda,\end{aligned}$$
(4.58)

with $\partial_\Pi = \left(\frac{\partial}{\partial\theta}, \frac{\partial}{\partial t}\right)$ and $D_\Pi = \left(\frac{\partial}{\partial\theta} - i\theta\frac{\partial}{\partial t}, \frac{\partial}{\partial t}\right)$. From (4.15)

$$\tilde{E}_M{}^\theta = E_M{}^\theta, \quad \tilde{E}_M{}^t = E_M{}^t - E_M{}^\theta i\theta$$
(4.59) $$\tilde{\Xi}^\theta = \Xi^\theta, \quad \tilde{\Xi}^t = \Xi^t - \Xi^\theta i\theta.$$

One finds by straightforward substitution

$$\begin{aligned}\delta E_M{}^\theta &= \delta\tilde{E}_M{}^\theta = \Xi^\Pi D_\Pi \tilde{E}_M{}^\theta - E_M{}^\Pi D_\Pi \tilde{\Xi}^\theta \\ &= \Xi^\Pi D_\Pi E_M{}^\theta - E_M{}^\Pi D_\Pi \Xi^\theta.\end{aligned}$$
(4.60)

This agrees with (4.36). However, in $\delta E_M{}^t$ there are extra terms beyond those which are present in the corresponding relation for $\tilde{E}_M{}^t$.

(4.61) $$\delta E_M{}^t = \delta(\tilde{E}_M{}^t + E_M{}^\theta i\theta) = \Xi^\Pi D_\Pi \tilde{E}_M{}^t - E_M{}^\Pi D_\Pi(\Xi^t - \Xi^\theta i\theta)$$
$$+(\Xi^\Pi D_\Pi E_M{}^\theta - E_M{}^\Pi D_\Pi \Xi^\theta)i\theta.$$

Various terms cancel and one obtains

(4.62) $$\delta E_M{}^t = \Xi^\Pi D_\Pi E_M{}^t - E_M{}^\Pi D_\Pi \Xi^t - 2i E_M{}^\theta \Xi^\theta.$$

The last term is due to the rigid torsion $T_{\theta\theta}{}^t = -2i$. Expanding the finite result for E' and X' in (4.52) to first order in $\hat{X} = \Xi^\theta$ we find agreement.

One may also check that

(4.63) $$\begin{aligned}\delta \mathcal{D} &= (e^{\hat{X}D} E_{\bar\theta}{}^\Lambda e^{-\hat{X}D})D_\Lambda + E_{\bar\theta}{}^\Lambda(e^{\hat{X}D} D_\Lambda e^{-\hat{X}D})\\ &= \delta E_{\bar\theta}{}^\Lambda D_\Lambda,\end{aligned}$$

but note that $e^{\hat{X}D} E_{\bar\theta}{}^\Lambda e^{-\hat{X}D}$ only yields the transport terms $E_{\bar\theta}{}^\Lambda(Z')$, while the terms with $e^{\hat{X}D} D_\Lambda e^{-\hat{X}D}$ yield the terms which are due to the index Λ of $E_{\bar\theta}{}^\Lambda$.

5. Extended rigid supersymmetries

We can also construct models with extended ($N > 1$) susy. We restrict ourselves to rigid supersymmetry, but $N > 1$ supergravities can also be constructed. We perform the analysis in x-space but one could also use $N > 1$ superspace.

Consider the action

(5.1) $$L = \frac{1}{2}\dot\phi\dot\phi + \frac{i}{2}\sum_{j=1}^N \lambda^j \dot\lambda^j.$$

It is invariant under the following rigid susy transformations

(5.2) $$\delta\phi = i\sum_{j=1}^N \epsilon^j \lambda^j, \delta\lambda^j = -\dot\phi \epsilon^j.$$

The proof is as before: $\delta L = \dot\phi \frac{d}{dt}(i\epsilon^j \lambda^j) + i\lambda^j \frac{d}{dt}(\dot\phi \epsilon^j)$ vanishes after partial integration.

There is something unusal about this action: it has more fermions (N) than bosons (one). Nevertheless it has the same number of fermionic and bosonic states because after quantization there are only zero modes ($\hat{x}_0$ and $\hat\lambda_0^j$) and their conjugate momenta ($\hat{p}$ and again $\hat\lambda_0^j$). The $\hat\lambda_0^j$ can be combined into $[\frac{N}{2}]$ annihilation and creation operators $(\hat\lambda_0^1 \pm i\hat\lambda_0^2)/\sqrt{2}$, ... If N is odd, the last $\hat\lambda_0^N$ yields projection operators $P_\pm = \frac{1}{2}(1 \pm \hat\lambda_0^N)$ and the Hilbert space $\mathcal{H}$ split into two spaces $\mathcal{H} = P_\pm \mathcal{H}$. No operator can bring one from a state in $\mathcal{H}_+$ to a state in $\mathcal{H}_-$ because operators either have no $\hat\lambda_0^N$, or if they have one $\hat\lambda_0^N$ then still $\hat\lambda_0^N \mathcal{H}_+ = \mathcal{H}_+$ because $\hat\lambda_0^N \frac{1}{2}(1+\hat\lambda_0^N) = \frac{1}{2}(1+\hat\lambda_0^N)$. The bosonic states are $e^{ip\hat{x}_0}\,|\,0\rangle =|\,p\rangle$ but also $\hat\lambda_0^1 \hat\lambda_0^2\,|\,p\rangle$ etc., and the fermionic states are $\hat\lambda_0^j\,|\,p\rangle$ but also $\hat\lambda_0^1 \hat\lambda_0^2 \hat\lambda_0^3\,|\,p\rangle$ etc. In fact there are 2^{N-1} bosonic states (with an even number of $\hat\lambda_0^j$) and 2^{N-1} fermionic states (with an odd number of λ_0^j). In higher dimensions there are also oscillators, and then one can count the number of bosonic and fermionic states by looking at the number of propagating fields.[21] For example, in $d = (3,1)$ one has the Wess-Zumino model with one Majorana spinor λ^α (2 states) and two real propagating massless scalars (again 2 states per oscillator).

In our model similar things happen. To close the algebra off-shell we need an equal number of bosonic and fermionic field components. Thus for $N = 2$ we need one real auxiliary field, which we call F. The action reads

$$(5.3) \qquad L = \frac{1}{2}\dot\varphi\dot\varphi - \frac{i}{2}(\lambda^1\dot\lambda^1 + \lambda^2\dot\lambda^2) + \frac{1}{2}F^2.$$

One can view (φ, λ^1) as one multiplet, and (λ^2, F) as another. The action for these multiplets reads in superspace

$$(5.4) \qquad \begin{aligned} L(\varphi, \lambda^1) &= \frac{1}{2}\dot\varphi\dot\varphi - \frac{i}{2}\lambda^1\dot\lambda^1 \sim \int d\theta D\Phi\dot\Phi, \quad \Phi = \varphi + i\theta\lambda^1, \\ L(\lambda^2, F) &= -\frac{i}{2}\lambda^2\dot\lambda^2 + \frac{1}{2}F^2 \sim \int d\theta \Psi D\Psi, \quad \Psi = \lambda^2 + \theta F. \end{aligned}$$

Clearly $\delta\varphi = i\epsilon^1\lambda^1$ and $\delta\lambda^1 = \dot\varphi\epsilon^1$ form a closed susy algebra, $\{Q, Q\} \sim P$. Also $\delta\lambda^2 = F\epsilon^2$ and $\delta F = i\epsilon^2\dot\lambda^2$ forms the same closed algebra. However, in x-space we can write down more general rules

$$(5.5) \qquad \begin{aligned} \delta\varphi &= \sum_{j=1}^{2} i\epsilon^j\lambda^j, \delta\lambda^j = \dot\varphi\epsilon^j + \alpha^{jk}\epsilon^k F, \\ \delta F &= i\epsilon^j\beta^{jk}\dot\lambda^k, \end{aligned}$$

where α and β are arbitrary real matrices. Invariance of the action requires

$$(5.6) \qquad \alpha^T = \beta.$$

The commutator algebra on φ yields

$$(5.7) \qquad \begin{aligned} [\delta(\epsilon_1), \delta(\epsilon_2)]\varphi &= \sum i\epsilon_2^j(\dot\varphi\epsilon_1^j + \alpha^{jk}\epsilon_1^k F) - (1 \leftrightarrow 2) \\ &= (2i\epsilon_2^j\epsilon_1^j)\dot\varphi + [i\epsilon_2(\alpha + \alpha^T)\epsilon_1]F. \end{aligned}$$

Clearly, for α antisymmetric (hence $\alpha^{jk} \sim \epsilon^{jk}$) we find only a translation of φ. For F one finds

$$(5.8) \qquad \begin{aligned} [\delta(\epsilon_1), \delta(\epsilon_2)]F &= i\epsilon_2^j\beta^{jk}\frac{d}{dt}(\dot\varphi\epsilon_1^k + \alpha^{kl}\epsilon_1^l F) - (1 \leftrightarrow 2) \\ &= [i\epsilon_2(\alpha + \alpha^T)\epsilon_1]\ddot\varphi + [2i\epsilon_2\alpha^T\alpha\epsilon_1]\dot F. \end{aligned}$$

We now recognize various symmetries.

(i) $\delta\varphi = zF$ and $\delta F = z\ddot\varphi$ with $z = i\epsilon_2(\alpha + \alpha^T)\epsilon_1$. This is an equation of motion symmetry. (These are symmetries of pairs of fields f and g of the form $\delta f = A\frac{\delta S}{\delta g}$ and $\delta g = -A\frac{\delta S}{\delta f}$. Obviously they leave the action invariant).
(ii) The usual translation.
(iii) An extra symmetry $\delta F = \sigma\dot F$ where σ is $2i\epsilon_2\alpha^T\alpha\epsilon_1 - 2i\epsilon_2\epsilon_1$. This is clearly a symmetry of $L = \frac{1}{2}F^2$.

---

[21]In $d = (1,1)$ one can divide the real scalar into a left-moving piece $\varphi(x+t)$ and a right-moving piece $\varphi(x-t)$, and susy requires then a two-component spinor $\lambda = \binom{\lambda_+}{\lambda_-}$ with $\lambda_+(x+y)$ right-moving on-shell and $\lambda_-(x-t)$ left-moving. In the zero mode sector there are always the same number of bosonic and fermionic states but in the nonzero mode sector one may drop λ_+ (or λ_-) and still have susy. Then λ_- transforms into φ and vice-versa, while λ_+ is inert.

For the fermions the susy commutator yields

$$[\delta(\epsilon_1), \delta(\epsilon_2)]\lambda^j = \delta(\epsilon_1)[\dot\varphi\epsilon_2^j + \alpha^{jk}\epsilon_2^k F] - (1 \leftrightarrow 2)$$
$$= -i(\epsilon_1^l \epsilon_2^j - \epsilon_2^l \epsilon_1^j)\dot\lambda^l + [(\alpha^{jk}\epsilon_2^k)(i\epsilon_1^l \beta^{lm}\dot\lambda^m) - (1 \leftrightarrow 2)]$$
(5.9) $$= [2i\epsilon_2^l \epsilon_1^l]\dot\lambda^j + i(\epsilon_2^l \epsilon_1^j \dot\lambda^l - \epsilon_2^l \epsilon_1^l \dot\lambda^j - (1 \leftrightarrow 2)) + [\cdots].$$

(iv) $\delta\lambda^j = s^{jl}\dot\lambda^l$ with symmetric and real s^{jl}. This is again an equation of motion symmetry.

The extra symmetries do not form a closed algebra. Their commutators generate new equation of motion symmetries, etc., etc. However, the minimal extension of the susy rules with F forms a closed algebra

$$\delta\varphi = i(\epsilon^1\lambda^1 + \epsilon^2\lambda^2), \quad \delta\lambda^1 = \dot\varphi\epsilon^1 - F\epsilon^2, \quad \delta\lambda^2 = \dot\varphi\epsilon^2 + F\epsilon^1,$$
(5.10) $$\delta F = i\epsilon^1\dot\lambda^2 - i\epsilon^2\dot\lambda^1.$$

We can try to construct and $N = 2$ superspace by introducing

$$\theta = \frac{\theta^1 + i\theta^2}{\sqrt{2}}, \quad \bar\theta = \frac{\theta^1 - i\theta^2}{\sqrt{2}}, \quad \bar D = \frac{\partial}{\partial\theta} + i\bar\theta\partial_t, \quad D = \frac{\partial}{\partial\bar\theta} + i\theta\partial_t,$$

$$\bar Q = \frac{\partial}{\partial\theta} - i\bar\theta\partial_t, \quad Q = \frac{\partial}{\partial\bar\theta} - i\theta\partial_t, \quad \lambda = \frac{\lambda^1 + i\lambda^2}{\sqrt{2}}, \quad \bar\lambda = \frac{\lambda^1 - i\lambda^2}{\sqrt{2}},$$

$$L = \frac{1}{2}\dot\varphi\dot\varphi - i\bar\lambda\dot\lambda + \frac{1}{2}F^2, \quad \Phi = \varphi + i\theta\lambda + i\bar\theta\bar\lambda + F\bar\theta\theta,$$

$$L = -\int d\theta d\bar\theta \bar D\Phi D\Phi = -\int d\theta d\bar\theta [i\bar\lambda - F\bar\theta + i\bar\theta\dot\varphi - \bar\theta\theta\dot{\bar\lambda}][i\lambda + F\theta + i\theta\dot\varphi - \theta\bar\theta\dot\lambda]$$

(5.11) $$= -i\bar\lambda\dot\lambda + i\dot{\bar\lambda}\lambda + \dot\varphi\dot\varphi + F^2.$$

The t-space components of Φ are given by $\Phi_| = \varphi, D\Phi_| = i\lambda, \bar D\Phi_| = i\bar\lambda$ and $\frac{1}{2}[\bar D, D] = F$. The susy rules are generated by $(\bar\epsilon D + \epsilon\bar D)$ acting on these components, using the D algebra.

The $N = 2$ model has a mass term and a Yukawa coupling

(5.12) $$L = \left[\frac{1}{2}\dot\varphi^2 - \frac{i}{2}\lambda_1\dot\lambda_1 - \frac{i}{2}\lambda_2\dot\lambda_2 + \frac{1}{2}F^2\right]$$
$$-m(F\varphi + i\lambda_1\lambda_2) + g\left(\frac{1}{2}F\varphi^2 + i\lambda_1\lambda_2\varphi\right).$$

The corresponding action in $N = 1$ superspace is

(5.13) $$S^{(0)} = \left(\frac{-i}{2}\right)\int dt d\theta [(D_\theta\phi)\partial_t\phi + i\psi D_\theta\psi],$$
$$\phi = \varphi + i\theta\lambda_1, \quad \psi = \lambda_2 + \theta F,$$

(5.14) $$S(m) = -\int dt d\theta m\phi\psi,$$
$$S(g) = \int dt d\theta \frac{1}{2}g\phi\phi\psi.$$

One can write more generally a superpotential term as

(5.15) $$S(W) = \int dt d\theta W(\phi)\psi = \int dt (WF + i\lambda_1 W'\lambda_2),$$

where $W(\phi) = -m\phi + \frac{1}{2}g\phi^2 + \cdots$.

One can also write down nonlinear sigma models for the $N = 1$ model

$$\begin{aligned}S(\text{nonl}) &= \frac{-i}{2} \int dt d\theta G(\phi) D_\theta \phi \partial_t \phi \\ &= \int dt \left[G(\varphi) \left(\frac{1}{2} \dot\varphi^2 - \frac{i}{2} \lambda_1 \dot\lambda_1 \right) + \frac{i}{2} \lambda_1 G'(\varphi) i \lambda_1 \dot\varphi \right] \\ &= \int dt G(\varphi) \left(\frac{1}{2} \dot\varphi^2 - \frac{i}{2} \lambda_1 \dot\lambda_1 \right). \end{aligned}$$
(5.16)

Similarly, one can find nonlinear σ models with $N = 2$ susy in $N = 1$ superspace. For all these $N = 2$ models in $N = 1$ superspace there is also a $N = 2$ superspace formulation.

The action in Hamiltonian form for the $N = 2$ model reads

$$\begin{aligned} L &= \dot\varphi p + \dot\lambda^1 \pi^1 + \dot\lambda^2 \pi^2 + \dot F p_F - \frac{1}{2} F^2 - \frac{1}{2} p^2, \\ \delta\varphi &= \frac{i}{2} \epsilon^j \left(\lambda^j + \frac{2}{i} \pi^j \right), \quad \delta p = 0, \quad \delta\pi^j = \frac{i}{2} p \epsilon^j, \\ \delta\lambda^1 &= p^1 \epsilon^1 + F \epsilon^2, \quad \delta\lambda^2 = p^2 \epsilon^2 - F \epsilon^1, \\ \delta p_F &= \epsilon^1 \pi^2 - \epsilon^2 \pi^1, \quad \delta F = 0. \end{aligned}$$
(5.17)

As in the $N = 1$ case, there are primary constraints $\pi^j - \frac{i}{2} \lambda^j = 0$ and $p_F = 0$, and secondary constraints $F = 0$. Since $\delta F \sim \dot\lambda$ in the Lagrangian formulation, but $\dot\lambda = 0$ is the full field equation for λ, we cannot express $\dot\lambda$ in terms of $\partial_x \lambda$ by means of its field equation, and this explains why $\delta F = 0$ and $\delta p = 0$. The susy Noether charges in the Hamiltonian approach are given by

$$\begin{aligned} Q_H^1 &= \left(i\pi^1 - \frac{1}{2} \lambda^1 \right) p - iF\pi^2, \\ Q_H^2 &= \left(i\pi^2 - \frac{1}{2} \lambda^2 \right) p + iF\pi^1. \end{aligned}$$
(5.18)

These changes reproduce the Hamiltonian susy laws *exactly* if one uses ordinary Poisson brackets. The reason is that in the Hamiltonian action there are no constraints. In the action of the Lagrangian formulation one finds Dirac brackets and the following hermitian susy charges

$$Q_L^j = -\lambda^j p.$$
(5.19)

The algebra reads in both cases

$$\begin{aligned} \{Q_H^i, Q_H^j\}_P &= \delta^{ij}(-p^2) = -2\delta^{ij} H, \\ \{Q_L^i, Q_L^j\}_D &= \delta^{ij}(-p^2) = -2\delta^{ij} H. \end{aligned}$$
(5.20)

Thus F and p_F do not transform under Q_L, but p_F transforms under Q_H. However, the algebra on all fields is the same. For example, $\epsilon Q_H \lambda \sim p\epsilon + F\epsilon = 0$ and $Q_L p = \epsilon$, in agreement with $H\lambda = 0$ for $H = H_L = H_H$.

The $N=2$ model can be used as a toy model for instanton physics. In Minkowski time t the action is

$$L = \frac{1}{2}\dot{\varphi}^2 - i\bar{\lambda}\dot{\lambda} - \bar{\lambda}\lambda(m - g\varphi) + \frac{1}{2}\left(F - m\varphi + \frac{1}{2}g\varphi^2\right)^2$$

(5.21)
$$-\frac{1}{8}g^2\left[\left(\varphi - \frac{m}{g}\right)^2 - \left(\frac{m}{g}\right)^2\right]^2.$$

This action is hermitian, and $\bar{\lambda} = (\lambda)^\dagger$. In Euclidean time τ, λ and $\bar{\lambda}$ become independent complex spinors. The Wick rotation is a complex Lorentz rotation (a $U(1)$ rotation) in the (t, τ) plane (see the joint paper with Waldron)

(5.22) $$t_\theta^1 = e^{i\theta}t, \quad t_{\theta=\pi/2}^1 \equiv \tau = it.$$

The spinor λ transforms then half as fast

(5.23) $$\lambda_\theta^1 = e^{i\theta/2}\lambda, \lambda_{\theta=\pi/2} \equiv \lambda_E = \sqrt{i}\lambda.$$

Making these substitutions, one automatically obtains a supersymmetric action for the Euclidean case.

It is a pleasure to thank Martin Roček for discussions about the covariant approach to superspace supergravity and the field redefinition in (4.57).

6. BRST quantization in a Hamiltonian approach

So far we have been discussing classical actions. The classical supergravity action was a gauge action, an action with two local symmetries in out toy model: diffeomorphisms and local supersymmetry. In the quantum theory, one uses a quantum action which is obtained by adding two more terms: a gauge fixing term and a ghost action. The local symmetry is then broken, but a rigid residual symmetry remains, the so-called BRST symmetry (due to Becchi, Rouet, Stora and Tyutin). The crucial property of BRST transformations is that they are nilpotent (see below). An infinitesimal BRST transformation has as parameter a purely imaginary anticommuting constant Λ. It is not the supersymmetry parameter (which is also anticommuting). Physicists use (at least) two ways to formulate the BRST formalism: a Lagrangian approach and a Hamiltonian approach. The infinitesimal BRST transformation rules in the former are written as $\delta_B \Phi$ (where Φ denotes any field), but in the Hamiltonian approach one uses operators and brackets. For example, the BRST transformations are generated by a BRST operatorial charge Q, and for every field there is a canonically conjugate field (called momentum by physicists) which satisfy equal-time canonical commutation rules or anticommutation rules. Then one has $\delta_B \Phi \sim \{\Phi, Q\}$. Nilpotency means $\delta_B \delta_B \Phi = 0$ in the Lagrangian approach, and $\{Q, Q\} = 0$ in the Hamiltonian approach. In this section we apply this general formalism to our quantum mechanical toy model. One word about terminology: we use the words real and hermitian (and purely imaginary and antihermitian) as equivalent.

To quantize the supergravity action using covariant quantization in the Lagrangian approach to BRST symmetry, we should add a gauge fixing term and a corresponding ghost action. We begin with the classical action in (3.10) although we could also start with the classical action in (3.43). We fix the gauge of general coordinate transformations by $h = 0$, and the gauge of local supersymmetry by $\psi = 0$. The corresponding gauge fixing terms in the action are then given by

(6.1) $$L_{\text{fix}} = dh + \Delta \psi.$$

The fields d and Δ are Lagrange multipliers which fix the gauges according to $h = 0$ and $\psi = 0$. Hermiticity of the action requires that d be real and Δ antihermitian. The ghost action is then given by

(6.2) $$L(\text{ghost})\Lambda = b\delta_B h + \beta\delta_B \psi,$$

where δ_B are the BRST transformations and Λ is the constant anticommuting imaginary BRST parameter. The fields b and β are the antighosts; b and β are antihermitian. (Mathematical physicists remove Λ on both sides of this equation and call it then a derivation. It is usually denoted by s, and the relation to δ_B is $\delta_B \phi = \Lambda s\phi$ for any field ϕ). We obtain thus for the full quantum action

(6.3) $$\begin{aligned} L &= L(\text{class}) + L(\text{fix}) + L\text{ghost}), \\ L(\text{class}) &= \frac{1}{2}\dot{\varphi}^2 + \frac{i}{2}\lambda\dot{\lambda} - h\dot{\varphi}^2 - i\psi\dot{\varphi}\lambda, \\ L(\text{fix}) &= dh + \Delta\psi, \quad L(\text{ghost})\Lambda = b\delta_B h + \beta\delta_B \psi. \end{aligned}$$

For the classical fields (h and ψ and φ and λ) the BRST transformations are just gauge transformations with a special choice of the parameters: $\xi = c\Lambda$ and $\epsilon = -i\gamma\Lambda$ where c and γ are real ghosts (Λ is imaginary, and ξ and ϵ are real). Because ξ is commuting, c is anticommuting, and because ϵ is anticommuting, γ is commuting. The BRST transformation rules for the classical fields follow from (3.10), (3.33) and (3.37)

(6.4) $$\begin{aligned} \delta_B h &= [(1 - 2h)\psi\gamma + \frac{1}{2}(1 - 2h)\dot{c} + \dot{h}c]\Lambda, \\ \delta_B \psi &= [-i(1 - 2h)\dot{\gamma} + \dot{\psi}c]\Lambda, \\ \delta_B \varphi &= (-\lambda\gamma + \dot{\varphi}c)\Lambda, \quad \delta_B \lambda = [i(1 - 2h)\dot{\varphi}\gamma + \psi\lambda\gamma + \dot{\lambda}c]\Lambda. \end{aligned}$$

The ghost action then becomes

(6.5) $$L(\text{ghost}) = b[(1 - 2h)\psi\gamma + \frac{1}{2}(1 - 2h)\dot{c} + \dot{h}c] + \beta[-i(1 - 2h)\dot{\gamma} + \dot{\psi}c].$$

It is clearly hermitian.

The BRST transformation rules of the ghosts follow from the structure constants of the classical gauge algebra or from the nilpotency of BRST transformations. One has uniformly on all classical fields (φ, λ, h and ψ), as discussed in (3.18)

(6.6) $$\begin{aligned} [\delta_s(\epsilon_2), \delta_s(\epsilon_1)] &= \delta_g(2i(1 - 2h)\epsilon_2\epsilon_1) + \delta_s(-2i\psi\epsilon_2\epsilon_1), \\ [\delta_g(\xi_2), \delta_g(\xi_1)] &= \delta_g(-\xi_2\dot{\xi}_1 + \xi_1\dot{\xi}_2), \\ [\delta_g(\xi), \delta_s(\epsilon)] &= \delta_s(-\xi\dot{\epsilon}). \end{aligned}$$

Thus the classical local gauge algebra "closes": the commutator of two local symmetries is a linear combination of local symmetries. New from a mathematical point of view is the appearance of fields (h and ψ) in the composite parameters, and thus in the structure constants. (One should thus rather speak of "structure functions"). One obtains

$$
\begin{aligned}
\delta_B c &= -c\dot{c}\Lambda + i(1-2h)\gamma\gamma\Lambda, \\
\delta_B \gamma &= c\dot{\gamma}\Lambda + \psi\gamma\gamma\Lambda.
\end{aligned}
\tag{6.7}
$$

An easier way to obtain these results is to use that the BRST variation of (6.4) should vanish. One may check that all BRST transformation rules preserve the reality properties of the fields.

The antighosts b and β transform into the auxiliary fields d and Δ, and the auxiliary fields are BRST invariant ("contractible pairs")

$$
\delta_B b = \Lambda d, \quad \delta_B \beta = \Lambda\Delta, \quad \delta_B d = 0, \quad \delta_B \Delta = 0.
\tag{6.8}
$$

All BRST transformation laws are nilpotent, and they leave the action $S = \int L dt$ invariant.

In the Lagrangian approach the BRST charge Q is the Noether charge for rigid BRST transformations but one does not use brackets. To obtain Q, one lets Λ become local (t-dependent), and one collects all terms in the variation of the action proportional to $\dot{\Lambda}$. (We used this procedure before to construct the supersymmetry charge). The BRST transformation rules of the fields themselves for local Λ do not contain by definition any $\dot{\Lambda}$; for example, $\delta_B \psi = (1-2h)(-i\dot{\gamma})\Lambda(t) + \ldots$, and not $\delta_B \psi = (1-2h)\frac{d}{dt}(-i\gamma\Lambda(t)) + \ldots$. The classical action yields the following terms proportional to $\dot{\Lambda}$

$$
\delta S(\text{class}) = \int [L(\text{class})c\dot{\Lambda} - (1-2h)\lambda\dot{\varphi}\gamma\dot{\Lambda}]dt.
\tag{6.9}
$$

The term with $c\dot{\Lambda}$ is expected from the result $\delta\mathcal{L} = \partial_\alpha(\xi^\alpha \mathcal{L})$ in general relativity, while the term with $\gamma\dot{\Lambda}$ comes from the variation with $\frac{d}{dt}\epsilon$ in $i\lambda\frac{d}{dt}\delta\lambda$ which is canceled by the Noether term. The gauge fixing term produces no terms with $\dot{\Lambda}$ because $\delta_B h$ and $\delta_B \psi$ do not contain $\dot{\Lambda}$ terms by definition, but the ghost action yields further terms proportional to $\dot{\Lambda}$

$$
\begin{aligned}
\delta S(\text{ghost}) = &\ b[\frac{1}{2}(1-2h)(-c\dot{c} + i(1-2h)\gamma\gamma)\dot{\Lambda} \\
&+ c((1-2h)\psi\gamma + \frac{1}{2}(1-2h)\dot{c} + \dot{h}c)\dot{\Lambda}] \\
&+ \beta[(-i)(1-2h)(c\dot{\gamma} + \psi\gamma\gamma)\dot{\Lambda} + ic(1-2h)\dot{\gamma}\dot{\Lambda}]
\end{aligned}
\tag{6.10}
$$

Several terms cancel in this expression.

The BRST charge in the Lagrangian approach is thus

$$
\begin{aligned}
Q =&\ cL(\text{class}) - (1-2h)\lambda\dot{\varphi}\gamma \\
&+ b[\frac{i}{2}(1-2h)^2\gamma\gamma + (1-2h)c\psi\gamma] \\
&- i\beta(1-2h)\psi\gamma\gamma.
\end{aligned}
\tag{6.11}
$$

In the Hamiltonian approach the BRST charge should contain terms of the form (we discuss this in more detail later)

$$
Q = c^\alpha \phi_\alpha - \frac{1}{2}c^\beta c^\alpha f_{\alpha\beta}{}^\gamma b_\gamma (-)^\beta,
\tag{6.12}
$$

with ghosts $c^\alpha = (c, \gamma)$, antighosts $b_\alpha = (b, \beta)$, and first class constraints $\phi_\alpha = (T, J)$ where T is the generator of diffeomorphisms and J the generator of supersymmetry. The sign $(-)^\beta$ is equal to $+1$ if the corresponding symmetry has a commuting parameter (and thus an anticommuting ghost); when the ghost is commuting $(-)^\beta$ equals -1. However, because the structure constants contain $\dot\xi$ and $\dot\epsilon$ we expect in Q terms with derivatives of c, namely $bc\dot c$ and $\beta\dot\gamma c$ terms. On the other hand, in a truly Hamiltonian approach no time derivatives of fields are allowed in the charges. This suggests that the b and β field equations have been used to eliminate $\dot c$

$$\text{(6.13)} \qquad \dot c = -2\psi\gamma - \frac{2}{1-2h}\dot h c; \qquad \dot\gamma = -\frac{i\dot\psi c}{1-2h}.$$

However, then one obtains time derivatives of h and ψ. Another problem is that we seem to have too many factors of $(1-2h)$ but this could be repaired by redefining fields. To resolve these issues we first construct Q by Hamiltonian methods. This is a very general approach which only uses as input the first-class constraints of the classical theory, and which provides a quantum action in phase space and a Hamiltonian BRST charge with in general many more fields than in the usual (Lagrangian) formulation. Eliminating nonpropagating fields we should regain the Lagrangian BRST charge Q in (6.11). Let's see how this works out.

In the Hamiltonian framework of Fradkin and Vilkovisky (and others) the quantum action is of the form

$$\text{(6.14)} \qquad L = \dot q^i p_i - H + \{Q_H, \psi_g\},$$

where q^i denotes all fields: classical fields (including the Lagrange multipliers h and ψ), ghosts and antighosts. The p_i are canonical momenta *for all of them*. (So, for example, there are canonical momenta for the ghosts, and separate canonical momenta for the antighosts). The BRST charge Q_H should be nilpotent

$$\text{(6.15)} \qquad \{Q_H, Q_H\} = 0.$$

The full quantum Hamiltonian H is constructed from Dirac's Hamiltonian H_D such that

$$\text{(6.16)} \qquad [H, Q_H] = 0.$$

Neither H_D nor H should depend on Lagrange multipliers, and will be constructed below. The BRST charge is in general given by

$$\text{(6.17)} \qquad Q_H = c^\alpha \varphi_\alpha + p^\mu(B) \pi_\mu(\lambda) - \frac{1}{2} c^\beta c^\alpha f_{\alpha\beta}{}^\gamma p_\gamma (-)^\beta,$$

where φ_α are the first-class constraints, $f_{\alpha\beta}{}^\gamma$ the structure functions defined by $\{\varphi_\alpha, \varphi_\beta\} = f_{\alpha\beta}{}^\gamma \varphi_\gamma$, and λ^μ are the Lagrange multipliers (classical fields which appears in the classical action without time derivatives), and B denotes the antighosts. The structure functions should only depend on p_i and q^i but not on λ^μ. Usually one has to take suitable linear combinations of local symmetries and add so-called equation-of-motion symmetries to achieve this. We shall demonstrate this in our model.

The BRST invariance of the action in (6.14) is almost obvious: each of the 3 terms is separately invariant due to the relations $\int \frac{d}{dt} Q_H dt = 0$, $[Q_H, H] = 0$ and $\{Q_H, Q_H\} = 0$.

We start again from the classical gauge invariant action in (3.10)

$$(6.18) \qquad L = \frac{1}{2}(1-2h)\dot{\varphi}^2 + \frac{i}{2}\lambda\dot{\lambda} - i\psi\dot{\varphi}\lambda.$$

The primary constraints are $p_h = 0$, $\pi_\psi = 0$ and $\pi_\lambda + \frac{i}{2}\lambda = 0$, and the naive Hamiltonian with all primary constraints added, reads

$$(6.19) \qquad H_{\text{naive}} = \frac{1}{1-2h}\frac{1}{2}(p+i\psi\lambda)^2 + ap_h + \alpha\pi_\psi + \eta(\pi_\lambda + \frac{i}{2}\lambda).$$

The functions $a(t)$, $\alpha(t)$ and $\eta(t)$ are arbitrary Lagrange multipliers which enforce the primary constraints. By a redefinition of η we can replace λ in the first term by the field $\frac{1}{2}\lambda + i\pi_\lambda$ which anticommutes with the constraint $\pi_\lambda + \frac{i}{2}\lambda$. This will simplify the analysis.

Conservation of the 3 primary constraints yields two secondary constraints and fixes one Lagrange multiplier

$$(6.20) \qquad p^2 = 0, \quad p(\pi_\lambda - \frac{i}{2}\lambda) = 0, \quad \eta(t) = 0.$$

Both secondary constraints are preserved in time because $\pi_\lambda - \frac{i}{2}\lambda$ anticommutes with the constraint $\pi_\lambda + \frac{i}{2}\lambda$. The Hamiltonian can be rewritten as

$$(6.21) \qquad H_{\text{naive}} = (1+2H)\frac{1}{2}p^2 + i\Psi(\frac{\lambda}{2} + i\pi_\lambda)p + ap_h + \alpha\pi_\psi.$$

Here

$$(6.22) \qquad \frac{1}{1-2h} = 1+2H \quad \text{and} \quad \frac{\psi}{1-2h} = \Psi,$$

so we encounter again the fields H and Ψ of (3.13). From now on we use $G \equiv 1+2H$ as gravitational field, and Ψ as gravitino. Hence, the Dirac Hamiltonian (the Hamiltonian on the constraint surface) vanishes

$$(6.23) \qquad H_D = 0.$$

and there are 5 constraints: one second class constraint

$$(6.24) \qquad \pi_\lambda + \frac{i}{2}\lambda = 0,$$

and four first-class constraints

$$(6.25) \qquad p_h = \pi_\psi = p^2 = p(\pi_\lambda - \frac{i}{2}\lambda) = 0.$$

Thus the Dirac brackets in (2.19) remain valid for supergravity.

The constraint $p^2 = 0$ generates diffeomorphisms. One might expect that they are given by

$$\delta\varphi = \xi\dot{\varphi}, \quad \delta p = \xi\dot{p}, \quad \delta\lambda = \xi\dot{\lambda}, \quad \delta\pi = \xi\dot{\pi},$$
$$(6.26) \qquad \delta H = \frac{d}{dt}\left[\frac{1}{2}(1+2H)\xi\right], \quad \delta\Psi = \frac{d}{dt}(\xi\Psi),$$

because these transformations leave (3.11) invariant. However in the Hamiltonian tranformation laws, no time derivatives are allowed. Moreover one expects that p^2 can only act on φ (and perhaps p and H), but not an λ, π and ψ. We now perform a series of modifications of the transformation rules which cast (6.26) into the expected form.

By adding terms proportional to the p and φ field equations to $\delta\varphi$ and δp, respectively (so-called equation of motion symmetries), namely, $\delta\phi = -\xi \frac{\partial S}{\partial p}$ and $\delta p = \xi \frac{\partial S}{\partial \phi}$, one obtains

$$\delta\varphi = (1+2H)\xi p - \xi\Psi(\pi - \frac{i}{2}\lambda),$$
(6.27)
$$\delta p = 0.$$

Similarly, adding terms proportional to the π and λ field equations to $\delta\lambda$ and $\delta\pi$ yields

(6.28) $$\delta\lambda = -\xi p\Psi, \quad \delta\pi_\lambda = \frac{i}{2}\xi p\Psi.$$

As a check of these last two results note that the constraint $\pi_\lambda + \frac{i}{2}\lambda$ is invariant. Finally we can remove the complicated last term in $\delta\varphi$ adding a local susy transformation in (3.11) with parameter $\epsilon = -\xi\Psi$ to **all** fields. This yields

$$\delta\varphi = \hat{\xi}p, \quad \delta p = 0, \quad \delta\lambda = 0, \quad \delta\pi_\lambda = 0,$$
(6.29)
$$\delta\Psi = 0, \quad \delta(1+2H) = \frac{d}{dt}\hat{\xi}, \quad \hat{\xi} = (1+2H)\xi.$$

These are the transformations of the matter fields generated by $\frac{1}{2}p^2$. The classical gauge fields are $(1+2H)$ and Ψ and they transform in general as

(6.30) $$\delta h^A = \frac{d}{dt}\epsilon^A + f^A{}_{BC} h^B_\mu \epsilon^C,$$

where ϵ^A and h^A_μ correspond to $\hat{\xi}$ and $1+2H$ in the case of diffeomorphisms. The same results should be obtained by taking the brackets with Q_H.

Next we consider the local susy generator in (6.20) which we already multiply with the classical susy gauge parameter $\epsilon(t)$, hence $\epsilon(\pi_\lambda - \frac{i}{2}\lambda)p$. It generates the following classical transformation laws, obtained using the Dirac brackets,

(6.31) $$\delta\varphi = -\epsilon(\pi_\lambda - \frac{i}{2}\lambda), \quad \delta p = 0, \quad \delta\lambda = -p\epsilon, \quad \delta\pi_\lambda = \frac{i}{2}p\epsilon.$$

These are the transformation laws of (3.11). The gauge fields should transform according to (6.30)

(6.32) $$\delta\Psi = \dot{\epsilon}, \quad \delta(1+2H) = -2i\epsilon\Psi.$$

(The factor 2 in $-2i\epsilon\Psi$ comes from the two terms in $f^A{}_{BC}h^B_\mu\epsilon^C$ with $h^B_\mu = \Psi$ and $\epsilon^C = \epsilon$, or vice-versa). Also these rules agree with (3.11). (Note that the local classical gauge algebra based on (6.29), (6.31) and (6.32) closes, and that only the commutator of two local supersymmetry transformations is nonzero).

The local gauge algebra of the transformation in (6.29) and (6.31), (6.32) has now structure functions which are independent of the Lagrange multipliers h and ψ (or H and Ψ), just as required for a Hamiltonian treatment. Having shown that the two first class constraints $\frac{1}{2}p^2$ and $p(\pi_\lambda - \frac{i}{2}\lambda)$ generate indeed the local symmetries of the classical phase space action in (3.11), we now proceed with the construction of the BRST charge Q_H and the quantum action.

The BRST charge is given by

(6.33) $$Q_H = \frac{1}{2}cp^2 - i\gamma p(\pi_\lambda - \frac{i}{2}\lambda) + p_b p_G + \pi_\beta \pi_\Psi - i\pi_c \gamma\gamma,$$

where $G = 1 + 2H$. We denote antihermitian conjugate momenta by π as in π_λ, π_Ψ, π_c, π_β, but hermitian momenta by p as in p_G, p_b, p_γ. By p in (6.33) we mean p_φ, as before. (Recall that all ghosts are real, but the antighosts b and β are antihermitian). The BRST charge is real and anticommuting. The first four terms contain the four first-class constraints, and the last term in (6.33) comes from the last term in (6.17). The coefficients of the terms with p_b and π_β are not fixed by nilpotency. It is easiest to fix the coefficients and signs by working out the transformation rules and fitting to the results obtained earlier, although in principle we need not follow this path since all terms are well defined. One can prove the nilpotency of the BRST laws by directly evaluating $\{Q_H, Q_H\}$, using the equal-time canonical (anti)commutations relations.

Defining BRST transformations by

(6.34) $$\delta_B \Phi = -i[\Lambda Q_H, \Phi] = -i[\Phi, Q_H \Lambda],$$

for any field Φ, we find the following results

(6.35)
$$\begin{aligned}
&\delta_B \varphi = cp\Lambda - i(\pi_\lambda - \tfrac{i}{2}\lambda)\gamma\Lambda, & &\delta_B p = 0, \\
&\delta_B \lambda = i\gamma p\Lambda, & &\delta_B \pi_\lambda = \tfrac{1}{2}\gamma p\Lambda, \\
&\delta_B G = p_b \Lambda, & &\delta_B \Psi = -\pi_\beta \Lambda, \\
&\delta_B c = i\gamma\gamma\Lambda, & &\delta_B b = -\Lambda p_G, \\
&\delta_B \gamma = 0, & &\delta_B \beta = \pi_\psi \Lambda, \\
&\delta_B \pi_c = -\tfrac{1}{2}p^2 \Lambda & &\delta_B p_\gamma = 2i\pi_c \gamma \Lambda + ip(\pi_\lambda - \tfrac{i}{2}\lambda). \\
&\delta_B p_G = \delta_B \pi_\psi = \delta_B p_b = \delta_B \pi_\beta = 0,
\end{aligned}$$

On the classical fields these rules agree with the classical gauge transformations with $\xi = c\Lambda$ and $\epsilon = -i\gamma\Lambda$. In principle one should use Dirac brackets to obtain these transformation rules. These Dirac brackets can be constructed from the second class constraint in (6.24), but because the second class constraint commutes with the first class constraints, the results are the same[22].

The rules in (6.35) are nilpotent. In fact, we could have used nilpotency of Q_H to derive $\delta_B c$ and $\delta_B \gamma$ from $\delta_B \varphi$, namely as follows

$$\delta_B^2 \varphi = 0 = \delta_B \left[cp - i\gamma(\pi_\lambda - \frac{i}{2}\lambda) \right]$$

(6.36) $$= (\delta_B c)p - i(\delta_B \gamma)(\pi_\lambda - \frac{i}{2}\lambda) - i\gamma(\gamma p \Lambda) \Rightarrow \delta_B c = i\gamma\gamma\Lambda, \quad \delta_B \gamma = 0.$$

The result $\delta_B \gamma = 0$ can also be derived from $\delta_B^2 \lambda = 0$.

Comparison of $\delta_B b$ and $\delta_B \beta$ in (6.35) and (6.8) reveals that the BRST auxiliary fields are the canonical momenta of the Lagrange multipliers

(6.37) $$p_G = -d, \qquad \pi_\Psi = -\Delta.$$

Finally we construct the quantum action in the Hamiltonian approach

(6.38) $$L = \dot{q}^i p_i - H + \{Q_H, \psi_g\}.$$

As "gauge-fixing fermion" we take the following hermitian anticommuting expression

(6.39) $$\psi_g = -i(b(G-1) + G\pi_c) + (-i\beta\Psi - \Psi p_\gamma).$$

---

[22]Only in the exceptional case that the Poisson bracket of a first and a second class constraint gives a square of a second class constraint, one would get a different answer, and in that case one would need to use Dirac brackets. This situation does not occur in our model.

The quantum Hamiltonian H_H which commutes with Q_H vanishes in our case (and in any gravitational theory) because $H_D = 0$ (see (6.23)). We denote the gravitational field $1 + 2H$ by G, and find then for the quantum Hamiltonian

(6.40) $\quad L = \dot{\varphi}p + \dot{\lambda}\pi_\lambda + \dot{G}p_G + \dot{\Psi}\pi_\Psi + \dot{c}\pi_c + \dot{b}p_b + \dot{\gamma}p_\gamma + \dot{\beta}\pi_\beta + \{Q_H, \psi_g\}.$

The evaluation of $\{Q_H, \psi_g\}$ is tedious but straightforward. For any pair of canonically conjugate variables we have $[p, q] = \frac{1}{i}$ or $\{\pi, q\} = \frac{1}{i}$. This yields

$$\{Q_H, -i(b + \pi_c)G + (-i\beta - p_\gamma)\Psi\} = -p_G(G - 1) - \frac{1}{2}Gp^2 + (b + \pi_c)p_b - \pi_\Psi \Psi$$

(6.41) $\qquad\qquad\qquad -p(\pi_\lambda - \frac{i}{2}\lambda)\Psi - 2\pi_c\gamma\Psi + \pi_\beta(-\beta + ip_\gamma).$

The transformation laws $\delta_B G = p_b \Lambda$ and $\delta_B \Psi = -\pi_\beta \Lambda$ which follow from (6.34) agree with the rules $\delta_B G = (\dot{c} + 2\gamma\Psi)\Lambda$ and $\delta\Psi = -i\dot{\gamma}\Lambda$ which we found from the classical transformation laws by substituting $\xi = c\Lambda$ and $\epsilon = -i\gamma\Lambda$, provided

(6.42) $\qquad\qquad -p_b + \dot{c} + 2\gamma\Psi = 0, \qquad i\pi_\beta + \dot{\gamma} = 0.$

These should be algebraic field equations, and indeed they are the field equations of π_c and p_γ. The relevant terms in the action are

(6.43) $\qquad\qquad L = (-p_b + \dot{c} + 2\gamma\Psi)\pi_c + (\dot{\gamma} + i\pi_\beta)p_\gamma.$

Integrating out π_c, p_b, p_γ and π_β imposes (6.42).

The terms $-p_G(G-1)$ and $-\pi_\Psi \Psi$ are the gauge fixing terms. Thus to compare with the action as obtained from the Lagrangian BRST formalism we should use $b(G-1) + \Delta\Psi$ in (6.3) as gauge fixing term. However, we now got terms $\dot{G}p_G - p_G G$ and $\dot{\Psi}\pi_\Psi - \pi_\Psi \Psi$ in the action. In Yang-Mills gauge theories one usually takes $\psi_g = b_a \chi^a + p_{c,a} A_0^a$ where A_0^a is the time-component of the classical gauge field, and $\chi^a = \partial^k A_k^a$. Then one makes a change of integration variables $p(A_0) = kp(A_0)'$ and $b_a = kb'_a$ where k is a constant. The superjacobian is unity, and taking the limit $k \to 0$ one arrives at the quantum action with relativistic unweighted gauge $\partial^\mu A_\mu^a = 0$ [4]. In our case the gauge choices were $G - 1 = 0$ and $\Psi = 0$. We used the same ψ_g but with both χ^a and A_0^a equal to $2H$ and Ψ. This led to (6.39), but it is clear that we cannot rescale $A_0^a \sim (G$ and $\Psi)$ but keep fixed $\chi \sim (G$ and $\Psi)$. In fact, we could have noticed before that there is something wrong with ψ_g in (6.39): the terms have different dimensions.

The resolution is also clear: drop the terms with b and β in (6.39). This removes the offending terms $-p_G G - \pi_\Psi \Psi$ and also the terms bp_b and $-\pi_\beta\beta$ in (6.41). The action now becomes after eliminating π_c, p_b, π_γ and π_b

(6.44) $\begin{aligned} L &= \dot{\varphi}p + \dot{\lambda}\pi_\lambda - \frac{1}{2}Gp^2 - p(\pi_\lambda - \frac{i}{2}\lambda)\Psi \\ &\quad + \dot{G}p_G + \dot{\Psi}\pi_\Psi \\ &\quad + \dot{b}(\dot{c} + 2\gamma\Psi) + \dot{\beta}(i\dot{\gamma}). \end{aligned}$

The first line is (3.11), the second line is the gauge fixing term, and the third line the ghost action. So the action in (6.40) from the Hamiltonian BRST formalism indeed agrees with the action from the Lagrangian BRST formalism if one chooses $\dot{G}$ and $\dot{\Psi}$ instead of G and Ψ as gauge fixing terms.

We have seen that there is a Lagrangian and a Hamiltonian approach. The latter contains conjugate momenta for all variables, hence brackets can be defined and charges constructed. We saw that the gauges $G - 1 = 0$ and $\Psi = 0$ in the Hamiltonian approach led to the same results after solving the algebraic field equations for the canonical momenta as the gauge $\dot{G} = 0$ and $\dot{\Psi} = 0$ in the Lagrangian approach. (We recall that $G = 1 + 2H = \frac{1}{1-2h}$ and $\Psi = \frac{\psi}{1-2h}$.) One might ask whether the Hamiltonian approach can also lead to the gauge $G - 1 = 0$ and $\Psi = 0$ in the Lagrangian approach (corresponding to $h = 0$ and $\psi = 0$ in (6.1)). This is indeed possible. One chooses as gauge fermion $\psi_g = 0$. The action becomes then very simple

(6.45) $\qquad L = \dot{\varphi}p + \dot{\lambda}\pi_\lambda + \dot{G}p_G + \dot{\Psi}p_\Psi + \dot{c}\pi_c + \dot{b}p_b + \dot{\gamma}p_\gamma + \dot{\beta}\pi_\beta.$

Next one factorizes the path integral into a minimal part with the fields (p, φ), (λ, π), (c, π_c) and (γ, p_γ), and a nonminimal part with the pairs (G, p_G), (Ψ, π_Ψ), (b, p_b) and (β, π_β). Finally one discards the latter, and *reinterprets* the momenta π_c and p_γ as the Lagrangian antighosts b and β, respectively. This yields then the action in (6.3) with d, Δ, h and ψ integrated out (removed by their algebraic field equations).

This concludes our discussion of BRST formalism applied to our simple model. We obtained one result which is a bit surprising (or interesting): the differentiated gauge choices $\dot{G} = \dot{\Psi} = 0$ in the Lagrangian approach correspond directly to the gauge choices $G - 1 = \Psi = 0$ in the Hamiltonian approach. On the other hand, the gauge choices $G - 1 = \Psi = 0$ in the Lagrangian approach did not correspond in a direct way to the Hamiltonian approach (we had to discard a sector). String theory uses Lagrangian gauge choices corresponding to $G - 1 = \Psi = 0$. Perhaps the corresponding differentiated gauge choices have advantages in certain respects.

References

[1] F.A. Berezin, The method of second quantization, Academic Press 1966. In this book manipulations with Grassmann variables are defined. See also F.A. Berezin, Introduction to Super Analysis, Reidel, 1987.

[2] R. Feynman and A. Hibbs, Quantum mechanics and path integrals, McGraw Hill 1965. Feynman pioneered the path integral in physics, and this is the standard reference.

[3] Two introductory texts which the reader may try to tackle after reading these lectures are J. Bagger and J. Wess, Supersymmetry and Supergravity, Princeton University Press, 1992, and M.F. Sohnius, Introducing supersymmetry, *Phys. Rep.* **128** (1985) 39. More advanced introductions are P. van Nieuwenhuizen, Supergravity, *Phys. Rep.* **68** (1981) 189, and S.J. Gates, M.T. Grisaru, M. Roček and W. Siegel, Superspace, One thousand and one lessons in supersymmetry.

[4] For a discussion of the Hamiltonian form of BRST symmetry, see M. Henneaux, *Phys. Rep.* **126** (1985) 1. A good introduction for mathematicians to BRST symmetry and general properties of gauge systems is M. Henneaux and C. Teitelboim, Quantization of Gauge Systems, Princeton University Press 1992. We used sections 16.5.6, 16.54b, 19.1.7 and 19.1.8 for discussions at the very end of chapter 6.

[5] Spinning point particles have been discussed in L. Brink, S. Deser, B. Zumino, P. Di Vecchia and P. Howe, *Phys. Lett.* **64B** (1976) 435; see also F.A. Berezin and M.S. Marinov, *Ann. Phys. (N.Y.)* **104** (1977) 336. A Hamiltonian path integral treatment of relativistic supersymmetric particles is given in M. Henneaux and C. Teitelboim, *Ann. Phys. (N.Y.)* **143** (1982) 127.

C.N. Yang Institute for Theoretical Physics, Stony Brook University, Stony Brook, NY, 11794-3840, USA

Titles in This Series

73 **Mikhail Lyubich and Leon Takhtajan, Editors,** Graphs and Patterns in Mathematics and Theoretical Physics (Stony Brook University, Stony Brook, NY, June 14–21, 2001)

72 **Michel L. Lapidus and Machiel van Frankenhuijsen, Editors,** Fractal geometry and applications: A jubilee of Benoît Mandelbrot, Parts 1 and 2 (San Diego, California, 2002 and École Normale Supérieure de Lyon, 2001)

71 **Gordana Matić and Clint McCrory, Editors,** Topology and Geometry of Manifolds (University of Georgia, Athens, Georgia, 2001)

70 **Michael D. Fried and Yasutaka Ihara, Editors,** Arithmetic fundamental groups and noncommutative algebra (Mathematical Sciences Research Institute, Berkeley, California, 1999)

69 **Anatole Katok, Rafael de la Llave, Yakov Pesin, and Howard Weiss, Editors,** Smooth ergodic theory and its applications (University of Washington, Seattle, 1999)

68 **Robert S. Doran and V. S. Varadarajan, Editors,** The mathematical legacy of Harish-Chandra: A celebration of representation theory and harmonic analysis (Baltimore, Maryland, 1998)

67 **Wayne Raskind and Charles Weibel, Editors,** Algebraic K-theory (University of Washington, Seattle, 1997)

66 **Robert S. Doran, Ze-Li Dou, and George T. Gilbert, Editors,** Automorphic forms, automorphic representations, and arithmetic (Texas Christian University, Fort Worth, 1996)

65 **M. Giaquinta, J. Shatah, and S. R. S. Varadhan, Editors,** Differential equations: La Pietra 1996 (Villa La Pietra, Florence, Italy, 1996)

64 **G. Ferreyra, R. Gardner, H. Hermes, and H. Sussmann, Editors,** Differential geometry and control (University of Colorado, Boulder, 1997)

63 **Alejandro Adem, Jon Carlson, Stewart Priddy, and Peter Webb, Editors,** Group representations: Cohomology, group actions and topology (University of Washington, Seattle, 1996)

62 **János Kollár, Robert Lazarsfeld, and David R. Morrison, Editors,** Algebraic geometry—Santa Cruz 1995 (University of California, Santa Cruz, July 1995)

61 **T. N. Bailey and A. W. Knapp, Editors,** Representation theory and automorphic forms (International Centre for Mathematical Sciences, Edinburgh, Scotland, March 1996)

60 **David Jerison, I. M. Singer, and Daniel W. Stroock, Editors,** The legacy of Norbert Wiener: A centennial symposium (Massachusetts Institute of Technology, Cambridge, October 1994)

59 **William Arveson, Thomas Branson, and Irving Segal, Editors,** Quantization, nonlinear partial differential equations, and operator algebra (Massachusetts Institute of Technology, Cambridge, June 1994)

58 **Bill Jacob and Alex Rosenberg, Editors,** K-theory and algebraic geometry: Connections with quadratic forms and division algebras (University of California, Santa Barbara, July 1992)

57 **Michael C. Cranston and Mark A. Pinsky, Editors,** Stochastic analysis (Cornell University, Ithaca, July 1993)

56 **William J. Haboush and Brian J. Parshall, Editors,** Algebraic groups and their generalizations (Pennsylvania State University, University Park, July 1991)

55 **Uwe Jannsen, Steven L. Kleiman, and Jean-Pierre Serre, Editors,** Motives (University of Washington, Seattle, July/August 1991)

54 **Robert Greene and S. T. Yau, Editors,** Differential geometry (University of California, Los Angeles, July 1990)

For a complete list of titles in this series, visit the
AMS Bookstore at **www.ams.org/bookstore/**.

ISBN 0-8218-3666-8

PSPUM/73

OHIO UNIVERSITY LIBRARY

Please ret___ ook as ___ on as you have
shed ___ _rde_ ___ _id a___ e it must
t___ l___ ed be-